Global Warming

The Complete Briefing • Fourth Edition

John Houghton's market-leading textbook is now in full colour and includes the latest IPCC findings and future energy scenarios from the International Energy Agency, making it the definitive guide to climate change. Written for students across a wide range of disciplines, its simple, logical flow of ideas gives an invaluable grounding in the science and impacts of climate change and highlights the need for action on global warming.

'The addition of colour serves the diagrams so they deliver the necessary message and information they intend ... to instructors and students in interdisciplinary programmes who need an accessible, broad-view text on the subject of climate change.'

YOCHANAN KUSHNIR, *Lamont-Doherty Earth Observatory of Columbia University*

'The new edition provides the most up-to-date and comprehensive coverage of climate change for teaching in an undergraduate class. It covers the latest on climate science, climate change impacts and adaptation, and approaches to slowing climate change through reducing emissions from energy use, transport, and deforestation. These complex issues are presented clearly and throughly, based on the recent Fourth Assessment Report of the Intergovernmental Panel on Climate Change and many other sources. The new edition has significantly expanded and updated sections on slowing and stabilising climate change and on energy and transport for the future, which complement the sections on climate science. The addition of colour adds clarity and emphasis to the many valuable figures. I will definitely be using this book in all my courses on climate change.'

PROF DAVID KAROLY, *University of Melbourne (formerly of the University of Oklahoma)*

'It is difficult to imagine how Houghton's exposition of this complex body of information might be substantially improved upon ... Seldom has such a complex topic been presented with such remarkable simplicity, directness and crystalline clarity... Houghton's complete briefing is without doubt the best briefing the concerned citizen could hope to find within the pages of a pocketable book.'

JOHN PERRY, *Bulletin of the American Meteorological Society*

'I can recommend (this book) to anyone who wants to get a better perspective on the topic of global warming . . . a very readable and comprehensive guide to the changes that are occurring now, and could occur in the future, as a result of human action . . . brings the global warming debate right up to date . . . .'

WILLIAM HARSTON, *The Independent*

'... a widely praised book on global warming and its consequences.'

The Economist

'I would thoroughly recommend this book to anyone concerned about global warming. It provides an excellent, essentially non-technical guide on scientific and political aspects of the subject. It is an essential briefing for students and science teachers.'

TONY WATERS, *The Observatory*

'For the non-technical reader, the best program guide to the political and scientific debate is John Houghton's book *Global Warming: The Complete Briefing*. With this book in hand you are ready to make sense of the debate and reach your own conclusions.'

ALAN HECHT, *Climate Change*

'This is a remarkable book ... It is a model of clear exposition and comprehensible writing ... Quite apart from its value as a background reader for science teachers and students, it would make a splendid basis for a college general course.'

ANDREW BISHOP, *Association for Science Education*

' ... a useful book for students and laymen to understand some of the complexities of the global warming issue. Questions and essay topics at the end of each chapter provide useful follow-up work and the range of material provided under one cover is impressive. At a student-friendly price, this is a book to buy for yourself and not rely on the library copy.'

ALLEN PERRY, *Holocene*

'This book is one of the best I have encountered, that deal with climate change and some of its anthropogenic causes. Well written, well organised, richly illustrated and referenced, it should be required reading for anybody concerned with the fate of our planet.'

ELMAR R. REITER, *Meteorology and Atmospheric Physics*

‘Sir John Houghton is one of the few people who can legitimately use the phrase “the complete briefing” as a subtitle for a book on global warming … Sir John has done us all a great favour in presenting such a wealth of material so clearly and accessibly and in drawing attention to the ethical underpinnings of our interpretation of this area of environmental science.’

Progress in Physical Geography

‘Throughout the book this argument is well developed and explained in a way that the average reader could understand – especially because there are many diagrams, tables, graphs and maps which are easy to interpret.’

SATYA

GLOBAL

WARMING

The Complete Briefing | **Fourth Edition**

Sir John Houghton

CAMBRIDGE UNIVERSITY PRESS
Cambridge, New York, Melbourne, Madrid, Cape Town, Singapore, São Paulo, Delhi

Cambridge University Press
The Edinburgh Building, Cambridge CB2 8RU, UK

Published in the United States of America by Cambridge University Press, New York

www.cambridge.org
Information on this title: www.cambridge.org/houghton

First published 1994 by Lion Publishing plc
Second edition published 1997 by Cambridge University Press
Third edition published 2004
Fourth edition published 2009

Printed in the United Kingdom at the University Press, Cambridge

A catalogue record for this publication is available from the British Library

ISBN 978-0-521-88256-9 hardback
ISBN 978-0-521-70916-3 paperback

To my grandchildren,
Daniel, Hannah, Esther, Max,
Jonathan, Jemima and Sam
and their generation

Contents

Preface

Global Warming is a topic that increasingly occupies the attention of the world. Is it really happening? If so, how much of it is due to human activities? How far will it be possible to adapt to changes of climate? What action to combat it can or should we take? How much will it cost? Or is it already too late for useful action? This book sets out to provide answers to all these questions by providing the best and latest information available.

I was privileged to chair or co-chair the Scientific Assessments for the Intergovernmental Panel on Climate Change (IPCC) from its inception in 1988 until 2002. During this period the IPCC published three major comprehensive reports – in 1990, 1995 and 2001 – that have influenced and informed those involved in climate change research and those concerned with the impacts of climate change. In 2007, a fourth assessment report was published. It is the extensive new material in this latest report that has provided the basis for the substantial revision necessary to update this fourth edition.

The IPCC reports have been widely recognised as the most authoritative and comprehensive assessments on a complex scientific subject ever produced by the world's scientific community. On the completion of the first assessment in 1990, I was asked to present it to Prime Minister Margaret Thatcher's cabinet – the first time an overhead projector had been used in the Cabinet Room in Number 10 Downing Street. In 2005, the work of the IPCC was cited in a joint statement urging action on climate change presented to the G8 meeting in that year by the Academies of Science of all G8 countries plus China, India and Brazil. The world's top scientists could not have provided stronger approval of the IPCC's work. An even wider endorsement came in 2007 when the IPCC was awarded a Nobel Peace Prize.

Many books have been published on global warming. My choice of material has been much influenced by the many lectures I have given in recent years to professional, student and general audiences.

The strengths of this book are that it is:

- **up-to-date with the latest reliable, accurate and understandable information** about all aspects of the global warming problem for students, professionals and interested or concerned citizens.
- **accessible** to both scientists and non-scientists. Although there are many numbers in the book – I believe quantification to be essential – there are no

mathematical equations. Some important technical material is included in boxes.

- **comprehensive**, as it moves through the basic science of global warming, impacts on human communities and ecosystems, economic, technological and ethical considerations and policy options for action both national and international.
- appropriate as a **general text for students,** from high-school level up to university graduate. Questions and problems for students to consider and to test their understanding of the material are included in each chapter.
- Its **simple and effective visual presentation of the vast quantities of data** available on climate change ensures that readers can see how conclusions are made, without being overwhelmed. Illustrations are available online.

Over the 20 years since the inception of the IPCC, our understanding of climate change has much increased and significant changes in climate due to human activities have been experienced. Further, studies of the feedbacks that determine the climate response have shown an increasing likelihood of enhanced response, so leading over these years to greater concern about the future impact of climate change on both human populations and ecosystems. Can much be done to alleviate the impact or mitigate future climate change? Later chapters of the book address this question and demonstrate that the technology is largely available to support urgent and affordable action. They also point to the many other benefits that will accrue to all sectors of society as the necessary action is taken. However, what seems lacking as yet is the will to take that action.

As I complete this revised edition I want to express my gratitude, first to those who inspired me and helped with the preparation of the earlier editions, with many of whom I was also involved in the work of the IPCC or of the Hadley Centre. I also acknowledge those who have assisted with the material for this edition or who have read and helpfully commented on my drafts, in particular, Fiona Carroll, Jim Coakley, Peter Cox, Simon Desjardin, Michael Hambery, Marc Humphreys, Chris Jones, Linda Livingstone, Jason Lowe, Tim Palmer, Martin Parry, Ralph Sims, Susan Solomon, Peter Smith, Chris West, Sue Whitehouse and Richard Wood. My thanks are also due to Catherine Flack, Matt Lloyd, Anna-Marie Lovett and Jo Endell-Cooper of Cambridge University Press for their competence and courtesy as they steered the book through its gestation and production.

Finally, I owe an especial debt to my wife, Sheila, who gave me strong encouragement to write the book in the first place, and who has continued her encouragement and support through the long hours of its production.

Global warming and climate change

1

Hurricane Wilma hit Florida's southern west coast on 24 October 2005.

THE PHRASE 'global warming' has become familiar to many people as one of the most important issues of our day. Many opinions have been expressed concerning it, from the doom-laden to the dismissive. This book aims to state the current scientific position on global warming clearly, so that we can make informed decisions on the facts.

Is the climate changing?

In the year 2060 my grandchildren will be approaching 70 years old; what will their world be like? Indeed, what will it be like during the 70 years or so of their normal lifespan? Many new things have happened in the last 70 years that could not have been predicted in the 1930s. The pace of change is such that even more novelty can be expected in the next 70. It seems certain that the world will be even more crowded and more connected. Will the increasing scale of human activities affect the environment? In particular, will the world be warmer? How is its climate likely to change?

Before addressing future climate changes, what can be said about climate changes in the past? In the more distant past there have been very large changes. The last million years has seen a succession of major ice ages interspersed with warmer periods. The last of these ice ages began to come to an end about 20 000 years ago and we are now in what is called an interglacial period. Chapter 4 will focus on these times far back in the past. But have there been changes in the very much shorter period of living memory – over the past few decades?

Variations in day-to-day weather are occurring all the time; they are very much part of our lives. The climate of a region is its average weather over a period that may be a few months, a season or a few years. Variations in climate are also very familiar to us. We describe summers as wet or dry, winters as mild, cold or stormy. In the British Isles, as in many parts of the world, no season is the same as the last or indeed the same as any previous season, nor will it be repeated in detail next time round. Most of these variations we take for granted; they add a lot of interest to our lives. Those we particularly notice are the extreme situations and the climate disasters (for instance, Figure 1.1 shows the significant climate events and disasters during the year 1998 – one of the warmest years on record). Most of the worst disasters in the world are, in fact, weather- or climate-related. Our news media are constantly bringing them to our notice as they occur in different parts of the world – tropical cyclones (called hurricanes or typhoons), windstorms, floods and tornadoes, also droughts whose effects occur more slowly, but which are probably the most damaging disasters of all.

The last 30 years

The closing decades of the twentieth century and the early years of the present century were unusually warm. Globally speaking, the last 30 years have been the warmest since accurate records began somewhat over 100 years ago. Twelve of the 13 years 1995 to 2007 rank among the 13 warmest in the instrumental record of global surface air temperature that began around 1850, the

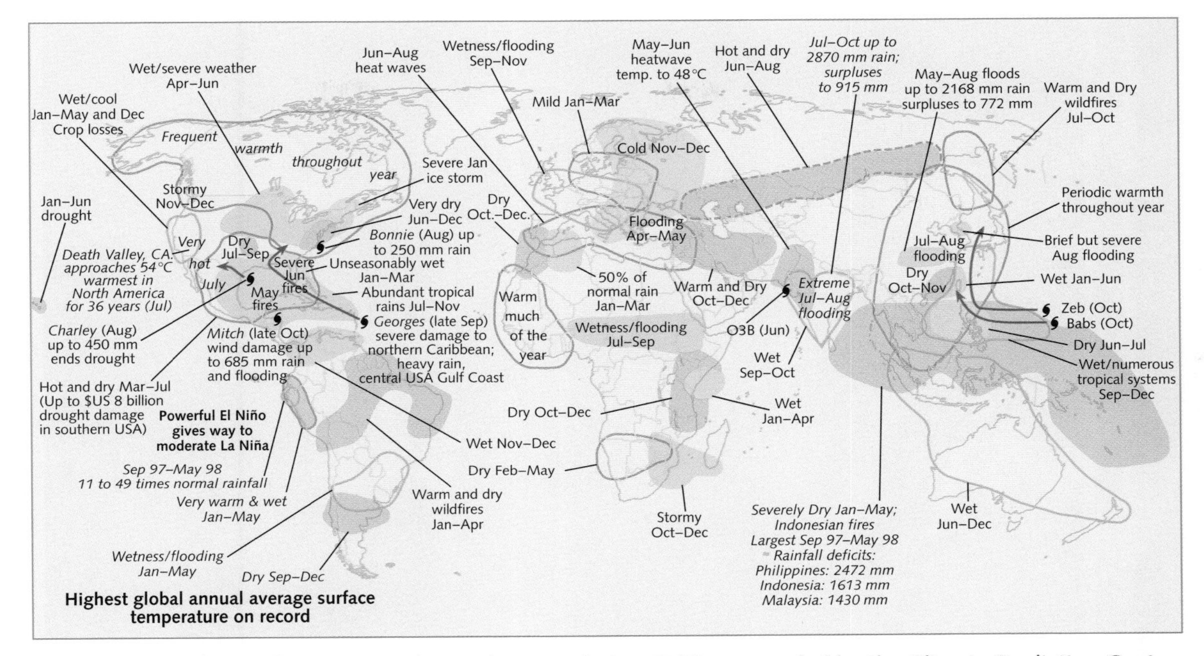

Figure 1.1 Significant climate anomalies and events during 1998 as recorded by the Climate Prediction Center of the National Oceanic and Atmospheric Administration (NOAA) of the United States.

years 1998 and 2005 being the warmest (different analyses disagree which is the warmer of the two). The Intergovernmental Panel on Climate Change in its 2007 Assessment[1] states:

> Warming of the climate system is unequivocal, as is now evident from observations of increases in global average air and ocean temperatures, widespread melting of snow and ice, and rising global average sea level.

The period has also been remarkable (just how remarkable will be considered later) for the frequency and intensity of extremes of weather and climate. Let me give a few examples. An extremely unusual heatwave in central Europe occurred in the summer of 2003 and led to the premature deaths of over 20 000 people (see Chapter 7, page 215). Periods of unusually strong winds have been experienced in western Europe. During the early hours of the morning of 16 October 1987, over 15 million trees were blown down in southeast England and the London area. The storm also hit northern France, Belgium and the Netherlands with ferocious intensity; it turned out to be the worst storm experienced in the area since 1703. Storm-force winds of similar or even greater intensity but covering a greater area of western Europe have struck since – on four occasions in 1990 and three occasions in December 1999.

Hurricane Mitch was one of the deadliest and most powerful hurricanes on record in the Atlantic basin, with maximum sustained winds of 180 mph (290 km h^{-1}). The storm was the thirteenth tropical storm, ninth hurricane and third major hurricane of the 1998 Atlantic hurricane season.

But those storms in Europe were mild by comparison with the much more intense and damaging storms other parts of the world have experienced during these years. About 80 hurricanes and typhoons – other names for tropical cyclones – occur around the tropical oceans each year, familiar enough to be given names: Hurricane Gilbert caused devastation on the island of Jamaica and the coast of Mexico in 1988, Typhoon Mireille hit Japan in 1991, Hurricane Andrew caused a great deal of damage in Florida and other regions of the southern United States in 1992, Hurricane Mitch caused great devastation in Honduras and other countries of central America in 1998 and Hurricane Katrina caused record damages as it hit the Gulf Coast of the United States in 2005 are notable recent examples. Low-lying areas such as Bangladesh are particularly vulnerable to the storm surges associated with tropical cyclones; the combined

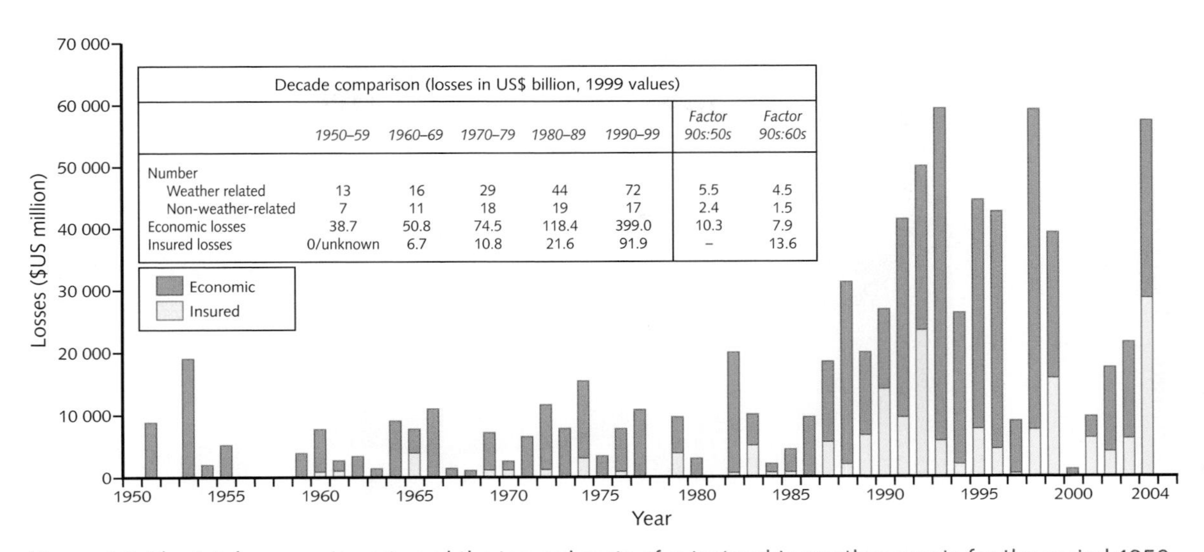

Decade comparison (losses in US$ billion, 1999 values)							
	1950–59	*1960–69*	*1970–79*	*1980–89*	*1990–99*	*Factor 90s:50s*	*Factor 90s:60s*
Number							
Weather related	13	16	29	44	72	5.5	4.5
Non-weather-related	7	11	18	19	17	2.4	1.5
Economic losses	38.7	50.8	74.5	118.4	399.0	10.3	7.9
Insured losses	0/unknown	6.7	10.8	21.6	91.9	–	13.6

Figure 1.2 The total economic costs and the insured costs of catastrophic weather events for the period 1950 to 2004 as recorded by the Munich Re insurance company. For 2005, because of Hurricane Katrina in the USA the figures are off the page – over $US200 billion for economic losses and over $US80 billion for insured losses. Both costs show a rapid upward trend in recent decades. The number of non-weather-related disasters is included for comparison. Tables 7.3 and 7.4 in Chapter 7 provide some regional detail and list some of the recent disasters with the greatest economic and insured losses.

effect of intensely low atmospheric pressure, extremely strong winds and high tides causes a surge of water which can reach far inland. In one of the worst such disasters in the twentieth century over 250 000 people were drowned in Bangladesh in 1970. The people of that country experienced another storm of similar proportions in 1999 as did the neighbouring Indian state of Orissa also in 1999, and smaller surges are a regular occurrence in that region.

The increase in storm intensity during recent years has been tracked by the insurance industry, which has been hit hard by recent disasters. Until the mid 1980s, it was widely thought that windstorms or hurricanes with insured losses exceeding $US1 billion (thousand million) were only possible, if at all, in the United States. But the gales that hit western Europe in October 1987 heralded a series of windstorm disasters that make losses of $US10 billion seem commonplace. Hurricane Andrew, for instance, left in its wake insured losses estimated at nearly $US21 billion (1999 prices) with estimated total economic losses of nearly $US37 billion. Figure 1.2 shows the costs of weather-related disasters[2] over the past 50 years as calculated by the insurance industry. It shows an increase in economic losses in such events by a factor of over 10 in real terms between the 1950s and the present day. Some of this increase can be attributed

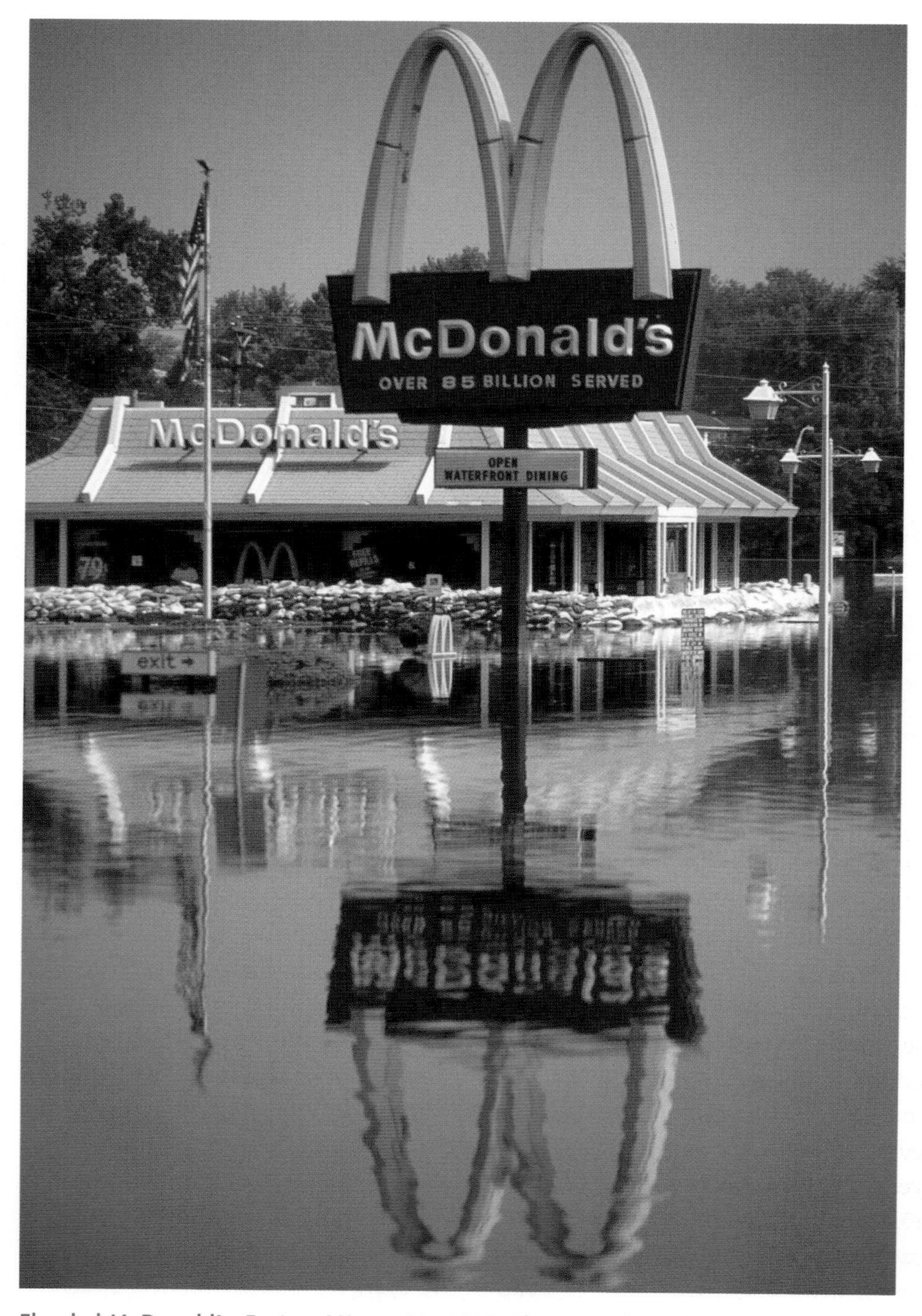

Flooded McDonald's, Festus, Missouri in 1993. The spot where this photo was taken is nearly 1.5 miles (2.5 km) and 30 feet (9 m) above the river.

to the growth in population in particularly vulnerable areas and to other social or economic factors; the world community has undoubtedly become more vulnerable to disasters. However, a significant part of it has also arisen from the increased storminess in the recent years compared with the 1950s.

Windstorms or hurricanes are by no means the only weather and climate extremes that cause disasters. Floods due to unusually intense or prolonged rainfall or droughts because of long periods of reduced rainfall (or its complete absence) can be even more devastating to human life and property. These events occur frequently in many parts of the world especially in the tropics and subtropics. There have been notable examples during the last two decades. Let me mention a few of the floods. In 1988, the highest flood levels ever recorded occurred in Bangladesh, and 80% of the entire country was affected; China experienced devastating floods affecting many millions of people in 1991, 1994–5 and 1998; in 1993, flood waters rose to levels higher than ever recorded in the region of the Mississippi and Missouri rivers in the United States, flooding an area equivalent in size to one of the Great Lakes; major floods in Venezuela in 1999 led to a large landslide and left 30 000 people dead; two widespread floods in Mozambique occurred within a year in 2000–1 leaving over half a million homeless; and in the summer of 2002 Europe experienced its worst floods for centuries. Droughts during these years have been particularly intense and prolonged in areas of Africa, both north and south. It is in Africa especially that they bear on the most vulnerable in the world, who have little resilience to major disasters. Figure 1.3 shows that in the 1980s droughts accounted for more deaths in Africa than all other disasters added together and illustrates the scale of the problem.

El Niño events

Rainfall patterns which lead to floods and droughts especially in tropical and semi-tropical areas are strongly influenced by the surface temperature of the oceans around the world, particularly the pattern of ocean surface temperature in the Pacific off the coast of South America (see Chapter 5 and Figure 5.9). About every three to five years a large area of warmer water appears and persists for a year or more. Because they usually occur around Christmas these are known as El Niño ('the boy child') events.[3] They have been well known for centuries to the countries along the coast of South America because of their devastating effect on the fishing industry; the warm top waters of the ocean prevent the nutrients from lower, colder levels required by the fish from reaching the surface.

A particularly intense El Niño, the second most intense in the twentieth century, occurred in 1982–3; the anomalous highs in ocean surface temperature

The Great Flood of 1993 occurred in the American Midwest, along the Mississippi and Missouri rivers from April to October 1993. The flood was among the most costly and devastating to ever occur in the United States, with $US15 billion in damages, and a flooded area of around 30000 square miles (80000 km^2). These images from Landsat-5 Thematic Mapper show the Mississippi near St Louis before and during the flood.

compared to the average reached 7 °C. Droughts and floods somewhere in almost all the continents were associated with that El Niño (Figure 1.4). Like many events associated with weather and climate, El Niños often differ very much in their detailed character; that has been particularly the case with the El Niño events of the 1990s. For instance, the El Niño event that began in 1990 and reached maturity early in 1992, apart from some weakening in mid 1992, continued to be dominated by the warm phase until 1995. The exceptional floods in the central United States and in the Andes and droughts in Australia and Africa

are probably linked with this unusually protracted El Niño. This, the longest El Niño of the twentieth century, was followed in 1997–8 by the century's most intense El Niño which brought exceptional floods to China and to the Indian sub-continent and drought to Indonesia – that in turn brought extensive forest fires creating an exceptional blanket of thick smog which was experienced over 1000 miles away (Figure 1.1).

Studies with computer models of the kind described later (in Chapter 5) provide a scientific basis for links between the El Niño and these extreme weather events; they also give some confidence that useful forecasts of such disasters will in due course be possible. A scientific question that is being urgently addressed is the possible link between the character and intensity of El Niño events and global warming due to human-induced climate change.

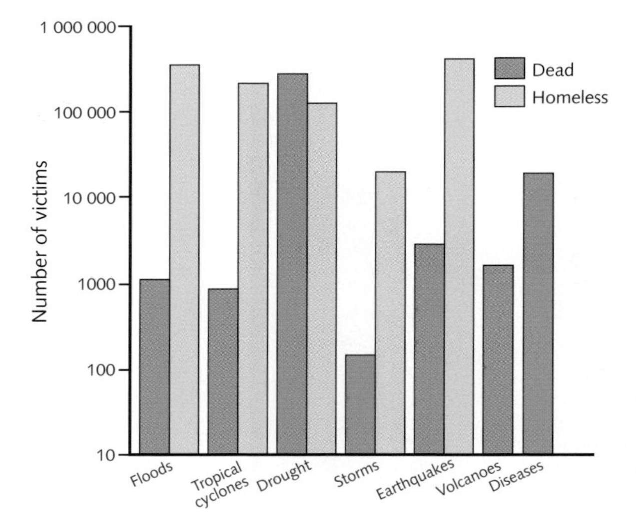

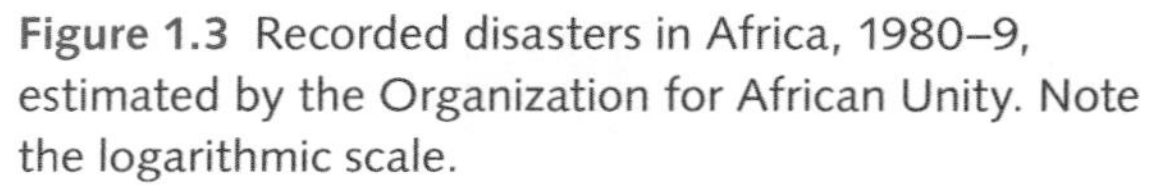

Figure 1.3 Recorded disasters in Africa, 1980–9, estimated by the Organization for African Unity. Note the logarithmic scale.

The effect of volcanic eruptions on temperature extremes

Natural events such as volcanoes can also affect the climate. Volcanoes inject enormous quantities of dust and gases into the upper atmosphere. Large amounts of sulphur dioxide are included, which through photochemical reactions using the Sun's energy are transformed to sulphuric acid and sulphate particles. Typically these particles remain in the stratosphere (the region of atmosphere above about 10 km in altitude) for several years before they fall into the lower atmosphere and are quickly washed out by rainfall. During this period they disperse around the whole globe and cut out some of the radiation from the Sun, thus tending to cool the lower atmosphere.

One of the largest volcanic eruptions in the twentieth century was that from Mount Pinatubo in the Philippines on 12 June 1991 which injected about 20 million tonnes of sulphur dioxide into the stratosphere together with enormous amounts of dust. This stratospheric dust caused spectacular sunsets around the world for many months following the eruption. The amount of radiation from the Sun reaching the lower atmosphere fell by about 2%. Global average temperatures lower by about a quarter of a degree Celsius were experienced for the following two years. There is also evidence that some of the unusual weather patterns of 1991 and 1992, for instance unusually cold winters in the Middle East and mild winters in western Europe, were linked with effects of the volcanic dust.

Vulnerability to change

Over the centuries, although different human communities have adapted to their particular climate, any large change to the average climate tends to bring stress of one kind or another. It is particularly the extreme climate events and climate disasters that emphasise the importance of climate to our lives and that demonstrate to countries around the world their vulnerability to climate change – a vulnerability that is enhanced by rapidly increasing world population and demands on resources.

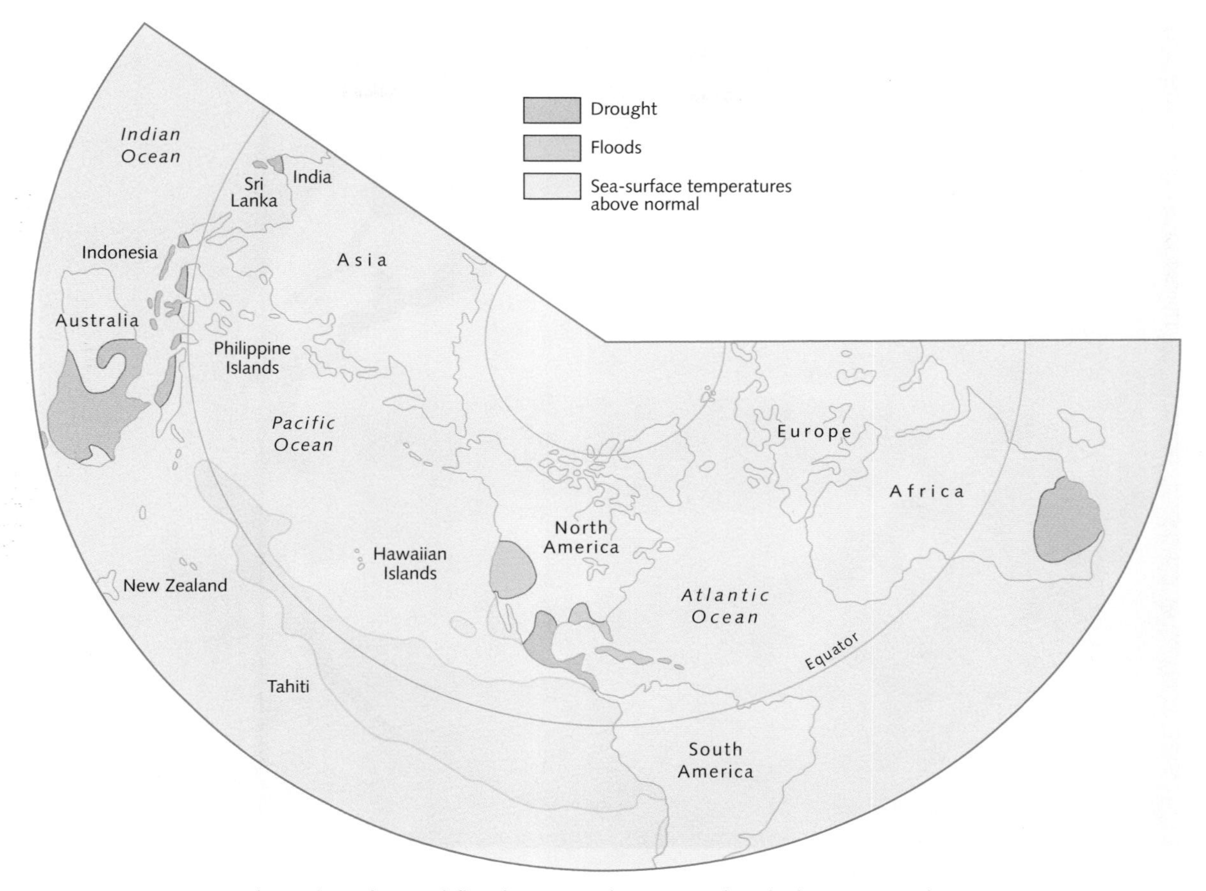

Figure 1.4 Regions where droughts and floods occurred associated with the 1982–3 El Niño.

But the question must be asked: how remarkable are these extreme events that I have been listing? Do they point to a changing climate due to human activities? Here a note of caution must be sounded. The range of normal natural climate variation is large. Climate extremes are nothing new. Climate records are continually being broken. In fact, a month without a broken record somewhere would itself be something of a record!

Many of us may remember the generally cold period over large areas of the world during the 1960s and early 1970s that caused speculation that the world was heading for an ice age. A British television programme about climate change called 'The ice age cometh' was prepared in the early 1970s and widely screened – but the cold trend soon came to an end. We must not be misled by our relatively short memories.

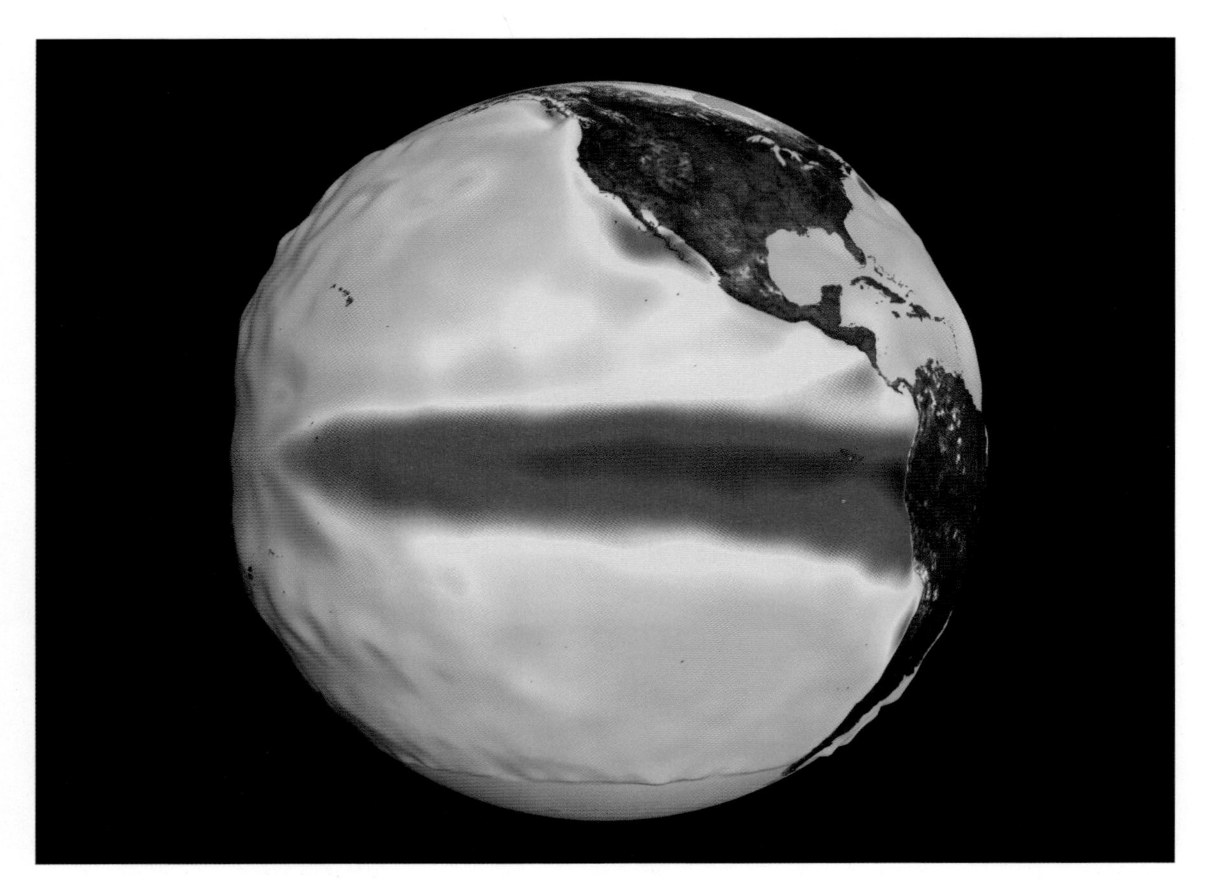

The El Niño event of 1997–8 is the most intense on record. One result was the drought that led to forest fires in Asia, which burned thousands of square miles of rainforest, plantations, conversion forest and scrubland in Indonesia alone. The above shows a superposition of sea surface temperature anomalies on anomalies of the sea surface elevation, showing warm water building up eastwards across the Pacific Ocean and reaching South America.

We may be sure about the warming that has occurred over the last few decades but do we have the evidence that this is linked with the development of human industry over the last 200 years? To identify climate change related to this development, we need to look for trends in global warming over similar lengths of time. They are long compared with both the memories of a generation and the period for which accurate and detailed records exist. Although, therefore, it can be ascertained that there was more storminess, for instance, in the region of the north Atlantic during the 1980s and 1990s than in the previous three decades, it is difficult to know just how exceptional those decades were compared with other periods in previous centuries. There is even more difficulty in tracking detailed climate trends in many other parts of the world, owing to the lack of adequate records; further, trends in the frequency of rare events are not easy to detect.

What is important is continually to make careful comparisons between practical observations of the climate and its changes and what scientific knowledge leads us to expect. During the last few years, as the occurrence of extreme events has made the public much more aware of environmental issues,[4] scientists in their turn have become more sure about just what human activities are doing to the climate. Later chapters will look in detail at the science of global warming and at the climate changes that we can expect, as well as investigating how these changes fit in with the recent climate record. First, however, I present a brief outline of our current scientific understanding.

What is global warming?

We know for sure that because of human activities, especially the burning of fossil fuels, coal, oil and gas, together with widespread deforestation, the gas carbon dioxide has been emitted into the atmosphere in increasing amounts over the past 200 years and more substantially over the past 50 years. Every year these emissions currently add to the carbon already present in the atmosphere a further 8000 million tonnes, much of which is likely to remain there for a period of 100 years or more. Because carbon dioxide is a good absorber of heat radiation coming from the Earth's surface, increased carbon dioxide acts like a blanket over the surface, keeping it warmer than it would otherwise be. With the increased temperature the amount of water vapour in the atmosphere also increases, providing more blanketing and causing it to be even warmer. The gas methane is also increasing because of different human activities, for instance mining and agriculture, and adding to the problem.

Being kept warmer may sound appealing to those of us who live in cool climates. However, an increase in global temperature will lead to global climate change. If the change were small and occurred slowly enough we would almost certainly be able to adapt to it. However, with rapid expansion taking place in the world's industry the change is unlikely to be either small or slow. The estimate I present in later chapters is that, in the absence of efforts to curb the rise in the emissions of carbon dioxide, the global average temperature will rise by about a third of a degree Celsius or more every ten years – or three or more degrees in a century.

This may not sound very much, especially when it is compared with normal temperature variations from day to night or between one day and the next. But it is not the temperature at one place but the temperature averaged over the whole globe. The predicted rate of change of 3 °C a century is probably faster than the global average temperature has changed at any time over the past 10 000 years. And as there is a difference in global average temperature of only about five or six degrees between the coldest part of an ice age and the warm periods in between ice ages (see Figure 4.6), we can see that a few degrees in this global average can represent a big change in climate. It is to this change and especially to the very rapid rate of change that many ecosystems and human communities (especially those in developing countries) will find it difficult to adapt.

Not all the climate changes will in the end be adverse. While some parts of the world experience more frequent or more severe droughts, floods or significant sea level rise, in other places crop yields may increase due to the fertilising effect of carbon dioxide. Other places, perhaps for instance in the sub-arctic, may become more habitable. Even there, though, the likely rate of change will cause problems: large damage to buildings will occur in regions of melting permafrost, and trees in sub-arctic forests like trees elsewhere will not have time to adapt to new climatic regimes.

Scientists are confident about the fact of global warming and climate change due to human activities. However, uncertainty remains about just how large the warming will be and what will be the patterns of change in different parts of the world. Although useful indications can be given, scientists cannot yet say in precise detail which regions will be most affected. Intensive research is needed to improve the confidence in scientific predictions.

Adaptation and mitigation

An integrated view of anthropogenic climate change is presented in Figure 1.5 where a complete cycle of cause and effect is shown. Begin in the box at the bottom where economic activity, both large and small scale, whether in developed or developing countries, results in emissions of greenhouse gases (of which carbon dioxide is the most important) and aerosols. Moving in a clockwise direction around the diagram, these emissions lead to changes in atmospheric concentrations of important constituents that alter the energy input and output of the climate system and hence cause changes in the climate. These climate changes impact both humans and natural ecosystems altering patterns of resource availability and affecting human livelihood and health. These impacts in their turn affect human development in all its aspects. Anticlockwise arrows illustrate possible development pathways and global emission constraints that would reduce the risk of future impacts that society may wish to avoid.

Figure 1.5 also shows how both causes and effects can be changed through *adaptation* and *mitigation*. In general adaptation is aimed at reducing the effects

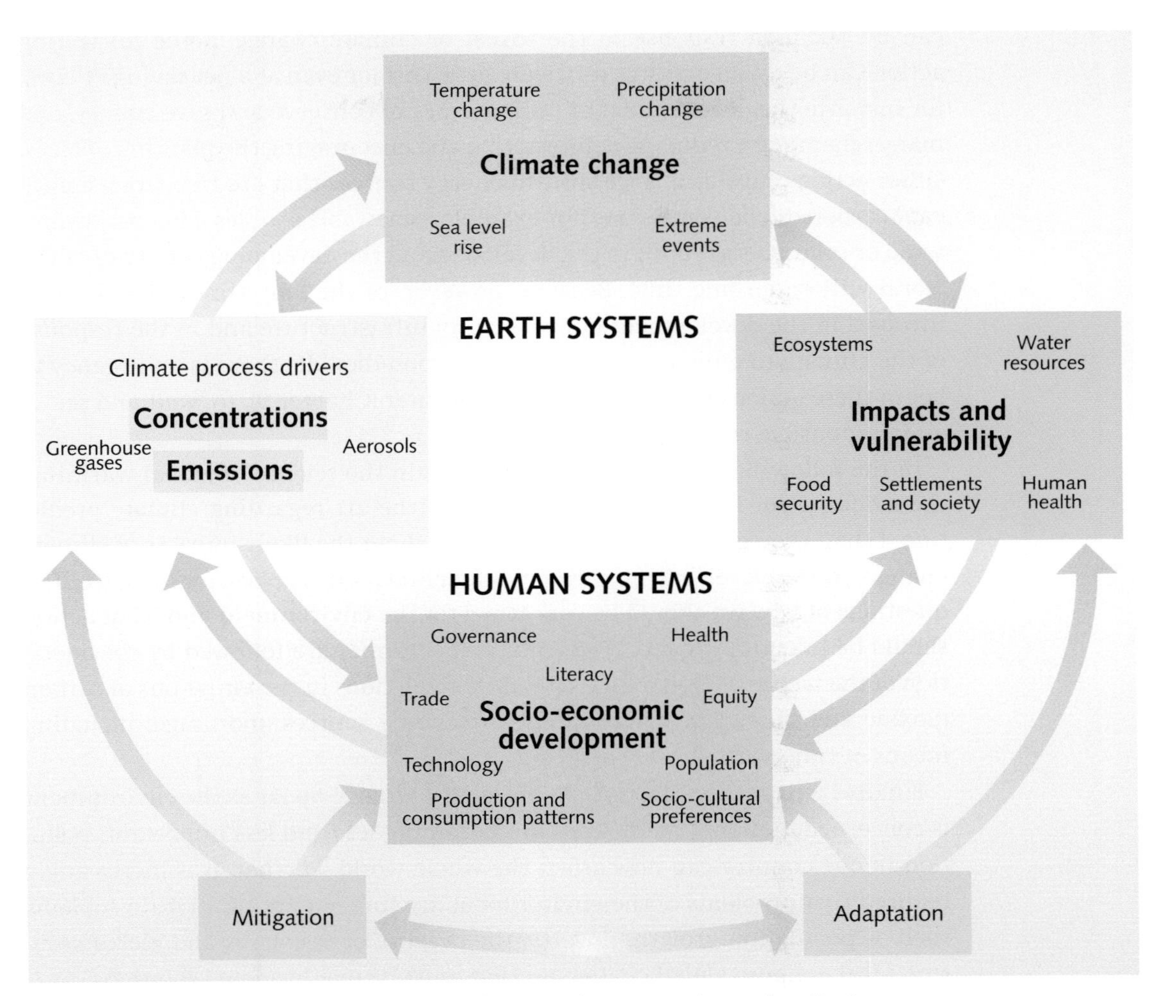

Figure 1.5 Climate change – an integrated framework (see text for explanation).

and mitigation is aimed at reducing the causes of climate change, in particular the emissions of the gases that give rise to it.

Uncertainty and response

Predictions of the future climate are surrounded with considerable uncertainty which arises from our imperfect knowledge both of the science of climate change and of the future scale of the human activities that are its cause. Politicians and others making decisions are therefore faced with the need to weigh all aspects of uncertainty against the desirability and the cost of the actions that can be taken in response to the threat of climate change. Some mitigating action can be taken easily at relatively little cost (or even at a net saving of cost), for instance the development of programmes to conserve and save energy, and many schemes for reducing deforestation and encouraging the planting of trees. Other actions such as a large shift to energy sources that are free from significant carbon dioxide emissions (for example, renewable sources: biomass, hydro, wind or solar energy) both in the developed and the developing countries of the world will take some time. Because, however, of the long timescales that are involved in the development of new energy infrastructure and in the response of the climate to emissions of gases like carbon dioxide, there is an urgency to begin these actions now. As we shall argue later (Chapter 9), to 'wait and see' is an irresponsible response.

In the following chapters I shall first explain the science of global warming, the evidence for it and the current state of the art regarding climate prediction. I shall then go on to say what is known about the likely impacts of climate change – on sea level, extreme events, water and food supplies, for instance. The questions of why we should be concerned for the environment and what action should be taken in the face of scientific uncertainty are followed by consideration of the technical possibilities for large reductions in the emissions of carbon dioxide and how these might affect our energy sources and usage, including means of transport.

Finally I will address the issue of the 'global village'. So far as the environment is concerned, national boundaries are becoming less and less important; pollution in one country can now affect the whole world. Further, it is increasingly realised that problems of the environment are linked to other global problems such as population growth, poverty, the overuse of resources and global security. All these pose global challenges that must be met by global solutions.

QUESTIONS

1. Look through recent copies of newspapers and magazines for articles that mention climate change, global warming or the greenhouse effect. How many of the statements made are accurate?
2. Make up a simple questionnaire about climate change, global warming and the greenhouse effect to find out how much people know about these subjects, their relevance and importance. Analyse results from responses to the questionnaire in terms of the background of the respondents. Suggest ways in which people could be better informed.

▶ FURTHER READING AND REFERENCE

Walker, Gabrielle and King, Sir David. 2008. *The Hot Topic*. London: Bloomsbury. A masterful paperback on climate change for the general reader covering the science, impacts, technology and political solutions.

NOTES FOR CHAPTER 1

1 Summary for policymakers, p. 5 in Solomon, S., Qin, D., Manning, M., Chen, Z., Marquis, M., Averyt, K.B., Tignor, M., Miller, H.L. (eds.) 2007. *Climate Change 2007: The Physical Science Basis. Contribution of Working Group 1 to the Fourth Assessment Report of the Intergovernmental Panel on Climate Change*. Cambridge: Cambridge University Press.

2 Including windstorms, hurricanes or typhoons, floods, tornadoes, hailstorms and blizzards but not including droughts because their impact is not immediate and occurs over an extended period.

3 A description of the variety of El Niño events and their impacts on different communities worldwide over centuries of human history can be found in a paperback by Ross Couiper-Johnston, *El Niño: The Weather Phenomenon that Changed the World*. 2000. London: Hodder and Stoughton.

4 A gripping account of some of the changes over recent decades can be found in a book by Mark Lynas, *High Tides: News from a Warming World*. 2004. London: Flamingo.

2 The greenhouse effect

This view of the rising Earth greeted the Apollo 8 astronauts as they came out from behind the Moon.

THE BASIC principle of global warming can be understood by considering the radiation energy from the Sun that warms the Earth's surface and the thermal radiation from the Earth and the atmosphere that is radiated out to space. On average these two radiation streams must balance. If the balance is disturbed (for instance by an increase in atmospheric carbon dioxide) it can be restored by an increase in the Earth's surface temperature.

How the Earth keeps warm

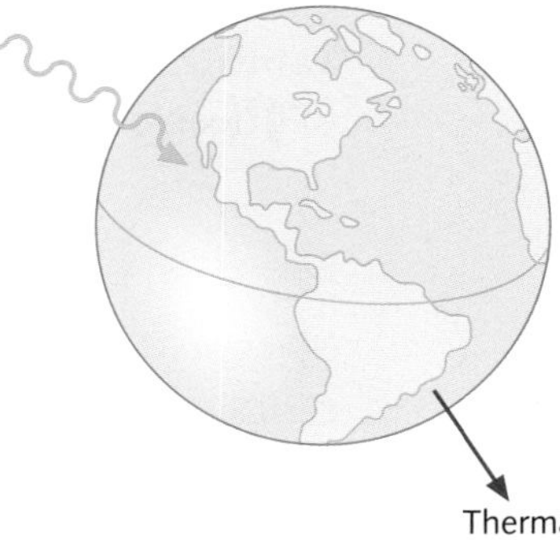

To explain the processes that warm the Earth and its atmosphere, I will begin with a very simplified Earth. Suppose we could, all of a sudden, remove from the atmosphere all the clouds, the water vapour, the carbon dioxide and all the other minor gases and the dust, leaving an atmosphere of nitrogen and oxygen only. Everything else remains the same. What, under these conditions, would happen to the atmospheric temperature?

The calculation is an easy one, involving a relatively simple radiation balance. Radiant energy from the Sun falls on a surface of one square metre in area outside the atmosphere and directly facing the Sun at a rate of about 1370 watts – about the power radiated by a reasonably sized domestic electric fire. However, few parts of the Earth's surface face the Sun directly and in any case for half the time they are pointing away from the Sun at night, so that the average energy falling on one square metre of a level surface outside the atmosphere is only one-quarter of this[1] or about 342 watts. As this radiation passes through the atmosphere a small amount, about 6%, is scattered back to space by atmospheric molecules. About 10% on average is reflected back to space from the land and ocean surface. The remaining 84%, or about 288 watts per square metre on average, remains actually to heat the surface – the power used by three good-sized incandescent electric light bulbs.

Figure 2.1 The radiation balance of planet Earth. The net incoming solar radiation is balanced on average by outgoing thermal radiation from the Earth.

To balance this incoming energy, the Earth itself must radiate on average the same amount of energy back to space (Figure 2.1) in the form of thermal radiation. All objects emit this kind of radiation; if they are hot enough we can see the radiation they emit. The Sun at a temperature of about 6000 °C looks white; an electric fire at 800 °C looks red. Cooler objects emit radiation that cannot be seen by our eyes and which lies at wavelengths beyond the red end of the spectrum – infrared radiation (sometimes called longwave radiation to distinguish it from the shortwave radiation from the Sun). On a clear, starry winter's night we are very aware of the cooling effect of this kind of radiation being emitted by the Earth's surface into space – it often leads to the formation of frost.

The amount of thermal radiation emitted by the Earth's surface depends on its temperature – the warmer it is, the more radiation is emitted. The amount of radiation also depends on how absorbing the surface is; the greater the

Table 2.1 The composition of the atmosphere, the main constituents (nitrogen and oxygen) and the greenhouse gases as in 2007

Gas	Mixing ratio or mole fraction[a] expressed as fraction* or parts per million (ppm)
Nitrogen (N_2)	0.78*
Oxygen (O_2)	0.21*
Water vapour (H_2O)	Variable (0–0.02*)
Carbon dioxide (CO_2)	380
Methane (CH_4)	1.8
Nitrous oxide (N_2O)	0.3
Chlorofluorocarbons	0.001
Ozone (O_3)	Variable (0–1000)

[a] For definition see Glossary.

absorption, the more the radiation. Most of the surfaces on the Earth, including ice and snow, would appear 'black' if we could see them at infrared wavelengths; that means that they absorb nearly all the thermal radiation which falls on them instead of reflecting it. It can be calculated[2] that the 288 W m^{-2} of incoming solar radiation received by the Earth's surface can be balanced by thermal radiation emitted by the surface at a temperature of −6 °C.[3] This is over 20 °C colder than is actually the case. In fact, an average of temperatures measured near the surface all over the Earth – over the oceans as well as the land – averaging, too, over the whole year, comes to about 15 °C. Some factor not yet taken into account is needed to explain this difference.

The greenhouse effect

The gases nitrogen and oxygen that make up the bulk of the atmosphere (Table 2.1 gives details of the atmosphere's composition) neither absorb nor emit thermal radiation. It is the water vapour, carbon dioxide and some other minor gases present in the atmosphere in much smaller quantities (Table 2.1) that absorb some of the thermal radiation leaving the surface, acting as a partial blanket for this radiation and causing the difference of 20 to 30 °C between the actual average surface temperature on the Earth of about 15 °C and the temperature that would apply if greenhouse gases were absent.[4] This blanketing is known as the *natural greenhouse effect* and the gases are

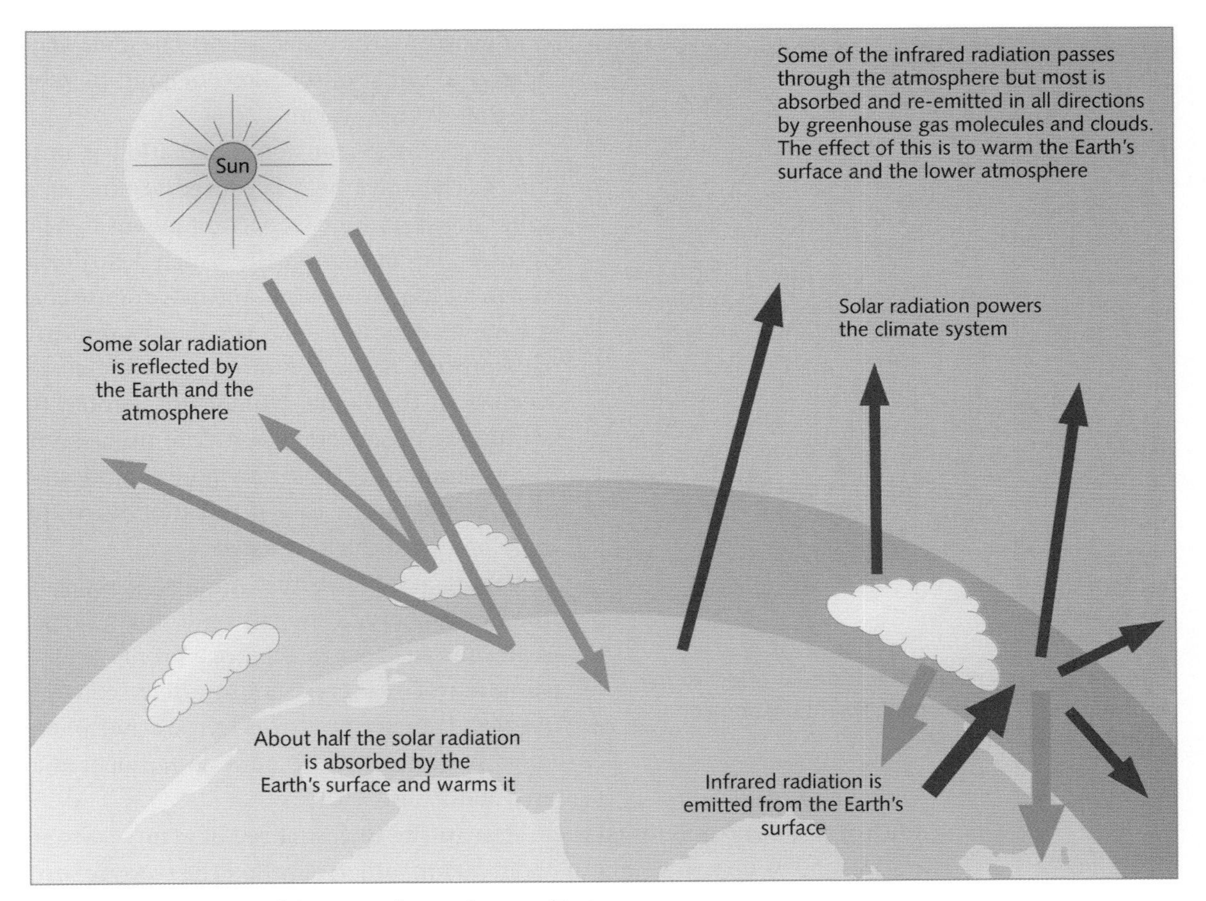

Figure 2.2 Schematic of the natural greenhouse effect.

known as greenhouse gases (Figure 2.2). It is called 'natural' because all the atmospheric gases (apart from the chlorofluorocarbons – CFCs) were there long before human beings came on the scene. Later on I will mention the *enhanced greenhouse effect*: the added effect caused by the gases present in the atmosphere due to human activities such as deforestation and the burning of fossil fuels.

The basic science of the greenhouse effect has been known since early in the nineteenth century (see box) when the similarity between the radiative properties of the Earth's atmosphere and of the glass in a greenhouse (Figure 2.3) was first pointed out – hence the name 'greenhouse effect'. In a greenhouse, visible radiation from the Sun passes almost unimpeded through the glass and is absorbed by the plants and the soil inside. The thermal radiation that is emitted by the plants and soil is, however, absorbed by the glass that re-emits some

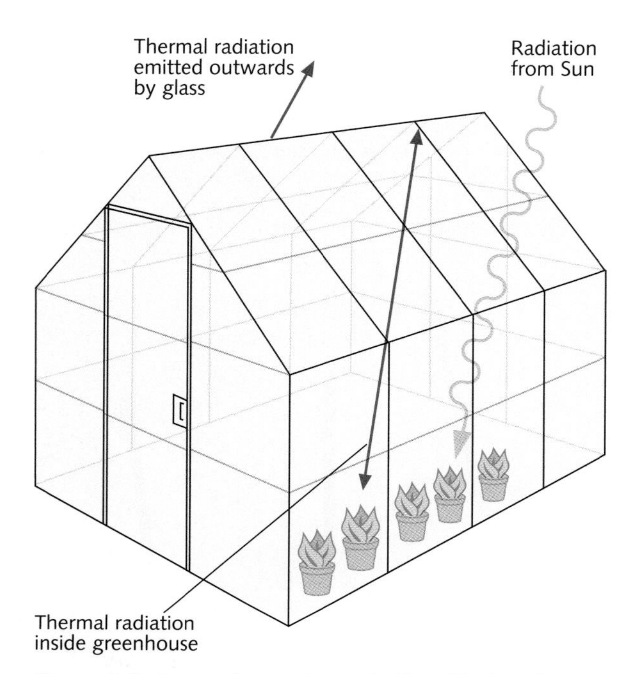

Figure 2.3 A greenhouse has a similar effect to the atmosphere on the incoming solar radiation and the emitted thermal radiation.

of it back into the greenhouse. The glass thus acts as a 'radiation blanket' helping to keep the greenhouse warm.

However, the transfer of radiation is only one of the ways heat is moved around in a greenhouse. A more important means of heat transfer is convection, in which less dense warm air moves upwards and more dense cold air moves downwards. A familiar example of this process is the use of convective electric heaters in the home, which heat a room by stimulating convection in it. The situation in the greenhouse is therefore more complicated than would be the case if radiation were the only process of heat transfer.

Mixing and convection are also present in the atmosphere, although on a much larger scale, and in order to achieve a proper understanding of the greenhouse effect, convective heat transfer processes in the atmosphere must be taken into account as well as radiative ones.

Within the atmosphere itself (at least in the lowest three-quarters or so of the atmosphere up to a height of about 10 km which is called the troposphere) convection is, in fact, the dominant process for transferring heat. It acts as follows. The surface of the Earth is warmed by the sunlight it absorbs. Air close to the surface is heated and rises because of its lower density. As the air rises it expands and cools – just as the air cools as it comes out of the valve of a tyre. As some air masses rise, other air masses descend, so the air is continually turning over as different movements balance each other out – a situation of convective equilibrium. Temperature in the troposphere falls with height at a rate determined by these convective processes; the fall with height (called the lapse rate) turns out on average to be about 6 °C per kilometre of height (Figure 2.4).

A picture of the transfer of radiation in the atmosphere may be obtained by looking at the thermal radiation emitted by the Earth and its atmosphere as observed from instruments on satellites orbiting the Earth (Figure 2.5). At some wavelengths in the infrared the atmosphere – in the absence of clouds – is largely transparent, just as it is in the visible part of the spectrum. If our eyes were sensitive at these wavelengths we would be able to peer through the

Pioneers of the science of the greenhouse effect[5]

The warming effect of the greenhouse gases in the atmosphere was first recognised in 1827 by the French scientist Jean-Baptiste Fourier, best known for his contributions to mathematics. He also pointed out the similarity between what happens in the atmosphere and in the glass of a greenhouse, which led to the name 'greenhouse effect'. The next step was taken by a British scientist, John Tyndall, who, around 1860, measured the absorption of infrared radiation by carbon dioxide and water vapour; he also suggested that a cause of the ice ages might be a decrease in the greenhouse effect of carbon dioxide. It was a Swedish chemist, Svante Arrhenius, in 1896, who calculated the effect of an increasing concentration of greenhouse gases; he estimated that doubling the concentration of carbon dioxide would increase the global average temperature by 5 to 6 °C, an estimate not too far from our present understanding.[6] Nearly 50 years later, around 1940, G. S. Callendar, working in England, was the first to calculate the warming due to the increasing carbon dioxide from the burning of fossil fuels.

Svante August Arrhenius (19 February 1859 – 2 October 1927).

The first expression of concern about the climate change that might be brought about by increasing greenhouse gases was in 1957, when Roger Revelle and Hans Suess of the Scripps Institute of Oceanography in California published a paper which pointed out that in the build-up of carbon dioxide in the atmosphere, human beings are carrying out a large-scale geophysical experiment. In the same year, routine measurements of carbon dioxide were started from the observatory on Mauna Kea in Hawaii. The rapidly increasing use of fossil fuels since then, together with growing interest in the environment, has led to the topic of global warming moving up the political agenda through the 1980s, and eventually to the Climate Convention signed in 1992 – of which more in later chapters.

atmosphere to the Sun, stars and Moon above, just as we can in the visible spectrum. At these wavelengths all the radiation originating from the Earth's surface leaves the atmosphere.

At other wavelengths radiation from the surface is strongly absorbed by some of the gases present in the atmosphere, in particular by water vapour and carbon dioxide.

Objects that are good absorbers of radiation are also good emitters of it. A black surface is both a good absorber and a good emitter, while a highly reflecting surface absorbs rather little and emits rather little too (which is why highly

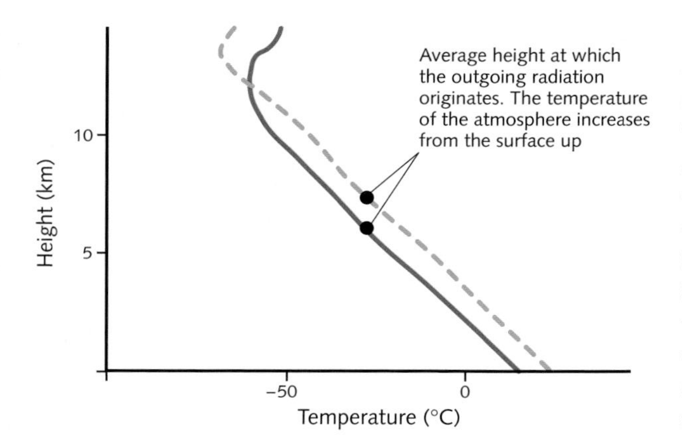

Figure 2.4 The distribution of temperature in a convective atmosphere (red line). The green line shows how the temperature increases when the amount of carbon dioxide present in the atmosphere is increased (in the diagram the difference between the lines is exaggerated – for instance, for doubled carbon dioxide in the absence of other effects the increase in temperature is about 1.2 °C). Also shown for the two cases are the average levels from which thermal radiation leaving the atmosphere originates (about 6 km for the unperturbed atmosphere).

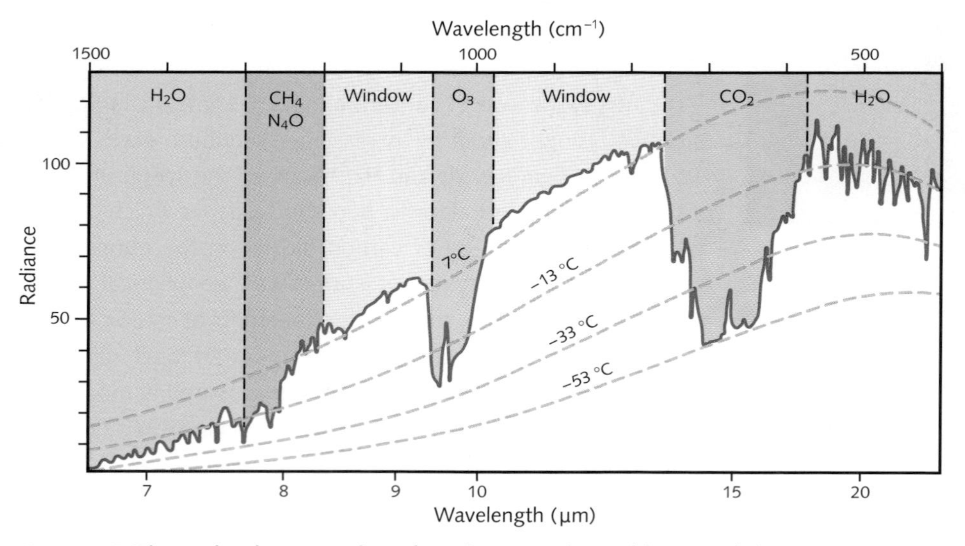

Figure 2.5 Thermal radiation in the infrared region (the visible part of the spectrum is between about 0.4 and 0.7 μm) emitted from the Earth's surface and atmosphere as observed over the Mediterranean Sea from a satellite instrument orbiting above the atmosphere, showing parts of the spectrum where different gases contribute to the radiation. Between the wavelengths of about 8 and 14 μm, apart from the ozone band, the atmosphere, in the absence of clouds, is substantially transparent; this is part of the spectrum called a 'window' region. Superimposed on the spectrum are curves of radiation from a black body at 7 °C, −13 °C, −33 °C and −53 °C. The units of radiance are watts per square metre per steradian per wavenumber.

Ice, oceans, land surfaces and clouds all play a role in determining how much incoming solar radiation the Earth reflects back into space.

reflecting foil is used to cover the surface of a vacuum flask and why it is placed above the insulation in the lofts of houses).

Absorbing gases in the atmosphere absorb some of the radiation emitted by the Earth's surface and in turn emit radiation out to space. The amount of thermal radiation they emit is dependent on their temperature.

Radiation is emitted out to space by these gases from levels somewhere near the top of the atmosphere – typically from between 5 and 10 km high (see Figure 2.5). Here, because of the convection processes mentioned earlier, the temperature is much colder – 30 to 50 °C or so colder – than at the surface. Because the gases are cold, they emit correspondingly less radiation. What these gases have to do, therefore, is absorb some of the radiation emitted by the Earth's surface but then to emit much less radiation out to space. They, therefore, act as a radiation blanket over the surface (note that the outer surface of a blanket is colder than inside the blanket) and help to keep it warmer than it would otherwise be (Figure 2.6).

There needs to be a balance between the radiation coming in and the radiation leaving the top of the atmosphere – as there was in the very simple model with which this chapter started. Figure 2.7 shows the various components of the radiation entering and leaving the top of the atmosphere for the real atmosphere situation. On average, 235 watts per square metre of solar radiation are absorbed by the atmosphere and the surface; this is less than the 288 watts mentioned at the beginning of the chapter, because now the effect of clouds is being taken into account. Clouds reflect some of the incident radiation from the Sun back out to space. However, they also absorb and emit thermal radiation and have a blanketing effect similar to that of the greenhouse gases. These two effects work in opposite senses: one (the reflection of solar radiation) tends

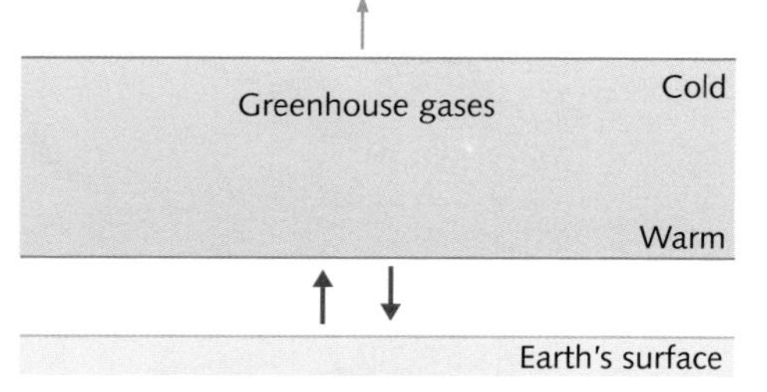

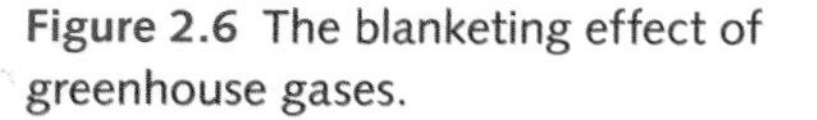

Figure 2.6 The blanketing effect of greenhouse gases.

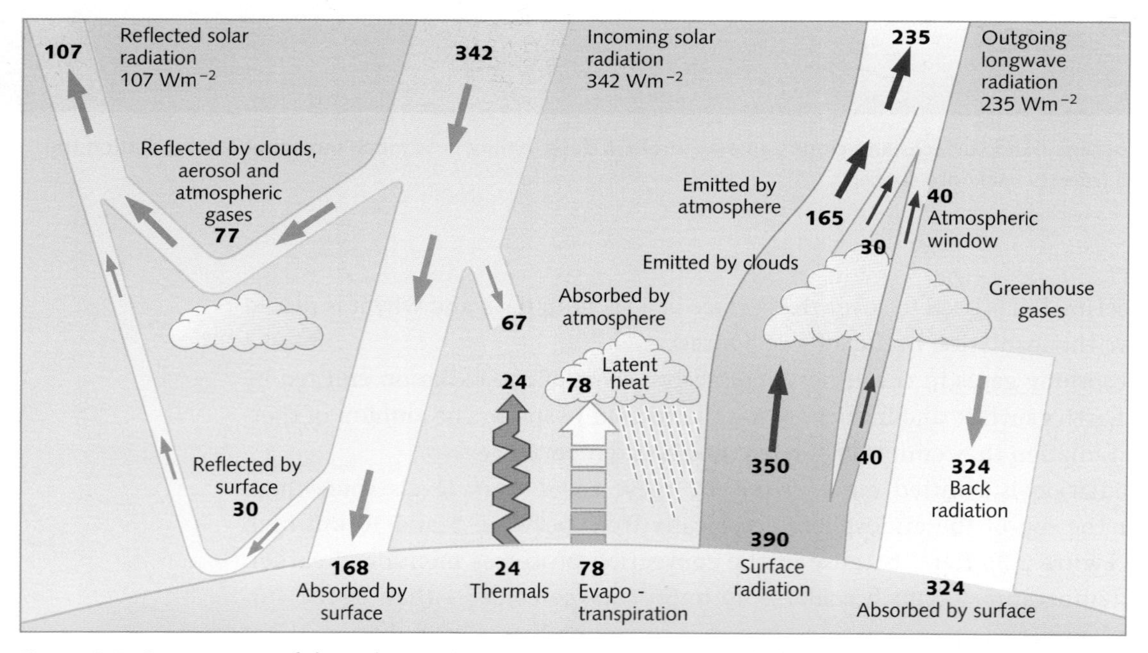

Figure 2.7 Components of the radiation (in watts per square metre) which on average enter and leave the Earth's atmosphere and make up the radiation budget for the atmosphere. About half of the incoming solar radiation is absorbed by the Earth's surface. This energy is transferred to the atmosphere by warming the air in contact with the surface (thermals), by evapotranspiration and by longwave radiation that is absorbed by clouds and greenhouse gases. The atmosphere in turn radiates longwave energy back to Earth as well as out to space.

to cool the Earth's surface and the other (the absorption of thermal radiation) tends to warm it. Careful consideration of these two effects shows that on average the net effect of clouds on the total budget of radiation results in a slight cooling of the Earth's surface.[7]

The numbers in Figure 2.7 demonstrate the required balance: 235 watts per square metre on average coming in and 235 watts per square metre on average going out. The temperature of the surface and hence of the atmosphere above adjusts itself to ensure that this balance is maintained. It is interesting to note that the greenhouse effect can only operate if there are colder temperatures in the higher atmosphere. Without the structure of decreasing temperature with height, therefore, there would be no greenhouse effect on the Earth.

Mars and Venus

Similar greenhouse effects also occur on our nearest planetary neighbours, Mars and Venus. Mars is smaller than the Earth and possesses, by Earth's standards, a very thin atmosphere. A barometer on the surface of Mars would record an atmospheric pressure less than 1% of that on the Earth. Its atmosphere, which consists almost entirely of carbon dioxide, contributes a small but significant greenhouse effect.

The planet Venus, which can often be seen fairly close to the Sun in the morning or evening sky, has a very different atmosphere to Mars. Venus is about the same size as the Earth. A barometer for use on Venus would need to survive very hostile conditions and would need to be able to measure a pressure about 100 times as great as that on the Earth. Within the Venus atmosphere, which consists very largely of carbon dioxide, deep clouds consisting of droplets of almost pure sulphuric acid completely cover the planet and prevent most of the sunlight from reaching the surface. Some Russian space probes that have landed there have recorded what would be dusk-like conditions on the Earth – only 1% or 2% of the sunlight present above the clouds penetrates that far. One might suppose, because of the small amount of solar energy available to keep the surface warm, that it would be rather cool; on the contrary, measurements from the same Russian space probes find a temperature there of about 525 °C – a dull red heat, in fact.

The reason for this very high temperature is the greenhouse effect. Because of the very thick absorbing atmosphere of carbon dioxide, little of the thermal radiation from the surface can get out. The atmosphere acts as such an effective radiation blanket that, although there is not much solar energy to warm the surface, the greenhouse effect amounts to nearly 500 °C.

The planets Mars, Earth and Venus have significant atmospheres. This diagram shows the approximate relative sizes of the terrestrial planets.

The 'runaway' greenhouse effect

What occurs on Venus is an example of what has been called the 'runaway' greenhouse effect. It can be explained by imagining the early history of the Venus atmosphere, which was formed by the release of gases from the interior of the planet. To start with it would contain a lot of water vapour, a powerful greenhouse gas (Figure 2.8). The greenhouse effect of the water vapour would cause the temperature at the surface to rise. The increased temperature would lead to more evaporation of water from the surface, giving more atmospheric water vapour, a larger greenhouse effect and therefore a further increased surface temperature. The process would continue until either the atmosphere became saturated with water vapour or all the available water had evaporated.

A runaway sequence something like this seems to have occurred on Venus. Why, we may ask, has it not happened on the Earth, a planet of about the same size as Venus and, so far as is known, of a similar initial chemical composition? The reason is that Venus is closer to the Sun than the Earth; the amount of solar energy per square metre falling on Venus is about twice that falling on the Earth. The surface of Venus, when there was no atmosphere, would have started off at a temperature of just over 50 °C (Figure 2.8). Throughout the sequence described above for Venus, water on the surface would have been continuously boiling. Because of the high temperature, the atmosphere would never have become saturated with water vapour. The Earth, however, would have started at a colder temperature; at each stage of the sequence it would have arrived at

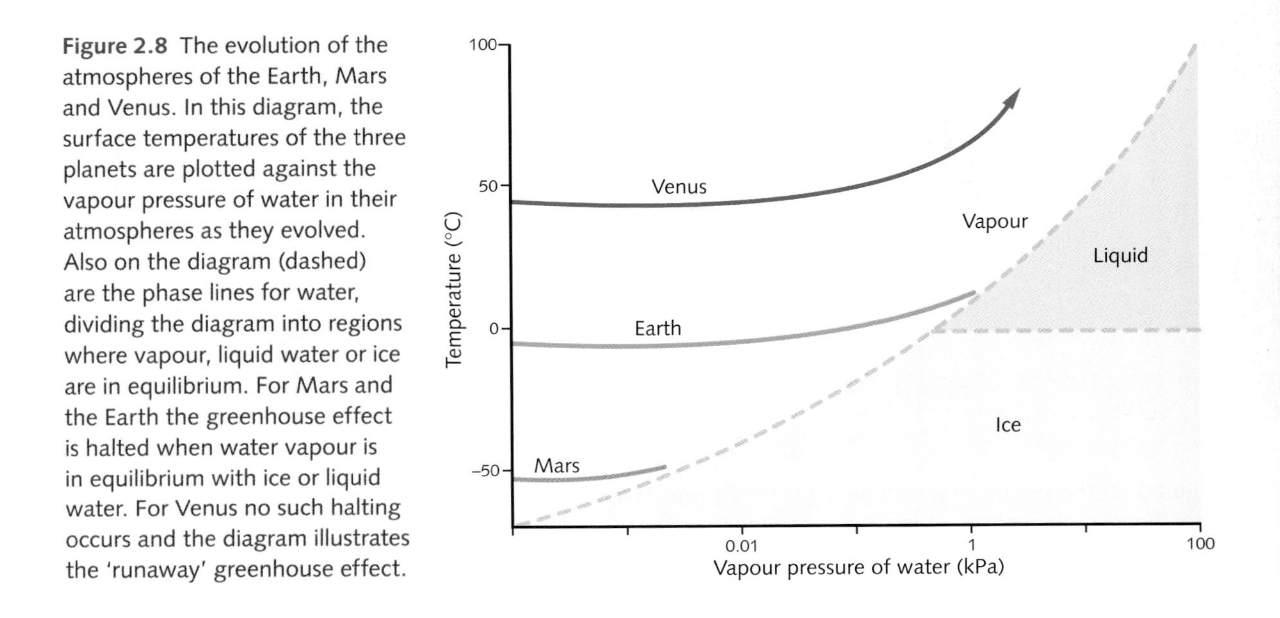

Figure 2.8 The evolution of the atmospheres of the Earth, Mars and Venus. In this diagram, the surface temperatures of the three planets are plotted against the vapour pressure of water in their atmospheres as they evolved. Also on the diagram (dashed) are the phase lines for water, dividing the diagram into regions where vapour, liquid water or ice are in equilibrium. For Mars and the Earth the greenhouse effect is halted when water vapour is in equilibrium with ice or liquid water. For Venus no such halting occurs and the diagram illustrates the 'runaway' greenhouse effect.

an equilibrium between the surface and an atmosphere saturated with water vapour. There is no possibility of such runaway greenhouse conditions occurring on the Earth.

The enhanced greenhouse effect

After our excursion to Mars and Venus, let us return to Earth! The natural greenhouse effect is due to the gases water vapour and carbon dioxide present in the atmosphere in their natural abundances as now on Earth. The amount of water vapour in our atmosphere depends mostly on the temperature of the surface of the oceans; most of it originates through evaporation from the ocean surface and is not influenced directly by human activity. Carbon dioxide is different. Its amount has changed substantially – by nearly 40% so far – since the Industrial Revolution, due to human industry and also because of the removal of forests (see Chapter 3). Future projections are that, in the absence of controlling factors, the rate of increase in atmospheric carbon dioxide will accelerate and that its atmospheric concentration will double from its pre-industrial value within the next 100 years (Figure 6.2).

This increased amount of carbon dioxide is leading to global warming of the Earth's surface because of its enhanced greenhouse effect. Let us imagine, for instance, that the amount of carbon dioxide in the atmosphere suddenly doubled, everything else remaining the same (Figure 2.9). What would happen

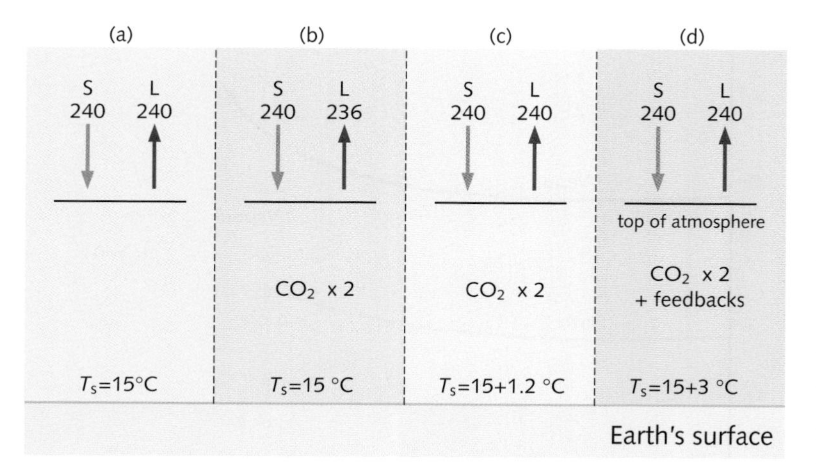

Figure 2.9 The enhanced greenhouse gas effect. Under natural conditions (a) the net solar radiation coming in (S = 240 watts per square metre) is balanced by thermal radiation (L) leaving the top of the atmosphere; average surface temperature (T_s) is 15°C. If the carbon dioxide concentration is suddenly doubled (b), L is decreased by 4 watts per square metre. Balance is restored if nothing else changes (c) apart from the temperature of the surface and lower atmosphere, which rises by 1.2 °C. If feedbacks are also taken into account (d), the average temperature of the surface rises by about 3 °C.

to the numbers in the radiation budget presented earlier (Figure 2.7). The solar radiation budget would not be affected. The greater amount of carbon dioxide in the atmosphere means that the thermal radiation emitted from it will originate on average from a higher and colder level than before (Figure 2.4). The thermal radiation budget will therefore be reduced, the amount of reduction being about 4 watts per square metre (a more precise value is 3.7).

This causes a net imbalance in the overall budget of 4 watts per square metre. More energy is coming in than going out. To restore the balance the surface and lower atmosphere will warm up. If nothing changes apart from the temperature – in other words, the clouds, the water vapour, the ice and snow cover and so on are all the same as before – the temperature change turns out to be about 1.2 °C.

In reality, of course, many of these other factors will change, some of them in ways that add to the warming (these are called positive feedbacks), others in ways that might reduce the warming (negative feedbacks). The situation is therefore much more complicated than this simple calculation. These complications will be considered in more detail in Chapter 5. Suffice it to say here that the best estimate at the present time of the increased average temperature of the Earth's surface if carbon dioxide levels were to be doubled is about twice that of the simple calculation: 3.0 °C. As the last chapter explained, for the global average temperature this is a large change. It is this global warming expected to result from the enhanced greenhouse effect that is the cause of current concern.

Having dealt with a doubling of the amount of carbon dioxide, it is interesting to ask what would happen if all the carbon dioxide were removed from the atmosphere. It is sometimes supposed that the outgoing radiation would be changed by 4 watts per square metre in the other direction and that the Earth would then cool by one or two degrees Celsius. In fact, that would happen if the carbon dioxide amount were to be halved. If it were to be removed altogether, the change in outgoing radiation would be around 25 watts per square metre – six times as big – and the temperature change would be similarly increased. The reason for this is that with the amount of carbon dioxide currently present in the atmosphere there is maximum carbon dioxide absorption over much of the region of the spectrum where it absorbs (Figure 2.5), so that a big change in gas concentration leads to a relatively small change in the amount of radiation it absorbs.[8] This is like the situation in a pool of water: when it is clear, a small amount of mud will make it appear muddy, but when it is muddy, adding more mud only makes a small difference.

An obvious question to ask is: has evidence of the enhanced greenhouse effect been seen in the recent climatic record? Chapter 4 will look at the record of temperature on the Earth during the last century or so, during which the Earth has warmed on average by about three-quarters of a degree Celsius. We shall see in Chapters 4 and 5 that there are good reasons for attributing most of this warming to the enhanced greenhouse effect, although because of the size of natural climate variability the exact amount of that attribution remains subject to some uncertainty.

SUMMARY

- No one doubts the reality of the natural greenhouse effect, which keeps us over 20 °C warmer than we would otherwise be. The science of it is well understood; it is similar science that applies to the enhanced greenhouse effect.
- Substantial greenhouse effects occur on our nearest planetary neighbours, Mars and Venus. Given the conditions that exist on those planets, the sizes of their greenhouse effects can be calculated, and good agreement has been found with those measurements that are available.
- Study of climates of the past gives some clues about the greenhouse effect, as Chapter 4 will show.

First, however, the greenhouse gases themselves must be considered. How does carbon dioxide get into the atmosphere, and what other gases affect global warming?

QUESTIONS

1. Carry out the calculation suggested in Note 4 (refer also to Note 2) to obtain an equilibrium average temperature for an Earth partially covered with clouds such that 30% of the incoming solar radiation is reflected. If clouds are assumed to cover half the Earth and if the reflectivity of the clouds increases by 1% what change will this make in the resulting equilibrium average temperature?
2. It is sometimes argued that the greenhouse effect of carbon dioxide is negligible because its absorption band in the infrared is so close to saturation that there is very little additional absorption of radiation emitted from the surface. What are the fallacies in this argument?
3. Use the information in Figure 2.5 to estimate approximately the surface temperature that would result if carbon dioxide were completely removed from the atmosphere. What is required is that the total energy radiated by the Earth plus atmosphere should remain the same, i.e. the area under the radiance curve in Figure 2.5 should be unaltered. On this basis construct a new curve with the carbon dioxide band absent.[9]
4. Using information from books or articles on climatology or meteorology describe why the presence of water vapour in the atmosphere is of such importance in determining the atmosphere's circulation.
5. Estimates of regional warming due to increased greenhouse gases are generally larger over land areas than over ocean areas. What might be the reasons for this?
6. (For students with a background in physics) What is meant by Local Thermodynamic Equilibrium (LTE),[10] a basic assumption underlying calculations of radiative transfer in the lower atmosphere appropriate to discussions of the greenhouse effect? Under what conditions does LTE apply?

▶ FURTHER READING AND REFERENCE

Historical overview of climate change science. Chapter 1, in Solomon, S., Qin, D., Manning, M., Chen, Z., Marquis, M., Averyt, K. B., Tignor, M., Miller, H. L. (eds.) 2007. *Climate Change 2007: The Physical Science Basis. Contribution of Working Group I to the Fourth Assessment Report of the Intergovernmental Panel on Climate Change*. Cambridge: Cambridge University Press.

Houghton J. 2002. *The Physics of Atmospheres*, third edition. Cambridge: Cambridge University Press, Chapters 1 and 14.

NOTES FOR CHAPTER 2

1 It is about one-quarter because the area of the Earth's surface is four times the area of the disc which is the projection of the Earth facing the Sun; see Figure 2.1.

2 The radiation by a black body is the Stefan–Boltzmann constant (5.67×10^{-8} J m^{-2} K^{-4} s^{-1}) multiplied by the fourth power of the body's absolute temperature in Kelvin. The absolute temperature is the temperature in degrees Celsius plus 273 (1 K = 1 °C).

3 These calculations using a simple model of an atmosphere containing nitrogen and oxygen only have been carried out to illustrate the effect of the other gases, especially water vapour and carbon dioxide. It is not, of course, a model that can exist in reality. All the water vapour could not be removed from the atmosphere above a water or ice surface. Further, with an average surface temperature of −6 °C, in a real situation the surface would have much more ice cover. The additional ice would reflect more solar energy out to space leading to a further lowering of the surface temperature.

4 The calculation I made giving a temperature of −6 °C for the average temperature of the Earth's surface if greenhouse gases are not present not only ignored the different reflectivity of ice compared with the present surface but also ignored the presence of clouds. Depending on the assumptions made regarding clouds and other factors, values ranging between 20 and 30 °C are quoted for the difference in surface temperature with and without greenhouse gases present.

5 Further details can be found in Mudge, F. B. 1997. The development of greenhouse theory of global climate change from Victorian times. *Weather*, **52**, 13–16.

6 A range of 2 to 4.5 °C is quoted in Chapter 6, page 143.

7 More detail of the radiative effects of clouds is given in Chapter 5; see Figures 5.14 and 5.15.

8 The dependence of the absorption on the concentration of gas is approximately logarithmic.

9 For some helpful diagrams and more information about the infrared spectrum of different greenhouse gases, see Harries, J. E. 1996. The greenhouse Earth: a view from space. *Quarterly Journal of the Royal Meteorological Society*, **122**, 799–818.

10 For information about LTE see, for instance, Houghton, J. T. 2002. *The Physics of Atmospheres*, third edition. Cambridge: Cambridge University Press.

3 The greenhouse gases

Industrial activity: a source of carbon dioxide and other gaseous and particulate pollution.

THE GREENHOUSE gases are those gases in the atmosphere which, by absorbing thermal radiation emitted by the Earth's surface, have a blanketing effect upon it. The most important of the greenhouse gases is water vapour, but its amount in the atmosphere is not changing directly because of human activities. The important greenhouse gases that are directly influenced by human activities are carbon dioxide, methane, nitrous oxide, the chlorofluorocarbons (CFCs) and ozone. This chapter will describe what is known about the origin of these gases, how their concentration in the atmosphere is changing and how it is controlled. Also considered will be particles in the atmosphere of anthropogenic origin, some of which can act to cool the surface.

Which are the most important greenhouse gases?

Figure 2.5 illustrated the regions of the infrared spectrum where the greenhouse gases absorb. Their importance as greenhouse gases depends both on their concentration in the atmosphere (Table 2.1) and on the strength of their absorption of infrared radiation. Both these quantities differ greatly for various gases.

Carbon dioxide is the most important of the greenhouse gases that are increasing in atmospheric concentration because of human activities. If, for the moment, we ignore the effects of the CFCs and of changes in ozone, which vary considerably over the globe and which are therefore more difficult to quantify, the increase in carbon dioxide (CO_2) has contributed about 72% of the enhanced greenhouse effect to date, methane (CH_4) about 21% and nitrous oxide (N_2O) about 7% (Figure 3.11).

Radiative forcing

In this chapter we shall use the concept of *radiative forcing* to compare the relative greenhouse effects of different atmospheric constituents. It is necessary therefore first to define radiative forcing.

In Chapter 2 we noted that, if the carbon dioxide in the atmosphere were suddenly doubled, everything else remaining the same, a net radiation imbalance near the top of the atmosphere of 3.7 W m^{-2} would result. This radiation imbalance is an example of radiation forcing, which is defined as the change in average net radiation at the top of the troposphere[1] (the lower atmosphere; for definition see Glossary) which occurs because of a change in the concentration of a greenhouse gas or because of some other change in the overall climate system; for instance, a change in the incoming solar radiation would constitute a radiative forcing. As we saw in the discussion in Chapter 2, over time the climate responds to restore the radiative balance between incoming and outgoing radiation. A positive radiative forcing tends on average to warm the surface and a negative radiative forcing tends on average to cool the surface.

Carbon dioxide and the carbon cycle

Carbon dioxide provides the dominant means through which carbon is transferred in nature between a number of natural carbon reservoirs – a process known as the carbon cycle. We contribute to this cycle every time we breathe. Using the oxygen we take in from the atmosphere, carbon from our food is burnt and turned into carbon dioxide that we then exhale; in this way we are provided with the energy we need to maintain our life. Animals contribute to atmospheric carbon dioxide in the same way; so do fires, rotting wood and decomposition of organic material in the soil and elsewhere. To offset these

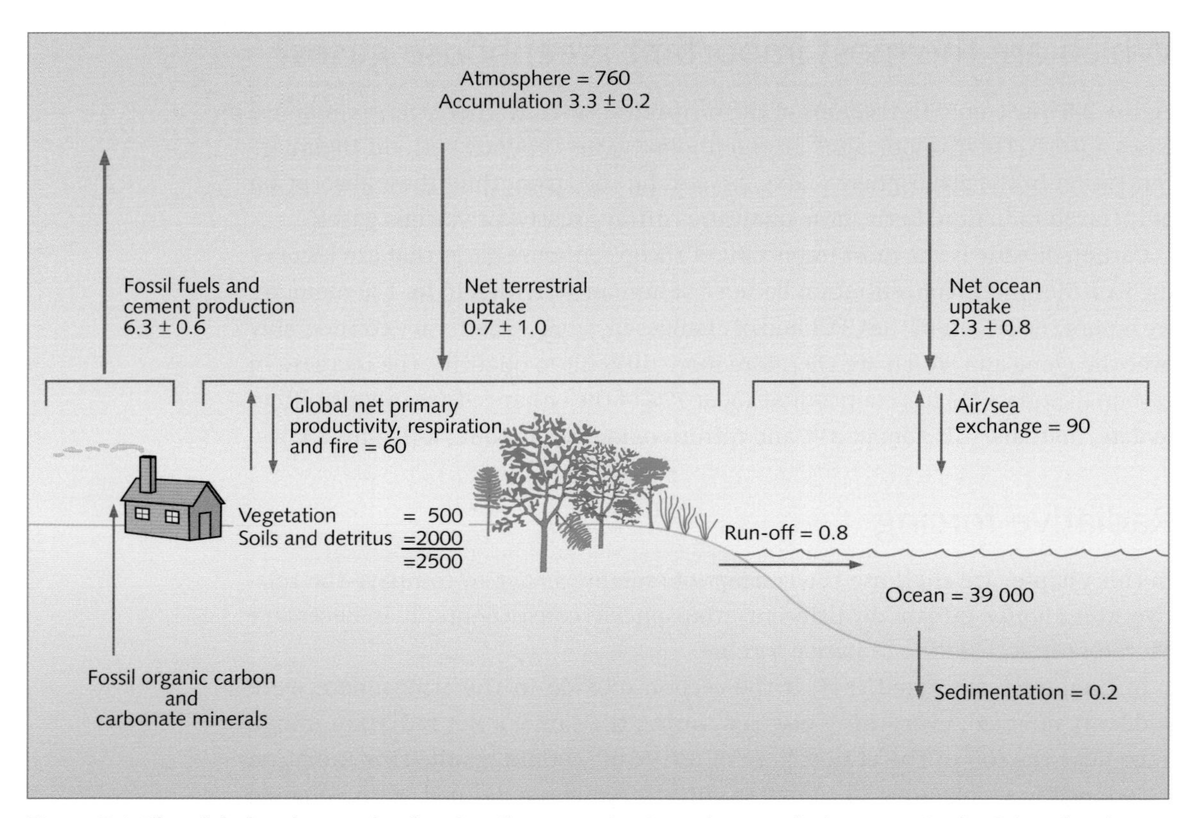

Figure 3.1 The global carbon cycle, showing the approximate carbon stocks in reservoirs (in Gt) and carbon flows (in Gt year^{-1}) relevant to the anthropogenic perturbation as annual averages over the decade from 1989 to 1998. Net ocean uptake of the anthropogenic perturbation equals the net air/sea input plus run-off minus sediment. The units are thousand millions of tonnes or gigatonnes (Gt). (More detail in Fig. 7.3 in Chapter 7 of IPCC AR4 WGI 2007.)

processes of respiration whereby carbon is turned into carbon dioxide, there are processes involving photosynthesis in plants and trees which work the opposite way; in the presence of light, they take in carbon dioxide, use the carbon for growth and return the oxygen back to the atmosphere. Both respiration and photosynthesis also occur in the ocean.

Figure 3.1 is a simple diagram of the way carbon cycles between the various reservoirs – the atmosphere, the oceans (including the ocean biota), the soil and the land biota (biota is a word that covers all living things – plants, trees, animals and so on – on land and in the ocean, which make up a whole known as the biosphere). The diagram shows that the movements of carbon (in the form of carbon dioxide) into and out of the atmosphere are quite large; about one-fifth of the total amount in the atmosphere is cycled in and out each year, part with the land biota and part through physical and chemical processes across the ocean surface. The land and ocean reservoirs are much larger than the amount in the atmosphere; small changes in these larger reservoirs could therefore have a large effect on the atmospheric concentration; the release of just 2% of the carbon stored in the oceans would double the amount of atmospheric carbon dioxide.

It is important to realise that on the timescales with which we are concerned anthropogenic carbon emitted into the atmosphere as carbon dioxide is not destroyed but redistributed among the various carbon reservoirs. Carbon dioxide is therefore different from other greenhouse gases that are destroyed by chemical action in the atmosphere. The carbon reservoirs exchange carbon between themselves on a wide range of timescales determined by their respective turnover times – which range from less than a year to decades (for exchange with the top layers of the ocean and the land biosphere) to millennia (for exchange with the deep ocean or long-lived soil pools). These timescales are generally much longer than the average time a particular carbon dioxide molecule spends in the atmosphere, which is only about four years. The large range of turnover times means that the time taken for a perturbation in the atmospheric carbon dioxide concentration to relax back to an equilibrium cannot be described by a single time constant. About 50% of an increase in atmospheric carbon dioxide will be removed within 30 years, a further 30% within a few centuries and the remaining 20% may remain in the atmosphere for many thousands of years.[2] Although a lifetime of about 100 years is often quoted for atmospheric carbon dioxide so as to provide some guide, use of a single lifetime can be very misleading.

Before human activities became a significant disturbance, and over periods short compared with geological timescales, the exchanges between the reservoirs were remarkably constant. For thousands of years before the beginning of industrialisation around 1750, a steady balance was maintained, such that the mixing ratio (or mole fraction; for definition see Glossary) of carbon dioxide in the atmosphere as measured from ice cores (see Chapter 4) kept within about 20 parts per million (ppm) of a mean value of about 280 ppm (Figure 3.2a).

The Industrial Revolution disturbed this balance and since its beginning over 600 thousand million tonnes (or gigatonnes, Gt) of carbon have been emitted into the atmosphere from fossil fuel burning. This has resulted in a concentration of carbon dioxide in the atmosphere that has increased by about 36%, from 280 ppm around 1700 to a value of over 380 ppm at the present day (Figure 3.2a), a greater concentration than for at least 650 000 years. Accurate measurements, which have been made since 1959 from an observatory near the summit of Mauna Loa in Hawaii, show that from 1995 to 2005 carbon dioxide increased

Didcot power station, near Oxford, UK.

on average each year by about 1.9 ppm (an increase from the average for the 1990s of about 1.5 ppm, although there are large variations from year to year (Figure 3.2b)). This increase spread through the atmosphere adds about 3.8 Gt to the atmospheric carbon reservoir each year.

It is easy to establish how much coal, oil and gas is being burnt worldwide each year. Most of it is to provide energy for human needs: for heating and domestic appliances, for industry and for transport (considered in detail in Chapter 11). The burning of these fossil fuels has increased rapidly since the Industrial Revolution (Table 3.1). Over the 1990s emissions rose about 0.7% per year; from 1999 to 2005 annual emissions rose systematically from 6.5 to 7.8 Gt of carbon (an annual increase averaging about 3%), nearly all of which enters the atmosphere as carbon dioxide. Another contribution to atmospheric carbon dioxide due to human activities comes from land-use change, in particular from tropical deforestation balanced in part by afforestation or forest regrowth. This contribution is not easy to quantify but some estimates are given in Table 3.1. For the 1990s (see Table 3.1), annual anthropogenic emissions from fossil fuel burning, cement manufacture (about 3% of the total) and land-use change amounted to about 8.0 Gt; over three-quarters of these resulted from fossil fuel burning. Since the annual net increase in the atmosphere was about 3.2 Gt, about 40% of the 8 Gt of new carbon remained to increase the atmospheric concentration

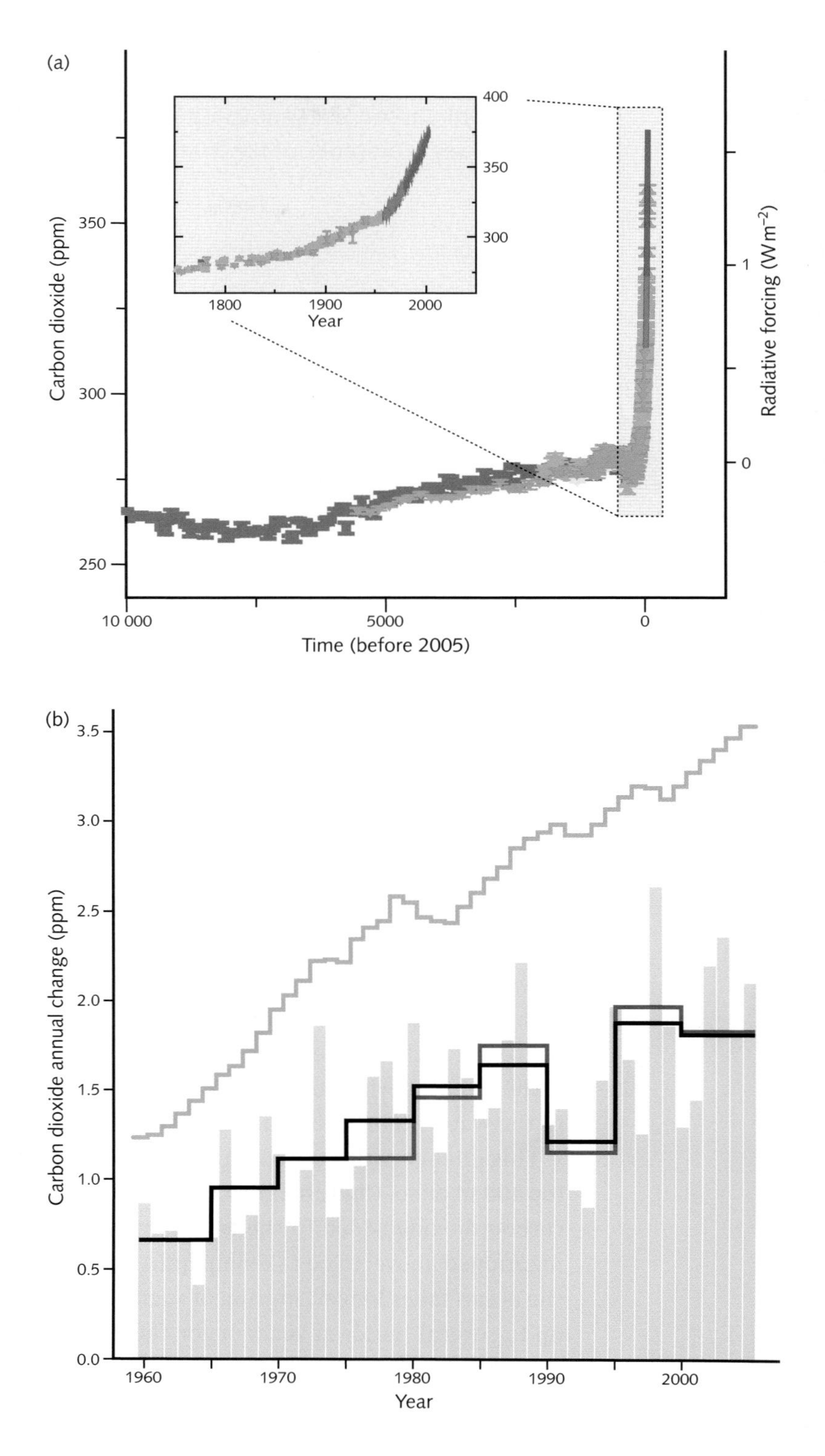

Figure 3.2 Atmospheric carbon dioxide concentration. (a) Over the last 10 000 years (inset since 1750) from various ice cores (symbols with different colours for different studies) and atmospheric samples (red lines). Corresponding radiative forcings shown on right-hand axis. (b) Annual changes in global mean and their five-year means from two different measurement networks (red and black stepped lines). The five-year means smooth out short-term perturbations associated with strong El Niño Southern Oscillation (ENSO) events in 1972, 1982, 1987 and 1997. The upper dark green line shows the annual increases that would occur if all fossil fuel emissions stayed in the atmosphere and there were no other emissions.

Table 3.1 Components of annual average global carbon budget for 1980s and 1990s in Gt of carbon per year (positive values are fluxes to the atmosphere, negative values represent uptake from the atmosphere)

	1980s	1990s	2000–2005
Emissions (fossil fuel, cement)	5.4 ± 0.3	6.4 ± 0.4	7.2 ± 0.3
Atmospheric increase	3.3 ± 0.1	3.2 ± 0.1	4.1 ± 0.1
Ocean–atmosphere flux	−1.8 ± 0.8	−2.2 ± 0.4	−2.2 ± 0.5
Land–atmosphere flux*	−0.3 ± 0.9	−1.0 ± 0.6	−0.9 ± 0.6
**partitioned as follows*			
Land-use change	1.4 (0.6 to 2.3)	1.6 (0.5 to 2.7)	not available
Residual terrestrial sink	−1.7 (−3.4 to 0.2)	−2.6 (−4.3 to −0.9)	not available

(Figure 3.2b). The other 60% was taken up between the other two reservoirs: the oceans and the land biota. Figure 3.5 shows that, as global average temperatures increase the fractions taken up by both land and ocean are likely to reduce.

About 95% of fossil fuel burning occurs in the northern hemisphere, so there is more carbon dioxide there than in the southern hemisphere. The difference is currently about 2 ppm (Figure 3.3) and, over the years, has grown in parallel with fossil fuel emissions, thus adding further compelling evidence that the atmospheric increase in carbon dioxide levels results from these emissions.

We now turn to what happens in the oceans. We know that carbon dioxide dissolves in water; carbonated drinks make use of that fact. Carbon dioxide is continually being exchanged with the air above the ocean across the whole ocean surface (about 90 Gt per year is so exchanged – Figure 3.1), particularly as waves break. An equilibrium is established between the concentration of carbon dioxide dissolved in the surface waters and the concentration in the air above the surface. The chemical laws governing this equilibrium are such that if the atmospheric concentration changes by 10% the concentration in solution in the water changes by only one-tenth of this: 1%.

This change will occur quite rapidly in the upper waters of the ocean, the top 100 m or so, thus enabling part of the anthropogenic (i.e. human-generated) carbon dioxide added to the atmosphere (most of the ocean's share of the 60% mentioned above) to be taken up quite rapidly. Absorption in the lower levels in the ocean takes longer; mixing of surface water with water at lower levels takes up to several hundred years or for the deep ocean over a thousand years. This process whereby carbon dioxide is gradually drawn from the atmosphere into the ocean's lower levels is sometimes known as the *solubility pump*.

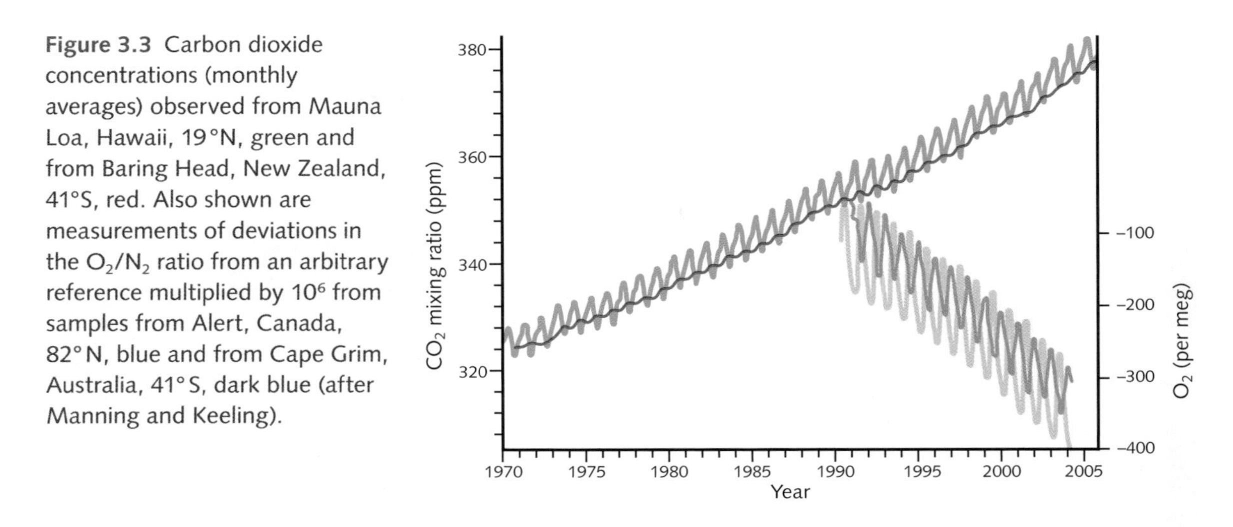

Figure 3.3 Carbon dioxide concentrations (monthly averages) observed from Mauna Loa, Hawaii, 19 °N, green and from Baring Head, New Zealand, 41°S, red. Also shown are measurements of deviations in the O_2/N_2 ratio from an arbitrary reference multiplied by 10^6 from samples from Alert, Canada, 82° N, blue and from Cape Grim, Australia, 41° S, dark blue (after Manning and Keeling).

So the oceans do not provide as immediate a sink for increased atmospheric carbon dioxide as might be suggested by the size of the exchanges with the large ocean reservoir. For short-term changes only the surface layers of water play a large part in the carbon cycle. Further, it is likely that a warmer regime will be associated on average with weaker overturning in the ocean and therefore with reduced carbon dioxide uptake.

Biological activity in the oceans also plays an important role. It may not be immediately apparent, but the oceans are literally teeming with life. Although the total mass of living matter within the oceans is not large, it has a high rate of turnover. Living material in the oceans is produced at some 30–40% of the rate of production on land. Most of this production is of plant and animal plankton which go through a rapid series of life cycles. As they die and decay some of the carbon they contain is carried downwards into lower levels of the ocean adding to the carbon content of those levels. Some is carried to the very deep water or to the ocean bottom where, so far as the carbon cycle is concerned, it is out of circulation for hundreds or thousands of years. This process, whose contribution to the carbon cycle is known as the *biological pump* (see box), was important in determining the changes of carbon dioxide concentration in both the atmosphere and the ocean during the ice ages (see Chapter 4).

Computer models – which calculate solutions for the mathematical equations describing a given physical situation, in order to predict its behaviour (see Chapter 5) – have been set up to describe in detail the exchanges of carbon between the atmosphere and different parts of the ocean. To test the validity of these models, they have also been applied to the dispersal in the ocean of the carbon isotope ^{14}C that entered the ocean after the nuclear tests of the 1950s;

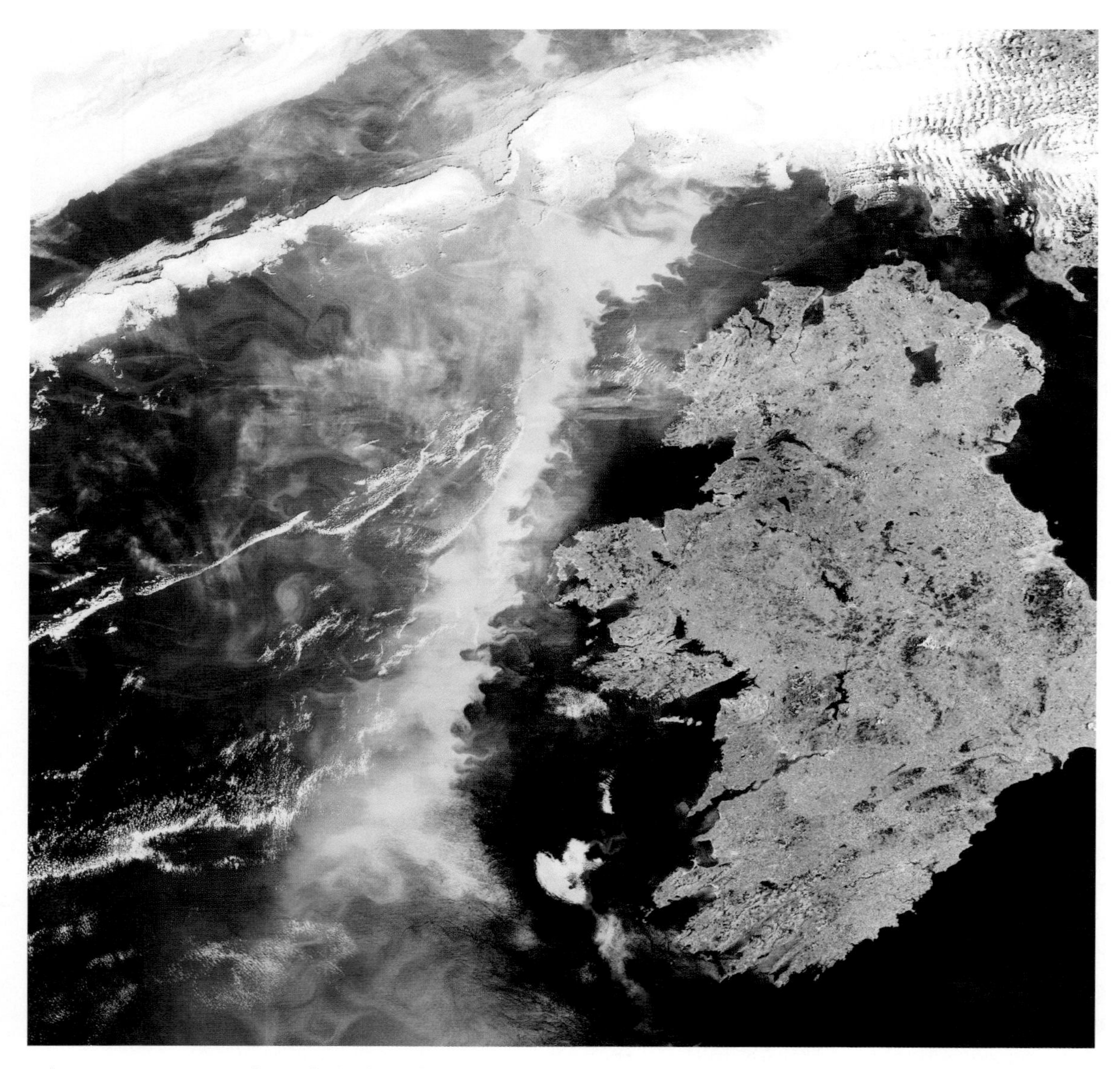

A large aquamarine-coloured plankton bloom is shown stretching across the length of Ireland in the North Atlantic Ocean in this image, captured on 6 June 2006 by Envisat's MERIS satellite, a dedicated ocean colour sensor able to identify phytoplankton concentrations.

the models simulate this dispersal quite well. From the model results, it is estimated that about 2 Gt ($\pm$ 0.8 Gt) of the carbon dioxide added to the atmosphere each year ends up in the oceans (see Table 3.1). Observations of the relative distribution of the other isotopes of carbon in the atmosphere and in the oceans also confirm this estimate (see box).

The *biological pump* in the oceans[3]

In temperate and high latitudes there is a peak each spring in ocean biological activity. During the winter, water rich in nutrients is transferred from deep water to levels near the surface. As sunlight increases in the spring an explosive growth of the plankton population occurs, known as the 'spring bloom'. Pictures of ocean colour taken from satellites demonstrate dramatically where this is happening.

Plankton are small plants (phytoplankton) and animals (zooplankton) that live in the surface waters of the ocean; they range in size between about 0.001mm across and the size of typical insects on land. Herbivorous zooplankton graze on phytoplankton; carnivorous zooplankton eat herbivorous zooplankton. Plant and animal debris from these living systems sinks in the ocean. While sinking, some decomposes and returns to the water as nutrients, some (perhaps about 1%) reaches the deep ocean or the ocean floor, where it is lost to the carbon cycle for hundreds, thousands or even millions of years. The net effect of the 'biological pump' is to move carbon from the surface waters to lower levels in the ocean. As the amount of carbon in the surface waters is reduced, more carbon dioxide from the atmosphere can be drawn down to restore the surface equilibrium. It is thought that the 'biological pump' has remained substantially constant in its operation during the last century unaffected by the increase in carbon dioxide.

Evidence of the importance of the 'biological pump' comes from the palaeoclimate record from ice cores (see Chapter 4). One of the constituents from the atmosphere trapped in bubbles in the ice is the gas methyl sulphonic acid, which originates from decaying ocean plankton; its concentration is therefore an indicator of plankton activity. As the global temperature began to increase when the last ice age receded nearly 20 000 years ago and as the carbon dioxide in the atmosphere began to increase (Figure 4.4), the methyl sulphonic acid concentration decreased. An interesting link is thereby provided between the carbon dioxide in the atmosphere and marine biological activity. During the cold periods of the ice ages, enhanced biological activity in the ocean could have been responsible for maintaining the atmospheric carbon dioxide at a lower level of concentration – the 'biological pump' was having an effect.

There is some evidence from the palaeo record of the biological activity in the ocean being stimulated by the presence of iron-containing dust blown over the oceans from the land surface. This has led to some proposals in recent years to enhance the 'biological pump' through artificially introducing iron over suitable parts of the ocean. While an interesting idea, it seems from careful studies that even a very large-scale operation would not have a large practical effect.

The question then remains as to why the ice ages were periods of greater marine biological activity than the warm periods in between. One possible contributing process is suggested by considering what happens in winter as nutrients are fed into the upper ocean ready for the spring bloom of biological activity. When there is less atmospheric carbon dioxide, cooling by radiation from the surface of the ocean increases. Since convection in the upper layers of the ocean is driven by cooling at the surface, increased cooling results in a greater depth of the mixed layer near the top of the ocean where all the biological activity occurs. This is an example of a positive biological feedback; a greater depth of layer means more plankton growth.[4]

What we can learn from carbon isotopes

Isotopes are chemically identical forms of the same element but with different atomic weights. Three isotopes of carbon are important in studies of the carbon cycle: the most abundant isotope ^{12}C which makes up 98.9% of ordinary carbon, ^{13}C present at about 1.1% and the radioactive isotope ^{14}C which is present only in very small quantities. About 10 kg of ^{14}C is produced in the atmosphere each year by the action of particle radiation from the Sun; half of this will decay into nitrogen over a period of 5730 years (the 'half-life' of ^{14}C).

When carbon in carbon dioxide is taken up by plants and other living things, less ^{13}C is taken up in proportion than ^{12}C. Fossil fuel such as coal and oil was originally living matter so also contains less ^{13}C (by about 18 parts per 1000) than the carbon dioxide in ordinary air in the atmosphere today. Adding carbon to the atmosphere from burning forests, decaying vegetation or fossil fuel will therefore tend to reduce the proportion of ^{13}C.

Because fossil fuel has been stored in the Earth for much longer than 5730 years (the half-life of ^{14}C), it contains no ^{14}C at all. Therefore, carbon from fossil fuel added to the atmosphere reduces the proportion of ^{14}C the atmosphere contains.

By studying the ratio of the different isotopes of carbon in the atmosphere, in the oceans, in gas trapped in ice cores and in tree rings, it is possible to find out where the additional carbon dioxide in the atmosphere has come from and also what amount has been transferred to the ocean. For instance, it has been possible to estimate for different times how much carbon dioxide has entered the atmosphere from the burning or decay of forests and other vegetation and how much from fossil fuels.

Similar isotopic measurements on the carbon in atmospheric methane provide information about how much methane from fossil fuel sources has entered the atmosphere at different times.

Further information regarding the broad partitioning of added atmospheric carbon dioxide between the atmosphere, the oceans and the land biota as presented in Table 3.1 comes from comparing the trends in atmospheric carbon dioxide concentration with the trends in very accurate measurements of the atmospheric oxygen/nitrogen ratio (Figures 3.3 and 3.4). This possibility arises because the relation between the exchanges of carbon dioxide and oxygen with the atmosphere over land is different from that over the ocean. On land, living organisms through photosynthesis take in carbon dioxide from the atmosphere and build up carbohydrates, returning the oxygen to the atmosphere. In the process of respiration they also take in oxygen from the atmosphere and convert it to carbon dioxide. In the ocean, by contrast, carbon dioxide taken from the atmosphere is dissolved, both the carbon and the oxygen in the molecules being removed. How such measurements can be interpreted for the period 1990–4 is shown in Figure 3.4. These data are consistent with budget for the 1990s shown in Table 3.1.

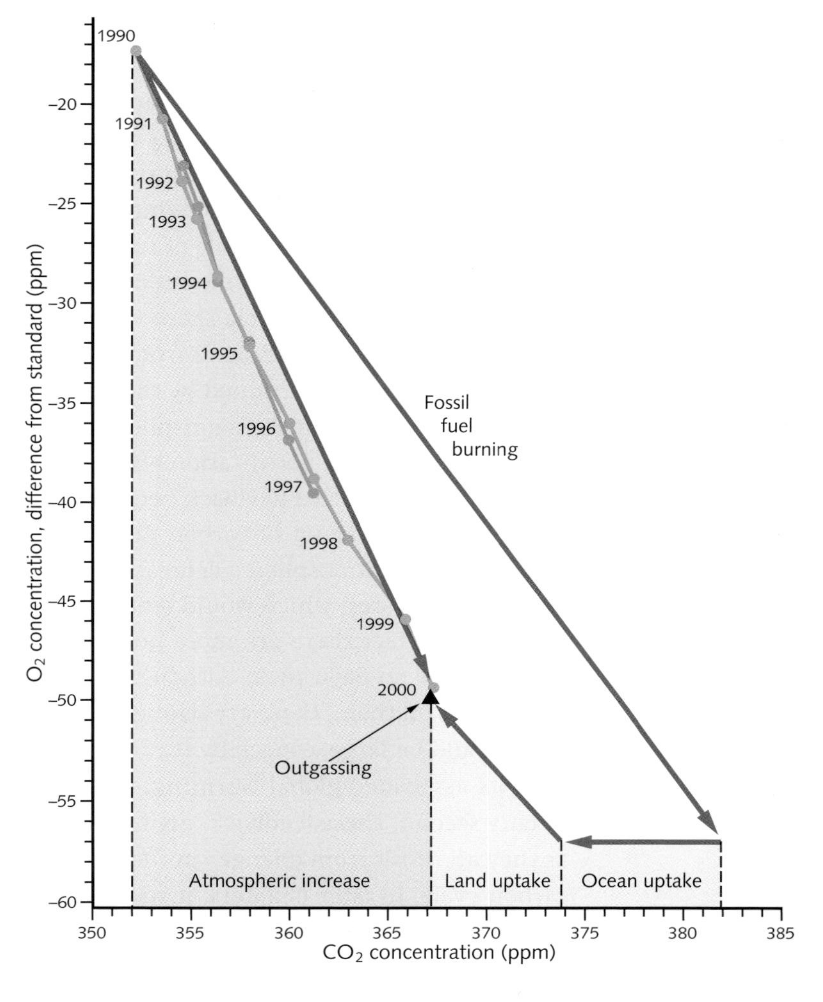

Figure 3.4 Partitioning of fossil fuel carbon dioxide uptake using oxygen measurements. Shown is the relationship between changes in carbon dioxide and oxygen concentrations. Observations are shown by solid circles. The arrow labelled 'fossil fuel burning' denotes the effect of the combustion of fossil fuels based on the $O_2:CO_2$ stoichiometric relation of the different fuel types. Uptake by land and ocean is constrained by the stoichiometric ratio associated with these processes, defining the slopes of the respective arrows.

The global land–atmosphere flux in Table 3.1 represents the balance of a net flux due to land-use changes which has generally been positive or a source of carbon to the atmosphere and a residual component that is, by inference, a negative flux or carbon sink. The estimates of land-use changes (Table 3.1) are dominated by deforestation in tropical regions although some uptake of carbon has occurred through forest regrowth in temperate regions of the northern hemisphere and other changes in land management. The main processes that contribute to the residual carbon sink are believed to be the carbon dioxide 'fertilisation' effect (increased carbon dioxide in the atmosphere leads to increased growth in some plants – see box in Chapter 7 on page 199), the effects of increased use of nitrogen fertilisers and of some changes in climate. The magnitudes of these contributions (Table 3.1) are difficult to estimate directly and are subject to much more uncertainty than their total, which can be inferred from the requirement to balance the overall carbon cycle budget.

A clue to the uptake of carbon by the land biosphere is provided from observations of the atmospheric concentration of carbon dioxide which, each year, show a regular cycle; the seasonal variation, for instance, at the observatory site at Mauna Loa in Hawaii approaches about 10 ppm (Figure 3.3). Carbon dioxide is removed from the atmosphere during the growing season and is returned as the vegetation dies away in the winter. In the northern hemisphere therefore a minimum in the annual cycle of carbon dioxide occurs in the northern summer. Since there is a larger amount of the terrestrial biosphere in the northern hemisphere the annual cycle there has a much greater amplitude than in the southern hemisphere. Estimates from carbon cycle models of the uptake by the land biosphere are constrained by these observations of the seasonal cycle and the difference between the hemispheres.[5]

The carbon dioxide fertilisation effect is an example of a biological feedback process. It is a negative feedback because, as carbon dioxide increases, it tends to increase the take-up of carbon dioxide by plants and therefore reduce the amount in the atmosphere, decreasing the rate of global warming. Positive feedback processes, which would tend to accelerate the rate of global warming, also exist; in fact there are more potentially positive processes than negative ones (see box on page 48–9). Although scientific knowledge cannot yet put precise figures on them, there are strong indications that some of the positive feedbacks could be large, especially if carbon dioxide were to continue to increase, with its associated global warming, through the twenty-first century into the twenty-second. These feedbacks are often called climate/carbon-cycle feedbacks as they all result from changes in the climate affecting the performance of the carbon cycle. In later chapters, it will be seen that these feedbacks can assert large influence on future concentrations of carbon dioxide and hence on future climate.

Carbon dioxide provides the largest single contribution to anthropogenic radiative forcing. Its radiative forcing from pre-industrial times to the present is shown in Figure 3.11. A useful formula for the radiative forcing R from atmospheric carbon dioxide when its atmospheric concentration is C ppm is: $R = 5.3 \ln(C/C_0)$ where C_0 is its pre-industrial concentration of 280 ppm.

Future emissions of carbon dioxide

To obtain information about future climate we need to estimate the future atmospheric concentrations of carbon dioxide, which depend on future anthropogenic emissions. In these estimates, the long time constants associated with the response of atmospheric carbon dioxide to change have important implications. Suppose, for instance, that all emissions into the atmosphere from human activities were suddenly halted. No sudden change would occur in the atmospheric concentration, which would decline only slowly. We could not expect it to approach its pre-industrial value for several hundred years.

But emissions of carbon dioxide are not halting, nor are they slowing; their increase is, in fact, becoming larger each year. The atmospheric concentration of carbon dioxide will therefore also increase more rapidly. Later chapters (especially Chapter 6) will present estimates of climate change during the twenty-first century due to the increase in greenhouse gases. A prerequisite for such estimates is knowledge of likely changes in carbon dioxide emissions. Estimating what will happen in the future is, of course, not easy. Because nearly everything we do has an influence on the emissions of carbon dioxide, it means estimating how human beings will behave and what their activities will be. For instance, assumptions have to be made about population growth, economic growth, energy use, the development of energy sources and the likely influence of pressures to preserve the environment. These assumptions are required for all countries of the world, both developing as well as developed ones. Further, since any assumptions made are unlikely to be fulfilled accurately in practice, it is necessary to make a variety of different assumptions, so that we can get some idea of the range of possibilities. Such possible futures are called *scenarios*.

In Chapter 6 and Chapter 11 are presented two sets of emission scenarios as developed respectively by the Intergovernmental Panel on Climate Change (IPCC) and the International Energy Agency (IEA). These emission scenarios are then turned into future projections of atmospheric carbon dioxide concentrations through the application of a computer model of the carbon cycle that includes descriptions of all the exchanges already mentioned. Further in Chapter 6, through the application of computer models of the climate (see Chapter 5), projections of the resulting climate change from different scenarios are also presented.

Feedbacks in the biosphere

As the greenhouse gases carbon dioxide and methane are added to the atmosphere because of human activities, biological or other feedback processes occurring in the biosphere (such as those arising from the climate change that has been induced) influence the rate of increase of the atmospheric concentration of these gases. These processes will tend either to add to the anthropogenic increase (positive feedbacks) or to subtract from it (negative feedbacks).

Two feedbacks, one positive (the plankton multiplier in the ocean) and one negative (carbon dioxide fertilisation), have already been mentioned in the text. Four other positive feedbacks are potentially important, although our knowledge is currently insufficient to quantify them precisely.

One is the effect of higher temperatures on respiration, especially through microbes in soils, leading to increased carbon dioxide emissions. Evidence regarding the magnitude of this effect has come from studies

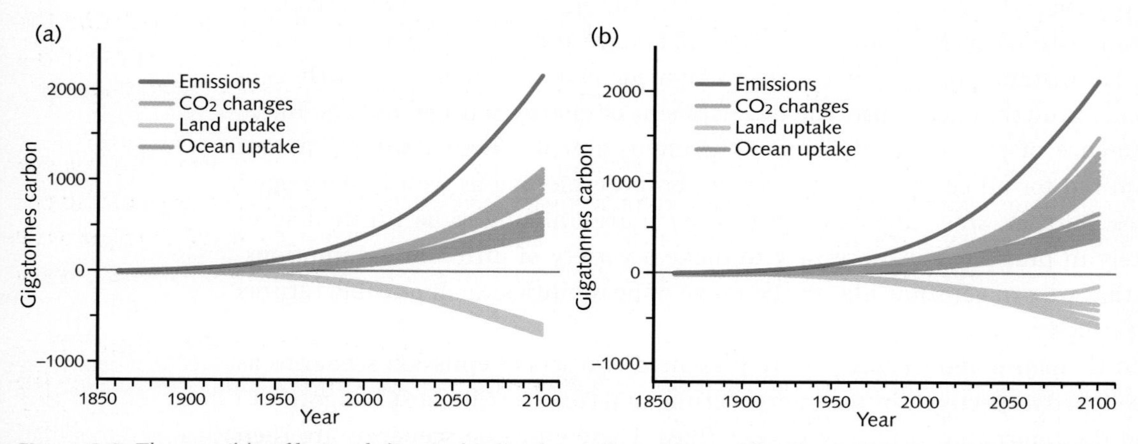

Figure 3.5 The possible effects of climate feedbacks on the carbon cycle. Shown are (1) accumulated fossil fuel CO_2 emissions from 1860 to the present and then projected to 2100 assuming the A2 SRES scenario (Figure 6.1) – in red (2) CO_2 from (1) absorbed into the ocean – in blue, (3) CO_2 taken up by the land (of the same sign as the other contributions but plotted below the axis for clarity) – in orange and (4) the residual CO_2 from (1) added to the atmosphere – in green. The nine different ocean and land budgets resulted from a study with nine coupled atmosphere–ocean general circulation climate models (AOGCMs – see chapter 5) organised internationally as part of a climate model intercomparison project (CMIP). (a) Shows results assuming no feedbacks from climate change into elements of the carbon cycle. (b) Shows results when climate feedbacks into the carbon cycle are included.

of the short-term variations of atmospheric carbon dioxide that have occurred during El Niño events and during the cooler period following the Pinatubo volcanic eruption in 1991. These studies, which covered variations over a few years, indicate a relation such that a change of 5 °C in average temperature leads to a 40% change in global average respiration rate[6] – a substantial effect. A question that needs to be resolved is whether this relation still holds over longer-term changes of the order of several decades to a century. A second positive feedback is the reduction of growth or the dieback especially in forests because of the stress caused by climate change, which may be particularly severe in Amazonia (see box in Chapter 7 on page 208).[7] As with the last effect, this will increase as the amount of climate change becomes larger. The combined result of these two feedbacks is that less carbon is taken up by the biosphere and more remains in the atmosphere.[8]

Figure 3.5 shows this combined result as estimated for the twenty-first century by nine different climate models that incorporate the relevant processes in both the ocean and the land biospheres. Note that one of the models in the ensemble, from the Hadley Centre, predicts the strongest values for the climate/carbon-cycle feedbacks mentioned above and projects the highest atmospheric carbon dioxide level by 2100.[9] The curve in Figure 3.5 for land uptake relating to this model begins to curve upwards from the middle of the century at which time the terrestrial biosphere changes from being a net sink of carbon (as in Table 3.1) to being a net source.

Taking the average of the nine models under the A2 scenario, about 50 Gt more carbon remains in the atmosphere in 2050 and 150 Gt in 2100 compared with what would occur in the absence of climate/carbon-cycle feedback. For the Hadley Centre model the numbers are about 50 Gt and 350 Gt respectively for 2050 and 2100. In terms of carbon dioxide concentration an additional 100 Gt means an additional 50 ppm.

A third positive feedback occurs through the release of greenhouse gases into the atmosphere due to the increase of fires in forested areas because of the drier conditions as climate warms or because of the dieback due to climate stress mentioned above[10].

The fourth positive feedback is the release of methane, as temperatures increase – from wetlands and from very large reservoirs of methane trapped in sediments in a hydrate form (tied to water molecules when under pressure) – mostly at high latitudes. Methane has been generated from the decomposition of organic matter present in these sediments over many millions of years. Because of the depth of the sediments this latter feedback is unlikely to become operative to a significant extent in the near future. However, were global warming to continue to increase unchecked for many decades, releases from hydrates could make a large contribution to methane emissions into the atmosphere and act as a large positive feedback on the climate.

Other greenhouse gases

Methane

Methane is the main component of natural gas. Its common name used to be marsh gas because it can be seen bubbling up from marshy areas where organic material is decomposing. Data from ice cores show that for at least 2000 years before 1800 its concentration in the atmosphere was about 700 ppb. Since then its concentration has more than doubled (Figure 3.6a) to a value that the ice core record shows is unprecedented over at least the last 650 000 years. During the 1980s it was increasing at about 10 ppb per year but during the 1990s the average rate of increase fell to around 5 ppb per year[11] and close to zero from 1999 to 2005. Although the concentration of methane in the atmosphere is much less than that of carbon dioxide (only 1.775 ppm in 2005 compared with about 380 ppm for carbon dioxide), its greenhouse effect is far from negligible. That is because the enhanced greenhouse effect caused by a molecule of methane is about eight times that of a molecule of carbon dioxide.[12]

The main natural source of methane is from wetlands. A variety of other sources result directly or indirectly from human activities, for instance from leakage from natural gas pipelines and from oil wells, from generation in rice paddy fields, from enteric fermentation (belching) from cattle and other livestock, from the decay of rubbish in landfill sites and from wood and peat burning. Details of estimates of the sizes of these sources during the 1990s are shown in Table 3.2. Attached to many of the numbers is a wide range of uncertainty. It is, for instance, difficult to estimate the amount produced in paddy fields averaged on a worldwide basis. The amount varies enormously during the rice growing season and also very widely from region to region. Similar problems arise when trying to estimate the amount produced by animals. Measurements of the proportions of the different isotopes of carbon (see box on page 44) in atmospheric methane assist considerably in helping to tie down the proportion that comes from fossil fuel sources, such as leakage from mines and from natural gas pipelines.

The main process for the removal of methane from the atmosphere is through chemical destruction. It reacts with hydroxyl (OH) radicals, which are present in the atmosphere because of processes involving sunlight, oxygen, ozone and water vapour. The average lifetime of methane in the atmosphere is determined by the rate of this loss process. At about 12 years[13] it is much shorter than the lifetime of carbon dioxide.

Although most methane sources cannot be identified very precisely, the largest sources apart from natural wetlands are closely associated with human activities.

Rice paddy fields have an adverse environmental impact because of the large quantities of methane gas they generate. World methane production due to paddy fields has been estimated to be in the range of 30 to 90 million tonnes per year.

It is interesting to note that the increase of atmospheric methane (Figure 3.6a) follows very closely the growth of human population since the Industrial Revolution. However, even without the introduction of deliberate measures to control human-related sources of methane because of the impact on climate change, it is not likely that this simple relationship with human population will continue. The IPCC Special Report on Emission Scenarios (SRES) presented in Chapter 6 include a wide range of estimates of the growth of human-related methane emissions during the twenty-first century – from approximately doubling over the century to reductions of about 25%. In Chapter 10 (page 305) ways are suggested in which

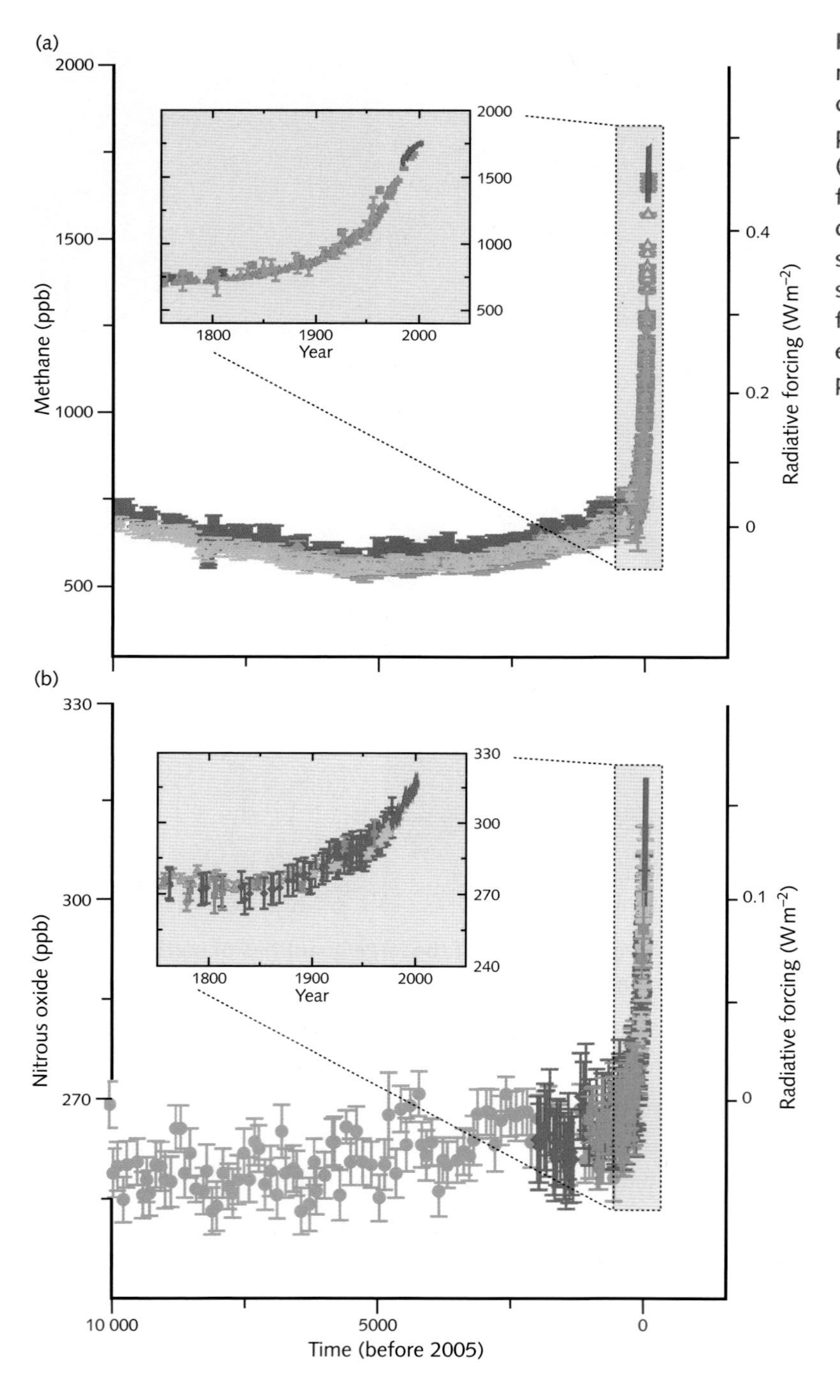

Figure 3.6 Change in (a) methane and (b) nitrous oxide concentration (mole fraction in ppb) over the last 10 000 years (insets from 1750) determined from ice cores (symbols with different colours from different studies) and atmospheric samples (red lines). Radiative forcing since the pre-industrial era due to the increases is plotted on the right-hand axes.

Table 3.2 Estimated sources and sinks of methane in millions of tonnes per year. The first column of data shows the best estimate from each source; the second column illustrates the uncertainty in the estimates by giving a range of values

	Best estimate	Uncertainty
Sources		
Natural		
Wetlands	150	(90–240)
Termites	20	(10–50)
Ocean	15	(5–50)
Other (including hydrates)	15	(10–40)
Human-generated		
Coal mining, natural gas, petroleum industry	100	(75–110)
Rice paddies	60	(30–90)
Enteric fermentation	90	(70–115)
Waste treatment	25	(15–70)
Landfills	40	(30–70)
Biomass burning	40	(20–60)
Sinks		
Atmospheric removal	545	(450–550)
Removal by soils	30	(15–45)
Atmospheric increase	22	(35–40)

For some more recent estimates see Table 7.7 in Denman, K. L., Brasseur, G. *et al.*, Chapter 7, in Solomon *et al.* (eds.) *Climate Change 2007: The Physical Science Basis.* The figure for atmospheric increase is an average for the 1990s; note that from 1999 to 2005 the increase was close to zero.

methane emissions could be reduced and methane concentrations in the atmosphere stabilised. Also see box on page 48–9 for possible destabilisation of methane emissions from methane hydrates especially at high latitudes.

Nitrous oxide

Nitrous oxide, used as a common anaesthetic and known as laughing gas, is another minor greenhouse gas. Its concentration in the atmosphere of about 0.3 ppm is rising at about 0.25% per year and is about 16% greater than in pre-industrial times (Figure 3.6b). The largest emissions to the atmosphere are associated with natural and agricultural ecosystems; those linked with human activities are probably due to increasing fertiliser use. Biomass burning and the chemical industry (for example, nylon production) also play some part. The sink of nitrous oxide is photodissociation in the stratosphere and reaction with electronically excited oxygen atoms, leading to an atmospheric lifetime of about 120 years.

Chlorofluorocarbons (CFCs) and ozone

The CFCs are man-made chemicals which, because they vaporise just below room temperature and because they are non-toxic and non-flammable, appear to be ideal for use in refrigerators, the manufacture of insulation and aerosol spray cans. Since they are so chemically unreactive, once they are released into the atmosphere they remain for a long time – 100 or 200 years – before being destroyed. As their use increased rapidly through the 1980s their concentration in the atmosphere has been building up so that they are now present (adding together all the different CFCs) in about 1 ppb (part per thousand million – or billion – by volume). This may not sound very much, but it is quite enough to cause two serious environmental problems.

The first problem is that they destroy ozone.[14] Ozone (O_3), a molecule consisting of three atoms of oxygen, is an extremely reactive gas present in small quantities in the stratosphere (a region of the atmosphere between about 10 km and 50 km in altitude). Ozone molecules are formed through the action of ultraviolet radiation from the Sun on molecules of oxygen. They are in turn destroyed by a natural process as they absorb solar ultraviolet radiation at slightly longer wavelengths – radiation that would otherwise be harmful to us and to other forms of life at the Earth's surface. The amount of ozone in the stratosphere is determined by the balance between these two processes, one forming ozone and one destroying it. What happens when CFC molecules move into the stratosphere is that some of the chlorine atoms they contain are stripped off, also by the action of ultraviolet sunlight. These chlorine atoms readily react with ozone, reducing it back to oxygen and adding to the rate of destruction of ozone. This occurs in a catalytic cycle – one chlorine atom can destroy many molecules of ozone.

The problem of ozone destruction was brought to world attention in 1985 when Joe Farman, Brian Gardiner and Jonathan Shanklin at the British Antarctic Survey discovered a region of the atmosphere over Antarctica where, during the southern spring, about half the ozone overhead disappeared. The existence of the 'ozone hole' was a great surprise to the scientists; it set off an intensive investigation into its causes. The chemistry and dynamics of its formation

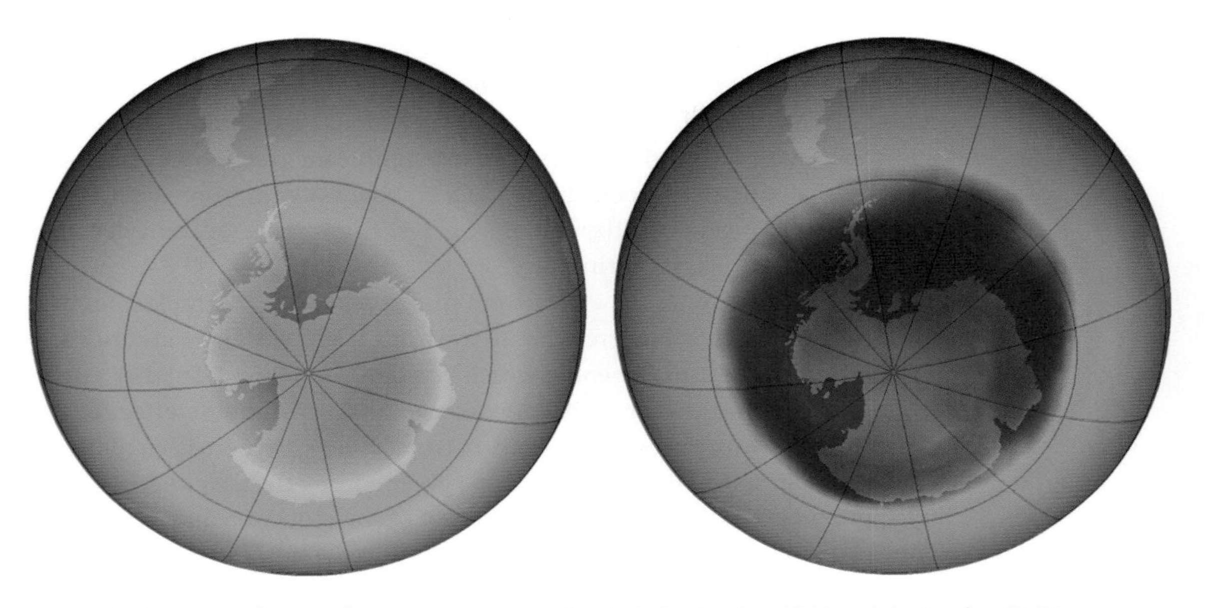

Ozone depletion can be seen by comparing ozone levels in September 1980 and September 2008. The dark blue and purple areas denote where the ozone layer is thinnest.

turned out to be complex. They have now been unravelled, at least as far as their main features are concerned, leaving no doubt that chlorine atoms introduced into the atmosphere by human activities are largely responsible. Not only is there depletion of ozone in the spring over Antarctica (and to a lesser extent over the Arctic) but also substantial reduction, of the order of 5%, of the total column of ozone – the amount above one square metre at a given point on the Earth's surface – at mid latitudes in both hemispheres.

Because of these serious consequences of the use of CFCs, international action has been taken. Many governments have signed the Montreal Protocol set up in 1987 which, together with the Amendments agreed in London in 1991 and in Copenhagen in 1992, required that manufacture of CFCs be phased out completely by the year 1996 in industrialised countries and by 2006 in developing countries. Because of this action the concentration of CFCs in the atmosphere is no longer increasing. However, since they possess a long life in the atmosphere, little decrease will be seen for some time and substantial quantities will be present well over 100 years from now.

So much for the problem of ozone destruction. The other problem with CFCs and ozone, the one which concerns us here, is that they are both greenhouse gases.[15] They possess absorption bands in the region known as the longwave atmospheric window (see Figure 2.5) where few other gases absorb. Because, as we have seen, the CFCs destroy some ozone, the greenhouse effect of the CFCs is partially compensated by the reduced greenhouse effect of atmospheric ozone.

First considering the CFCs on their own, a CFC molecule added to the atmosphere has a greenhouse effect 5000 to 10 000 times greater than an added molecule of carbon dioxide. Thus, despite their very small concentration compared, for instance, with carbon dioxide, they have a significant greenhouse effect. It is estimated that radiative forcing due to CFCs is about 0.3 W m^{-2} – or about 12% of the radiative forcing due to all greenhouse gases. This forcing will only decrease slowly in the twenty-first century.

Turning now to ozone, the effect from ozone depletion is complex because the amount by which ozone greenhouse warming is reduced depends critically on the height in the atmosphere at which it is being destroyed. Further, ozone depletion is concentrated at high latitudes while the greenhouse effect of the CFCs is uniformly spread over the globe. In tropical regions there is virtually no ozone depletion so no change in the ozone greenhouse effect. At mid latitudes, very approximately, the greenhouse effects of ozone reduction and of the CFCs compensate for each other. In polar regions the reduction in the greenhouse effect of ozone more than compensates for the greenhouse warming effect of the CFCs.[16]

As CFCs are phased out, they are being replaced to some degree by other halocarbons – hydrochlorofluorocarbons (HCFCs) and hydrofluorocarbons (HFCs). In Copenhagen in 1992, the international community decided that HCFCs would also be phased out by the year 2030. While being less destructive to ozone than the CFCs, they are still greenhouse gases. The HFCs contain no chlorine or bromine, so they do not destroy ozone and are not covered by the Montreal Protocol. Because of their shorter lifetime, typically tens rather than hundreds of years, the concentration in the atmosphere of both the HCFCs and the HFCs, and therefore their contribution to global warming for a given rate of emission, will be less than for the CFCs. However, since their rate of manufacture could increase substantially their potential contribution to greenhouse warming is being included alongside other greenhouse gases (see Chapter 10, page 296).

Concern has also extended to some other related compounds which are greenhouse gases, the perfluorocarbons (e.g. CF_4, C_2F_6) and sulphur hexafluoride (SF_6), which are produced in some industrial processes. Because they possess very long atmospheric lifetimes, probably more than 1000 years, all emissions of these gases accumulate in the atmosphere and will continue to influence climate for thousands of years. They are also therefore being included as potentially important greenhouse gases.

Ozone is also present in the lower atmosphere or troposphere, where some of it is transferred downwards from the stratosphere and where some is generated by chemical action, particularly as a result of the action of sunlight on the oxides of nitrogen. It is especially noticeable in polluted atmospheres near the surface; if present in high enough concentration, it can become a health hazard. In the northern hemisphere the limited observations available together with model simulations of the chemical reactions leading to ozone formation suggest that ozone concentrations in the troposphere have doubled since pre-industrial times – an increase which is estimated to have led to a global average radiative forcing of about 0.35 W m^{-2} (Figure 3.11). Ozone is also generated at levels in the upper troposphere as a result of the nitrogen oxides emitted from aircraft exhausts; nitrogen oxides emitted from aircraft are more effective at producing ozone in the upper troposphere than are equivalent emissions at the surface. The radiative forcing in northern mid latitudes from aircraft due to this additional ozone[17] is of similar magnitude to that from the carbon dioxide emitted from the combustion of aviation fuel which currently is about 3% of current global fossil fuel consumption.

Gases with an indirect greenhouse effect

I have so far described all the gases present in the atmosphere that have a direct greenhouse effect. There are also gases which through their chemical action on greenhouse gases, for instance on methane or on lower atmospheric ozone, have an influence on the overall size of greenhouse warming. Carbon monoxide (CO) and the nitrogen oxides (NO and NO_2) emitted, for instance, by motor vehicles and aircraft are some of these. Carbon monoxide has no direct greenhouse effect of its own but, as a result of chemical reactions, it forms carbon dioxide. These reactions also affect the amount of the hydroxyl radical (OH) which in turn affects the concentration of methane. Emissions of nitrogen oxides, for instance, result in a small reduction in atmospheric methane which partially compensates for the increase in ozone due to aircraft mentioned in the last paragraph. Substantial research has been carried out on the chemical processes in the atmosphere that lead to these indirect effects on greenhouse gases. It is of course important to take them properly into account, but it is also important to recognise that their combined effect is much less than that of the major contributors to human-generated greenhouse warming, namely carbon dioxide and methane.

Particles in the atmosphere

Small particles suspended in the atmosphere (often known as *aerosol*; see Glossary) affect its energy balance because they both absorb radiation from the Sun and scatter it back to space. We can easily see the effect of this on a bright day in the summer with a light wind when downwind of an industrial area. Although no cloud appears to be present, the Sun appears hazy. We call it 'industrial haze'. Under these conditions a significant proportion of the sunlight incident at the top of the atmosphere is being lost as it is scattered back and out of the atmosphere by the millions of small particles (typically between 0.001 and 0.01 mm in diameter) in the haze. The effect of particles can also be seen often when flying over or near industrial or densely populated areas for instance in Asia when although no cloud is present, it is too hazy to see the ground.[18]

Atmospheric particles come from a variety of sources. They arise partially from natural causes; they are blown off the land surface, especially in desert areas, they result from forest fires and they come from sea spray. From time to time large quantities of particles are injected into the upper atmosphere from volcanoes – the Pinatubo volcano which erupted in 1991 provides a good example (see Chapter 5). Some particles are also formed in the atmosphere itself, for instance sulphate particles from the sulphur-containing gases emitted from volcanoes. Other particles arise from human activities. Over the past ten years a large number of observations especially from satellite-borne instruments have provided much needed information about the aerosol distribution from both natural and anthropogenic sources in both space and time (Figure 3.7a).

The most important of the aerosols from anthropogenic sources are sulphate particles that are formed as a result of chemical action on sulphur dioxide, a gas that is produced in large quantities by power stations and other industries in which coal and oil (both of which contain sulphur in varying quantities) are burnt. Because these particles remain in the atmosphere only for about five days on average, their effect is mainly confined to regions near the sources of the particles, i.e. the major industrial regions of the northern hemisphere (Figure 3.7b). Sulphate particles scatter sunlight and provide a negative forcing, globally averaged estimated as -0.4 ± 0.2 W m^{-2}. Over limited regions of the northern hemisphere the radiative effect of these particles is comparable in size, although opposite in effect, to that of human-generated greenhouse gases up to the present time. Figure 3.8 illustrates a model estimate of the substantial effect on global atmospheric temperature of removing all sulphate aerosol in the year 2000.

An important factor that will influence the future concentrations of sulphate particles is 'acid rain' pollution, caused mainly by the sulphur dioxide emissions. This leads to the degradation of forests and fish stocks in lakes especially in regions downwind of major industrial areas. Serious efforts are therefore under way, especially in Europe and North America, to curb these emissions to a substantial degree. Although the amount of sulphur-rich coal being burnt elsewhere in the world, for instance in Asia, is increasing rapidly, the damaging

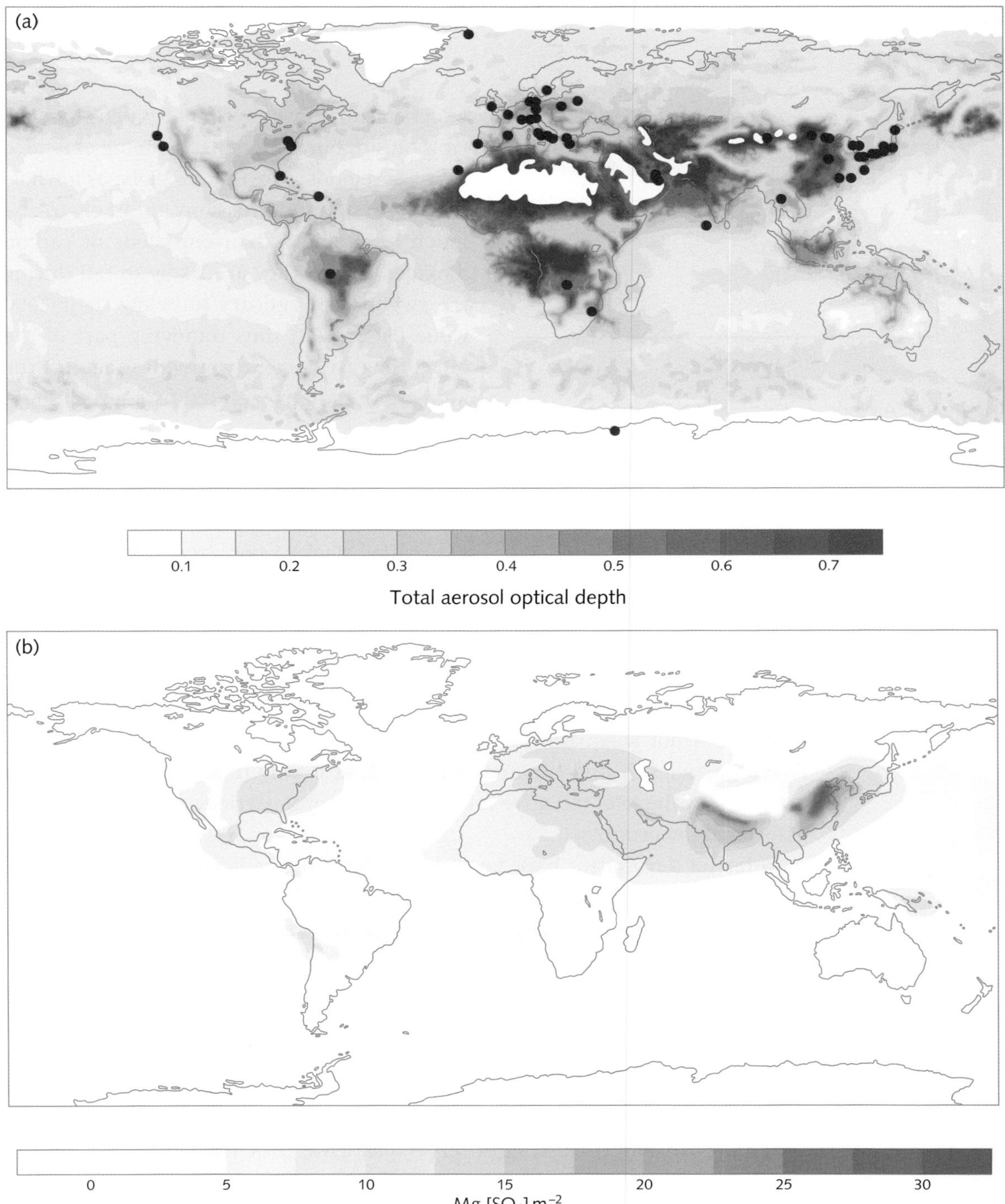

Figure 3.7 Distribution of atmospheric aerosols. (a) Total aerosol optical depth at a mid-visible wavelength (for definition see Glossary) due to natural plus anthropogenic aerosols determined from observations by the satellite instrument MODIS, averaged from August to October 2001. Also indicated are the locations of aerosol lidar network sites (red circles). (b) The amount of sulphate (SO_4) aerosol in the atmosphere in mg[SO_4] m^{-2} from human activities, 'background' non-explosive volcanoes and natural di-methyl sulphate (DMS) from ocean plankton, averaged over the decade of the 1990s calculated by the Hadley Centre model HadGEM1.

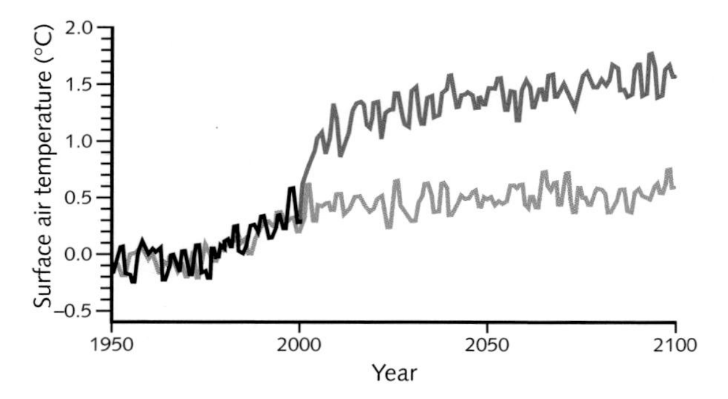

Figure 3.8 A model calculation of the effect on global mean surface air temperature of removing all sulphate aerosols in the year 2000 (red line) compared with maintaining the global burden of sulphate aerosols at the 2000 level for the twenty-first century (blue line).

effects of sulphur pollution are such that controls on sulphur emissions are being extended to these regions also. For the globe as a whole therefore, sulphur emissions are likely to rise much less rapidly than emissions of carbon dioxide. In fact, they are likely to fall during the twenty-first century to below their 2000 value (Figure 6.1) thus removing part of the offset they are currently providing against the increase in radiative forcing from greenhouse gases.

The radiative forcing from particles can be positive or negative depending on the nature of the particles. For instance, soot particles (also called black carbon) from fossil fuel burning absorb sunlight and possess a positive forcing globally averaged estimated as 0.2 ± 0.15 W m^{-2}. Other smaller anthropogenic contributions to aerosol radiative forcing come from biomass burning (e.g. the burning of forests), organic carbon particles from fossil fuel and nitrate and mineral dust particles. Because of the interactions that occur between particles from different sources and with clouds it is not adequate to add simply estimated radiative forcing from different particles to find the total forcing. That is why in Figure 3.11 estimates are given of the *total* radiative forcing from aerosol particles together with the associated uncertainty.

So far for aerosol we have been describing *direct* radiative forcing. There is a further way by which particles in the atmosphere could influence the climate; that is through their effect on cloud formation that is described as *indirect* radiative forcing. The mechanism of indirect forcing that is best understood arises from the influence of the number of particles and their size on cloud radiative properties (Figure 3.9). If particles are present in large numbers when clouds are forming, the resulting cloud consists of a large number of smaller drops – smaller than would otherwise be the case – similar to what happens as polluted fogs form in cities. Such a cloud will be more highly reflecting to sunlight than one consisting of larger particles, thus further increasing the energy loss resulting from the presence of the particles. Further the droplet size and number influence the precipitation efficiency, the lifetime of clouds and hence the geographic extent of cloudiness. Figure 3.10 is an illustration of the effect as it applies in the wakes of ships where clouds form possessing drop sizes much smaller than those pertaining to other clouds in the vicinity. There is now substantial observational evidence for these mechanisms but the processes

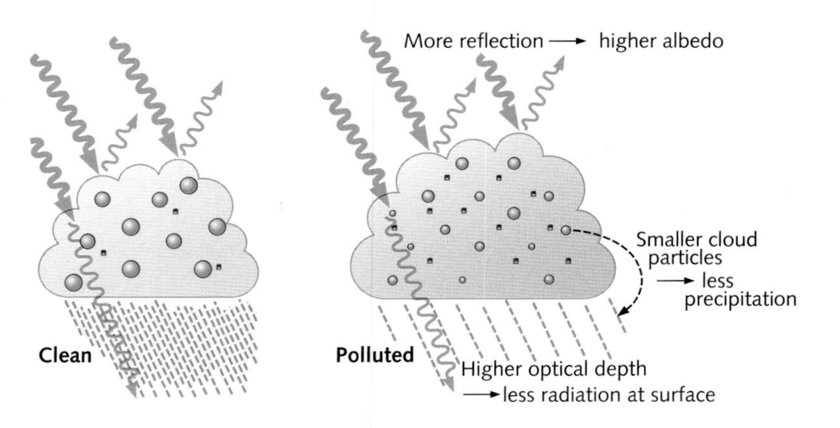

Figure 3.9 Schematic illustrating the cloud albedo and lifetime indirect effect on radiative forcing. Larger numbers of smaller particles in polluted clouds lead to more reflection of solar radiation from the cloud top, less radiation at the surface, less precipitation and a longer cloud lifetime.

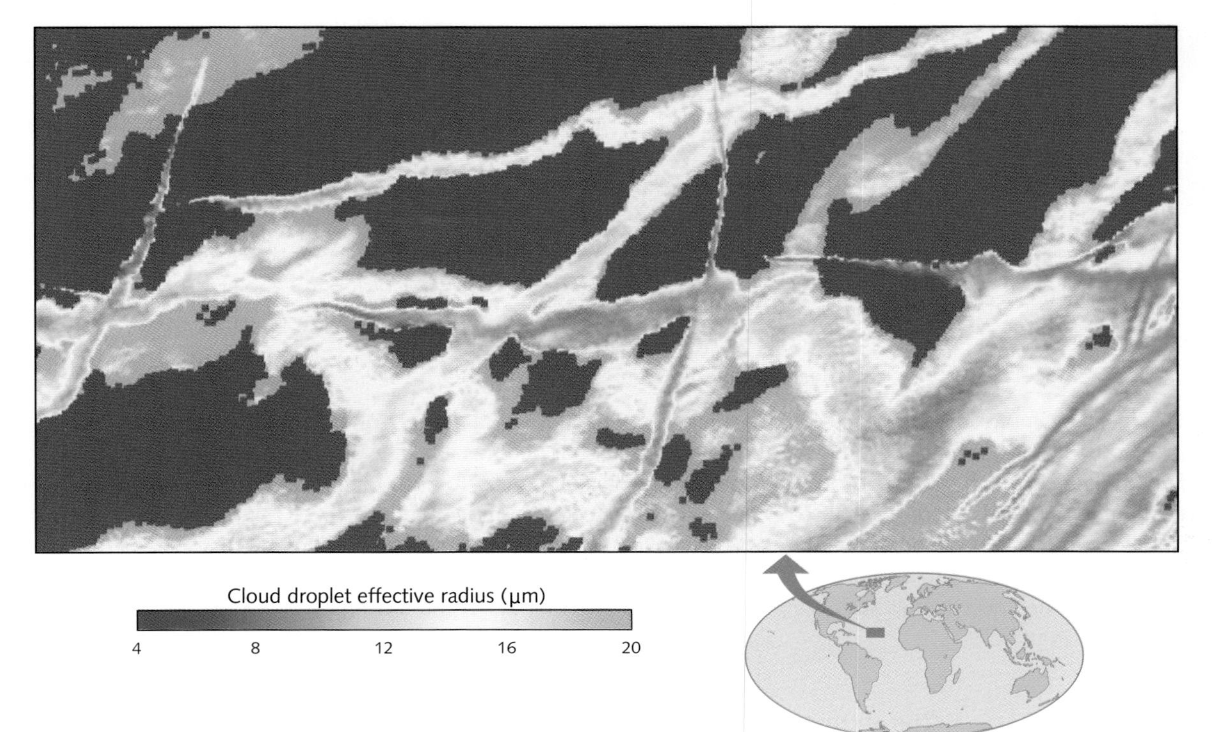

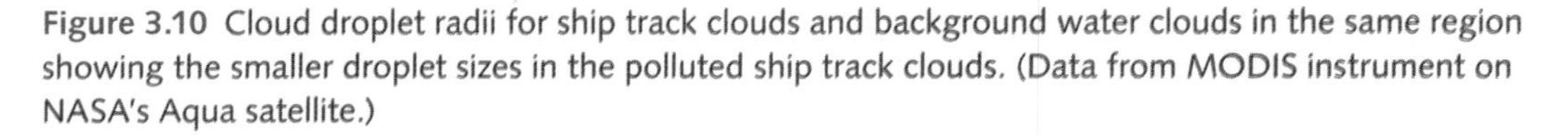

Figure 3.10 Cloud droplet radii for ship track clouds and background water clouds in the same region showing the smaller droplet sizes in the polluted ship track clouds. (Data from MODIS instrument on NASA's Aqua satellite.)

involved are not easy to model and will vary a great deal with the particular situation. Substantial uncertainty therefore remains in estimates of their magnitude as shown in Figure 3.11[19]. To refine these estimates, more studies are required especially through making careful measurements on suitable clouds.

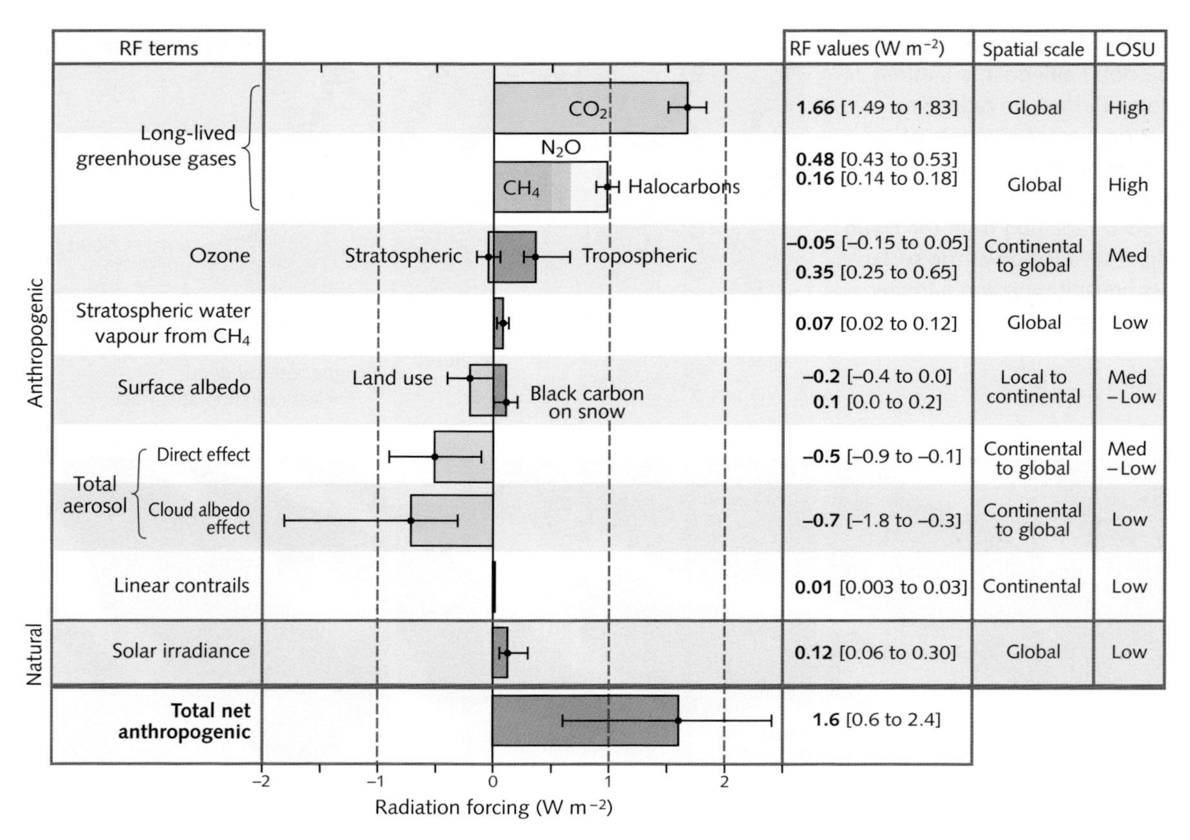

Figure 3.11 Global, annual mean radiative forcings (W m^{-2}) due to a number of agents for the period from pre-industrial (1750) to 2005. The size of the rectangular bar denotes a best estimate value; the horizontal lines indicate estimates of the uncertainty (90% confidence) ranges. To each forcing an indication is given of the geographical extent (spatial scale) and a 'level of scientific understanding' (LOSU) index is accorded. This latter represents a judgement about the reliability of the forcing estimate involving factors such as the assumptions necessary to evaluate the forcing, the degree of knowledge of the mechanisms determining the forcing and the uncertainties surrounding the quantitative estimate of the forcing.

The estimates for the radiative effects of particles as in Figure 3.11 can be compared with the global average radiative forcing to date due to the increase in greenhouse gases of about 2.6 W m^{-2}. Comparing global average forcings, however, is not the whole story. Although the effects of particles on the global climate are well indicated by using globally averaged forcing estimates, for their effects on regional climate, information about their regional distribution (Figure 3.7) has also to be included (see Chapter 6).

A particular effect on cloudiness arises from aircraft flying in the upper troposphere which influence high cloud cover through their emissions of water vapour and of particles that can act as nuclei on which condensation can occur. As we shall see in Chapter 5 (page 111) high cloud provides a blanketing effect on the Earth's surface similar to that of greenhouse gases and therefore leads to positive radiative forcing. Extensive formation of contrails in the upper troposphere by aircraft frequently occurs; an estimate of radiative forcing from this cause is included in Figure 3.11. Persistent contrails also tend to lead to increased overall cloudiness in the region where the contrails have formed. This is called *aviation induced cloudiness* and is difficult to quantify. Because of these effects of aircraft and also the effect of increased ozone (reduced by methane reduction) mentioned on page 57, the overall greenhouse effect of aircraft has been estimated as the equivalent of two or possibly up to four times the effect of their carbon dioxide emissions.[20]

Global warming potentials

It is useful to be able to compare the radiative forcing generated by the different greenhouse gases. Because of their different lifetimes, the future profile of radiative forcing due to releases of greenhouse gases varies from gas to gas. An index called the global warming potential (GWP)[21] has been defined for greenhouse gases that takes the ratio of the time-integrated radiative forcing from the instantaneous release of 1 kg of a given gas to that from the release of 1 kg of carbon dioxide. A time horizon has also to be specified for the period over which the integration is carried out. The GWPs of the six greenhouse gases included in the Kyoto Protocol are listed in Table 10.2. Applying the GWPs to the emissions from a mixture of greenhouse gases enables the mixture to be considered in terms of an equivalent amount of carbon dioxide. However, because the GWPs for different time horizons are very different, GWPs are of limited application and must be used with care.

Estimates of radiative forcing

This chapter has summarised current scientific knowledge about the sources and sinks of the main greenhouse gases and aerosol particles, the natural balances that are maintained between the different components and the way in which these balances are being disturbed by human-generated emissions.

This information has been employed together with information about the absorption by the different gases of radiation in different parts of the spectrum (see Chapter 2) to calculate the effect of increases in gases and particles on the amount of net solar radiation entering the atmosphere and of net thermal radiation leaving. Estimates of the radiative forcing from 1750 to 2005

This view over the Chicago and Lake Michigan area was taken (30 November 2003) by a crewmember on the International Space Station. Aircraft contrails are clearly visible.

for the different greenhouse gases and for tropospheric aerosols of different origins are brought together in Figure 3.11, together with estimates of their uncertainty.

SUMMARY

From Figure 3.11 it will be seen that:

- The dominant forcing for climate change over the last two centuries has been that from the increase of long-lived greenhouse gases, especially carbon dioxide.
- Since the mid twentieth century, significant offset to the positive forcing from greenhouse gases has arisen from negative forcing due to aerosols, especially from sulphates.

- Other smaller forcings are due to changes in ozone (stratospheric and tropospheric), stratospheric water vapour and land surface albedo (for definition see Glossary) (Figure 3.12) and persistent contrails from aircraft.
- Estimates of changes in solar irradiance are smaller than estimated in the 2001 IPCC report (see Chapter 6, page 166).
- Significant progress has been made in the understanding and estimating of indirect aerosol forcing since the 2001 IPCC report – although substantial uncertainties remain.

For the future, different assumptions about future emissions of greenhouse gases and aerosols are used to generate emission scenarios. From these scenarios estimates are made (for carbon dioxide, for instance, using a computer model of the carbon cycle) of likely increases in greenhouse gas concentrations in the future. Details of radiative forcings projected for the twenty-first century are presented in Chapter 6 (page 142). Chapters 5 and 6 will explain how estimates of radiative forcing can be incorporated into computer climate models so as to predict the climate change that is likely to occur because of human activities. However, before considering predictions of future climate change, it is helpful to gain perspective by looking at some of the climate changes that have occurred in the past.

QUESTIONS

1 The lifetime of a carbon dioxide molecule in the atmosphere before it is exchanged with the ocean is typically less than a year, while the time taken for an increase in carbon dioxide concentration from fossil fuel burning to diminish substantially is typically many years. Explain the reasons for this difference.

2 Estimate how much carbon dioxide you emit each year through breathing.

3 Estimate the size of your share of carbon dioxide emissions from the burning of fossil fuels.

4 A typical city in the developed world with a population of about one million produces about half a million tonnes of municipal waste each year. Suppose the waste is buried in a landfill site where the waste decays producing equal quantities of carbon dioxide and methane. Making assumptions about the likely carbon content of the waste and the proportion that eventually decays, estimate the annual production of methane. If all the methane leaks away, using the information in Note 12, compare the greenhouse effect of the

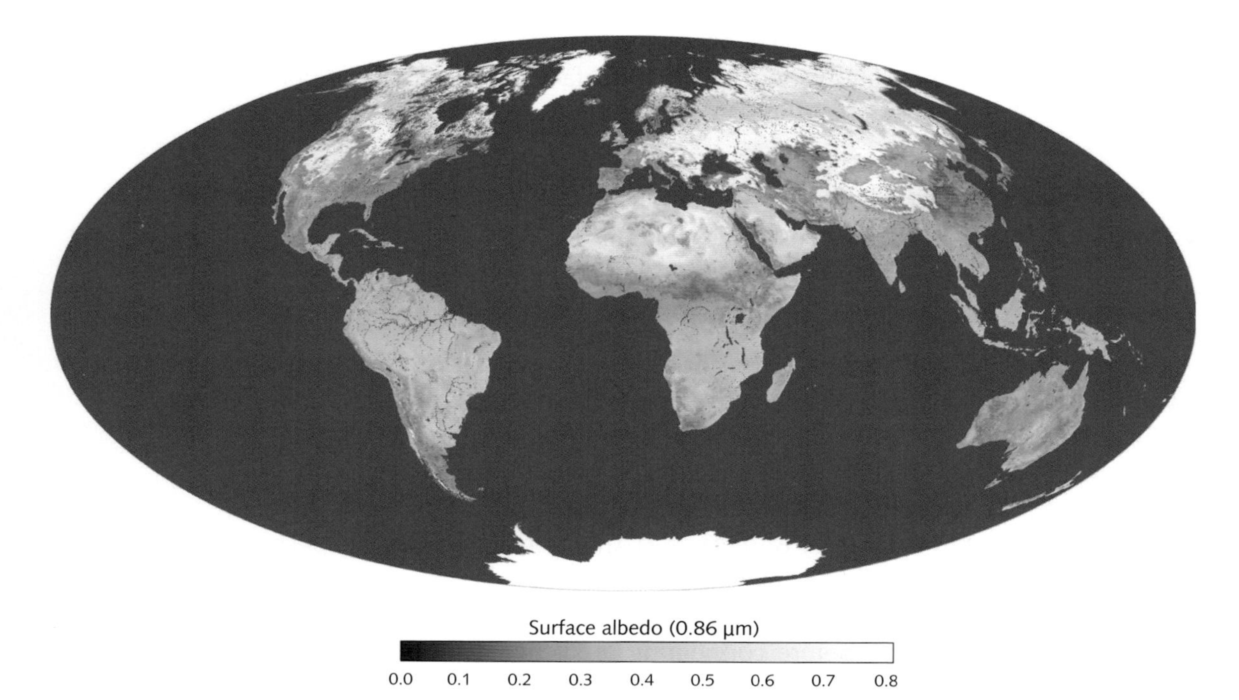

Figure 3.12 Global surface albedo for the period 1–16 January 2002, showing the proportion of incoming solar radiation that is reflected from the Earth's surface at a wavelength of 0.86 μm. (Data from MODIS instrument on NASA's Terra satellite; visualisation by Eric Moody, RS Information Systems Inc.)

carbon dioxide and methane produced from the landfill site with that of the carbon dioxide produced if the waste were incinerated instead. Discuss how far waste is a 'renewable' energy resource.

5. A new forest is planted containing 1 million trees which will mature in 40 years. Estimate the amount of carbon sequestered per year by the forest.
6. Find figures for the amount of fuel used by a typical aircraft and the size of fleets of the world's airlines and airforces and estimate the carbon dioxide emitted globally each year by the world's aircraft.
7. Search for information about the ozone hole and explain why it occurs mainly in the Antarctic.
8. What are the main uses of CFCs? Suggest ways in which the emissions of CFCs to the atmosphere could be reduced more rapidly.
9. Evidence is sometimes presented suggesting that the variations in global average temperature over the last century or more can all be explained as due to variations in the energy output of the Sun. There is therefore

nothing left to attribute to the increase in greenhouse gases. What is the fallacy in this argument?

10 With the use of the formula given in the text, calculate the radiative forcing due to carbon dioxide for atmospheric concentrations of 150, 280, 450, 560 and 1000 ppm.

11 Making approximate assumptions about particle size and scattering properties, estimate the optical depth equivalent to 25 mg[SO_4] m^{-2} in Figure 3.7b. Then estimate global average radiative forcing due to sulphate aerosol.

FURTHER READING AND REFERENCE

Solomon, S., Qin, D., Manning, M., Chen, Z., Marquis, M., Averyt, K.B., Tignor, M., Miller, H.L. (eds.) 2007. *Climate Change 2007: The Physical Science Basis. Contribution of Working Group I to the Fourth Assessment Report of the Intergovernmental Panel on Climate Change*. Cambridge: Cambridge University Press.

Technical Summary (summarises basic information about greenhouse gases, carbon cycle and aerosols)

Chapter 2 Changes in atmospheric constituents and radiative forcing

Chapter 7 Couplings between changes in the climate system and biogeochemistry

World Resources Institute www.wri.org – valuable for its catalogue of climate data (e.g. greenhouse gas emissions).

NOTES FOR CHAPTER 3

1 It is convenient to define radiative forcing as the radiative imbalance at the top of the troposphere rather than at the top of the whole atmosphere. The formal definition of radiative forcing as in the IPCC reports is 'the change in net (down minus up) irradiance (solar plus longwave in $W m^{-2}$) at the tropopause after allowing for stratospheric temperatures to readjust to radiative equilibrium, but with surface and tropospheric temperatures and state held fixed at the unperturbed values'.

2 See Denman, K.L., Brasseur, G. *et al.* 2007. Chapter 7, Section 7.3.12, in Solomon *et al.* (eds.) *Climate Change 2007: The Physical Science Basis*.

3 See Jansen, E., Overpeck, J. *et al.* 2007. Chapter 6 (especially Box 6.2), in Solomon *et al.* (eds.) *Climate Change 2007: The Physical Science Basis*.

4 The process has been called the 'plankton multiplier': Woods, J., Barkmann, W. 1993. The plankton multiplier: positive feedback in the greenhouse. *Journal of Plankton Research*, **15**, 1053–74.

5 For more details see House, J.I. *et al.* 2003. Reconciling apparent inconsistencies in estimates of terrestrial CO_2 sources and sinks. *Tellus*, **55B**, 345–63.

6 See Jones, C.D., Cox, P.M. 2001. Atmospheric science letters. doi.1006/asle. 2001.0041; the respiration rate varies approximately with a factor $2^{T-10}/10$.

7 Cox, P.M., Betts, R.A., Collins, M., Harris, P., Huntingford, C., Jones, C.D. 2004. Amazon dieback under climate-carbon cycle projections for the 21st century. *Theoretical and Applied Climatology*, **78**, 137–56.

8 Some recent papers have pointed out the need for urgent research into aspects of the climate/carbon-cycle feedback, for instance the importance of

linking together all components of the ecosystem chemistry and dynamics including, for instance, the influences of water and nitrogen. See Heimann, M., Reichstein, M. 2008. *Nature*, **451**, 289–92; Gruber, N., Galloway, J.N. 2008. *Nature*, **451**, 293–6; and Friedlingstein, P. 2008. *Nature*, **451**, 297–8.

9 In the absence of climate feedbacks the Hadley model is similar to the other models with slightly above average land carbon uptake and slightly below average ocean uptake.

10 The extensive and protracted forest fires in Indonesia and neighbouring areas in 1997–8 have been estimated to have resulted in the emission to the atmosphere of 0.8 to 2.6 GtC; this may be one of the reasons for particularly high growth in atmospheric carbon dioxide in 1998.

11 In the 1990s the rate of increase substantially slowed. The reason for this is not known but one suggestion is that, because of the collapse of the Russian economy, the leakage from Siberian natural gas pipelines was much reduced.

12 The ratio of the enhanced greenhouse effect of a molecule of methane compared to a molecule of carbon dioxide is known as its global warming potential (GWP); a definition of GWP is given later in the chapter. The figure of about 8 given here for the GWP of methane is for a time horizon of 100 years – see Lelieveld, J., Crutzen, P.J. 1992. *Nature*, **355**, 339–41; see also Prather *et al.*, Chapter 4, and Ramaswamy *et al.*, Chapter 6, in, Houghton, J.T., Ding, Y., Griggs, D.J., Noguer, M., van der Linden, P., Dai, X., Maskell, K., Johnson, C.A. (eds.) 2001. *Climate Change 2001: The Scientific Basis. Contribution of Working Group 1 to the Third Assessment Report of the Intergovernmental Panel on Climate Change*. Cambridge: Cambridge University Press. The GWP is also often expressed as the ratio of the effect for unit mass of each gas in which case the GWP for methane (whose molecular mass is 0.36 of that of carbon dioxide) becomes about 23 for the 100-year time horizon. About 75% of the contribution of methane to the greenhouse effect is because of its direct effect on the outgoing thermal radiation. The other 25% arises because of its influence on the overall chemistry of the atmosphere. Increased methane eventually results in small increases in water vapour in the upper atmosphere, in tropospheric ozone and in carbon dioxide, all of which in turn add to the greenhouse effect.

13 Taking into account the loss processes due to reaction with OH in the troposphere, chemical reactions and soil loss lead to a lifetime of about ten years. However, the effective lifetime of methane against a perturbation in concentration in the atmosphere (the number quoted here) is complex because it depends on the methane concentration. This is because the concentration of the radical OH (interaction with which is the main cause of methane destruction), due to chemical feedbacks, is itself dependent on the methane concentration (more details in Prather *et al.* 2001. Chapter 4, in Houghton *et al.* (eds.), *Climate Change 2001: The Scientific Basis*).

14 For more detail, see *Scientific Assessment of Ozone Depletion: 2003*. Geneva: World Meteorological Organization.

15 Prather *et al.*, Chapter 4, in Houghton *et al.* (eds.), *Climate Change 2001: The Scientific Basis.*

16 More detail on this and the radiative effects of minor gases and particles can be found in Ramaswamy *et al.* 2001, Chapter 6, in Houghton *et al.* (eds.), *Climate Change 2001: The Scientific Basis.*

17 More detail in Penner, J.E. *et al.* (eds.), 1999. *Aviation and the Global Atmosphere*. An IPCC Special Report. Cambridge: Cambridge University Press.

18 See for instance Ramanathan, V. *et al.* 2007, Warming trends in Asia amplified by brown cloud solar absorption. *Nature*, **448**, 575–8.

19 For variations of radioactive forcing from different forcing agents over the period 1880–2004, see James Hansen *et al.*, *Science* **308**, 1431.

20 More detail in Penner *et al.* (eds.), *Aviation and the Global Atmosphere.*

21 More detail on GWPs can be found in Ramaswamy *et al.*, Chapter 6, in Houghton *et al.* (eds.), *Climate Change 2001: The Scientific Basis.*

Climates of the past

4

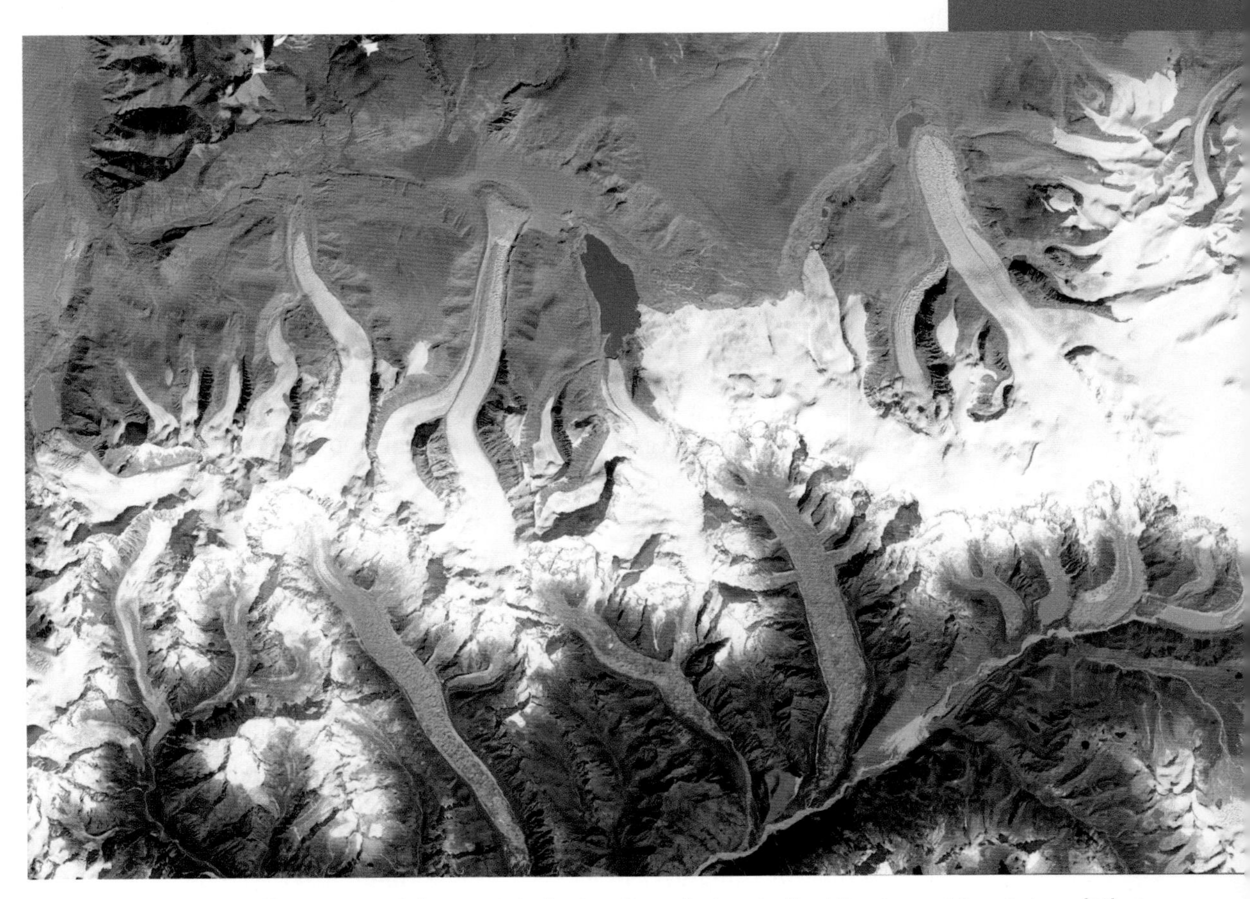

Satellite image of the termini of retreating glaciers in the Himalayan Mountains of Bhutan.

TO OBTAIN some perspective against which to view future climate change, it is helpful to look at some of the climate changes that have occurred in the past. This chapter will briefly consider climatic records and climate changes in three periods: the last hundred years, then the last thousand years and finally the last million years. At the end of the chapter some interesting evidence for the existence of relatively rapid climate change at various times during the past one or two hundred thousand years will be presented.

The last hundred years

The 1980s and 1990s and the early years of the twenty-first century have brought unusually warm years for the globe as a whole as is illustrated in Figure 4.1, which shows the global average temperature since 1850, the period for which the instrumental record is available with good accuracy and coverage. An increase over this period has taken place of $0.76 \pm 0.19\,°C$ (Figure 4.1a). The two warmest years in the record are 1998 and 2005, 1998 ranking highest on one estimate and 2005 highest on two other estimates. Also 12 of the 13 years 1995 to 2007 rank amongst the 13 warmest years in the whole record. A further striking statistic is that each of the first eight months of 1998 was *very likely*[1] the warmest of those months in the record up to that date. Although there is a distinct trend in the record, the increase is by no means a uniform one. In fact, some periods of cooling as well as warming have occurred and an obvious feature of the record is the degree of variability from year to year and from decade to decade.

Note also that there has been little if any average increase in warming during the years 2001–2006. Some have tried to argue that this shows the warming is over. However, as the figure illustrates, seven years of record is too short a period to establish a trend. Although the year 2007 was slightly cooler that 2006, the first seven years of the twenty-first century were on average nearly 0.2 °C warmer than the last seven years of the twentieth century, even though 1998 was the warmest year so far. Further, studies of interannual variability in the record demonstrate the strong influence of variations in El Niño and suggest that interannual variability may continue to offset anthropogenic warming until around 2009.[2]

Shown also in Figure 4.1b are the patterns of recent warming at the surface and averaged over the troposphere. Warming within the atmosphere is more spatially uniform than the surface record which shows more warming over land than over the ocean (see also Figure 4.2).

A sceptic may wonder how diagrams like those in Figure 4.1 can be prepared and whether any reliance can be placed upon them. After all, temperature varies from place to place, from season to season and from day to day by many tens of degrees. But here we are not considering changes in local temperature but in the average over the whole globe. A change of a few tenths of a degree in that average is a large change.

First of all, just how is a change in global average temperature estimated from a combination of records of changes in the near-surface temperature over land and changes in the temperature of the sea surface? To estimate the changes over land, weather stations are chosen where consistent observations have been taken from the same location over a substantial proportion of the whole

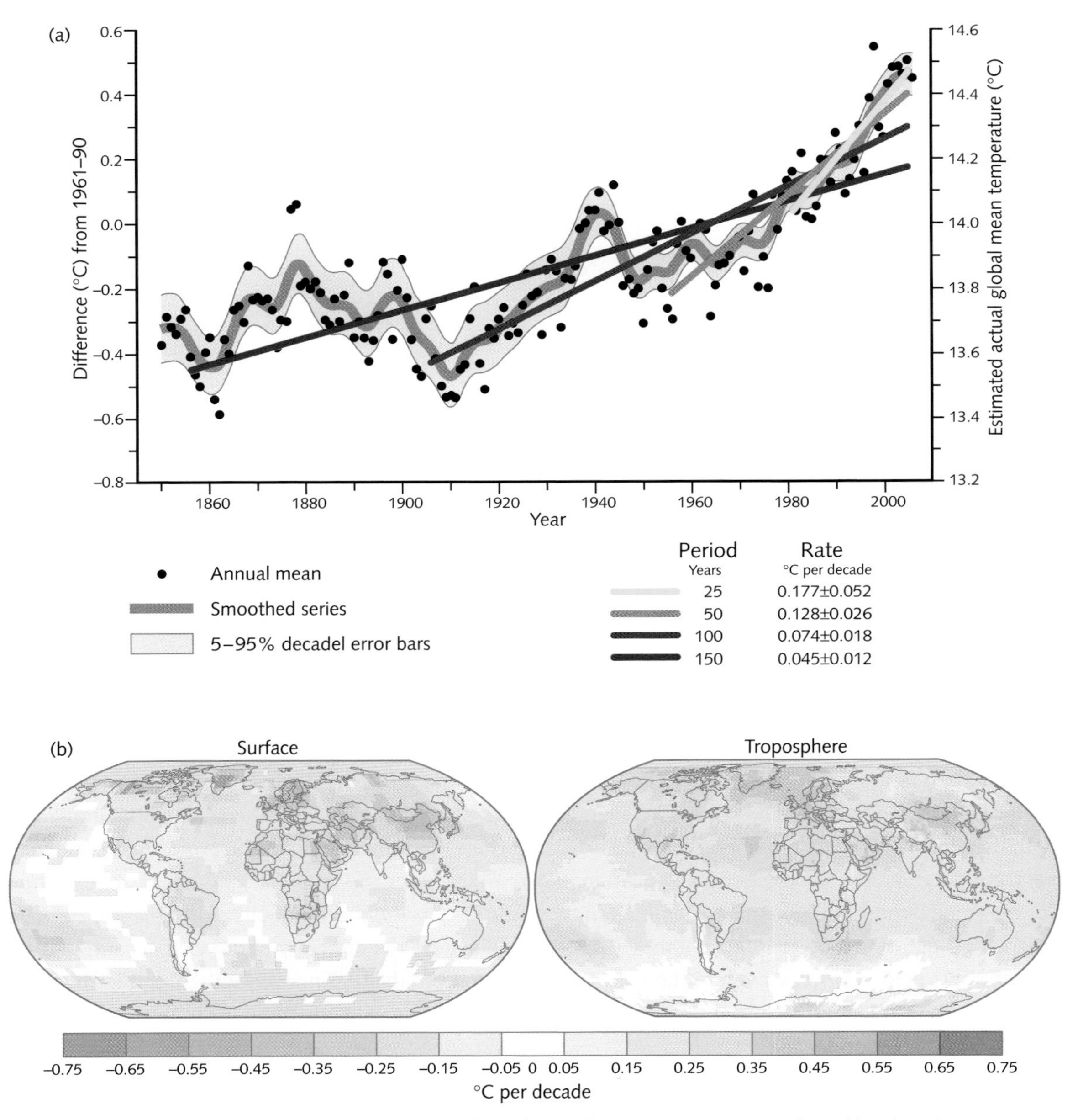

Figure 4.1 (a) Variations of the globally averaged Earth's surface temperature (combined land surface air temperature and sea surface temperature) for 1850–2006 relative to the 1961–90 mean. The black dots show annual means and the right-hand axis shows the estimated actual average temperature. Linear trend fits are shown to the last 25, 50, 100 and 150 years indicating accelerated warming. (b) Patterns of linear global temperature trends from 1979 to 2005 estimated at the surface (left), and for the troposphere (right) from the surface to about 10 km altitude from satellite records. Grey areas indicate incomplete data.

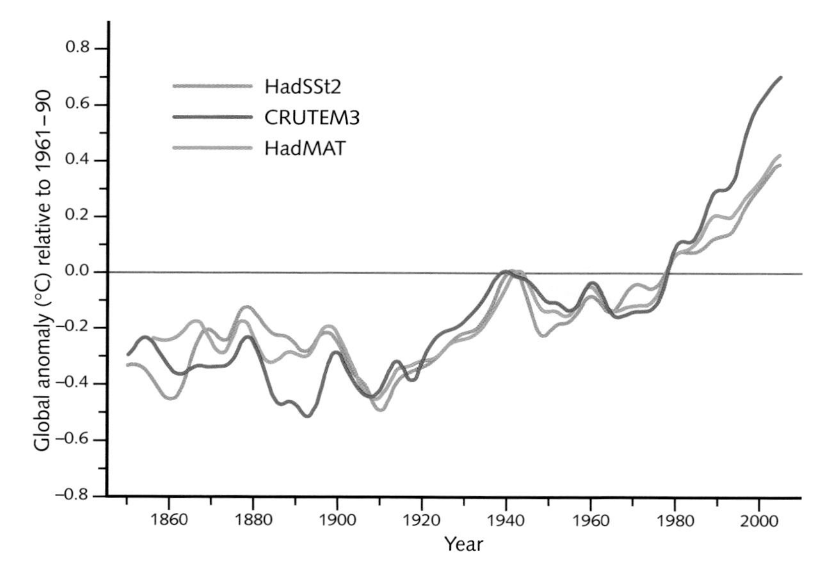

Figure 4.2 Decadally smoothed annual anomalies of global average sea surface temperature (blue), night marine air temperature (green) and land surface air temperature (red) relative to their 1961–90 means.

Atmospheric temperature observed by satellites

Since 1979 meteorological satellites flown by the National Oceanic and Atmospheric Administration (NOAA) of the United States have carried a microwave instrument, the Microwave Sounding Unit (MSU), for the remote observation of the average temperature of the lower part of the atmosphere up to about 7 km in altitude.

Figure 4.3b shows the record of global average temperature deduced from the MSU and compares it with data from sounding instruments carried on balloons for the same region of the atmosphere, showing very good agreement for the period of overlap. Figure 4.3c shows the record of surface air temperature for the same period. All three measurements show similar variability, the variations at the surface tracking well with those in the lower troposphere. The plots also illustrate the difficulty of deriving accurate trends from a short period of record where there is also substantial variability. Since 1979 the trend in the MSU observations of 0.12 to 0.19 °C per decade shows good agreement with the trend in surface observations of 0.16 to 0.18 °C per decade.

In the stratosphere the temperature trends are reversed (Figure 4.3a) ranging from a decrease of about 0.5 °C per decade in the lower stratosphere to 2.5 °C per decade in the upper stratosphere.

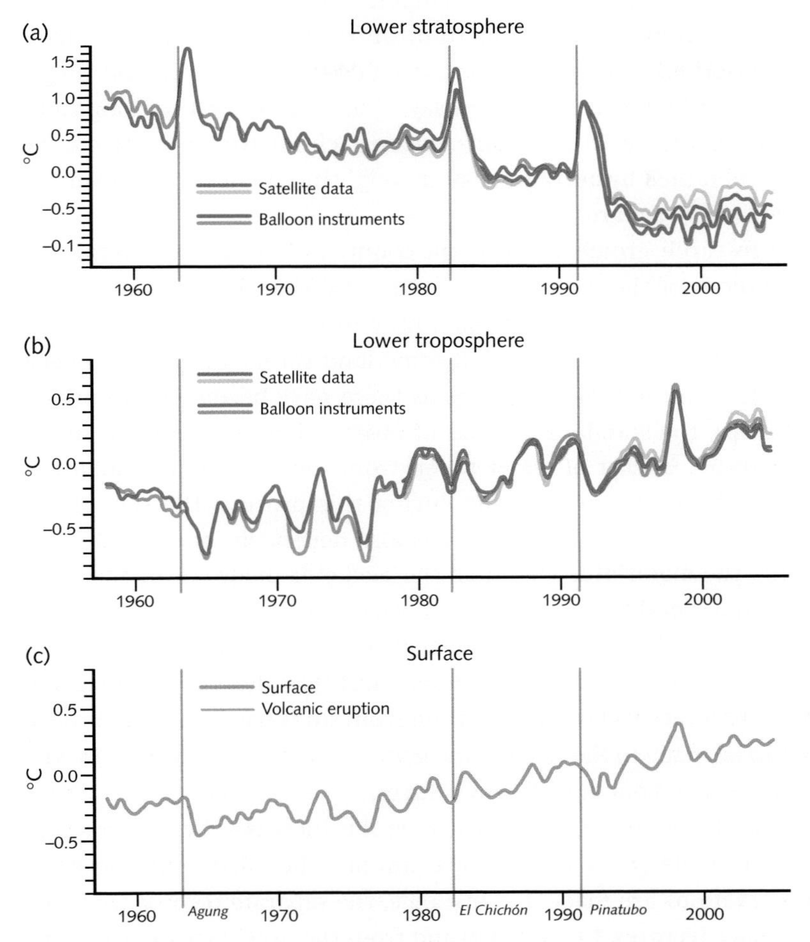

Figure 4.3 Time series of analyses of observations of global average temperature (°C) (relative to average for 1979–97) (a) for the lower stratosphere (~13 to 20 km) from balloon instruments (blue and red) and since 1979 from satellite MSUs (purple and brown); (b) for the lower troposphere (up to ~7 km) from balloon instruments (blue and red) and since 1979 from satellite MSUs (purple and brown); (c) for the surface. All time series are monthly mean anomalies relative to 1979 to 1997 smoothed with a seven-month running mean filter. Times of major volcanic eruptions are indicated by vertical lines.

140-year period. Changes in sea surface temperature have been estimated by processing over 60 million observations from ships – mostly merchant ships – over the same period. All the observations, from land stations and from ships, are then located within a grid of squares, say 1° of latitude by 1° of longitude, covering the Earth's surface. Observations within each square are averaged; the global average is obtained by averaging (after weighting them by area) over the averages for each of the squares.

A number of research groups in different countries have made careful and independent analyses of these observations. In somewhat different ways they have made allowances for factors that could have introduced artificial changes in the records. For instance, the record at some land stations could have been affected by changes in their surroundings as these have become more urban. In the case of ships, the standard method of observation used to be to insert a thermometer into a bucket of water taken from the sea. Small changes of temperature have been shown to occur during this process; the size of the changes varies between day and night and is also dependent on several other factors including the material from which the bucket is made – over the years wooden, canvas and metal buckets have been variously employed. Nowadays, a large proportion of the observations are made by measuring the temperature of the water entering the engine cooling system. Careful analysis of the effects of these details on observations both on land and from ships has enabled appropriate corrections to be made to the record, and good agreement has been achieved between analyses carried out at different centres.

Confidence that the observed variations are real is increased by noticing that the trend and the shape of the changes are similar when different selections of the total observations are made. For instance, the separate records from the land and sea surface (Figures 4.1b and 4.2) and from the northern and southern hemispheres are closely in accord. Further indirect indicators such as changes in borehole temperatures and sub-surface ocean temperatures, decrease in snow cover and glacier shrinkage provide independent support for the observed warming (Table 4.1.)

During the last 30 years or so observations have been available from satellites orbiting around the Earth. Their great advantage is that they automatically provide data with global coverage, which are often lacking in other data sets. The length of the record from satellites, however, is less than 30 years, a comparatively short period in climate terms. At the time of the IPCC 2001 assessment there were suggestions that the satellite measurements of lower atmospheric temperature since 1979 showed a substantially smaller warming trend than the surface observations. However, more careful analyses since 2001 of the satellite observations now bring the two trends into agreement within their respective uncertainties (see box).

The most obvious feature of the climate record illustrated in Figure 4.1 is that of considerable variability, not just from year to year, but from decade to decade. Some of this variability will have arisen through causes external to the atmosphere and the oceans, for instance as a result of volcanic eruptions such as those of Krakatoa in 1883 or of Pinatubo in the Philippines in 1991 (the low global average temperature in 1992 and 1993, compared with neighbouring years, is almost certainly due to the Pinatubo volcano). But there is no need to invoke volcanoes or other external causes to explain all the variations in the record. Many of them result from internal variations within the total climate system, for instance between different parts of the ocean (see Chapter 5 for more details).

The warming during the twentieth century has not been uniform over the globe. For instance, the recent warming has been greatest over northern hemisphere continents at mid to high latitudes. There have also been areas of cooling, for instance over some parts of the North Atlantic ocean associated with changes in ocean circulation (see Chapter 5). Some of the regional patterns of temperature change are related to different phases of atmosphere–ocean oscillations, such as the El Niño Southern Oscillation (ENSO) and the North Atlantic Oscillation (NAO). The positive phase of the NAO, with high pressure over the sub-tropical Atlantic and southern Europe and mild winters over northwest Europe, has tended to be dominant since the mid 1980s.

An interesting feature of the increasing temperature during the last few decades has been that, in the daily cycle of temperature, minimum temperatures over land have increased about twice as much as maximum temperatures. A likely explanation for this, in addition to the effects of enhanced greenhouse gases, is an increase in cloud cover which has been observed in many of the areas with reduced temperature range. An increase in cloud tends to obstruct daytime sunshine and tends also to reduce the escape of terrestrial radiation at night.

As might be expected the increases in temperature have led on average to increases in precipitation, although precipitation shows even more variability in both space and time than temperature. The increases have been particularly noticeable in the northern hemisphere in mid to high latitudes, often appearing particularly as increases in heavy rainfall events (see Table 4.1).

The broad features of these changes in temperature and precipitation are consistent with what is expected because of the influence of increased greenhouse gases (see Chapter 5), although there is much variability in the record that arises for reasons not associated with human activities. For instance, the particular increase from 1910 to 1940 (Figure 4.1a) is too rapid to have been due to the rather

Table 4.1 Twentieth-century changes in the Earth's atmosphere, climate and biophysical system

Indicator	Observed changes
Concentration indicators	
Atmospheric concentration of CO_2	280 ppm for the period 1000–1750 to 368 ppm in year 2000 (31 ± 4% increase) – 380 ppm in 2006
Terrestrial biospheric CO_2 exchange	Cumulative source of about 30 GtC between the years 1800 and 2000; but during the 1990s a net sink of about 10 ± 6 GtC
Atmospheric concentration of CH_4	700 ppb for the period 1000–1750 to 1750 ppb in year 2000 (151 ± 25% increase) – 1775 ppb in 2005
Atmospheric concentration of N_2O	270 ppb for the period 1000–1750 to 316 ppb in the year 2000 (17 ± 5% increase) – 319 ppb in 2005
Tropospheric concentration of O_3	Increased by 35 ± 15% from the years 1750 to 2000, varies with region
Stratospheric concentration of O_3	Decreased since 1970, varies with altitude and latitude
Atmospheric concentrations of HFCs, PFCs and SF_6	Increased globally over the last 50 years
Weather indicators	
Global mean surface temperature	Increased by 0.6 ± 0.2 °C over the twentieth century – 0.74 ± 0.18 over 100 years 1906–2005; land areas warmed more than the oceans (*very likely*)
Northern hemisphere surface temperature	Increase over the twentieth century greater than during any other century in the last 1000 years; 1990s warmest decade of the millennium (*likely*)
Diurnal surface temperature range	Decreased over the years 1950 to 2000 over land; night-time minimum temperatures increased at twice the rate of daytime maximum temperatures (*likely*)
Hot days/heat index	Increased (*likely*)
Cold/frost days	Decreased for nearly all land areas during the twentieth century (*very likely*)
Continental precipitation	Increased by 5–10% over the twentieth century in the northern hemisphere (*very likely*), although decreased in some regions (e.g. north and west Africa and parts of the Mediterranean)
Heavy precipitation events	Increased at mid and high northern latitudes (*likely*)
Drought	Increased summer drying and associated incidence of drought in a few areas (*likely*). Since 1970s, increase in total area affected in many regions of the world (*likely*)

Table 4.1 (Cont.)

Tropical cyclones Intense extratropical storms	Since 1970s, trend towards longer lifetimes and greater storm intensity but no trend in frequency (*likely*) Since 1950s, net increase in frequency/intensity and poleward shift in track (*likely*)
Biological and physical indicators	
Global mean sea level	Increased at an average annual rate of 1–2 mm during the twentieth century – rising to about 3 mm from 1993–2003
Duration of ice cover of rivers and lakes	Decreased by about two weeks over the twentieth century in mid and high latitudes of the northern hemisphere (*very likely*)
Arctic sea-ice extent and thickness	Thinned by 40% in recent decades in late summer to early autumn (*likely*) and decreased in extent by 10–15% since the 1950s in spring and summer
Non-polar glaciers	Widespread retreat during the twentieth century
Snow cover	Decreased in area by 10% since global observations became available from satellites in the 1960s (*very likely*)
Permafrost	Thawed, warmed and degraded in parts of the polar, sub-polar and mountainous regions
El Niño events	Became more frequent, persistent and intense during the last 30 years compared to the previous 100 years
Growing season	Lengthened by about one to four days per decade during the last 50 years in the northern hemisphere, especially at higher latitudes
Plant and animal ranges	Shifted poleward and up in elevation for plants, insects, birds and fish
Breeding, flowering and migration	Earlier plant flowering, earlier bird arrival, earlier dates of breeding season and earlier emergence of insects in the northern hemisphere
Coral reef bleaching	Increased frequency, especially during El Niño events
Economic indicators	
Weather-related economic losses	Global inflation-adjusted losses rose by an order of magnitude over the last 50 years. Part of the observed upward trend is linked to socio-economic factors and part is linked to climatic factors

Note: This table provides examples of key observed changes and is not an exhaustive list. It includes both changes attributable to anthropogenic climate change and those that may be caused by natural variations or anthropogenic climate change. Confidence levels (for explanation see Note 1) are reported where they are explicitly assessed by the relevant Working Group of the IPCC.

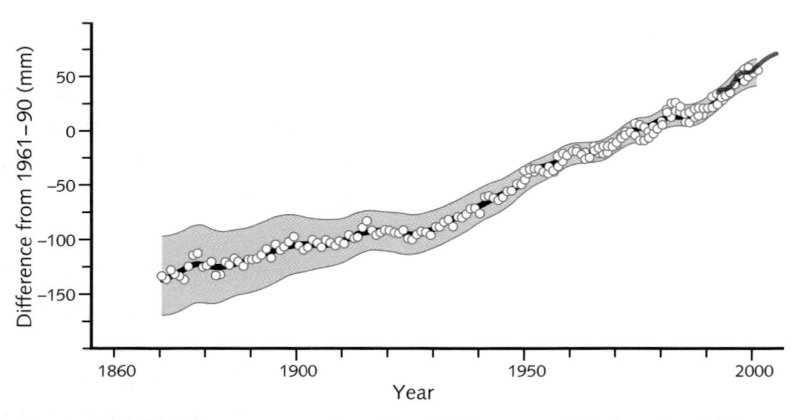

Figure 4.4 Global average sea level from tide gauge (blue) and satellite (red) data relative to the 1961–90 mean.

small increase in greenhouse gases during that period. The particular reasons for this will be discussed in the next chapter where comparisons of observed temperatures with simulations from climate models for the whole of the twentieth century will be presented, not just as they concern the global mean but also the regional patterns of change. We conclude therefore that, although the expected signal is still emerging from the noise of natural variability, most of the observed warming over the last 50 years is *very likely* to have been due to the increase in greenhouse gas concentrations.

Significant cooling of the lower stratosphere (the region at altitudes between about 10 and 30 km) has been observed over the last two decades (Figure 4.3a). This is to be expected both because of the decrease in the concentration of ozone (which absorbs solar radiation) and because of the increased carbon dioxide concentration which leads to increased cooling at these levels (see Chapter 2). Because warming of the troposphere and cooling of the stratosphere are occurring because of the increase in greenhouse gases, an increase on average in the height of the tropopause, the boundary between the troposphere and the stratosphere, is also expected. There is now observational evidence for this increase.

A further source of information regarding climate change comes from measurements of change in sea level (Figure 4.4). Over the twentieth century sea level rose by 17 ± 5 cm. The rate of rise increased to 3.1 ± 0.7 cm over the decade 1993 to 2003 of which 1.6 ± 0.5 cm is estimated to be from thermal expansion of the ocean as its average temperature increased and 1.2 ± 0.4 cm from the melting of glaciers that have generally been retreating over the last century. The net contribution from the Greenland and Antarctic ice caps is more uncertain but is believed to be small.

In Chapter 1, we mentioned the increasing vulnerability of human populations to climate extremes, which has brought about more awareness of recent extremes in the forms of floods, droughts, tropical cyclones and windstorms. It is therefore of great importance to know whether there is evidence of an increase in the frequency or severity of these and other extreme events. The available evidence regarding how these and other relevant parameters have

Nukuoro Atoll, Federated States of Micronesia is home to 900 people around the 6-km wide lagoon. It is part of an island chain that stretches northeast of Papua New Guinea in the western Pacific. It was reported in November 2005 that the islands have progressively become uninhabitable, with an estimate of their total submersion by 2015. Storm surges and high tides continue to wash away homes, destroy vegetable gardens and contaminate fresh water supplies. Photographed from the International Space Station.

changed during the twentieth century is summarised in Table 4.1 in terms of different indicators: concentrations of greenhouse gases; temperature, hydrological and storm-related indicators; and biological and physical indicators. To what extent these changes are expected to continue or to intensify during the twenty-first century will be addressed in Chapter 6.

The last thousand years

The detailed systematic record of weather parameters such as temperature, rainfall, cloudiness and the like presented above, covering a good proportion of the globe over the last 140 years, is not available for earlier periods. Further back, the record is more sparse and doubt arises over the consistency of the instruments

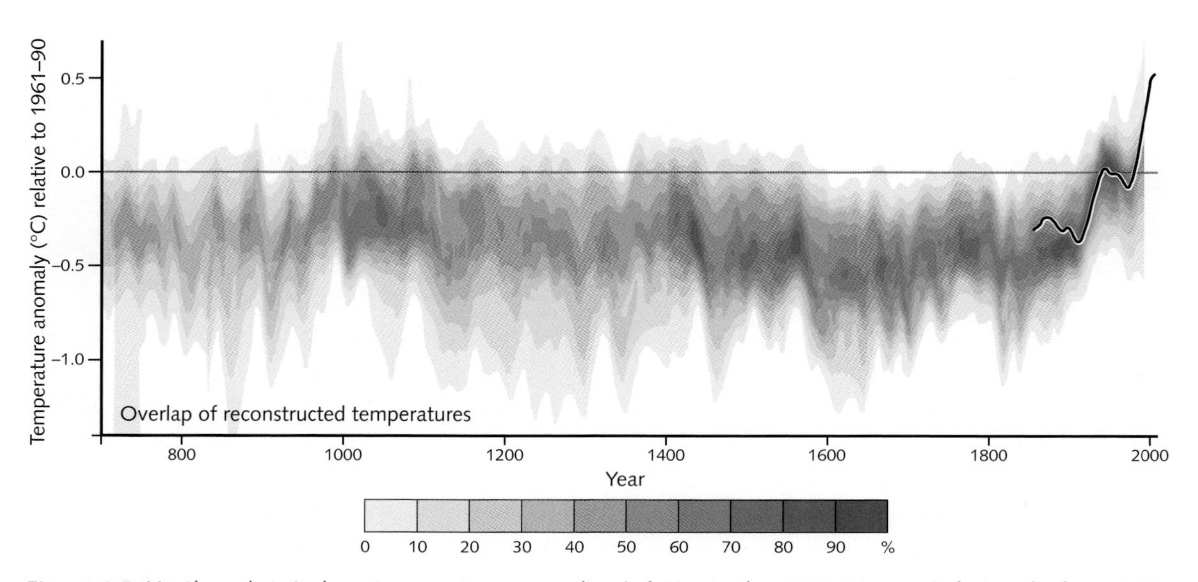

Figure 4.5 Northern hemisphere temperature anomalies (relative to the 1961–90 mean) during the last 1300 years from ten published overlapping reconstructions from proxy records – e.g. tree rings, corals, ice cores and historical records – (shadings), and from 1860 from instrumental data (black line). For the percentage shading scale, temperatures within ±1 standard error (SE) of a reconstruction score 10% and regions within the 5–95% uncertainty range score 5%. The maximum of 100% is obtained only for temperatures that fall within ±1 SE of all ten reconstructions.

used for observation. Most thermometers in use 200 years ago were not well calibrated or carefully exposed. However, many diarists and writers kept records at different times; from a wide variety of sources weather and climate information can be pieced together. Indirect sources, such as are provided by ice cores, tree rings and records of lake levels, of glacier advance and retreat, and of pollen distribution (found in sediments in lakes for instance), can also yield information to assist in building up the whole climatic story. From a variety of sources, for instance, it has been possible to put together for China a systematic atlas of weather patterns covering the last 500 years.

Similarly, from direct and indirect sources, it has been possible to deduce the average temperature over the northern hemisphere for the last millennium (Figure 4.5). Sufficient data are not available for the same reconstruction to be carried out over the southern hemisphere. Because of the uncertainties underlying the precise interpretation of proxy data and because of the sparsity of data and coverage especially for earlier periods, there are large uncertainties associated with the reconstructions shown in Figure 4.5 – as is illustrated also by the range within the ten different reconstructions. However, it is just possible to identify the 'Medieval Warm Period' associated with the eleventh to fourteenth centuries and a relatively cool period, the 'Little Ice Age', associated with the fifteenth to nineteenth centuries. There has been much debate about the extent of these particular periods that only affected part of the northern hemisphere and are therefore more prominent in local records, for instance those from central England. The increase in temperature over the twentieth century is particularly striking and the 1990s are *likely* to have been the warmest decade of the millennium in the northern hemisphere.

Although there is as yet no complete explanation for the variations that occurred between 1000 and 1900, it is clear that greenhouse gases such as carbon dioxide and methane cannot have been the cause of change. For the millennium before 1800 their concentration in the atmosphere was rather stable, the carbon dioxide concentration, for instance, varying by less than 3%. However, the influence of variations in volcanic activity can be identified especially in some of the downturns of temperature in the record of Figure 4.3. For instance, one of the largest eruptions during the period was that of Tambora in Indonesia in April 1815, which was followed in many places by two exceptionally cold years; and 1816 was described in New England and Canada as the 'year without a summer'. Although the effect on the climate even of an eruption of the magnitude of Tambora only lasts a few years, variations in *average* volcanic activity have a longer-term effect. It is likely also variations in the output of energy from the Sun provide some part of the explanation.[3] Although accurate direct measurements of total solar radiation are not available (apart from those made during the last two decades from satellite instruments), other evidence suggests that the solar output could have varied significantly in the past. For instance, compared with its value today it may have been somewhat lower (by a few tenths of a watt per square metre) during the Maunder Minimum in the seventeenth century (a period when almost no sunspots were recorded; see also box on page 166). There is no need, however, to invoke volcanoes or variations in solar output as the cause of all the climate variations over this period. As with the shorter-term changes mentioned earlier, such variations of climate can arise naturally from internal variations within the atmosphere and the ocean and in the two-way relationship – coupling – between them.

The millennial record of Figure 4.5 is particularly important because it provides an indication of the range and character of climate variability that arises from natural causes. As we shall see in the next chapter, climate models also provide some information on natural climate variability. Careful assessments of these observational and model results confirm that natural variability (the combination of internal variability and naturally forced, e.g. by volcanoes or change in solar output) is *very unlikely* to explain the warming in the latter half of the twentieth century.

Scientists in Antarctica use a hand drill to take 10-metre ice cores. Chemical analyses of the cores will reveal changes in climate and the composition of the atmosphere.

The past million years

To go back before recorded human history, scientists have to rely on indirect methods to unravel much of the story of the past climate. A particularly valuable information source is the record stored in the ice that caps Greenland and the Antarctic continent. These ice caps are several thousands of metres thick. Snow deposited on their surface gradually becomes compacted as further snow falls, becoming solid ice. The ice moves steadily downwards, eventually flowing outwards at the bottom of the ice-sheet. Ice near the top of the layer will have been deposited fairly recently; ice near the bottom will have fallen on the surface many tens or hundreds of thousands of years ago. Analysis of the ice at different levels can, therefore, provide information about the conditions prevailing at different times in the past.

Deep cores have been drilled out of the ice at several locations in both Greenland and Antarctica. At Russia's Vostok station in east Antarctica, for instance, drilling has been carried out for over 25 years. The longest and most recent core reached a depth of over 3.5 km; the ice at the bottom of the hole fell as snow on the surface of the Antarctic continent well over half a million years ago (Figure 4.6b).

Small bubbles of air are trapped within the ice. Analysis of the composition of that air shows what was present in the atmosphere for the time at which the ice

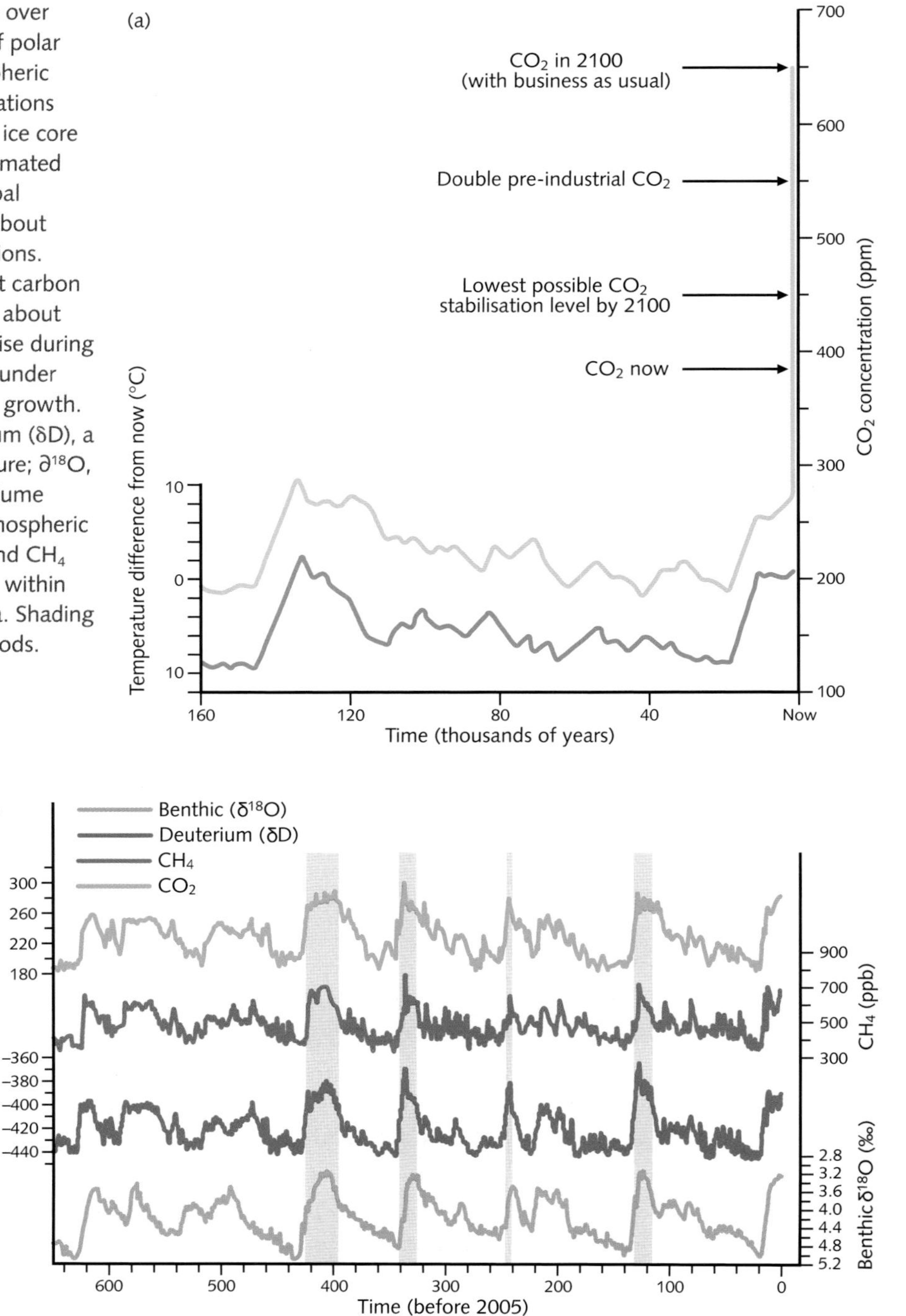

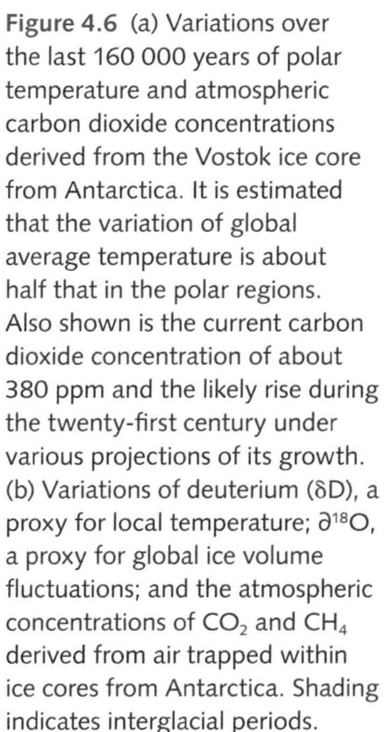

Figure 4.6 (a) Variations over the last 160 000 years of polar temperature and atmospheric carbon dioxide concentrations derived from the Vostok ice core from Antarctica. It is estimated that the variation of global average temperature is about half that in the polar regions. Also shown is the current carbon dioxide concentration of about 380 ppm and the likely rise during the twenty-first century under various projections of its growth. (b) Variations of deuterium (δD), a proxy for local temperature; $\partial^{18}O$, a proxy for global ice volume fluctuations; and the atmospheric concentrations of CO_2 and CH_4 derived from air trapped within ice cores from Antarctica. Shading indicates interglacial periods.

Palaeoclimate reconstruction from isotope data

The isotope ^{18}O is present in natural oxygen at a concentration of about 1 part in 500 compared with the more abundant isotope ^{16}O. When water evaporates, water containing the lighter isotope is more easily vaporised, so that water vapour in the atmosphere contains less ^{18}O compared with sea water. Similar separation occurs in the process of condensation when ice crystals form in clouds. The amount of separation between the two oxygen isotopes in these processes depends on the temperatures at which evaporation and condensation occur. Measurements on snowfall in different places can be used to calibrate the method; it is found that the concentration of ^{18}O varies by about 0.7 of a part per 1000 for each degree of change in average temperature at the surface. Information is therefore available in the ice cores taken from polar ice caps concerning the variation in atmospheric temperature in polar regions during the whole period when the ice core was laid down.

Since the ice caps are formed from accumulated snowfall which contains less ^{18}O compared with sea water, the concentration of ^{18}O in water from the oceans provides a measure of the total volume of the ice in the ice caps; it changes by about 1 part in 1000 between the maximum ice extent of the ice ages and the warm periods in between. Information about the ^{18}O content of ocean water at different times is locked up in corals and in cores of sediment taken from the ocean bottom, which contain carbonates from fossils of plankton and small sea creatures from past centuries and millennia. Measurements of radioactive isotopes, such as the carbon isotope ^{14}C, and correlations with other significant past events enable the corals and sediment cores to be dated. Since the separation between the oxygen isotopes which occurs as these creatures are formed also depends on the temperature of the sea water (although the dependence is weaker than the other dependencies considered above) information is also available about the distribution of ocean surface temperature at different times in the past.

was formed – gases such as carbon dioxide or methane. Dust particles that may have come from volcanoes or from the sea surface are also contained within the ice. Further information is provided by analysis of the ice itself. Small quantities of different oxygen isotopes and of the heavy isotope of hydrogen (deuterium) are contained in the ice. The ratios of these isotopes that are present depend sensitively on the temperatures at which evaporation and condensation took place for the water in the clouds from which the ice originated (see box). These in turn are dependent on the average temperature near the surface of the Earth. A temperature record for the polar regions can therefore be constructed from analyses of the ice cores. The associated changes in global average temperature are estimated to be about half the changes in the polar regions.

Such a reconstruction from a Vostok core for the temperature and the carbon dioxide content is shown in Figure 4.6a for the past 160 000 years, which includes the last major ice age that began about 120 000 years ago and began to come to an end about 20 000 years ago. Figure 4.6b extends the record to

650 000 years ago. The close connections that exist between temperature, carbon dioxide and methane concentrations are evident in Figure 4.6. Note also from Figure 4.6 the likely growth of atmospheric carbon dioxide during the twenty-first century, taking it to levels that are *unlikely* to have been exceeded during the past 20 million years.

Further information over the past million years is available from investigations of the composition of ocean sediments. Fossils of plankton and other small sea creatures deposited in these sediments also contain different isotopes of oxygen. In particular the amount of the heavier isotope of oxygen (^{18}O) compared with the more abundant isotope (^{16}O) is sensitive both to the temperature at which the fossils were formed and to the total volume of ice in the world's ice caps at the time of the fossils' formation that is linked to the global sea level. For instance, from such data it can be deduced that the sea level at the last glacial maximum, 20 000 years ago, was about 120 m lower than today and that during the last interglacial period, about 125 000 years ago, it was likely between 4 and 6 m higher than today due to some melting of the polar ice caps in both Greenland and Antarctica.

From the variety of palaeoclimate data available, variations in the volume of ice in the ice caps can be reconstructed over the greater part of the last million years (Figures 4.6b, lower curve, and 4.7c). In this record six or seven major ice ages can be identified with warmer periods in between, the period between these major ice ages being approximately 100 000 years. Other cycles are also evident in the record.

The most obvious place to look for the cause of regular cycles in climate is outside the Earth, in the Sun's radiation. Has this varied in the past in a cyclic way? So far as is known the output of the Sun itself has not changed to any large extent over the last million years or so. But because of variations in the Earth's orbit, the distribution of solar radiation has varied in a more or less regular way during this period.

Three regular variations occur in the orbit of the Earth around the Sun (Figure 4.7a). The Earth's orbit, although nearly circular, is actually an ellipse. The eccentricity of the ellipse (which is related to the ratio between the greatest and the least diameters) varies with a period of about 100 000 years; that is the slowest of the three variations. The Earth also spins on its own axis, the axis of spin being tilted with respect to the axis of the Earth's orbit, the angle of tilt varying between 21.6° and 24.5° (currently it is 23.5°) with a period of about 41 000 years. The third variation is of the time of year when the Earth is closest to the Sun (the Earth's perihelion). The time of perihelion moves through the months of the year with a period of about 23 000 years (see also Figure 5.19); in the present configuration, the Earth is closest to the Sun in January.

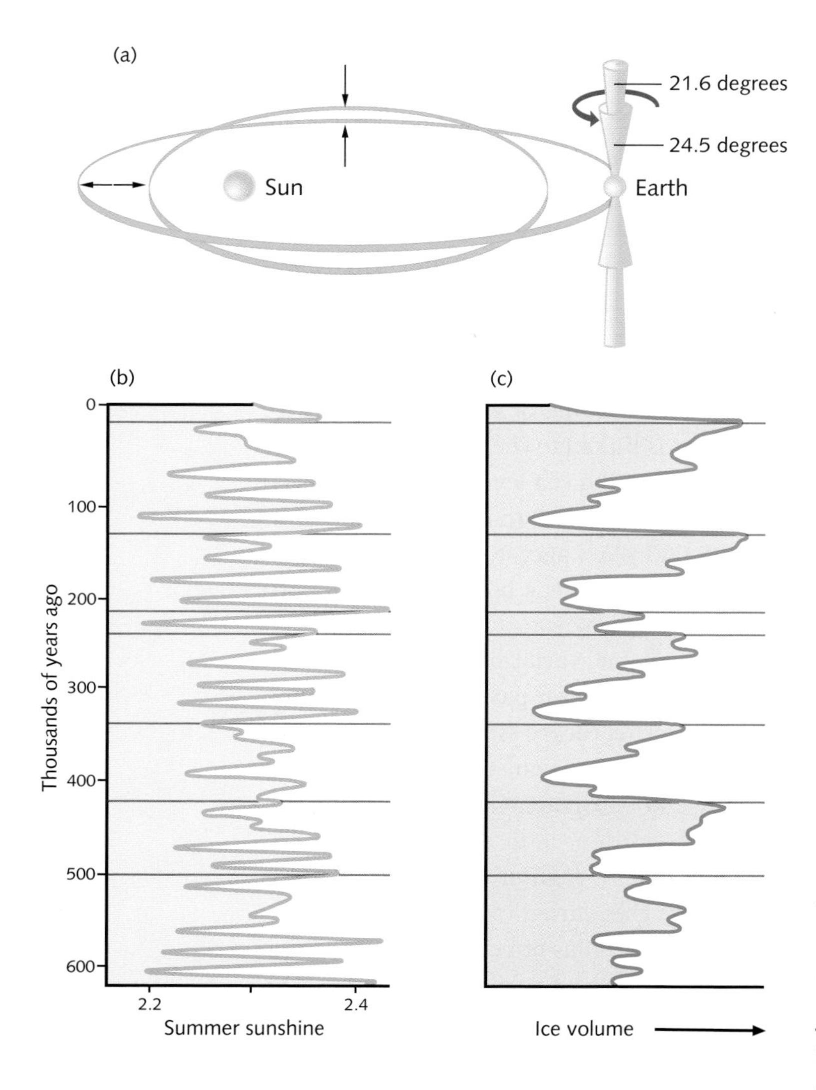

Figure 4.7 Variations in the Earth's orbit (a), in its eccentricity, the orientation of its spin axis (between 21.6° and 24.5°) and the longitude of perihelion (i.e. the time of year when the Earth is closest to the Sun; see also Figure 5.19), cause changes in the average amount of summer sunshine (in millions of joules per square metre per day) near the poles (b). These changes appear as cycles in the climate record in terms of the volume of ice in the ice caps (c).

As the Earth's orbit changes its relationship to the Sun, although the total quantity of solar radiation reaching the Earth varies very little, the distribution of that radiation with latitude and season over the Earth's surface changes considerably. The changes are especially large in polar regions where the variations in summer sunshine, for instance, reach about 10% (Figure 4.7b). James Croll, a British scientist, first pointed out in 1867 that the major ice ages of the past might be linked with these regular variations in the seasonal distribution of solar radiation reaching the Earth. His ideas were developed in 1920 by Milutin Milankovitch, a climatologist from Yugoslavia, whose name is usually linked with the theory. Inspection by eye of the relationship between the variations of polar summer sunshine and global ice volume shown in Figure 4.7 suggests a significant connection. Careful study of the correlation between the two curves confirms this and demonstrates that 60% of the variance in the climatic record of global ice volume falls close to the three frequencies of regular variations in the Earth's orbit, thus providing support for the Milankovitch theory.[4]

More careful study of the relationship between the ice ages and the Earth's orbital variations shows that the size of the climate changes is larger than might be expected from forcing by the radiation changes alone. Other processes that enhance the effect of the radiation changes (in other words, positive feedback processes) have to be introduced to explain the climate variations. One such feedback arises from the changes in carbon dioxide influencing atmospheric temperature through the greenhouse effect, illustrated by the strong correlation observed in the climatic record between average atmospheric temperature and carbon dioxide concentration (Figure 4.6). Such a correlation does not, of course, prove the existence of the greenhouse feedback; in fact part of the correlation arises because the atmospheric carbon dioxide concentration is itself influenced, through biological feedbacks (see Chapter 3), by factors related to the average global temperature.[5] Further, since Antarctic temperature started to rise several centuries before atmospheric carbon dioxide during past glacial terminations, it is clear that carbon dioxide variations have not provided the trigger for the end of glacial periods. However, as we shall see in Chapter 5, climates of the past cannot be modelled successfully without taking the greenhouse feedback into account.[6]

An obvious question to ask is when, on the Milankovitch theory, is the next ice age due? It so happens that we are currently in a period of relatively small solar radiation variations and the best projections for the long term are of a longer than normal interglacial period leading to the beginning of a new ice age perhaps in 50 000 years' time.[7]

How stable has past climate been?

The major climate changes considered so far in this chapter have taken place relatively slowly. The growth and recession of the large polar ice-sheets between the ice ages and the intervening warmer interglacial periods have taken on average many thousands of years. However, the ice core records such as those in Figures 4.6 and 4.8 show evidence of large and relatively rapid fluctuations. Ice cores from Greenland provide more detailed evidence of these than those from Antarctica. This is because at the summit of the Greenland ice cap, the rate of accumulation of snow has been higher than that at the Antarctica drilling locations. For a given period in the past, the relevant part of the Greenland ice core is longer and more detail of variations over relatively short periods is therefore available.

The data show that the last 8000 years have been unusually stable compared with earlier epochs. In fact, as judged from the Vostok (Figure 4.6) and the Greenland records (Figure 4.8) this long stable period in the Holocene is a

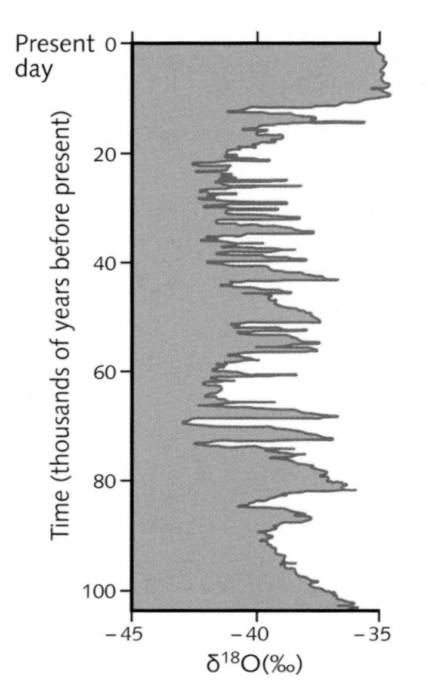

Figure 4.8 Variations in Arctic temperature over the past 100 000 years as deduced from oxygen isotope measurements (in terms of $\delta^{18}O$) from the 'Summit' ice core in Greenland. The quantity $\delta^{18}O$ plotted in Figures 4.6 and 4.7 is the difference (in parts per thousand) between the $^{18}O/^{16}O$ ratio in the sample and the same ratio in a laboratory standard. The overall shape of the record is similar to that from the Vostok ice core shown in Figure 4.6 but much more detail is apparent in the 'Summit' record's stable period over the last 8000 years. A change of 5 parts per 1000 in $\delta^{18}O$ in the ice core corresponds to about a 7 °C change in temperature.

unique feature of climate during the past 420 000 years. It has been suggested that this had profound implications for the development of civilisations.[8] Model simulations (see Chapter 5) indicate that the detail of long-term changes during the Holocene is consistent with the influence of orbital forcing (Figure 4.7). For instance, some northern hemisphere glaciers retreated between 11 000 and 5000 years ago and were smaller during the later part of this period than they are today. The present-day glacier retreat cannot be attributed to the same natural causes as the decrease in summer insolation due to orbital forcing during the past few millennia has tended to glacier increase.

It is also interesting to inspect the rate of temperature change during the recovery period from the last glacial maximum about 20 000 years ago and compare it with recent temperature changes. The data indicate an average warming rate of about 0.2 °C per century between 20 000 and 10 000 years before present (BP) over Greenland, with lower rates for other regions. Compare this with a temperature rise during the twentieth century of about 0.6 °C and the rates of change of a few degrees Celsius per century projected to occur during the twenty-first century because of human activities (see Chapter 6).

The ice core data (Figure 4.8) demonstrate that a series of rapid warm and cold oscillations called Dansgaard–Oeschger events punctuated the last glaciation. Comparison between the results from ice cores drilled at different locations within the Greenland ice cap confirm the details up to about 100 000 years ago. Comparison with data from Antarctica suggests that the fluctuations of temperature over Greenland (perhaps up to 16 °C) have been larger than those over Antarctica. Similar large and relatively rapid variations are evident from North Atlantic deep sea sediment cores.

Another particularly interesting period of climatic history, more recently, is the Younger Dryas event (so called because it was marked by the spread of an arctic flower, *Dryas octopetala*), which occurred over a period of about 1500 years between about 12 000 and 10 700 years ago. For 6000 years before the start of this event the Earth had been warming up after the end of the last ice age. But

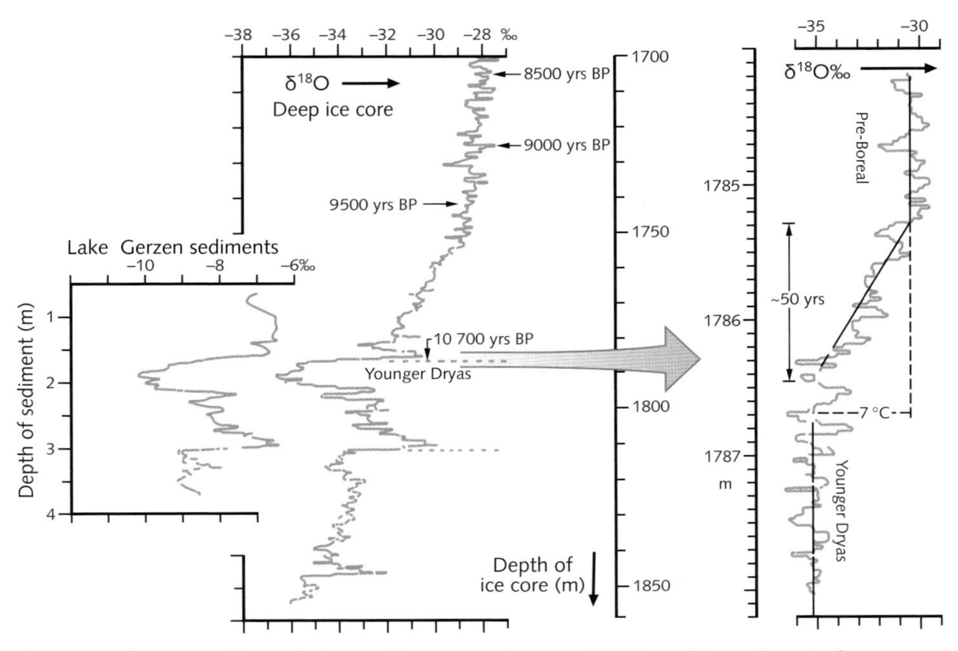

Figure 4.9 Records of the variations of the oxygen isotope $\delta^{18}O$ from lake sediments from Lake Gerzen in Switzerland and from the Greenland ice core 'Dye 3' showing the Younger Dryas event and its rapid end about 10 700 years ago. Dating of the ice core was by counting the annual layers down from the surface; dating of the lake sediment was by the ^{14}C method. A change of 5 parts per 1000 in $\delta^{18}O$ in the ice core corresponds to about a 7 °C change in temperature.

then during the Younger Dryas period, as demonstrated from many different sources of palaeoclimatic data, the climate swung back again into much colder conditions similar to those at the end of the last ice age (Figure 4.9). The ice core record shows that at the end of the event, 10 700 years ago, the warming in the Arctic of about 7 °C occurred over only about 50 years and was associated with decreased storminess (shown by a dramatic fall in the amount of dust in the ice core) and an increase in precipitation of about 50%.

Two main reasons for these rapid variations in the past have been suggested. One reason particularly applicable to ice age conditions is that, as the ice-sheets over Greenland and eastern Canada have built up, major break-ups have occurred from time to time, releasing massive numbers of icebergs into the North Atlantic in what are called Heinrich events. The second possibility is that the ocean circulation in the North Atlantic region has been strongly affected by injections of fresh water from the melting of ice. At present the ocean circulation in this region is strongly influenced by cold salty water sinking to deep ocean levels because its saltiness makes it dense; this sinking process is part of the 'conveyor belt' which is the major feature of the circulation of deep ocean water around the world (see Figure 5.18). Large quantities of fresh water from the melting of ice would make the water less salty, preventing it from sinking and thereby altering the whole Atlantic circulation.

This link between the melting of ice and the ocean circulation is a key feature of the explanation put forward by Professor Wallace Broecker for the Younger Dryas event.[9] As the great ice-sheet over North America began to melt at the end of the last ice age, the melt water at first drained through the Mississippi into the Gulf of Mexico. Eventually, however, the retreat of the ice opened up a channel for the water in the region of the St Lawrence River. This influx of fresh water into the North Atlantic reduced its saltiness, thus, Broecker postulates, cutting off the formation of deep water and that part of the ocean 'conveyor belt'.[10] Warm water was therefore prevented from flowing northward, resulting in a reversal to much colder conditions. The suggestion is also that a reversal of this process with the starting up of the Atlantic 'conveyor belt' could lead to a sudden onset of warmer conditions.

Although debate continues regarding the details of the Younger Dryas event, there is considerable evidence from palaeodata, especially those from ocean sediments, for the main elements of the Broecker explanation which involve the deep ocean circulation. It is also clear from palaeodata that large changes have occurred at different times in the past in the formation of deep water and in the deep ocean circulation. Chapter 3 mentioned the possibility of such changes being induced by global warming through the growth of greenhouse gas concentrations. Our perspective regarding the possibilities of future climate change needs to take into account the rapid climate changes that have occurred in the past.

SUMMARY

In this chapter we have learnt that:

- Records of temperature, atmospheric composition and sea level from ice cores from Greenland and Antarctica, from ocean and lake sediment cores and other proxy records have provided a wealth of information about past climates over much of the past million years.
- The current levels of carbon dioxide and methane concentrations in the atmosphere and their rate of increase is unprecedented in the palaeoclimate record over the last half million years and probably for much longer.

- It is *likely* that the 50-year period of the second half of the twentieth century was the warmest northern hemisphere period in the last 1300 years.
- It is *very likely* that the warming of 4 to 7 °C since the last glacial maximum 18 000 years ago occurred at an average rate about ten times slower than the warming of the twentieth century.
- The main trigger mechanism for the series of ice ages over the last million years or more has been the variations in the distribution of solar radiation especially in the polar regions arising from the regular variations in parameters of the Earth's orbit around the Sun – called Milankovitch cycles. Variations in greenhouse gases have served to add a positive feedback to this forcing. These orbital variations are well understood and the next ice age is not expected to begin for at least 30000 years.
- For the first half of the last interglacial period (~130 000–123 000 years ago), a large increase in summer solar radiation due to orbital forcing led to higher temperatures in polar regions 3 to 5 °C warmer than today and melting in polar regions that led to 4 to 6 metres higher sea level than today.
- Some of the abrupt events during the last 100 000 years that have been identified in the records from ice core and other data may have been associated with large fresh water inputs into the ocean due for instance to large ice discharges that resulted in large-scale changes to the ocean circulation.

Having now in these early chapters set the scene, by describing the basic science of global warming, the greenhouse gases and their origins and the current state of knowledge regarding past climates, I move on in the next chapter to describe how, through computer models of the climate, predictions can be made about what climate change can be expected in the future.

QUESTIONS

1. Given that the sea level at the end of the last glacial maximum was 120 m lower than that today, estimate the volume of ice in the ice-sheets that covered the northern parts of the American and Eurasian continents.
2. How much energy would be required to melt the volume of ice you have calculated in Question 1? Compare this with the extra summer sunshine north of latitude 60° which might have been available between 18 000 and 6000 years before the present according to the data in Figure 4.7. Does your answer support the Milankovitch theory?

3 It is sometimes suggested that the large reserves of fossil fuels on Earth should be preserved until the onset of the next ice age is closer so that some of its impact can be postponed. From what you know of the greenhouse effect and of the behaviour of carbon dioxide in the atmosphere and the oceans, consider the influences that human burning of the known reserves of fossil fuels (see Figure 11.2) could have on the onset of the next ice age.

FURTHER READING AND REFERENCE

James Hansen *et al.*, Climate Change and Trace Gases, *Phil. Trans. R. Soc.A* (2007), **365**, 1925–1954. Summarises influence of greenhouse gases on paleoclimates of different epochs.

Solomon, S., Qin, D., Manning, M., Chen, Z., Marquis, M., Averyt, R.B., Tignor, M., Miller, H.L. (eds.) 2007. *Climate Change 2007: The Physical Science Basis. Contribution of Working Group I to the Fourth Assessment Report of the Intergovernmental Panel on Climate Change.* Cambridge: Cambridge University Press.

Technical Summary (summarises basic information about greenhouse gases and observations of the present and past climates)

Chapter 3 Observations: Surface and atmospheric climate change

Chapter 4 Observations: Changes in snow, ice and frozen ground

Chapter 5 Observations: Oceanic climate change and sea level

Chapter 6 Palaeoclimate

NOTES FOR CHAPTER 4

1 In Solomon, S. *et al*, (eds.) *Climate Change 2007: The Physical Science Basis* expressions of certainty such as *very likely* were related so far as possible to quantitative statements of confidence as follows: *virtually certain* >99% probability of occurrence, *extremely likely* >95%, *very likely* >90%, *likely* >66%, *more likely than not* >50%, *unlikely* <33%, *very unlikely* <10%, *extremely unlikely* <5%. When these 'likely' words, employed in this way, appear in the text they are italicised.

2 Smith, D.M. *et al.*, 2007. *Science*, **317**, 796–9.

3 See, for instance, Crowley, T.J. 2000. Causes of climate change over the past 1000 years. *Science*, **289**, 270–7.

4 Raymo, M.E., Huybers, P. 2008. *Nature*, **451**, 284–5.

5 For a discussion of what caused the low CO_2 concentrations during glacial times see Box 6.2, Chapter 6, in Solomon *et al.* (eds.) *Climate Change 2007: The Physical Science Basis.*

6 See, for instance, James Hansen, Bjerknes Lecture at American Geophysical Union, 17 December 2008 at www.columbia.edu/njeh1/2008/AGUBjerknes_2008/217.pdf

7 Berger, A., Loutre, M.F. 2002. *Science*, **297**, 1287–8. The *IPCC 2007 Report* states that 'it is *very unlikely* that the Earth would naturally enter another ice age for at least 30,000 years'. (Chapter 6 Summary, in Solomon *et al.* (eds.) *Climate Change 2007: The Physical Science Basis*).

8 Petit, J.R. *et al.* 1999. *Nature*, **399**, 429–36.

9 Broecker, W.S., Denton, G.H. 1990. What drives glacial cycles? *Scientific American*, **262**, 43–50.

10 More information in Chapter 5, see especially Figure 5.18.

5 Modelling the climate

This supercell thunderstorm (the largest and most severe class of thunderstorm) caused widespread damage in northwest Colorado, June 2006.

CHAPTER 2 looked at the greenhouse effect in terms of a simple radiation balance. That gave an estimate of the rise in the average temperature at the surface of the Earth as greenhouse gases increase. But any change in climate will not be distributed uniformly everywhere; the climate system is much more complicated than that. More detail in climate change prediction requires very much more elaborate calculations using computers. The problem is so vast that the fastest and largest computers available are needed. But before computers can be set to work on the calculation, a model of the climate must be set up for them to use. A model of the weather as used for weather forecasting will be used to explain what is meant by a numerical model on a computer, followed by a description of the increase in elaboration required to include all parts of the climate system in the model.

Modelling the weather

An English mathematician, Lewis Fry Richardson, set up the first numerical model of the weather. During his spare moments while working for the Friends' Ambulance Unit (he was a Quaker) in France during the First World War he carried out the first numerical weather forecast. With much painstaking calculation with his slide rule, he solved the appropriate equations and produced a six-hour forecast. It took him six months – and then it was not a very good result. But his basic methods, described in a book published in 1922,[1] were correct. To apply his methods to real forecasts, Richardson imagined the possibility of a very large concert hall filled with people, each person carrying out part of the calculation, so that the integration of the numerical model could keep up with the weather. But he was many years before his time! It was not until some forty years later that, essentially using Richardson's methods, the first operational weather forecast was produced on an electronic computer. Computers more than one trillion times faster than the one used for that first forecast (Figure 5.1) now run the numerical models that are the basis of all weather forecasts.

Lewis Fry Richardson (11 October 1881–30 September 1953).

Numerical models of the weather and the climate are based on the fundamental mathematical equations that describe the physics and dynamics of the movements and processes taking place in the atmosphere, the ocean, the ice and on the land. Although they include empirical information, they are not based to any large degree on empirical relationships – unlike numerical models of many other systems, for instance in the social sciences.

Setting up a model of the atmosphere for a weather forecast (see Figure 5.2) requires a mathematical description of the way in which energy from the Sun enters the atmosphere from above, some being reflected by the surface or by clouds and some being absorbed at the surface or in the atmosphere (see Figure 2.7). The exchange of energy and water vapour between the atmosphere and the surface must also be described.

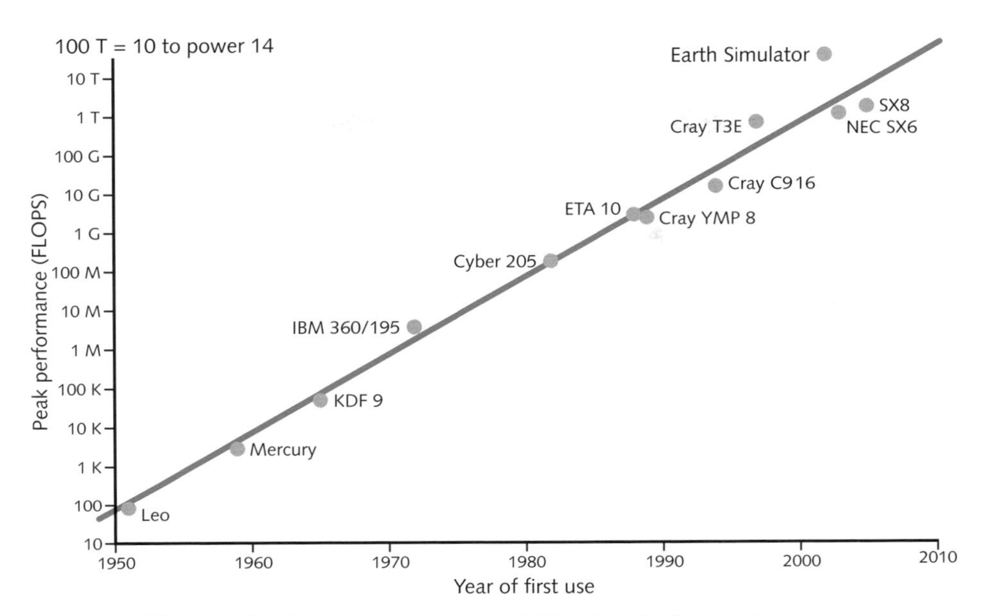

Figure 5.1 The growth of computer power available at major forecasting centres. The computers are those used by the UK Met Office for numerical weather prediction research, since 1965 for operational weather forecasting and most recently for research into climate prediction. Richardson's dream computer of a large 'human' computer mentioned at the beginning of the chapter would possess a performance of perhaps 500 FLOPS (floating point operations per second). The largest computer on which meteorological or climate models are run in 2007 is the Earth Simulator in Japan. The straight line illustrates a rate of increase in performance of a factor of 10 every five years.

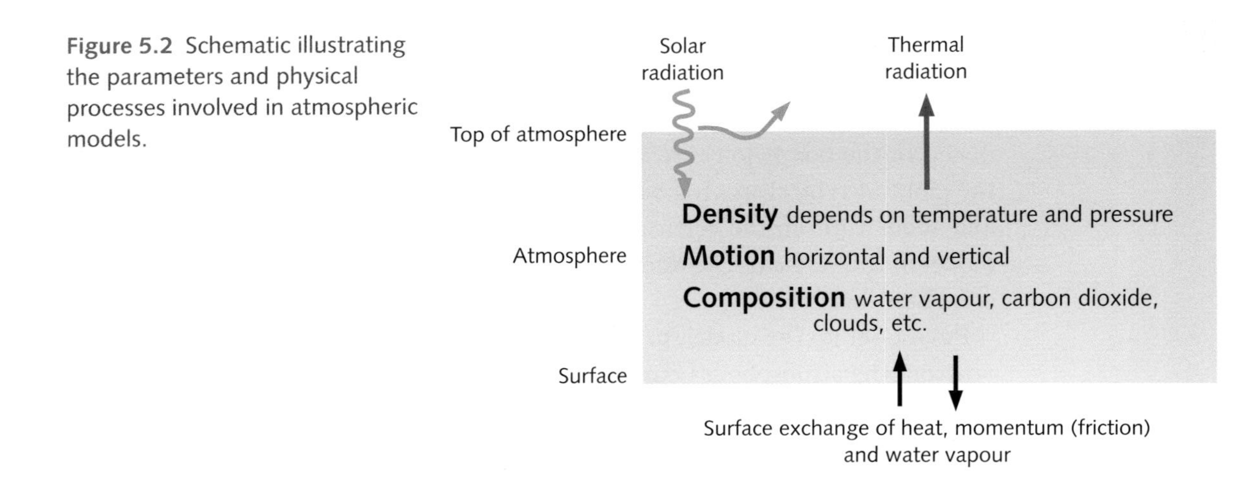

Figure 5.2 Schematic illustrating the parameters and physical processes involved in atmospheric models.

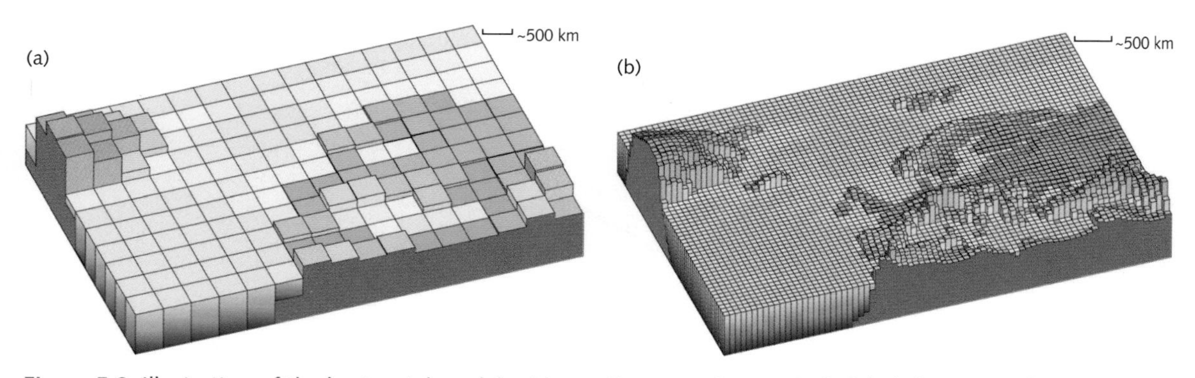

Figure 5.3 Illustration of the horizontal model grid over Europe as in a typical global climate model (a) in 1990 employed for the *IPCC 1st Assessment Report* and (b) in 2007 as employed for the *IPCC 4th Assessment Report*. Note the large improvement from the coarse grid of 1990.

Water vapour is important because of its associated latent heat (in other words, it gives out heat when it condenses) and also because the condensation of water vapour results in cloud formation, which modifies substantially the interaction of the atmosphere with the incoming energy from the Sun. Variations in both these energy inputs modify the atmospheric temperature structure, causing changes in atmospheric density (since warmed gases expand and are therefore less dense). It is these density changes that drive atmospheric motions such as winds and air currents, which in their turn alter and feed back on atmospheric density and composition. More details of the model formulation are given in the box below.

To forecast the weather for several days ahead a model covering the whole globe is required; for example, the southern hemisphere circulation today will affect northern hemisphere weather within a few days and vice versa. In a global forecasting model, the parameters (i.e. pressure, temperature, humidity, wind velocity and so on) that are needed to describe the dynamics and physics (listed in the box below) are specified at a grid of points (Figure 5.3) covering the globe. A typical spacing between points in the horizontal would be 100 km and about 1 km in the vertical; typically there would be 20 or 30 levels in the model in the vertical. The fineness of the spacing is limited by the power of the computers available.

Having set up the model, to generate a forecast from the present, it is started off from the atmosphere's current state and then the equations are integrated forward in time (see box below) to provide new descriptions of the atmospheric circulation and structure up to six or more days ahead. For a description of the atmosphere's current state, data from a wide variety of sources (see box below) have to be brought together and fed into the model.

Setting up a numerical atmospheric model

A numerical model of the atmosphere contains descriptions, in appropriate computer form and with necessary approximations, of the basic dynamics and physics of the different components of the atmosphere and their interactions.[2] When a physical process is described in terms of an algorithm (a process of step-by-step calculation) and simple parameters (the quantities that are included in a mathematical equation), the process is said to have been parameterised.

The dynamical equations are:

- The horizontal momentum equations (Newton's Second Law of Motion). In these, the horizontal acceleration of a volume of air is balanced by the horizontal pressure gradient and the friction. Because the Earth is rotating, this acceleration includes the Coriolis acceleration. The 'friction' in the model mainly arises from motions smaller than the grid spacing, which have to be parameterised.
- The hydrostatic equation. The pressure at a point is given by the mass of the atmosphere above that point. Vertical accelerations are neglected.
- The continuity equation. This ensures conservation of mass.

The model's physics consists of:

- The equation of state. This connects the quantities of pressure, volume and temperature for the atmosphere.
- The thermodynamic equation (the law of conservation of energy).
- Parameterisation of moist processes (such as evaporation, condensation, formation and dispersal of clouds).
- Parameterisation of absorption, emission and reflection of solar radiation and of thermal radiation.
- Parameterisation of convective processes.
- Parameterisation of exchange of momentum (in other words, friction), heat and water vapour at the surface.

Most of the equations in the model are differential equations, which means they describe the way in which quantities such as pressure and wind velocity change with time and with location. If the rate of change of a quantity such as wind velocity and its value at a given time are known, then its value at a later time can be calculated. Constant repetition of this procedure is called integration. Integration of the equations is the process whereby new values of all necessary quantities are calculated at later times, providing the model's predictive powers.

Since computer models for weather forecasting were first introduced, their forecast skill has improved to an extent beyond any envisaged by those involved in the development of the early models. As improvements have been made in the model formulation, in the accuracy or coverage of the data

Data to initialise the model

At a major global weather forecasting centre, data from many sources are collected and fed into the model. This process is called initialisation. Figure 5.4 illustrates some of the sources of data for the forecast

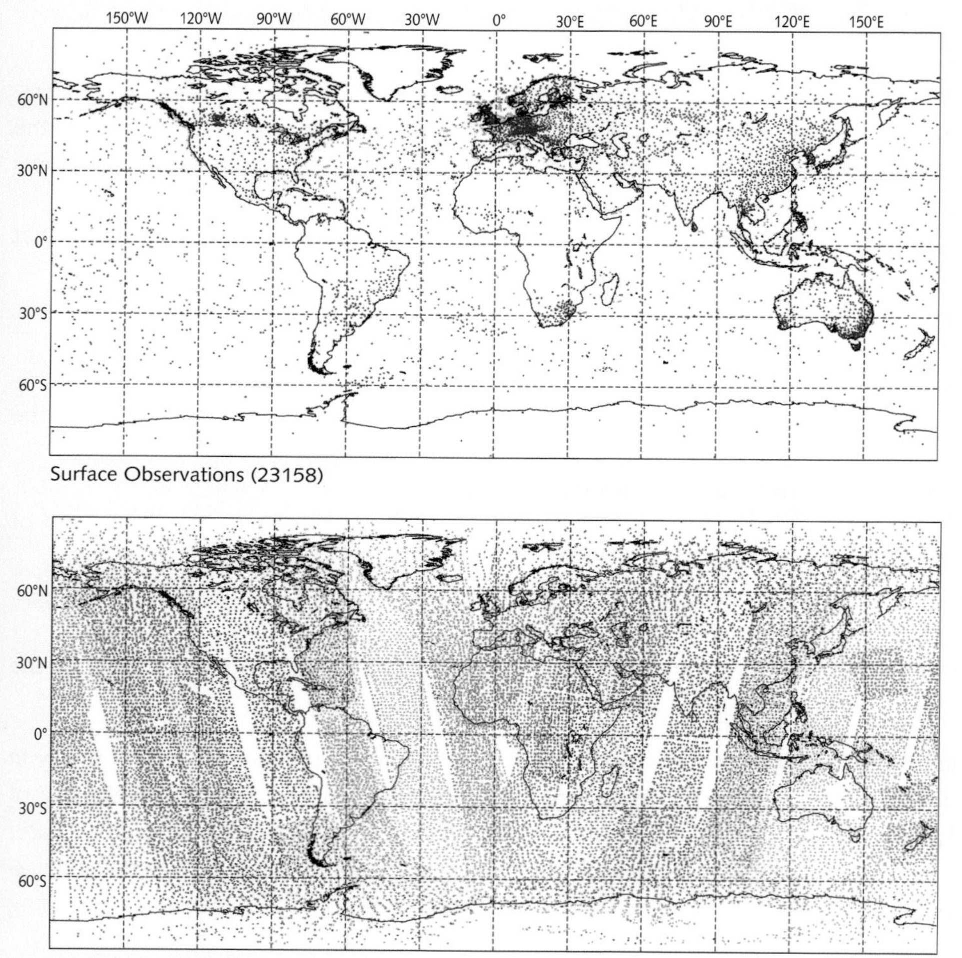

Figure 5.4 Some of the sources of data for input into the UK Met Office global weather forecasting model on a typical day. Surface observations are from land observing stations (manned and unmanned), from ships and from buoys. Radiosonde balloons make observations at up to 30 km altitude from land and from ship-borne stations. Satellite soundings are of temperature and humidity at different atmospheric levels deduced from observations of infrared or microwave radiation. Satellite cloud-track winds are derived from observing the motion of clouds in images from geostationary satellites. The number of observations of each type is given in brackets.

beginning at 0000 hours Universal Time (UT) on 20 May 2008. To ensure the timely receipt of data from around the world a dedicated communication network has been set up, used solely for this purpose. Great care needs to be taken with the methods for assimilation of the data into the model as well as with the data's quality and accuracy.

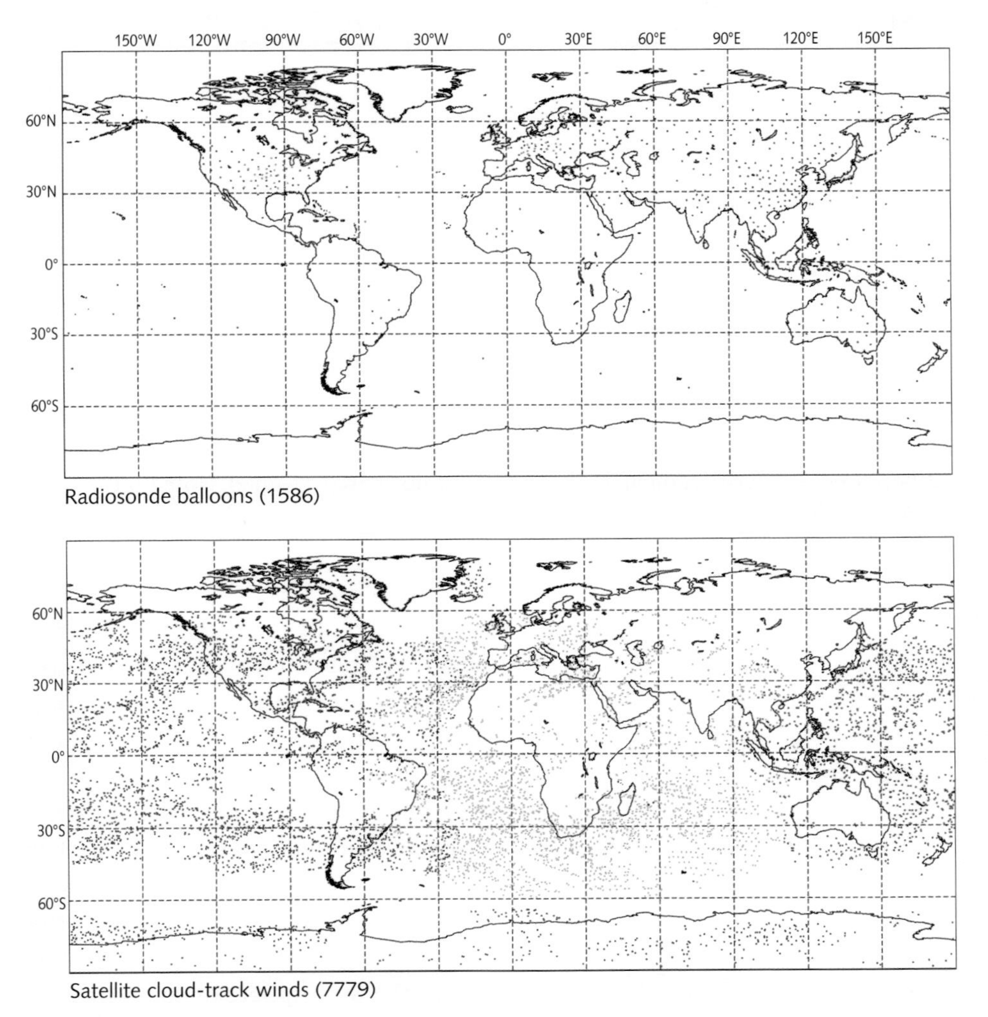

Satellite cloud-track winds (7779)

Figure 5.4 Continued

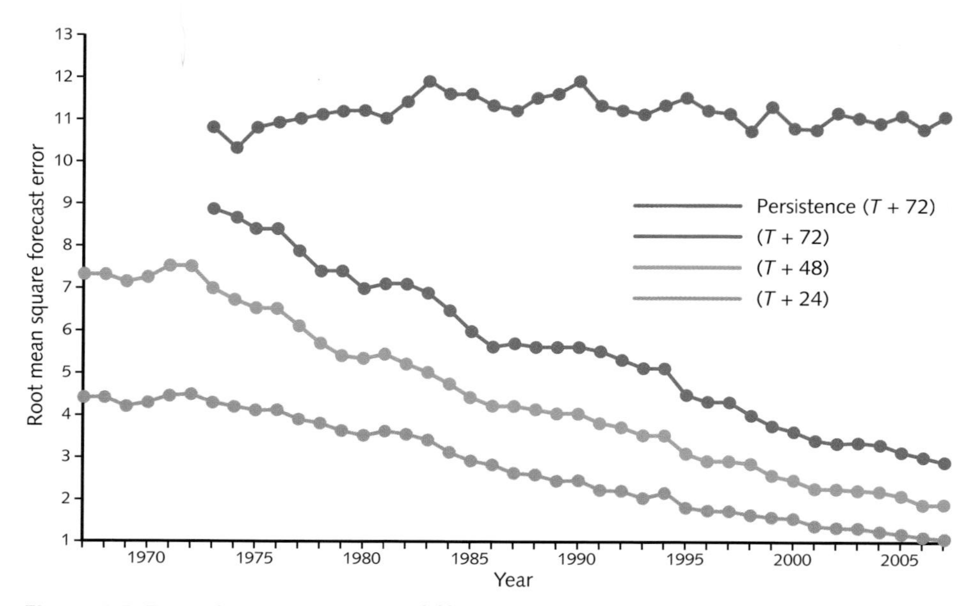

Figure 5.5 Errors (root mean square differences of forecasts of surface pressure in hPa compared with analyses) of UK Met Office forecasting models for the north Atlantic and Western Europe since 1966 for 24-hour (blue), 48-hour (green) and 72-hour (red) forecasts compared with assuming no change (purple). Note that 1 hPa = 1 mbar.

used for initialisation (see box) or in the resolution of the model (the distance between grid points), the resulting forecast skill has increased. For instance, for the British Isles, three-day forecasts of surface pressure today are as skilful on average as two-day forecasts of ten years ago, as can be seen from Figure 5.5.

When looking at the continued improvement in weather forecasts, the question obviously arises as to whether the improvement will continue or whether there is a limit to the predictability we can expect. Because the atmosphere is a partially chaotic system (see box below), even if perfect observations of the atmospheric state and circulation could be provided, there would be a limit to our ability to forecast the detailed state of the atmosphere at some time in the future. In Figure 5.6 current forecast skill is compared with the best estimate of the limit of the forecast skill for the British Isles (similar results would be obtained with any other mid latitude situation) with a perfect model and near-perfect data. According to that estimate, the limit of significant future skill is about 20 days ahead.

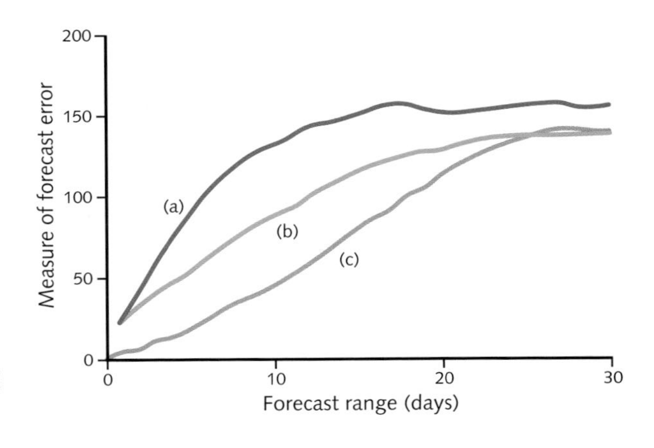

Figure 5.6 Potential improvements in forecast skill if there were better data or a better model. The ordinate (vertical axis) is a measure of the error of model forecasts (it is the root mean square differences of forecasts of the 500 hPa height field compared with analyses). Curve (a) is the error of 1990 UK Met Office forecasts as a function of forecast range. Curve (b) is an estimate showing how, with the same initial data, the error would be reduced if a perfect model could be used. Curve (c) is an estimate showing the further improvement which might be expected if near-perfect data could be provided for the initial state. After a sufficiently long period, all the curves approach a saturation value of the average root mean square difference between any forecasts chosen at random.

Forecast skill varies considerably between different weather situations or patterns. In other words some situations are more 'chaotic' (in the technical sense in which that word is used – see box below) than others. One way of identifying the skill that might be achieved in a given situation is to employ *ensemble* forecasting in which an ensemble of forecasts is run from a cluster of initial states that are generated by adding to an initial state small perturbations that are within the range of observational or analysis errors. The forecasts provided from the means of such ensembles show significant improvement compared with individual forecasts. Further, ensemble forecasts where the spread amongst the ensemble is low possess more skill than those where the spread in the ensemble is comparatively high (Figure 5.8).[3]

Seasonal forecasting

So far short-term forecasts of detailed weather have been considered. After 20 days or so they run out of skill. What about further into the future? Although we cannot expect to forecast the weather in detail, is there any possibility of predicting the average weather, say, a few months ahead? As this section shows, it is possible for some parts of the world, because of the influence of the distribution of ocean surface temperatures on the atmosphere's behaviour. For seasonal forecasting it is no longer the initial state of the atmosphere about which detailed knowledge is required. Rather, we need to know the conditions at the surface and how they might be changing.

Weather forecasting and chaos

The science of chaos has developed rapidly since the 1960s (when a meteorologist, Edward Lorenz, was one of its pioneers) along with the power of electronic computers. In this context, chaos[4] is a term with a particular technical meaning (see Glossary). A chaotic system is one whose behaviour is so highly sensitive to the initial conditions from which it started that precise future prediction is not possible. Even quite simple systems can exhibit chaos under some conditions. For instance, the motion of a simple pendulum (Figure 5.7) can be 'chaotic' under some circumstances, and, because of its extreme sensitivity to small disturbances, its detailed motion is not then predictable.

A condition for chaotic behaviour is that the relationship between the quantities which govern the motion of the system be non-linear; in other words, a description of the relationship on a graph would be a curve rather than a straight line.[5] Since the appropriate relationships for the atmosphere are non-linear it can be expected to show chaotic behaviour. This is illustrated in Figure 5.6, which shows the improvement in predictability that can be expected if the data describing the initial state are improved. However, even with virtually perfect initial data, the predictability in terms of days ahead only moves from about six days to about 20 days, because the atmosphere is a chaotic system.

For the simple pendulum not all situations are chaotic (Figure 5.7). Not surprisingly, therefore, in a system as complex as the atmosphere, some occasions are more predictable than others. A good illustration of an occasion with particular sensitivity to the initial data is provided by the exceptionally severe storm Lothar that crossed northern France in December 1999. It blew down hundreds of millions of trees and led to economic losses estimated at over 5 billion euros. Figure 5.8 shows an ensemble of forecasts carried out

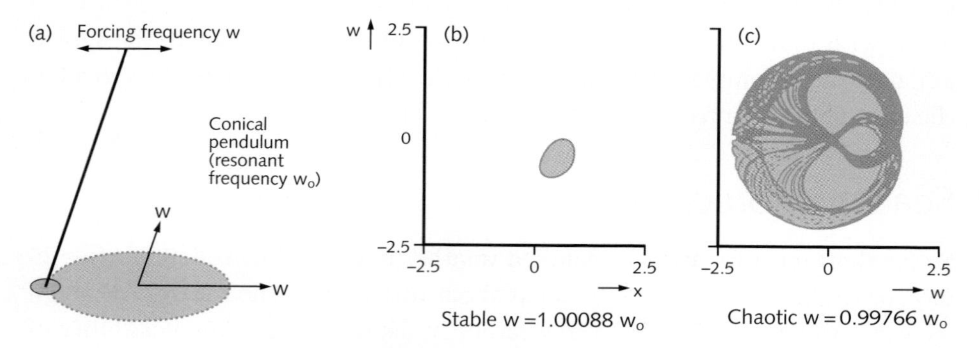

Figure 5.7 (a) A simple pendulum consisting of a bob at the end of a string of length 10 cm attached to a point of suspension which is moved with a linear oscillatory forcing motion at frequencies near the pendulum's resonance frequency f_0. (b) and (c) show plots of the bob's motion on a horizontal plane, the scale being in centimetres. (b) For a forcing frequency just above f_0 the motion of the bob settles down to a simple, regular pattern. (c) For a forcing frequency just below f_0 the bob shows 'chaotic' motion (although contained within a given region) which varies randomly and discontinuously as a function of the initial conditions.

by the European Centre for Medium Range Forecasting (ECMWF) starting from a set of slightly varying initial conditions 42 hours earlier.[6] The best-guidance deterministic forecast only predicts a weak trough in surface pressure which is supported by a number of members of the ensemble. However, a minority of the ensemble members show an intense vortex over France similar to what actually occurred, demonstrating the value of the ensemble in its prediction of the risk of the severe event even though a precise deterministic forecast was not possible. It is interesting that more recent deterministic reforecasts with an improved model have failed to predict this storm.

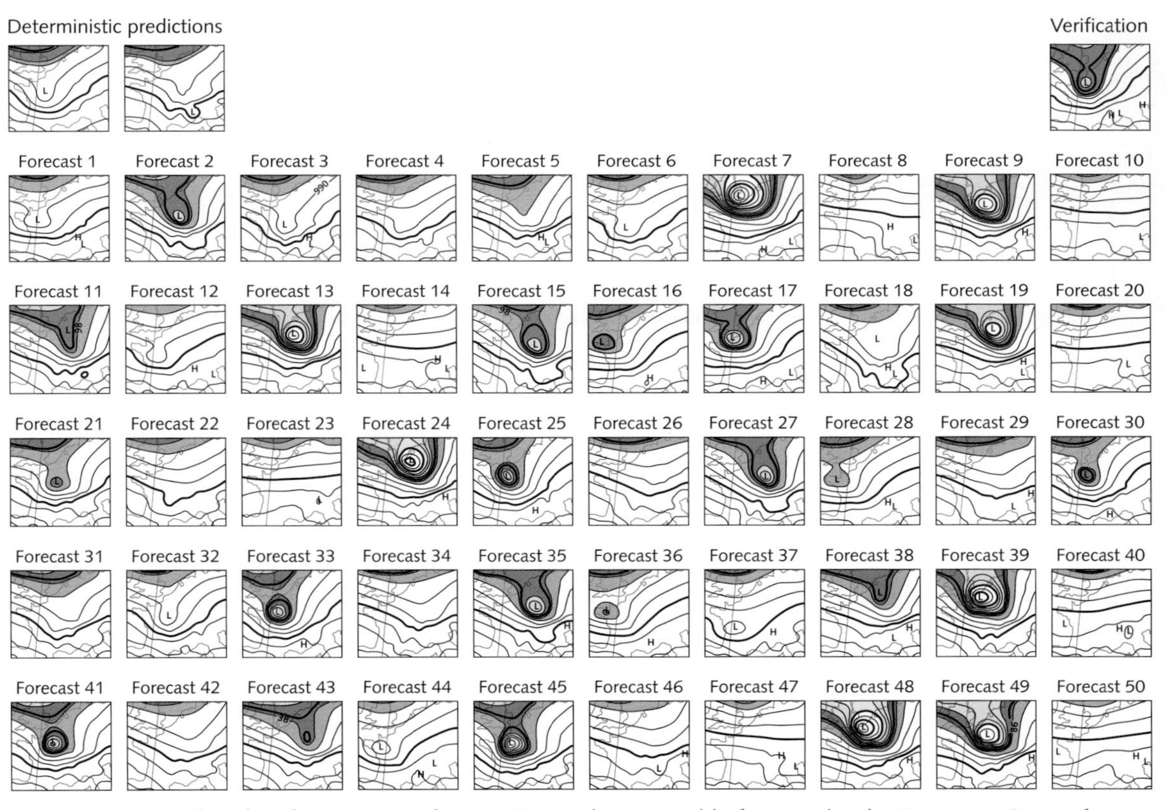

Figure 5.8 Isopleths of surface pressure from a 51-member ensemble forecast by the European Centre for Medium Range Forecasting (ECMWF) of the storm Lothar based on initial conditions 42 hours before the storm crossed northern France on 26 December 1999. The isobars are 5 mb apart and the thicker 1000 mb isobar runs across the middle of the figures. The top left shows forecasts made from the best estimate of the initial conditions that did not indicate the presence of a severe storm. Nor did many members of the ensemble. However, some of the ensemble members show an intense vortex indicating significant risk of its occurrence. The top right shows the situation at the end of the forecast period.

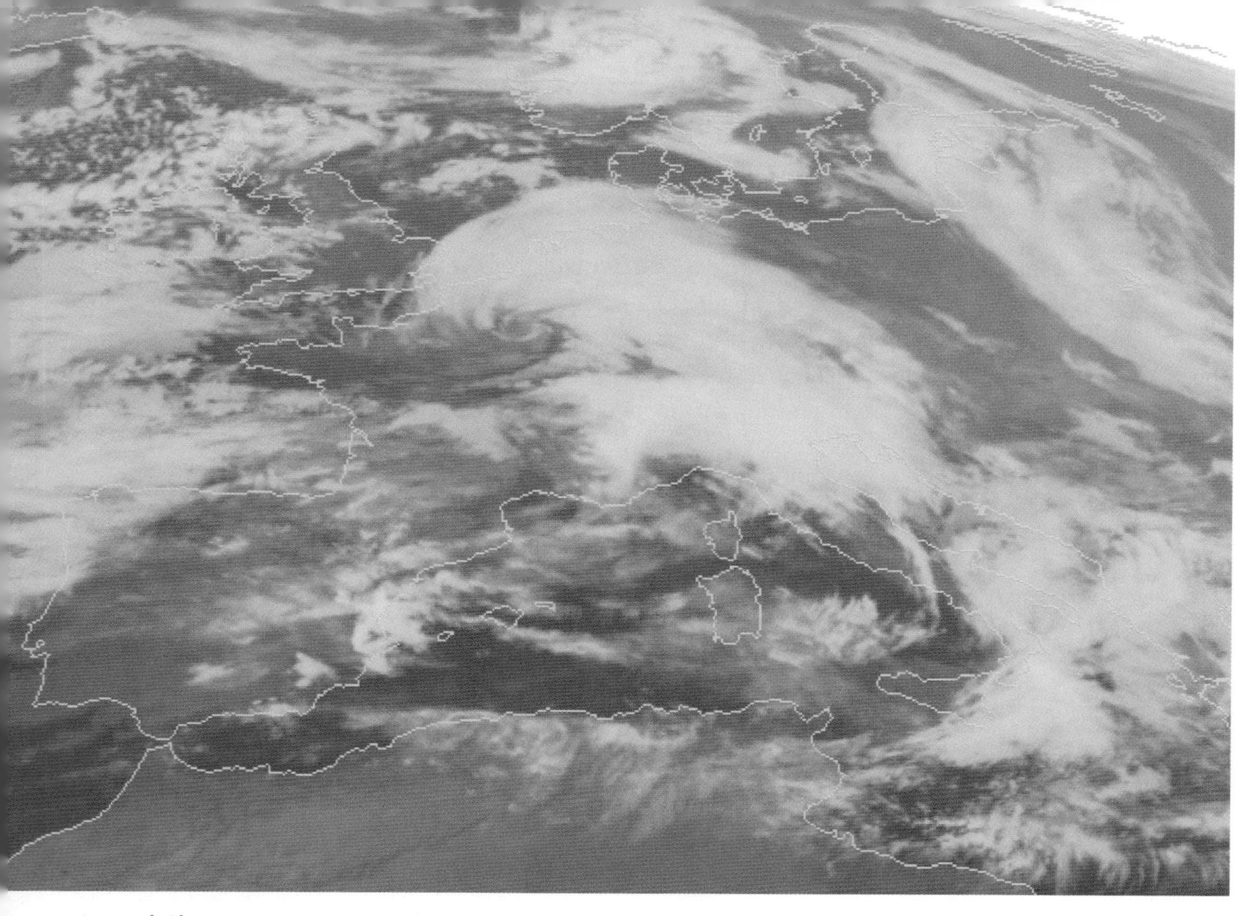

Around Christmas 1999, storm front Lothar raced across France, Switzerland and Germany, and 100 people died. (Also see Figure 5.8).

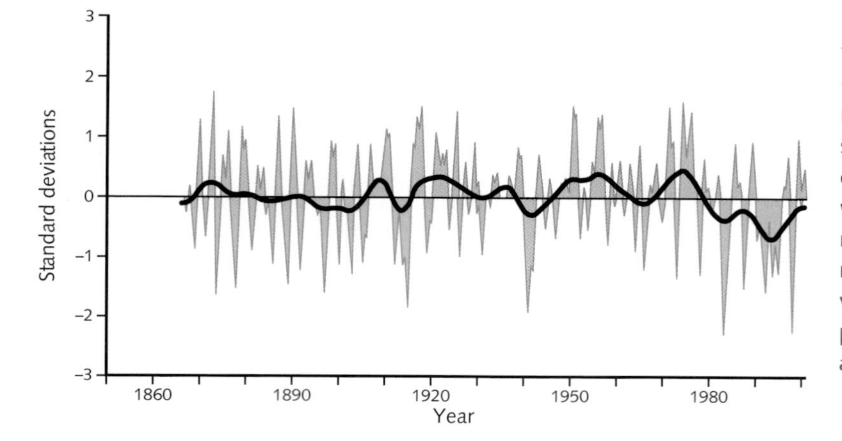

Figure 5.9 Monthly values of the Southern Oscillation Index (SOI) based on normalised Tahiti minus Darwin sea level pressures. An 11-point low pass filter effectively removes fluctuations with periods of less than eight months. The thick black line represents a decadal filter. Negative values indicate positive sea level pressure anomalies at Darwin and thus El Niño conditions.

In the tropics, the atmosphere is particularly sensitive to sea surface temperature. This is not surprising because the largest contribution to the heat input to the atmosphere is due to evaporation of water vapour from the ocean surface and its subsequent condensation in the atmosphere, releasing its latent heat. Because the saturation water vapour pressure increases rapidly with

A simple model of the El Niño

El Niño events are good examples of the strong coupling which occurs between the circulations of ocean and atmosphere. The stress exerted by atmospheric circulation – the wind – on the ocean surface is a main driver for the ocean circulation. Also, as we have seen, the heat input to the atmosphere from the ocean, especially that arising from evaporation, has a big influence on the atmospheric circulation.

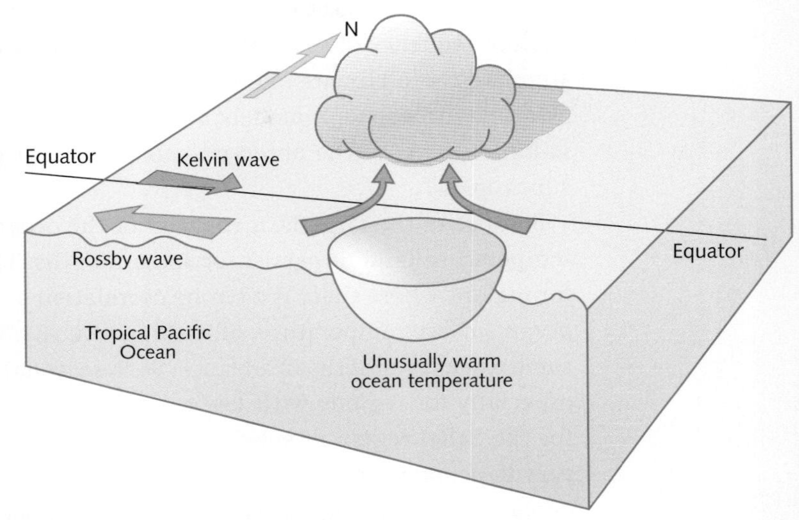

Figure 5.10 Schematic to illustrate El Niño oscillation.

A simple model of an El Niño event that shows the effect of different kinds of wave motions that can propagate within the ocean is illustrated in Figure 5.10. In this model a wave in the ocean, known as a Rossby wave, propagates westwards from a warm anomaly in ocean surface temperature near the equator. When it reaches the ocean's western boundary it is reflected as a different sort of wave, known as a Kelvin wave, which travels eastward. This Kelvin wave cancels and reverses the sign of the original warm anomaly, so triggering a cold event. The time taken for this half-cycle of the whole El Niño process is determined by the speed with which the waves propagate in the ocean; it takes about two years. It is essentially driven by ocean dynamics, the associated atmospheric changes being determined by the patterns of ocean surface temperature (and in turn reinforcing those patterns) that result from the ocean dynamics. Expressed in terms of this simple model, some of the characteristics of the El Niño process appear to be essentially predictable.

temperature – at higher temperatures more water can be evaporated before the atmosphere is saturated – evaporation from the surface and hence the heat input to the atmosphere is particularly large in the tropics.

The most prominent examples of interactions between atmosphere and ocean circulations are associated with the El Niño Southern Oscillation (ENSO)[7] that was first identified in the late nineteenth century as 'seesaw' surface pressure variations between South America and Indonesia (see box). It is during El Niño events associated with ENSO in the east tropical Pacific (see Figure 5.9) that the largest variations are found in ocean temperature.

Anomalies in the circulation and rainfall in all tropical regions and to a lesser extent at mid latitudes are associated with these El Niño events (see Figure 1.4). A good test of the atmospheric models described above is to run them with an El Niño sequence of sea surface temperatures and see whether they are able to simulate these climate anomalies. This has now been done with a number of different atmospheric models; they have shown considerable skill in the simulation of many of the observed anomalies, especially those in the tropics and sub-tropics.

Because of the large heat capacity of the oceans, anomalies of ocean surface temperature tend to persist for some months. The possibility therefore exists, for regions where there is a strong correlation between weather and patterns of ocean surface temperature, of making forecasts of climate (or average weather) some weeks or months in advance. Such seasonal forecasts have been attempted especially for regions with low rainfall; for instance, for northeast Brazil and for the Sahel region of sub-Saharan Africa, a region where human survival is very dependent on the marginal rainfall (see box). To make seasonal forecasts depends on the ability to forecast changes in ocean surface temperature. To do that requires understanding of, and the ability to model, the ocean circulation and the way it is coupled to the atmospheric circulation. Because the largest changes in ocean surface temperature occur in the tropics and because there are reasons to suppose that the ocean may be more predictable in the tropics than elsewhere, most emphasis on the prediction of ocean surface temperature has been placed in tropical regions, in particular on the prediction of the El Niño events themselves.

Later on in this chapter the coupling of atmospheric models and ocean models is described. For the moment it will suffice to say that, using coupled models together with detailed observations of both atmosphere and ocean in the Pacific region, significant skill in the prediction of El Niño events has been achieved for months and up to a year in advance (Figure 5.12) (see also Chapter 7).

The climate system

So far the forecasting of detailed weather over a few days and of average weather for a month or so, up to perhaps a season ahead, has been described in order to introduce the science and technology of modelling, and also because some of the scientific confidence in the more elaborate climate models arises from their ability to describe and forecast the processes involved in day-to-day weather.

Forecasting for the African Sahel region

The Sahel region of Africa forms a band about 500 km wide along the southern edge of the Sahara Desert that gets most of its rainfall during northern hemisphere summer (particularly July to September). Rainfall has decreased in this region since the 1960s. Particularly pronounced periods of drought occurred during the 1970s and 1980s with devastating impact on the local economy (Figure 5.12). Variations in Sahel rainfall are linked to fluctuations in patterns of sea surface temperature (SST), a connection that has provided since the 1980s a basis for generating seasonal forecasts of rainfall for the region.[8] The advent of climate models (see next sections) has brought substantial improvements in such forecasting (Figure 5.12) which is now limited by the accuracy of model predictions of sea surface temperature – possible for months ahead but not, as yet, for years.[9] Such forecasts can also benefit by including within the models variations in the characteristics of land surface vegetation and soil that also influence local rainfall.[10]

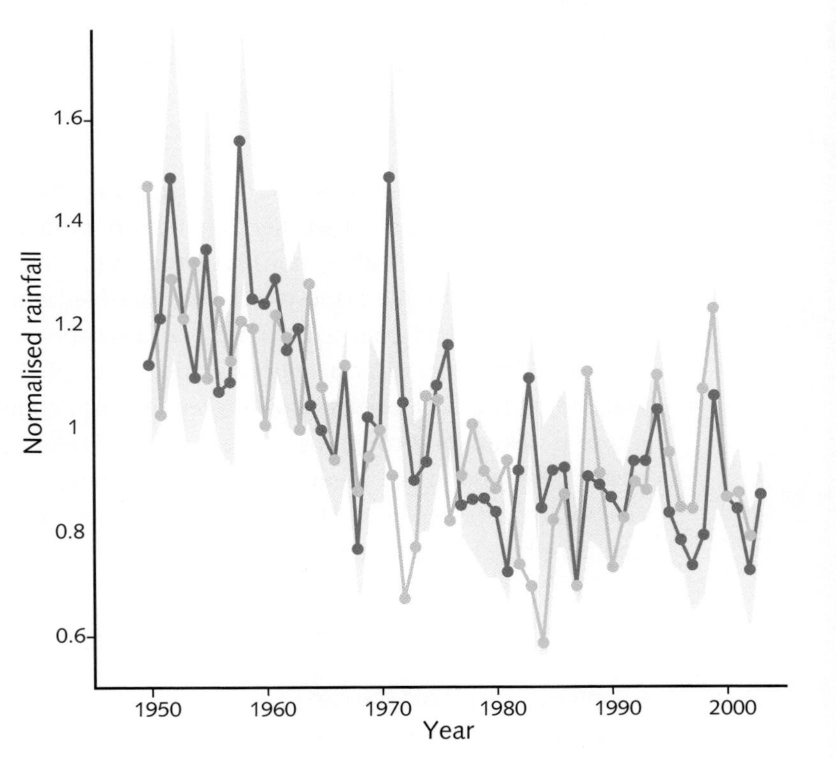

Figure 5.11 Observed Sahel July–September rainfall for each year (orange) compared to an ensemble mean of ten simulations with a climate model (GFDL-CM2.0) forced with observed sea surface temperatures (red). Both model and observations are normalised to unit mean over 1950–2000. The pink band represents ±1 standard deviation of intra-ensemble variability.

Climate is concerned with substantially longer periods of time, from a few years to perhaps a decade or longer. A description of the climate over a period involves the averages of appropriate components of the weather (for example, temperature and rainfall) over that period together with the statistical

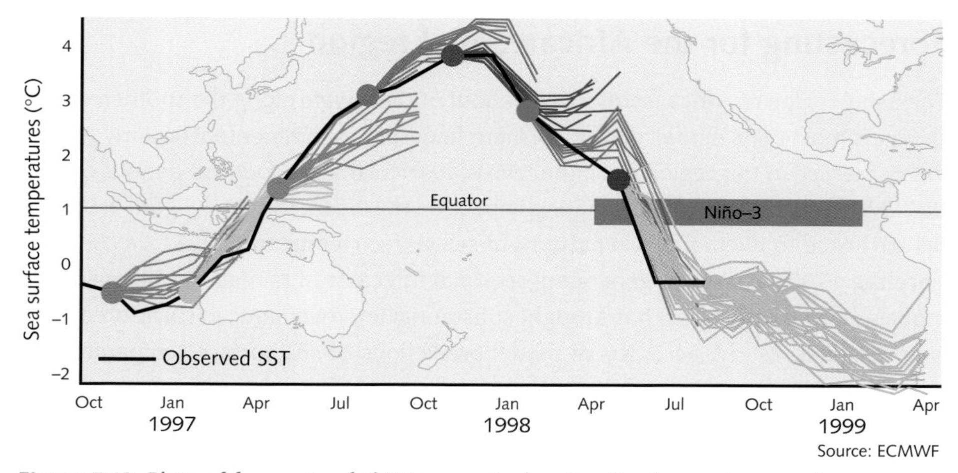

Source: ECMWF

Figure 5.12 Plots of forecasts of El Niño events showing the forecast sea surface temperature in the Niño-3 region from various start times throughout the large 1997–8 El Niño. Different lines of the same colour indicate different ensemble members. The background indicates the location of Niño-3.

variations of those components. In considering the effect of human activities such as the burning of fossil fuels, changes in climate over periods of decades up to a century or two ahead must be predicted.

Since we live in the atmosphere the variables commonly used to describe climate are mainly concerned with the atmosphere. But climate cannot be described in terms of atmosphere alone. Atmospheric processes are strongly coupled to the oceans (see above); they are also coupled to the land surface. There is also strong coupling to those parts of the Earth covered with ice (the cryosphere) and to the vegetation and other living systems on the land and in the ocean (the biosphere). These five components – atmosphere, ocean, land, ice and biosphere – together make up the climate system (Figure 5.13).

Feedbacks in the climate system[11]

Chapter 2 considered the rise in global average temperature which would result from the doubling of the concentration of atmospheric carbon dioxide assuming that no other changes occurred apart from the increased temperature at the surface and in the lower atmosphere. The rise in temperature was found to be 1.2 °C and results from what is often called *temperature feedback* (see box following this section). However, it was also established that, because of other feedbacks (which may be positive or negative) associated with the temperature

A Saharan dust storm originating in Mali blew off the west coast of Africa on 6 June 2006. Although partially hidden by the dust storm, the differences of the underlying landscape are still apparent as the sands of the Sahara give way to vegetation of the south. The Sahel is particularly vulnerable to desertification – land degradation from climate change and/or human activity that transforms a region to a desert.

increase, the actual rise in global average temperature was *likely* to be more than doubled to about 3.0°C. This section lists the most important of these feedbacks.

Water vapour feedback

This is the most important.[12] With a warmer atmosphere more evaporation occurs from the ocean and from wet land surfaces. On average, therefore, a warmer atmosphere will be a wetter one; it will possess a higher water vapour

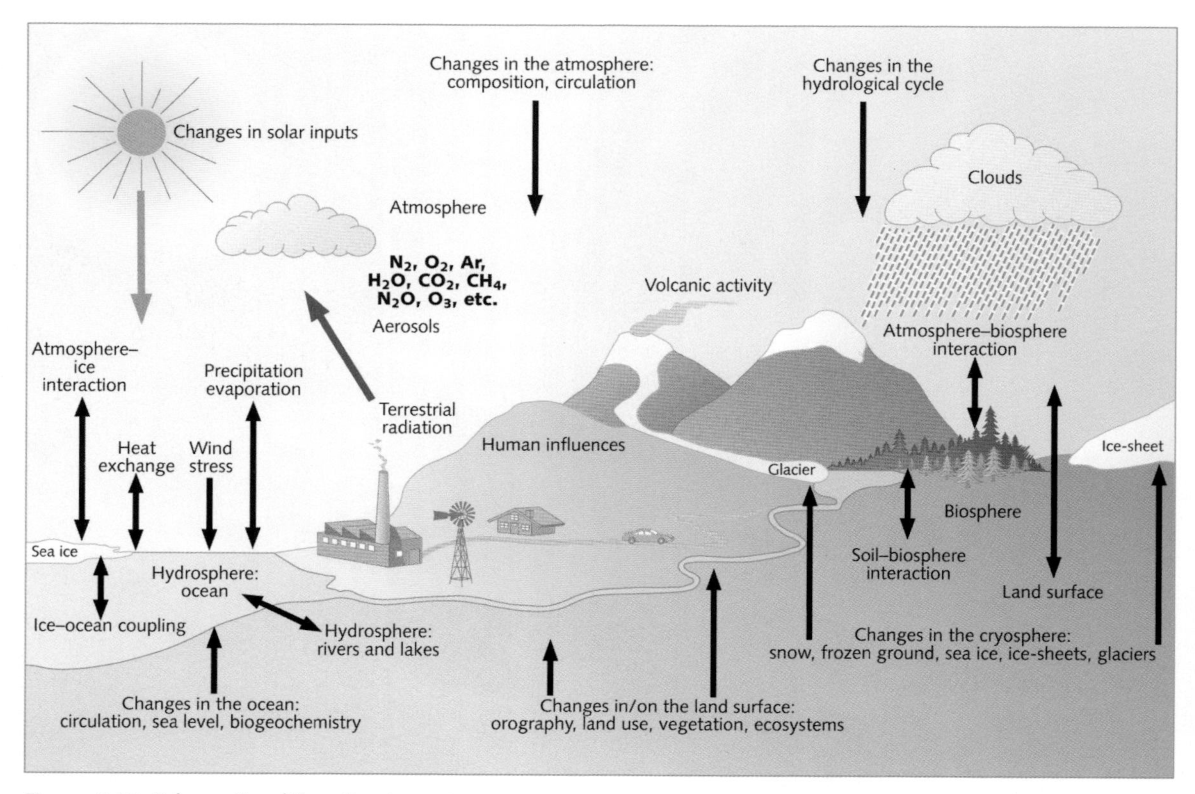

Figure 5.13 Schematic of the climate system.

content. Since water vapour is a powerful greenhouse gas, its potential feedback has been very thoroughly studied. It is found to provide on average a positive feedback of a magnitude that models estimate to approximately double the increase in the global average temperature that would arise with fixed water vapour.[13]

Cloud-radiation feedback

This is more complicated as several processes are involved. Clouds interfere with the transfer of radiation in the atmosphere in two ways (Figure 5.14). Firstly, they reflect a certain proportion of solar radiation back to space, so reducing the total energy available to the system. Secondly, they act as blankets to thermal radiation from the Earth's surface in a similar way to greenhouse gases. By absorbing thermal radiation emitted by the Earth's surface below, and by themselves emitting thermal radiation, they act to reduce the heat loss to space by the surface.

The effect that dominates for any particular cloud depends on the cloud temperature (and hence on the cloud height) and on its detailed optical properties (those properties which determine its reflectivity to solar radiation and its interaction with thermal radiation). The latter depend on whether the cloud is of water or ice, on its liquid or solid water content (how thick or thin it is) and on the average size of the cloud particles. In general for low clouds the reflectivity effect wins so they tend to cool the Earth–atmosphere system; for high clouds, by contrast, the blanketing effect is dominant and they tend to warm the system. The overall feedback effect of clouds, therefore, can be either positive or negative (see box below).

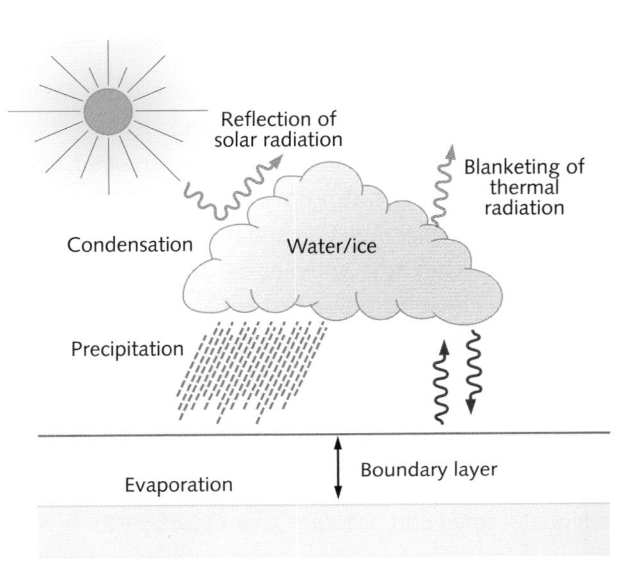

Figure 5.14 Schematic of the physical processes associated with clouds.

Climate is very sensitive to possible changes in cloud amount or structure, as can be seen from the results of models discussed in later chapters. To illustrate this, Table 5.1 shows that the hypothetical effect on the climate of a small percentage change in cloud cover is comparable with the expected changes due to a doubling of the carbon dioxide concentration.

Ocean-circulation feedback

The oceans play a large part in determining the existing climate of the Earth; they are likely therefore to have an important influence on climate change due to human activities.

The oceans act on the climate in four important ways. Firstly, there are close interactions between the ocean and the atmosphere; they behave as a strongly coupled system. As we have already noted, evaporation from the oceans provides the main source of atmospheric water vapour which, through its latent heat of condensation in clouds, provides the largest single heat source for the atmosphere. The atmosphere in its turn acts through wind stress on the ocean surface as the main driver of the ocean circulation.

Secondly, they possess a large heat capacity compared with the atmosphere; in other words a large quantity of heat is needed to raise the temperature of the oceans only slightly. In comparison, the entire heat capacity of the atmosphere is equivalent to less than 3 m depth of water. That means that in a world

Cloud radiative forcing

A concept helpful in distinguishing between the two effects of clouds mentioned in the text is that of cloud radiative forcing (CRF). Take the radiation leaving the top of the atmosphere above a cloud; suppose it has a value R. Now imagine the cloud to be removed, leaving everything else the same; suppose the radiation leaving the top of the atmosphere is now R'. The difference $R' - R$ is the cloud radiative forcing. It can be separated into solar radiation and thermal radiation components that generally act in opposite senses, each typically of magnitude between 50 and 100 W m^{-2}. On average, it is found that clouds tend slightly to cool the Earth–atmosphere system.

A map of cloud radiative forcing (Figure 5.15a) deduced from satellite observations illustrates the large variability in CRF over the globe with both positive and negative values. It is also helpful to study separately the shortwave and longwave components of the atmosphere's radiation budget (Figure 5.15b), the variations of which are dominated by variations in cloud cover and type. Model simulations are able to capture the overall pattern of these variations; the big challenge is to simulate the changing pattern with adequate detail and accuracy (see Question 8). It is through careful comparisons with observations that progress in the understanding of cloud feedback will be achieved.

(a)

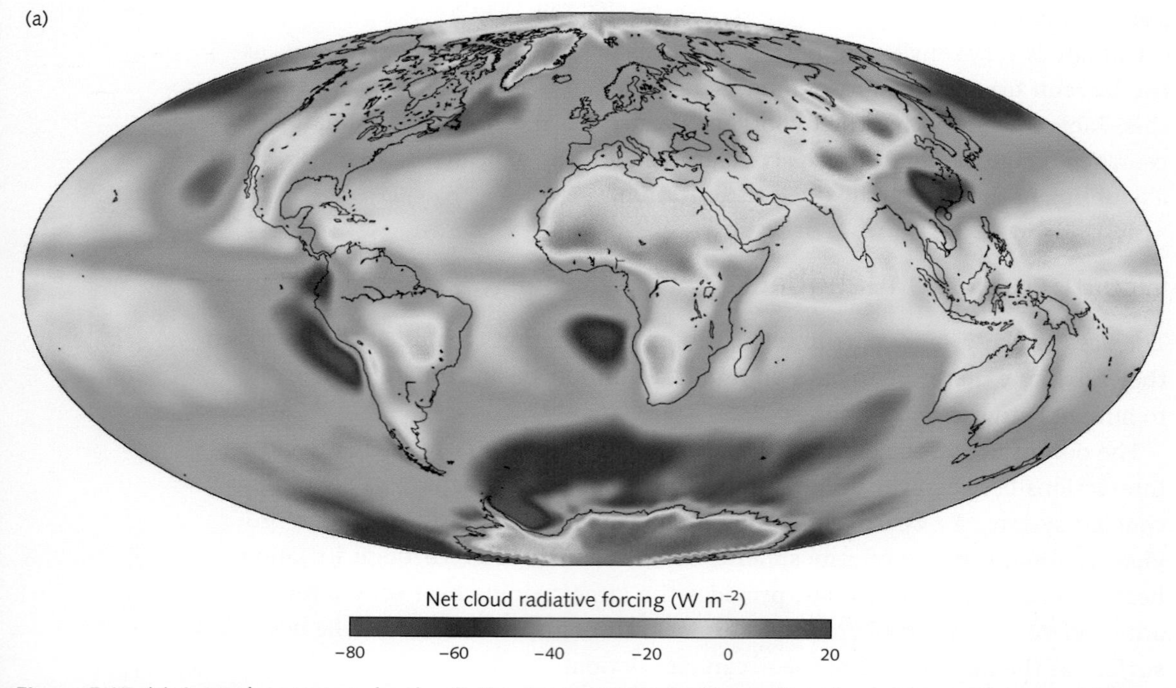

Figure 5.15 (a) Annual mean net cloud radiative forcing (CRF) for the period March 2000 to February 2001 as observed by the CERES instrument on the NASA Terra satellite. (b) Comparison of the observed longwave (pink/red), shortwave (orange) and net radiation at the top of the atmosphere for the tropics (20° N–20° S) as deviation from the mean for 1985–90; data from the ERBE instrument on the ERBS satellite and the CERES instrument on the TRMM satellite. Note the influence of the eruption of Pinatubo volcano.

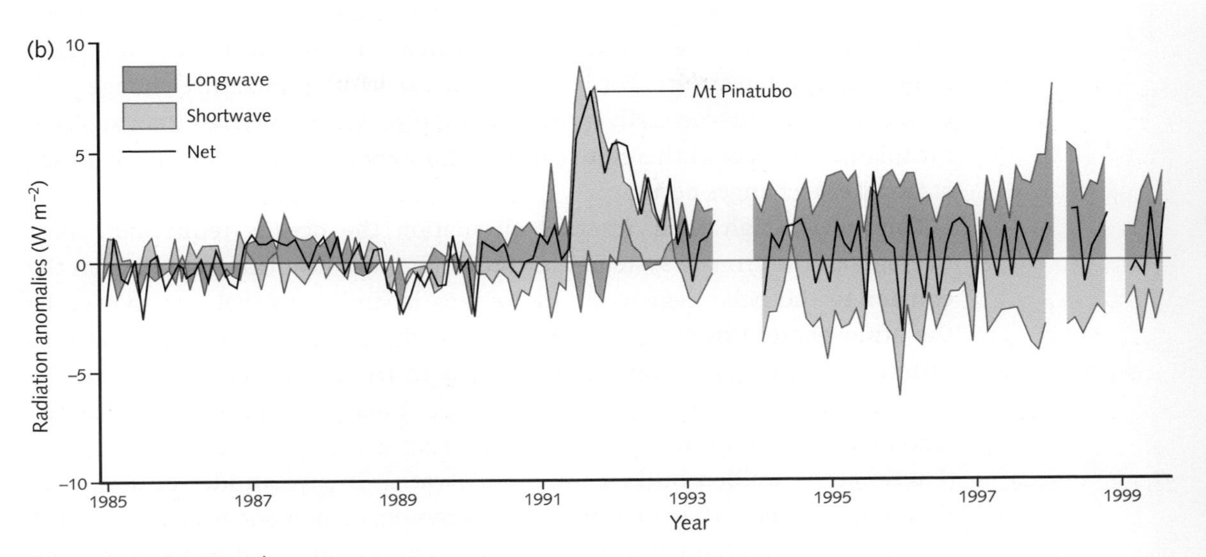

Figure 5.15 Continued

Table 5.1 Estimates of global average temperature changes under different assumptions about changes in greenhouse gases and clouds

Greenhouse gases	Clouds	Change (in °C) from current average global surface temperature of 15 °C
As now	As now	0
None	As now	–32
None	None	–21
As now	None	4
As now	As now but +3% high cloud	0.3
As now	As now but +3% low cloud	–1.0
Doubled CO_2 concentration otherwise as now	As now (no additional cloud feedback)	1.2
Doubled CO_2 concentration + best estimate of feedbacks	Cloud feedback included	3

that is warming, the oceans warm much more slowly than the atmosphere. We experience this effect of the oceans as they tend to reduce the extremes of atmospheric temperature. For instance, the range of temperature change both during the day and seasonally is much less at places near the coast than at places far inland. The oceans therefore exert a dominant control on the rate at which atmospheric changes occur.

Thirdly, through their internal circulation the oceans redistribute heat throughout the climate system. The total amount of heat transported from the equator to the polar regions by the oceans is similar to that transported by the atmosphere. However, the regional distribution of that transport is very different (Figure 5.16). Even small changes in the regional heat transport by the oceans could have large implications for climate change. For instance, the amount of heat transported by the north Atlantic Ocean is over 1000 terawatts (1 terawatt = 1 million million watts = 10^{12} watts). To give an idea of how large this is, we can note that a large power station puts out about 1000 million (10^{9}) watts and the total amount of commercial energy produced globally is about 12 terawatts. To put it further in context, considering the region of the north Atlantic Ocean between northwest Europe and Iceland, the heat input (Figure 5.16) carried by the ocean circulation is of similar magnitude to that reaching the ocean surface there from the incident solar radiation. Any accurate simulation of likely climate change, therefore, especially of its regional variations, must include a description of ocean structure and dynamics.

Ice-albedo feedback

An ice or snow surface is a powerful reflector of solar radiation (the albedo is a measure of its reflectivity). As some ice melts, therefore, at the warmer surface, solar radiation which had previously been reflected back to space by the ice or snow is absorbed, leading to further increased warming. This is another positive feedback which on its own would increase the global average temperature rise due to doubled carbon dioxide by about 20%.

In addition to the basic temperature feedback, four feedbacks have been identified, all of which play a large part in the determination of climate, especially its regional distribution. It is therefore necessary to introduce them into climate models. Because the global models allow for regional variation and also include the important non-linear processes in their formulation, they are able in principle to provide a full description of the effect of these feedbacks (see Figure 5.17). They are, in fact, the only tools available with this potential capability. It is to a description of climate prediction models that we now turn.

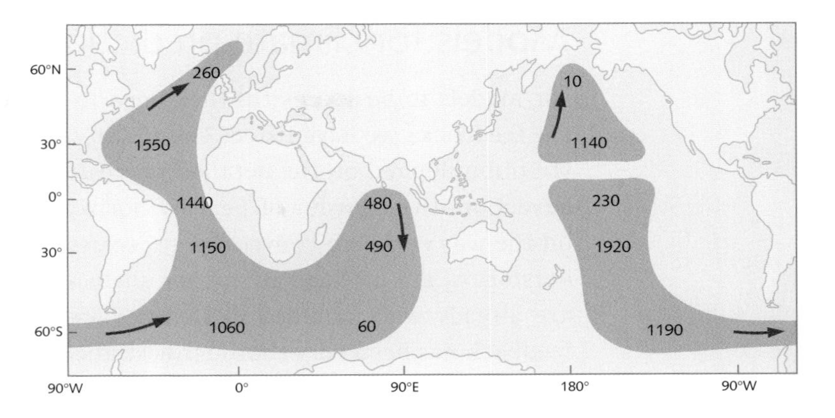

Figure 5.16 Estimates of transport of heat by the oceans. Units are terawatts (10^{12} W or 1 million million watts). Note the linkages between the oceans and that some of the heat transported by the North Atlantic originates in the Pacific.

Climate feedback comparisons

Climate feedbacks affect the *sensitivity* of the climate in terms of the temperature change ΔT_s at the surface that occurs for a given change ΔQ in the amount of net radiation at the top of the troposphere (known as the radiative forcing[14]). ΔQ and ΔT_s are related by a *feedback parameter* f (units $Wm^{-2}K^{-1}$) according to

$$\Delta Q = f\ \Delta T_s$$

If nothing changes other than the temperature (see fig 2.8), f is just the basic temperature feedback parameter $f_0 = 3.2\ W m^{-2} K^{-1}$ (i.e the change in radiation at the top of the troposphere that leads to a 1 °C change at the surface).

However, as we have seen other changes occur that result in feedbacks. The total feedback parameter f allows all the feedbacks to be added together

$$f = f_0 + f_1 + f_2 + f_3 + \ldots$$

where f_1, f_2, f_3 etc. are the feedback parameters describing water vapour, cloud, ice-albedo feedbacks, etc.

The amplification a of the temperature change ΔT_s that occurs with a total feedback parameter f compared with the basic temperature feedback f_0 is

$$a = f_0/f$$

Estimates of the feedback parameters for the main feedbacks from different climate models are:[15]

Water vapour (including lapse rate feedback – see Note 12)	– 1.2 ± 0.5
Cloud	– 0.6 ± 0.7
Ice albedo	– 0.3 ± 0.3
Total feedback parameter (sum of f_0 and the three above[16])	– 1.1 ± 0.5

Note that with this total feedback parameter the amplification factor is about 2.9 and the resulting climate sensitivity to doubled carbon dioxide a little over 3 °C.

Models for climate prediction

For models to be successful they need to include an adequate description of the feedbacks we have listed. The water vapour feedback and its regional distribution depend on the detailed processes of evaporation, condensation and advection (the transfer of heat by horizontal air flow) of water vapour, and on the way in which convection processes (responsible for showers and thunderstorms) are affected by higher surface temperatures. All these processes are already well included in weather forecasting models and water vapour feedback has been very thoroughly studied. The most important of the others are cloud-radiation feedback and ocean-circulation feedback. How are these incorporated into the models?

For modelling purposes, clouds divide into two types – *layer* clouds present on scales larger than the grid size and *convective* clouds generally on smaller scales than a grid box. For the introduction of layer clouds, early weather forecasting and climate models employed comparatively simple schemes. A typical scheme would generate cloud at specified levels whenever the relative humidity exceeded a critical value, chosen for broad agreement between model-generated cloud cover and that observed from climatological records. More recent models parameterise the processes of condensation, freezing, precipitation and cloud formation much more completely. They also take into account detailed cloud properties (e.g. water droplets or ice crystals and droplet number and size) that enable their radiative properties (e.g. their reflectivity and transmissivity) to be specified sufficiently well for the influence of clouds on the atmosphere's overall energy budget to be properly described. The most sophisticated models also include allowance for the effect of aerosols on cloud properties (Figure 3.9) – denoted the indirect aerosol effect in Figure 3.11. The effects of convective clouds are incorporated as part of the model's scheme for the parameterisation of convection.

The amount and sign (positive or negative) of the average cloud-radiation feedback in a particular climate model is dependent on many aspects of the model's formulation as well as on the particular scheme used for the description of cloud formation. Different climate models, therefore, can show average cloud-radiation feedback which can be either positive or negative (see box); further, the feedback can show substantial regional variation. For instance, models differ in their treatment of low cloud such that in some models the amount of low cloud increases with increased greenhouse gases, but in other models it decreases. Although considerable progress has been made in recent years in the observations of clouds and their representation in models, uncertainty

Snow-covered surfaces like the Arctic and Antarctica reflect 70% of the sunlight that hits them, but the polar regions don't have a large impact on the overall albedo of the Earth because the high latitudes get little sunlight to start with. Snow covering North America and Eurasia in the springtime, as the Sun returns in full force, has a much greater effect on the climate.

regarding cloud-radiation feedback continues to be the main reason for the wide uncertainty range in what is called *climate sensitivity* (see Chapter 6) or the change in global average surface temperature due to a doubling of carbon dioxide concentration.

The remaining feedback that is of great importance is that due to the effects of the ocean circulation. Compared with a global atmospheric model for weather forecasting, the most important elaboration of a climate model is the inclusion of the effects of the ocean. Early climate models only included the ocean very crudely; they represented it by a simple slab some 50 or 100 m deep, the approximate depth of the 'mixed layer' of ocean which responds to the seasonal heating and cooling at the Earth's surface. In such models, adjustments had to be made to allow for the transport of heat by ocean currents. When running the model

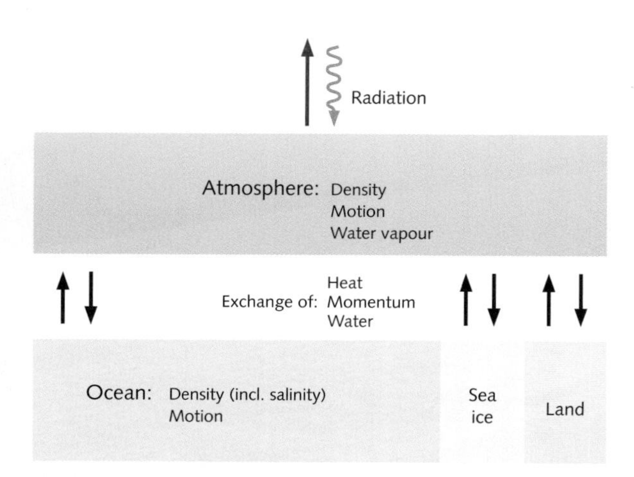

Figure 5.17 Component elements and parameters of a coupled atmosphere–ocean model including the exchanges at the atmosphere–ocean interface.

with a perturbation such as increased carbon dioxide, it was not possible to make allowance for any changes in that transport which might occur. Such models therefore possessed severe limitations.

For an adequate description of the influence of the ocean it is necessary to model the ocean circulation and its coupling to the atmospheric circulation. Figure 5.17 shows the ingredients of such a model. For the atmospheric part of the model, in order to accommodate long runs on available computers, the size of the grid has to be substantially larger, typically 100–300 km in the horizontal. Otherwise it is essentially the same as the global model for weather forecasting described earlier. The formulation of the dynamics and physics of the ocean part of the model is similar to that of its atmospheric counterpart. The effects of water vapour are of course peculiar to the atmosphere, but the salinity (the salt content) of the oceans has to be included as a parameter together with its considerable effects on the water density. Because dynamical systems, e.g. large scale eddies in the oceans, are of smaller scale than their atmospheric counterparts, the grid size of the ocean component is typically about half that of the atmospheric component. On the other hand, because changes in the ocean are slower, the time step for model integration can be greater for the ocean component.

At the ocean–atmosphere interface there are exchanges of heat, water and momentum (exchange of momentum leads to friction) between the two fluids. The importance of water in the atmosphere and its influence on the atmospheric circulation have already been shown. The distribution of fresh water precipitated from the atmosphere as rain or snow also has a large influence on the ocean's circulation through its effect on the distribution of salt in the ocean, which in turn affects the ocean density. It is not surprising, therefore, to find that the 'climate' described by the model is quite sensitive to the size and the distribution of water exchanges at the interface.

Before the model can be used for prediction it has to be run for a considerable time until it reaches a steady 'climate'. The 'climate' of the model, when it is run unperturbed by increasing greenhouse gases, should be as close as possible to the current actual climate. If the exchanges listed above are not correctly described, this will not be the case. Much effort has gone into model descriptions of these exchanges. Until about the year 2000, many coupled models introduced artificial adjustments to the fluxes at the surface of heat, water and momentum so as to ensure that the model's 'climate' was as identical as possible to the current climate. However, since that time the ocean component of the model has been improved especially through introducing higher resolution (100 km or less), so that models are now able to provide an adequate description of the climate with no such adjustments.

Before leaving the oceans, there is a particular feedback that should be mentioned between the hydrological cycle and the deep ocean circulation (see box below). Changes in rainfall, by altering the ocean salinity, can interact with the ocean circulation. This could affect the climate, particularly of the North Atlantic region; it may also have been responsible for some dramatic climate changes in the past (see Chapter 4).

The most important feedbacks belong to the atmospheric and the ocean components of the model. They are the largest components, and, because they are both fluids and have to be dynamically coupled together, their incorporation into the model is highly demanding. However, another feedback to be modelled is the ice-albedo feedback, which arises from the variations of sea-ice and of snow.

Sea-ice covers a large part of the polar regions in the winter. It is moved about by both the surface wind and the ocean circulation. So that the ice-albedo feedback can be properly described, the growth, decay and dynamics of sea-ice have to be included in the model. Land ice is also included, essentially as a boundary condition – a fixed quantity – because its coverage changes little from year to year. However, the model needs to show whether there are likely to be changes in ice volume, even though these are small, in order to find out their effect on sea level (Chapter 7 considers the impacts of sea-level change).

Interactions with the land surface must also be adequately described. The most important properties for the model are land surface wetness or, more precisely, soil moisture content (which will determine the amount of evaporation) and albedo (reflectivity to solar radiation). The models keep track of the changes in soil moisture through evaporation and precipitation. The albedo depends on soil type, vegetation, snow cover and surface wetness.

Validation of the model

In discussing various aspects of modelling we have already indicated how some validation of the components of climate models may be carried out.[17] The successful predictions of weather forecasting models provide validation of important aspects of the atmospheric component, as do the simulations mentioned earlier in the chapter of the connections between sea surface temperature

The ocean's deep circulation

For climate change over periods up to a decade, only the upper layers of the ocean have any substantial interaction with the atmosphere. For longer periods, however, links with the deep ocean circulation become important. The effects of changes in the deep circulation are of particular importance.

Experiments using chemical tracers, for instance those illustrated in Figure 5.20 (see next box), have been helpful in indicating the regions where strong coupling to the deep ocean occurs. To sink to the deep ocean, water needs to be particularly dense, in other words both cold and salty. There are two main regions where such dense water sinks down to the deep ocean, namely in the north Atlantic Ocean (in the Greenland Sea between Scandinavia and Greenland and the Labrador Sea west of Greenland) and in the region of Antarctica. Salt-laden deep water formed in this way contributes to a deep ocean circulation that involves all the oceans (Figure 5.18) and is known as the *thermohaline circulation* (THC).

In Chapter 4 we mentioned the link between the THC and the melting of ice. Increases in the ice melt can lead to the ocean surface water becoming less salty and therefore less dense. It will not sink so readily, the deep water formation will be inhibited and the THC is weakened. In Chapter 6, the link between the THC and the hydrological (water) cycle in the atmosphere is mentioned. Increased precipitation in the North Atlantic region, for instance, can lead to a weakening of the THC.

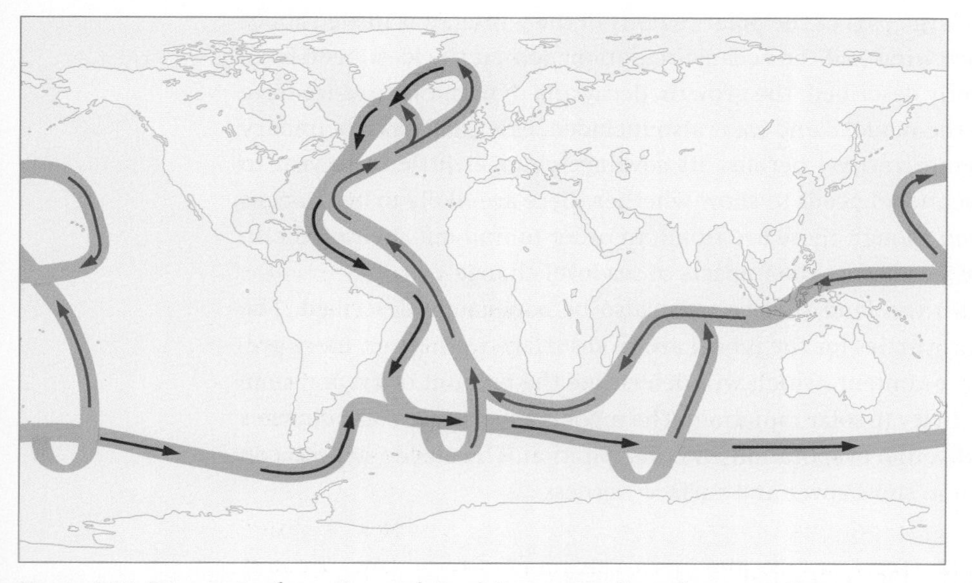

Figure 5.18 Deep water formation and circulation – sometimes known as the ocean 'conveyor belt' – connecting the oceans together. The deep salty current (blue) largely originates in the Nordic Seas and the Labrador Sea where northward flowing water (red) near the surface that is unusually salty becomes cooler and even more salty through evaporation, so increasing its density causing it to sink. Regions of upwelling in the southern ocean feed into the warm surface current (red).

anomalies and precipitation patterns in some parts of the world. Various tests have also been carried out of the ocean component of climate models; for instance, through comparisons between the simulation and observation of the movement of chemical tracers (see box below).

Once a comprehensive climate model has been formulated it can be tested in three main ways. Firstly, it can be run for a number of years of simulated time and the climate generated by the model compared in detail to the current climate. For the model to be seen as a valid one, the average distribution and the seasonal variations of appropriate parameters such as surface pressure, temperature and rainfall have to compare well with observation. In the same way, the variability of the model's climate must be similar to the observed variability. Climate models that are currently employed for climate prediction stand up well to such comparisons.

Recent progress in model performance has been evident in improved simulations of modes of climate variability on the large scale and from intraseasonal to interdecadal timescales. This is of particular importance because of the links that are likely between variations in modes such as the northern and southern annular modes (NAM and SAM) and the ENSO (El Niño Southern Oscillation) and the growth of atmospheric greenhouse gases.[18] Progress with the prediction of ENSO events and associated climate anomalies was mentioned earlier in the chapter.

Secondly, models can be compared against simulations of past climates when the distribution of key variables was substantially different from that at present; for example, the period around 9000 years ago when the configuration of the Earth's orbit around the Sun was different (see Figure 5.19). The perihelion (minimum Earth–Sun distance) was in July rather than in January as it is now; also the tilt of the Earth's axis was slightly different from its current value (24° rather than 23.5°). Resulting from these orbital differences (see Chapter 4), there were significant differences in the distribution of solar radiation throughout the year. The incoming solar energy when averaged over the northern hemisphere was about 7% greater in July and correspondingly less in January.

When these altered parameters are incorporated into a model, a different climate results. For instance, northern continents are warmer in summer and colder in winter. In summer a significantly expanded low-pressure region develops over north Africa and south Asia because of the increased land–ocean temperature contrast. The summer monsoons in these regions are strengthened and there is increased rainfall. These simulated changes are in qualitative agreement with palaeoclimate data; for example, these data provide evidence for that period (around 9000 years ago) of lakes and

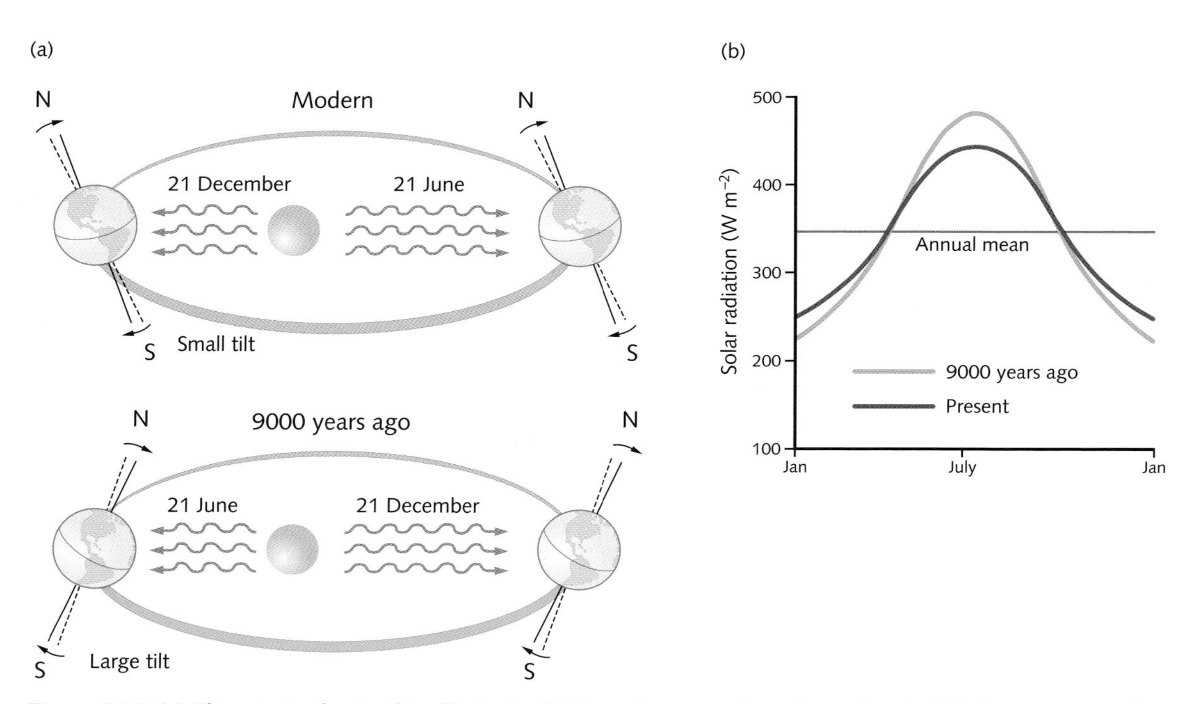

Figure 5.19 (a) Changes in the Earth's elliptical orbit from the present configuration to 9000 years ago and (b) changes in the average solar radiation during the year over the northern hemisphere.

vegetation in the southern Sahara about 1000 km north of the present limits of vegetation.

The accuracy and the coverage of data available for these past periods are limited. However, the model simulations for 9000 years ago, described above, and those for other periods in the past have demonstrated the value of such studies in the validation of climate models.[19]

A third way in which models can be validated is to use them to predict the effect of large perturbations on the climate such as occurs, for instance, with volcanic eruptions, the effects of which were mentioned in Chapter 1. Several climate models have been run in which the amount of incoming solar radiation has been modified to allow for the effect of the volcanic dust from Mount Pinatubo, which erupted in 1991 (Figure 5.20). Successful simulation of some of the regional anomalies of climate which followed that eruption, for instance the unusually cold winters in the Middle East and the mild winters in western Europe, has also been achieved by the models.[20]

In these three ways, which cover a range of timescales, confidence has been built in the ability of models to predict climate change due to human activities.

The 12 June 1991 eruption column from Mount Pinatubo taken from the east side of Clark Air Base.

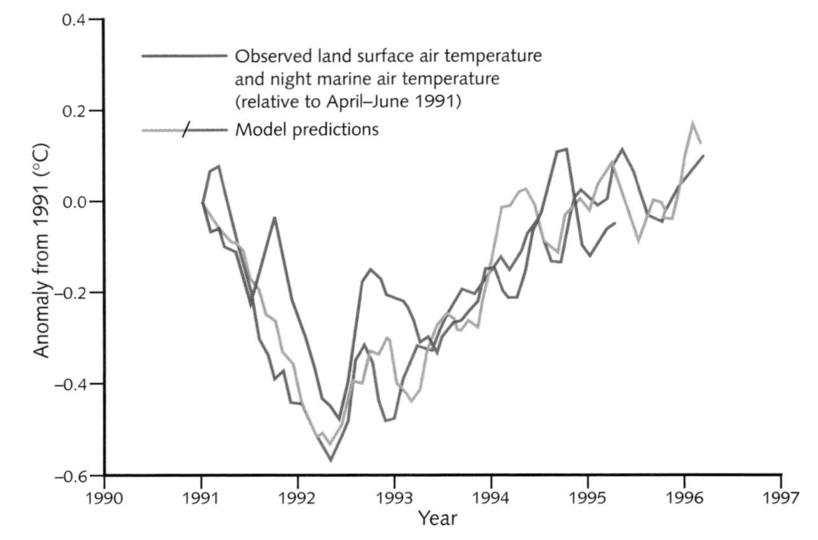

Figure 5.20 The predicted and observed changes in global land and ocean surface air temperature after the eruption of Mount Pinatubo, in terms of three-month running averages from April to June 1991 to March to May 1995.

Modelling of tracers in the ocean

A test that assists in validating the ocean component of the model is to compare the distribution of a chemical tracer as observed and as simulated by the model. In the 1950s radioactive tritium (an isotope of hydrogen) released in the major atomic bomb tests entered the oceans and was distributed by the ocean circulation and by mixing.

Figure 5.21 shows good agreement between the observed distribution of tritium (in tritium units) in a section of the western North Atlantic Ocean about a decade after the major bomb tests and the distribution as simulated by a 12-level ocean model. Similar comparisons have been made more recently of the measured uptake of one of the freons CFC-11, whose emissions into the atmosphere have increased rapidly since the 1950s, compared with the modelled uptake.

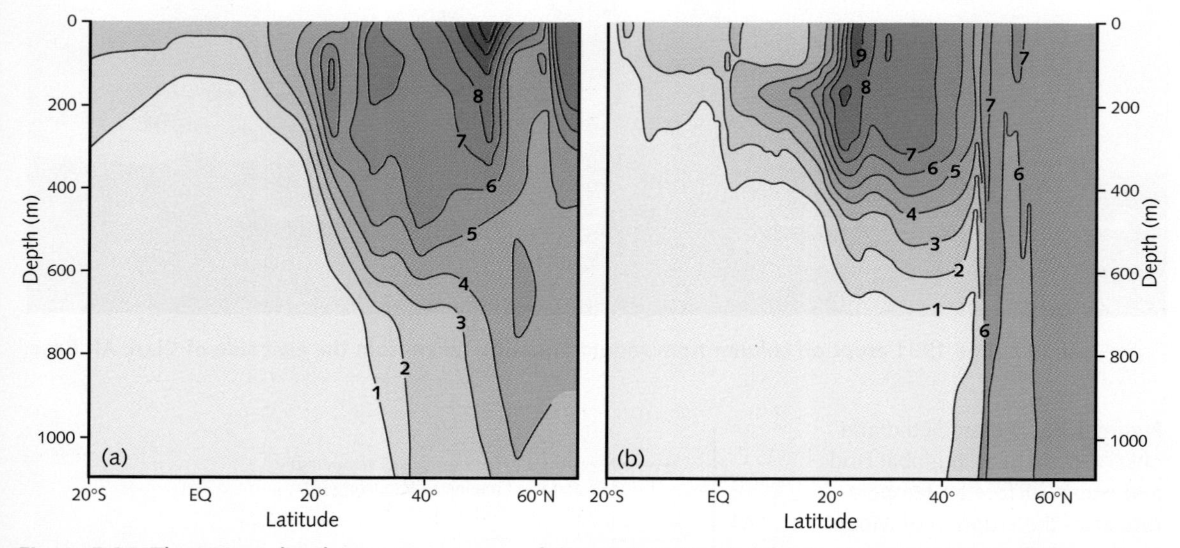

Figure 5.21 The tritium distribution in a section of the western North Atlantic Ocean approximately one decade after the major atomic bomb tests, as observed in the GEOSECS programme (a) and as modelled (b).

Comparison with observations

More than 20 centres in the world located in more than ten countries are currently running climate models of the kind I have described in which the circulations of the atmosphere and the ocean are fully coupled together. Some of these models have been employed to simulate the climate of the last 150 years allowing for variations in aspects of natural forcing (e.g. solar variations and volcanoes) and anthropogenic forcings (i.e. increases in the concentrations of greenhouse gases and aerosols).

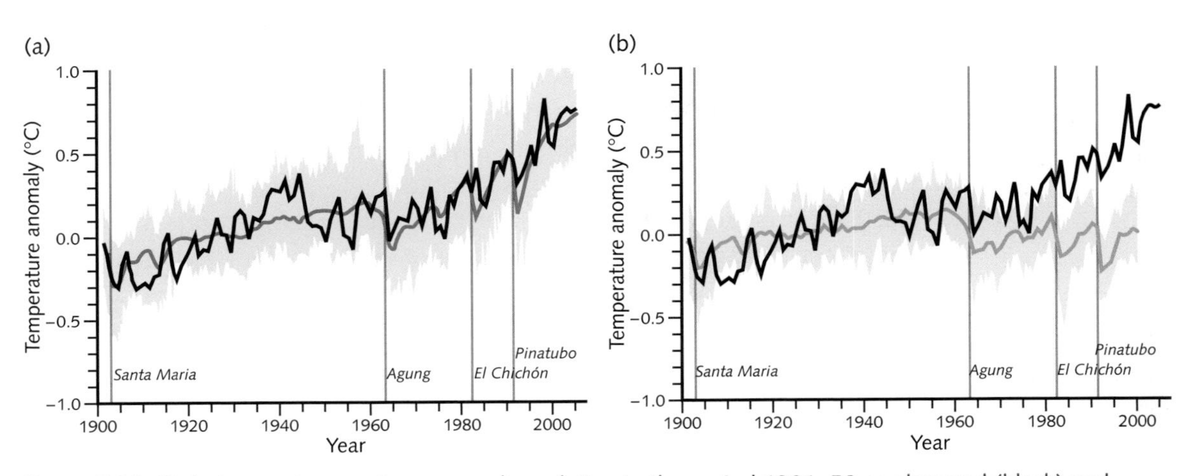

Figure 5.22 Global mean temperature anomalies relative to the period 1901–50 as observed (black) and from an ensemble of 58 simulations by 14 models with individual simulations shown by very thin lines. Simulations are shown with both anthropogenic and natural forcings (a) and with natural forcings only (b). The multimodel ensemble means are shown in red (a) and blue (b). Vertical lines indicate the timing of major volcanic events.

Examples of such simulations are shown in Figure 5.22, where the observed record of global average surface air temperature is compared with model simulations taking into account the combination of natural and anthropogenic forcings and natural forcings on their own.

Three interesting features of Figure 5.22 can be noted. Firstly, that the inclusion of both natural and anthropogenic forcings provides a plausible explanation for a large part of the observed temperature changes over the last century, and that the inclusion of anthropogenic factors is essential to explain the rapid increase in temperature over the last 40 years. Further, it is likely that changes in solar output and the comparative absence of volcanic activity were the most important variations in natural forcing factors during the first part of the twentieth century. Secondly, the model simulations show variability up to a tenth of a degree Celsius or more over periods of a few years up to decades. This variability is due to internal exchanges in the model between different parts of the climate system, and is not dissimilar to that which appears in the observed record. Thirdly, due to the slowing effect of the oceans on climate change, the warming observed or modelled so far is less than would be expected if the climate system were in equilibrium under the amount of radiative forcing due to the current increase in greenhouse gases and aerosols.

Because of the large amount of natural variability in both the observations and the simulations, much debate has taken place over the last two decades about the strength of the evidence that global warming due to the increase in greenhouse gases has actually been observed in the climate record. In other words has the 'signal' attributed to global warming risen sufficiently above the 'noise' of natural variability? The Intergovernmental Panel on Climate Change (IPCC) has been much involved in this debate.

The IPCC's first Report published in 1990[21] made a carefully worded statement to the effect that, although the size of the observed warming is broadly consistent with the predictions of climate models, it is also of similar magnitude to natural climate variability. An unequivocal statement that anthropogenic climate change had been detected could not therefore be made at that time. By 1995 more evidence was available and the IPCC 1995 Report[22] reached the cautious conclusion as follows.

> Our ability to quantify the human influence on global climate is currently limited because the expected signal is still emerging from the noise of natural climate variability, and because there are uncertainties in key factors. These include the magnitude and patterns of long term natural variability and the time-evolving pattern of forcing by, and response to, changes in the concentrations of greenhouse gases and aerosols, and land surface changes. Nevertheless, the balance of evidence suggests a discernible influence on global climate.

Since 1995 a large number of studies have addressed the problems of *detection* and *attribution*[23] of climate change. Better estimates of natural variability have been made, especially using models, and the conclusion reached that the warming over the last 100 years is *very unlikely* to be due to natural variability alone. In addition to studies using globally averaged parameters, there have been detailed statistical studies using pattern correlations based on optimum detection techniques applied to both model results and observations. The conclusion reached in the IPCC 2001 Report:[24]

> In the light of new evidence and taking into account the remaining uncertainties, most of the observed warming over the last 50 years is *likely* to have been due to the increase in greenhouse gas concentrations.

was strengthened in the IPCC 2007 Report[25] which summarised its conclusion as follows:

> It is *very likely* that anthropogenic greenhouse gas increases caused most of the observed increase in globally averaged temperatures since the mid-20th century. Discernible human influences now extend to other aspects

of climate, including continental-average temperatures, atmospheric circulation patterns and some types of extremes.

The 2007 Report made four further summary points:

- It is *likely* that greenhouse gases alone would have caused more warming than observed because volcanic and anthropogenic aerosols have offset some warming that would otherwise have taken place.
- The observed widespread warming of the atmosphere and ocean, together with ice mass loss, support the conclusion that it is *extremely unlikely* that global climate change of the past 50 years was caused by unforced variability alone.
- Warming of the climate system has been detected and attributed to anthropogenic forcing in surface and free atmosphere temperatures, in temperatures of the upper several hundred metres of the ocean and in contributions to sea-level rise. The observed pattern of tropospheric warming and stratospheric cooling can be largely attributed to the combined influences of greenhouse gas increases and stratospheric ozone depletion.
- It is *likely* that there has been significant anthropogenic warming over the past 50 years averaged over each continent except Antarctica. The observed patterns of warming, including greater warming over land than over the ocean and their changes over time, are simulated by models that include anthropogenic forcing.

Confidence having been established in climate models in the ways we have outlined in the last two sections, these models can now be used to generate projections of the likely climate change in the future due to human activities. Details of such projections will be presented in the next chapter.

Before leaving comparison with observations, I should mention observations of the warming of the ocean that add further confirmation to the picture that has been presented. In Chapter 2, the effect of an increase of greenhouse gases was expressed in terms of radiative forcing or, in other words, a net input of heat energy into the earth–atmosphere system. Most of this extra energy is stored in the ocean. From large numbers of measurements of the temperature increase in the ocean at different locations and depths down to 3 km, it has been estimated that, over the period 1961–2003, the ocean has been absorbing energy at a rate of 0.21 ± 0.04 W m^{-2} globally averaged over the Earth's surface.[26] Two-thirds of this energy is stored in the upper 700 m of the ocean. Within the limits of uncertainty, it agrees well with model estimates of ocean heat uptake.[27]

More detail from observations and models of heat penetration into the oceans is shown in Figure 5.23[28] that demonstrates that natural forcing due to solar variations and volcanic eruptions cannot explain the observed warming but that the addition of human induced greenhouse gas forcings brings observations and model simulations into good agreement for all three oceans.

Is the climate chaotic?

Throughout this chapter the implicit assumption has been made that climate change is predictable and that models can be used to provide predictions of climate change due to human activities. Before leaving this chapter I want to consider whether this assumption is justified.

The capability of the models themselves has been demonstrated so far as weather forecasting is concerned. They also possess some skill in seasonal forecasting. They can provide a good description of the current climate and its seasonal variations. Further, they provide predictions that on the whole are reproducible and that show substantial consistency between different models – although it might be argued, some of this consistency could be a property of the models rather than of the climate. Further over several recent decades for which comparison with observations is possible the predictions show good agreement with observations. But is there other evidence to support the view that climate change is predictable, particularly for the longer term?

A good place to look for further evidence is in the record of climates of the past, presented in Chapter 4. Correlation between the Milankovitch cycles in the Earth's orbital parameters and the cycles of climate change over the past half million years (see Figures 4.7 and 5.19) provides strong evidence to substantiate the Earth's orbital variations as the main factor responsible for the triggering of major climate changes – although the nature of the feedbacks which control the very different amplitudes of response to the three orbital variations still need to be understood. Some 60 ± 10% of the variance in the record of global average temperature from palaeontological sources over the past million years occurs close to frequencies identified in the Milankovitch theory. The existence of this surprising amount of regularity suggests that the climate system is not strongly chaotic so far as these large changes are concerned, but responds in a largely predictable way to the Milankovitch forcing.

This Milankovitch forcing arises from changes in the distribution of solar radiation over the Earth because of variations in the Earth's orbit. Changes in climate as a result of the increase of greenhouse gases are also driven by changes in the radiative regime at the top of the atmosphere. These changes are not dissimilar in kind (although different in distribution) from the changes that provide the Milankovitch forcing. It can be argued therefore that the increases in greenhouse gases will also result in a largely predictable response.

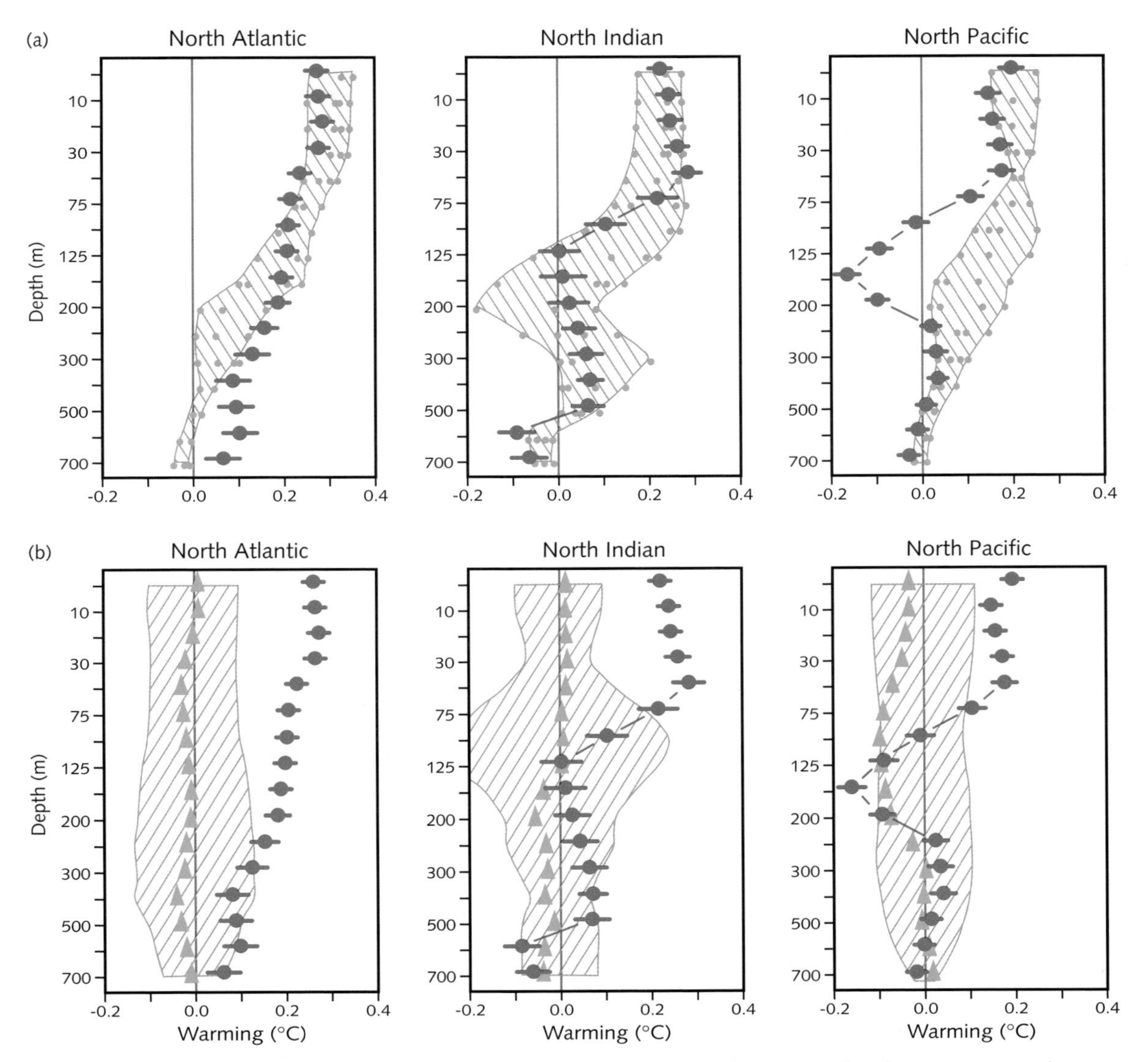

Figure 5.23 Warming of the top 700 m of three of the oceans resulting from natural (solar variations and volcanic eruptions) and anthropogenic (greenhouse gases and aerosols) forcing at the surface over the period 1961–2003. Observations (red dots) of warming compared with modelled changes from both natural and anthropogenic forcing (hatched regions in (a)) and from natural forcing only (green triangles in (b)). The hatched regions in (b) represent the 90% confidence limits of natural internal variability. The ranges of model estimates in (a) are taken from 4 runs (denoted by green dots) of the HadCM3 model at the UK Hadley Centre.

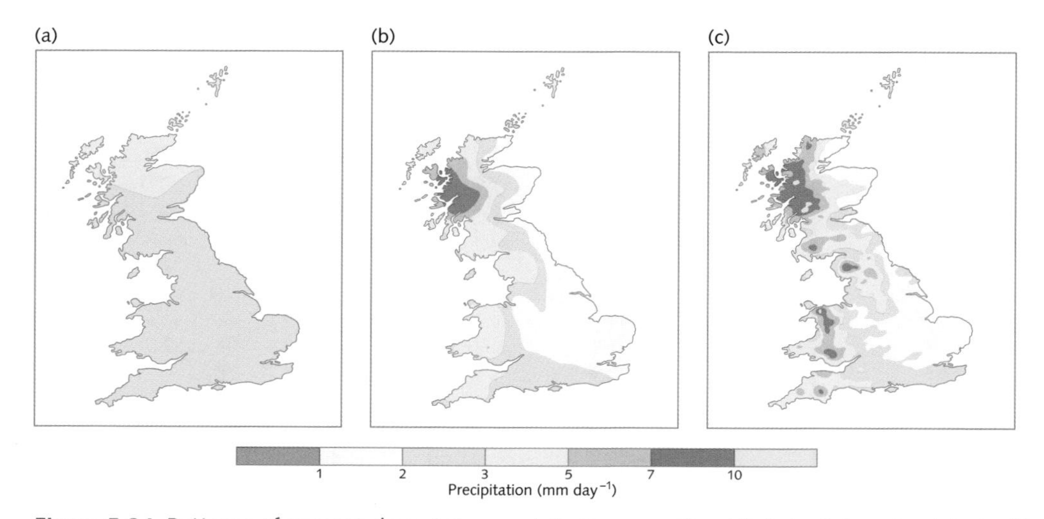

Figure 5.24 Patterns of present-day winter precipitation over Great Britain, (a) as simulated with a 300-km resolution global model, (b) with 50-km resolution regional model, (c) as observed with 10-km resolution.

Regional climate modelling

The simulations we have so far described in this chapter are with global circulation models (GCM) that typically possess a horizontal resolution (grid size) of 200 or 300 km – the size being limited primarily by the availability of computer power. Weather and climate on scales large compared with the grid size are described reasonably well. However, at scales comparable with the grid size, described as the regional scale,[29] the results from global models possess serious limitations. The effects of forcings and circulations that exist on the regional scale need to be properly represented. For instance, patterns of precipitation depend critically on the major variations in orography and surface characteristics that occur on this scale (see Figure 5.24). Patterns generated by a global model therefore will be a poor representation of what actually occurs on the regional scale.

To overcome these limitations regional modelling techniques have been developed.[30] That most readily applicable to climate simulation and prediction is the Regional Climate Model (RCM). A model covering an appropriate region at a horizontal resolution of say 25 or 50 km can be 'nested' in a global model. The global model provides information about the response of the global circulation to large-scale forcings and the evolution of boundary information for the RCM. Within the region, physical information, for instance concerning forcings, is entered on the scale of the regional grid and the evolution of the detailed circulation is developed within the RCM. The RCM is able to account for forcings on smaller scales than are included in the GCM (e.g. due to topography or land cover inhomogeneity; see Figure 5.24) and can also simulate atmospheric circulations and climate variables on these smaller scales.

A limitation of the regional modelling technique we have described is that, although the global model provides the boundary inputs for the RCM, the RCM provides no interaction back on to the global model. As larger computers become available it will be possible to run global models at substantially increased resolution so that this limitation becomes less serious; at the same time RCMs will acquire an ability to deal with detail on even smaller scales. Some examples of regional model simulations are shown in Figures 6.13 and 7.9.

Another technique is that of *statistical downscaling* which has been widely employed in weather forecasting. This uses statistical methods to relate large-scale climate variables (or 'predictors') to regional or local variables. The predictors from a global circulation climate model can be fed into the statistical model to estimate the corresponding regional climate characteristics. The advantage of this technique is that it can easily be applied. Its disadvantage from the point of view of simulating climate change is that it is not possible to be sure how far the statistical relations apply to a climate-changed situation.

The future of climate modelling

Very little has been said in this chapter about the biosphere. Chapter 3 referred to comparatively simple models of the carbon cycle which include chemical and biological processes and simple non-interactive descriptions of atmospheric processes and ocean transport. The large three-dimensional global circulation climate models described in this chapter contain a lot of dynamics and physics but no interactive chemistry or biology. As the power of computers has increased, it has become possible to incorporate into the physical and dynamical models some of the biological and chemical processes that make up the carbon cycle and the chemistry of other gases. These are enabling studies to be made of the detailed processes and interactions that occur in the complete climate system.

Climate modelling continues to be a rapidly growing science. Although useful attempts at simple climate models were made with early computers it is only during the last 20 years that computers have been powerful enough for coupled atmosphere–ocean models to be employed for climate prediction and that their results have been sufficiently comprehensive and credible for them to be taken seriously by policymakers. The climate models which have been developed are probably the most elaborate and sophisticated of computer models developed in any area of natural science. Further, climate models that describe the natural science of climate are now being coupled with socio-economic information in integrated assessment models (see box in Chapter 9, page 280).

As the power of computers increases it becomes more possible to investigate the sensitivity of models by running a variety of ensembles that include different initial conditions, model parameterisations and formulations. A particularly interesting project[31] involves thousands of computer users around the world in running state-of-the-art climate prediction models on their home, school or work computers. By collating data from thousands of models it is generating the world's largest climate modelling prediction experiment.

This chapter has mainly concentrated on modelling the climate response to anthropogenic forcing up to a century or so into the future taking into account what have been called the *fast feedbacks*. However, questions are increasingly being asked about what is likely to happen to the climate in the longer term as a result of human activities now. Much of the response to change by three components of the climate system, namely the oceans, ice sheets and land surface (e.g. vegetation), occurs over longer time scales than a century (cf Table 7.5). Associated with these are *slow feedbacks*[32] that tend to be non-linear. Even if the atmospheric composition is stabilised, warming of the deeper oceans will occur up to at least 1000 years and major changes of the polar ice caps could occur over many millennia. Because of the magnitude of these changes, their non-linearity and their large impact even in the relatively short term, for instance over sea level rise, it is vital that they are better understood. Much future research on past climates, modelling and observations will be concerned with the characteristics of both fast and slow responses within the climate system.

SUMMARY

This chapter has described the basis, assumptions, methods and development of computer numerical modelling of the atmosphere and the climate. Over the past 30 years alongside the rapid development in the performance and speed of computers, there has been enormous development in the sophistication, skill and performance of atmosphere–ocean coupled general circulation models of the climate. Crucial has been the careful incorporation of the variety of positive and negative feedback processes. Confidence in the ability of models to provide useful projections of future climate is based on model simulations that have been validated against:

- detailed observations of current and recent climate of both oceans and atmosphere
- detailed observations of particular climatic cycles such as El Niño events

- observations of perturbations arising from particular events such as volcanic eruptions
- palaeoclimate information from past climates under different orbital forcing.

A great deal remains to be done to narrow the uncertainty of model predictions. The modelling of cloud feedback processes remains the source of the largest uncertainty. Other priorities are to improve the modelling of ocean processes and the ocean–atmosphere interaction. Larger and faster computers continue to be required for these and also to improve the resolution of regional models. More thorough observations of all components of the climate system also continue to be necessary, so that more accurate validation of the model formulations can be achieved. Very substantial national and international programmes are under way to address all these issues.

QUESTIONS

1. Make an estimate of the speed in operations per second of Richardson's 'people' computer. Where does it fall in Figure 5.1?
2. If the spacing between the grid points in a model is 100 km and there are 20 levels in the vertical, what is the total number of grid points in a global model? If the distance between grid points in the horizontal is halved, how much longer will a given forecast take to run on the computer?
3. Take your local weather forecasts over a week and describe their accuracy for 12, 24 and 48 hours ahead.
4. Estimate the average energy received from the Sun over a square region of the ocean surface, one side of the square being a line between northern Europe and Iceland. Compare with the average transport of energy into the region by the North Atlantic Ocean (Figure 5.16).
5. Take a hypothetical situation in which a completely absorbing planetary surface at a temperature of 280 K is covered by a non-absorbing and non-emitting atmosphere. If a cloud which is non-absorbing in the visible part of the spectrum but completely absorbing in the thermal infrared is present above the surface, show that its equilibrium temperature will be 235 K ($=280/2^{0.25}$ K).[33] Show also that if the cloud reflects 50% of solar radiation, the rest being transmitted, the planet's surface will receive the same amount of energy as when the cloud is absent. Can you substantiate the statement that the presence of low clouds tends to cool the Earth while high clouds tend towards warming of it?

6 Associated with the melting of sea-ice which results in increased evaporation from the water surface, additional low cloud can appear. How does this affect the ice-albedo feedback? Does it tend to make it more or less positive?
7 Estimate from information in Chapter 3 the average net radiative forcing from 1960 to 2003. Compare this with the average heating rate at the Earth's surface deduced from measurements of the energy absorbed by the ocean as detailed on page 127. Comment on your results.
8 A change in radiative forcing at the top of the atmosphere of about 3 $W m^{-2}$ leads to a change in surface temperature of around 1 °C providing nothing else changes. Consider the plots shown in Figure 5.15b. What surface temperature change would be expected following the Pinatubo volcano? Compare with the information in Fig 5.21 and comment on your comparison. It is through the analysis of data of the kind illustrated in Figure 5.15a and b that the magnitude and sign of cloud-radiation feedback can be studied. Specify the requirements of a programme of measurements in terms of accuracy (in $W m^{-2}$) and coverage in both space and time if meaningful conclusions about cloud-radiation feedback are to be obtained.
9 It is sometimes argued that weather and climate models are the most sophisticated and soundly based models in natural science. Compare them (e.g. in their assumptions, their scientific basis, their potential accuracy, etc.) with other computer models with which you are familiar both in natural science and social science (e.g. models of the economy).

FURTHER READING AND REFERENCE

Solomon, S., Qin, D., Manning M., Chen, Z., Marquis, M., Averyt, K.B., Tigor, M., Miller, H.L. (eds.) 2007. *Climate Change 2007: The Physical Science Basis. Contribution of Working Group I to the Fourth Assessment Report of the Intergovernmental Panel on Climate Change*. Cambridge: Cambridge University Press.

Technical Summary (summarises basic information about modelling and its applications)

Chapter 1 Historical overview of climate change science

Chapter 8 Climate models and their evaluation

Chapter 9 Understanding and attributing climate change

Chapter 10 Global climate projections

Chapter 11 Regional climate projections

McGuffie, K., Henderson-Sellers, A. 2005. *A Climate Modeling Primer*, third edition. New York: Wiley.

Houghton, J.T. 2002. *The Physics of Atmospheres*, third edition. Cambridge: Cambridge University Press.

Palmer, T., Hagedorn, R. (eds.) 2006. *Predictability of Weather and Climate*. Cambridge: Cambridge University Press.

NOTES FOR CHAPTER 5

1 Richardson, L.F. 1922. *Weather Prediction by Numerical Processes*. Cambridge: Cambridge University Press. Reprinted by Dover, New York, 1965.

2 For more details see, for instance, Houghton, *The Physics of Atmospheres*.

3 See Palmer, T.N. 2006, Chapter 1 in Palmer, T., Hagedorn, R. (eds.) 2006 *Predictability of Weather and Climate*. Cambridge: Cambridge University Press.

4 For more detail see: Chapter 13 in Houghton, *The Physics of Atmospheres*; Palmer, T.N. 1993. A nonlinear perspective on climate change. *Weather*, **48**, 314–26; Palmer and Hagedorn (eds.), *Predictability of Weather and Climate*.

5 An equation such as $y = ax + b$ is linear; a plot of y against x is a straight line. Examples of non-linear equations are $y = ax^2 + b$ or $y + xy = ax + b$; plots of y against x for these equations would not be straight lines. In the case of the pendulum, the equations describing the motion are only approximately linear for very small angles from the vertical where the sine of the angle is approximately equal to the angle; at larger angles this approximation becomes much less accurate and the equations are non-linear.

6 Palmer, T.N., Chapter 1 in Palmer and Hagedorn (eds.), *Predictability of Weather and Climate*.

7 Named 'Southern Oscillation' by Sir Gilbert Walker in 1928.

8 Folland, C.K., Owen, J., Ward, M.N., Colman, A. 1991. Prediction of seasonal rainfall in the Sahel region using empirical and dynamical methods. *Journal of Forecasting*, **10**, 21–56.

9 See for instance Shukla, J., Kinter III, J.L., Chapter 12 in Palmer and Hagedorn (eds.), *Predictability of Weather and Climate*.

10 Xue, Y. 1997. Biospheric feedback on regional climate in tropical north Africa. *Quarterly Journal of the Royal Meteorological Society*, **123**, 1483–515.

11 For a review of climate feedback processes see Bony, S. *et al.*, 2006. How well do we understand and evaluate climate change feedback processes? *Journal of Climate*, **19**, 3445–82.

12 Associated with water vapour feedback is also *lapse rate feedback* which occurs because, associated with changes of temperature and water vapour content in the troposphere, are changes in the average lapse rate (the rate of fall of temperature with height). Such changes lead to this further feedback, which is generally much smaller in magnitude than water vapour feedback but of the opposite sign, i.e. negative instead of positive. Frequently, when overall values for water vapour feedback are quoted the lapse rate feedback has been included. For more details see Houghton, *The Physics of Atmospheres*.

13 Lindzen, R.S. 1990. Some coolness concerning global warming. *Bulletin of the American Meteorological Society*, **71**, 288–99. In this paper, Lindzen queries the magnitude and sign of the feedback due to water vapour, especially in the upper troposphere, and suggests that it could be much less positive than predicted by models and could even be slightly negative. Much has been done through observational and modelling studies to investigate the likely magnitude of water vapour feedback. More detail can be found in Stocker, T.F. *et al.*, Physical climate processes and feedbacks, Chapter 7 in Houghton *et al.* (eds.), *Climate Change 2001: The Science Basis*. The conclusion of that chapter, whose authors include Lindzen, is that 'the balance of evidence favours a positive clear-sky water vapour feedback of a magnitude comparable to that found in simulations'.

14 See Figure 2.8 and the definition of radiative forcing at the beginning of Chapter 3.

15 From Figure 8.14 in Randall, D., Wood, R.A. *et al.* *Climate Models and their Evaluation*, Chapter 8 in Solomon et al. (eds.) *Climate Change 2007: The Physical Science Basis*.

16 Note that the variance in the total is less than the sum of the variances of the three parameters. The total is obtained by first adding the values of the parameters from individual model runs.

17 For a description of a recent model and how it performs see Pope, V. *et al.* 2007. The Met Office Hadley Centre climate modelling capability: the competing requirements for improved resolution, complexity and dealing with uncertainty. *Philosophical Transactions of the Royal Society A*, **365**, 2635–2657.

18 Randall *et al.* Chapter 8, in Solomon *et al.* (eds.) *Climate Change 2007: The Physical Science Basis*.

19 For a recent review see Cane, M.A. *et al.* 2006. Progress in paleoclimate modeling. *Journal of Climate* **19**, 5031–57.
20 Graf, H.-E. *et al.* 1993. Pinatubo eruption winter climate effects: model versus observations. *Climate Dynamics*, **9**, 61–73.
21 See Policymakers' summary. In Houghton, J.T., Jenkins, G.J., Ephraums, J.J. (eds.) 1990. *Climate Change: The IPCC Scientific Assessment.* Cambridge: Cambridge University Press.
22 See Summary for policymakers. In Houghton, J.T., Meira Filho, L.G., Callander, B.A., Harris, N., Kattenberg, A., Maskell, K. (eds.) 1996. *Climate Change 1995: The Science of Climate Change.* Cambridge: Cambridge University Press.
23 Detection is the process demonstrating that an observed change is significantly different (in a statistical sense) than can be explained by natural variability. Attribution is the process of establishing cause and effect with some defined level of confidence, including the assessment of competing hypotheses. For further information about detection and attribution studies see Mitchell, J.F.B., Karoly, D.J. *et al.* 2001. Detection of climate change and attribution of causes, Chapter 12 in Houghton *et al.* (eds.), *Climate Change 2001: The Scientific Basis*, and Hegerl, G.C., Zwiers, F.W. *et al.* Understanding and attributing climate change, Chapter 9, in Solomon *et al.* (eds.) *Climate Change 2007: The Physical Science Basis.*
24 Summary for policymakers. In Houghton *et al.* (eds.) *Climate Change 2001: The Scientific Basis.*
25 Summary for policymakers, in Solomon *et al.* (eds.) *Climate Changes 2007: The Physical Science Basis.* The definitions of *likely, very likely*, etc. are given in Note 1 to Chapter 4.
26 Bindoff, N. Willebrand, J. *et al.* 2007. Observations: Oceanic climate change and sea level, Chapter 5, in Solomon *et al.* (eds.) *Climate Change 2007: The Physical Science Basis.*
27 See Gregory, J. *et al.* 2002. *Journal of Climate*, **15**, 3117–21.
28 From Barnett, T. P. et al, 2005, *Science* **309**, 284–287; modelling simulations from the Hadley Centre UK.
29 The regional scale is defined as describing the range of 10^4 to 10^7 km^2. The upper end of the range (10^7 km^2) is often described as a typical sub-continental scale. Circulations at larger than the sub-continental scale are on the planetary scale.
30 For more information see Giorgi, F., Hewitson, B. *et al.* 2001, Regional climate information – evaluation and projections. Chapter 10, in Houghton *et al.* (eds.), *Climate Change 2007: The Scientific Basis.*
31 See www.climateprediction.net
32 The terminology of fast and slow feedbacks has been introduced by James Hansen – see his Bjerknes Lecture at American Geophysical Union, 17 December 2008 at www.columbia.edu/~jeh1/2008/AGUBjerknes_20081217.pdf.
33 Hint: recall Stefan's blackbody radiation law that the energy emitted is proportional to the fourth power of the temperature.

Climate change in the twenty-first century and beyond

6

NASA'S 2006 CloudSat (artist's rendition) studies the role of clouds and aerosols in regulating the Earth's weather, climate and air quality

THE LAST chapter explained that the most effective tool we possess for the prediction of future climate change due to human activities is the climate model. This chapter will describe the predictions of models for likely climate change during the twenty-first century. It will also consider other factors that might lead to climate change and assess their importance relative to the effect of greenhouse gases.

Emission scenarios

A principal reason for the development of climate models is to learn about the detail of the likely climate change this century and beyond. Because model simulations into the future depend on assumptions regarding future anthropogenic emissions of greenhouse gases, which in turn depend on assumptions about many factors involving human behaviour, it has been thought inappropriate and possibly misleading to call the simulations of future climate so far ahead 'predictions'. They are therefore generally called 'projections' to emphasise that what is being done is to explore likely future climates which arise from a range of assumptions regarding human activities.

A starting point for any projections of climate change into the future is a set of descriptions of likely future global emissions of greenhouse gases. These will depend on a variety of assumptions regarding human behaviour and activities, including population, economic growth, energy use and the sources of energy generation. As was mentioned in Chapter 3, such descriptions of future emissions are called *scenarios*. A wide range of scenarios was developed by the IPCC in a Special Report on Emission Scenarios (SRES)[1] in preparation for its 2001 Report (see box below). It is these scenarios that have been used in developing the projections of future climate presented in this chapter. In addition, because it has been widely used in modelling studies, results are also presented using a scenario (IS 92a) taken from a set developed by the IPCC in 1992 and widely described as representative of 'business-as-usual'.[2] Details of these scenarios are presented in Figure 6.1.

The storylines on which the SRES scenarios are based incorporate a wide range of different assumptions regarding population, economic growth, technological innovation and attitudes to social and environmental sustainability. None of them, however, takes account of deliberate action to combat climate change and reduce greenhouse gases. Scenarios including such action will be presented in Chapters 10 and 11 where the possibilities for stabilisation of carbon dioxide concentration in the atmosphere is considered.

The SRES scenarios include estimates of greenhouse gas emissions resulting from all sources including land-use change. Estimates in the different scenarios begin from the current values for land-use change including deforestation (see Table 3.1). Assumptions in different scenarios vary, from continued deforestation, although reducing as less forest remains available for clearance, to substantial afforestation leading to an increased carbon sink. The next stage in the development of projections of climate change is to turn the emission profiles of

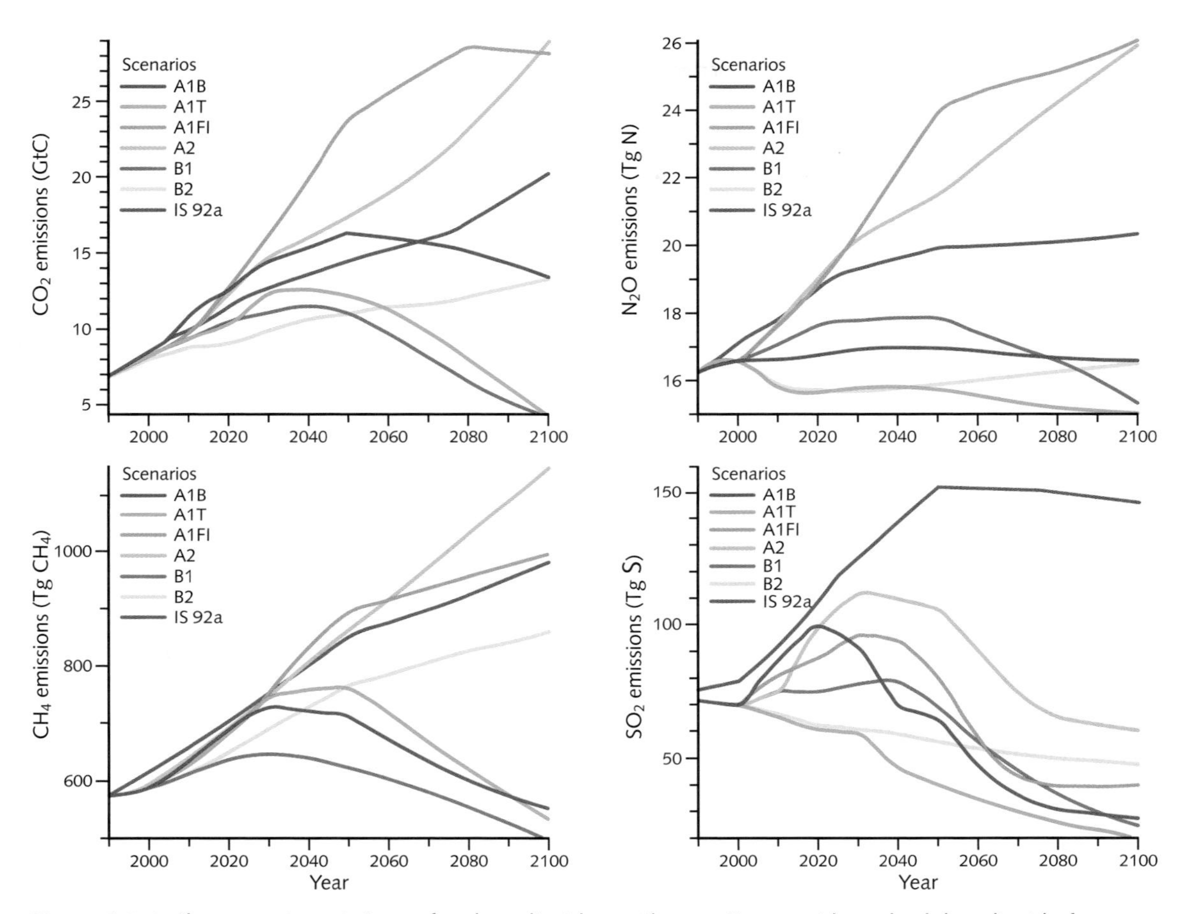

Figure 6.1 Anthropogenic emissions of carbon dioxide, methane, nitrous oxide and sulphur dioxide for the six illustrative SRES scenarios, A1B, AIT, A1FI, A2, B1 and B2. For comparison the IS 92a scenario is also shown.

greenhouse gases into greenhouse gas concentrations (Figure 6.2) and then into radiative forcing (Table 6.1). The methods by which these are done are described in Chapter 3, where the main sources of uncertainty are also mentioned. For the carbon dioxide concentration scenarios these uncertainties, especially those concerning the magnitude of the climate feedback from the terrestrial biosphere (see box on page 48–9), amount to a range of about –10% to +30% in 2100 for each profile.[3]

For most scenarios, emissions and concentrations of the main greenhouse gases increase during the twenty-first century. However, despite the increases projected in fossil fuel burning – very large increases in some cases – emissions of sulphur dioxide (Figure 6.1) and hence the concentrations of sulphate

The emission scenarios of the Special Report on Emission Scenarios (SRES)

The SRES scenarios are based on a set of four different storylines within each of which a family of scenarios has been developed – leading to a total of 35 scenarios.[4]

A1 storyline

The A1 storyline and scenario family describes a future world of very rapid economic growth, a global population that peaks in mid century and declines thereafter, and the rapid introduction of new and more efficient technologies. Major underlying themes are convergence among regions, capacity building and increased cultural and social interactions, with a substantial reduction in regional differences in per capita income. The A1 scenario family develops into three groups which describe alternative directions of technological change in the energy system. The three groups are distinguished by their technological emphasis: fossil fuel intensive (A1FI), non-fossil fuel energy sources (A1T) or a balance across all sources (A1B) – where balance is defined as not relying too heavily on one particular energy source, on the assumption that similar improvement rates apply to all energy-supply and end-use technologies.

A2 storyline

The A2 storyline and scenario family describes a very heterogeneous world. The underlying theme is self-reliance and preservation of local identities. Fertility patterns across regions converge very slowly, which results in a continuously increasing population. Economic development is primarily regionally oriented and per capita economic growth and technological change more fragmented and slower than other storylines.

B1 storyline

The B1 storyline and scenario family describes a convergent world, with the same global population that peaks in mid century and declines thereafter as in the A1 storyline, but with rapid change in economic structures towards a service and information economy, with reductions in material intensity and the introduction of clean and resource-efficient technologies. The emphasis is on global solutions to economic, social and environmental sustainability, including improved equity, but without additional climate-related initiatives.

B2 storyline

The B2 storyline and scenario family describes a world in which the emphasis is on local solutions to economic, social and environmental sustainability. It is a world with a continuously increasing global population, at a rate lower than in A2, intermediate levels of economic development and less rapid and more diverse technological change than in the B1 and A1 storylines. While the storyline is also oriented towards environmental protection and social equity, it focuses on local and regional levels.

From the total set of 35 scenarios, an illustrative scenario was chosen for each of the six scenario groups A1B, A1FI, A1T, A2, B1 and B2. All should be considered equally sound. It is mostly for this set of six illustrative scenarios that data are presented in this chapter.

The SRES scenarios do not include additional climate initiatives, which means that no scenarios are included that explicitly assume implementation of the United Nations Framework Convention on Climate Change or the emissions targets of the Kyoto Protocol.

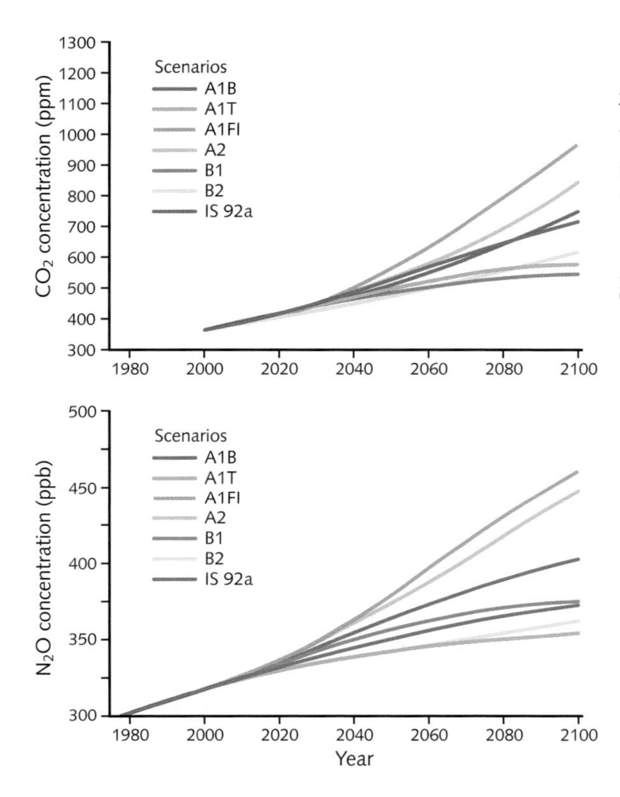

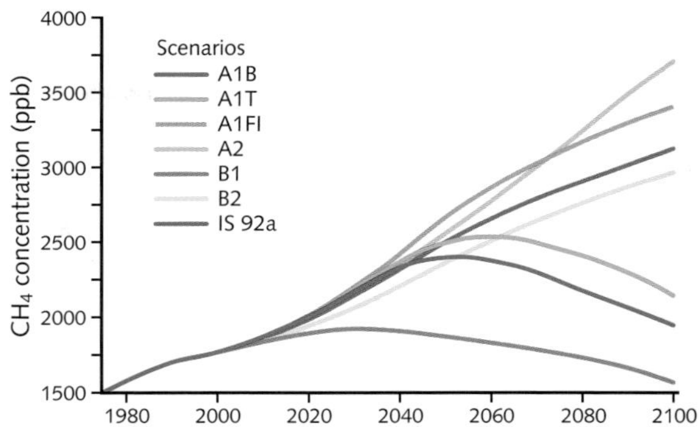

Figure 6.2 Atmospheric concentrations of carbon dioxide, methane and nitrous oxide resulting from the six illustrative SRES scenarios and from the IS 92a scenario. Uncertainties for each profile, especially those due to possible carbon feedbacks, have been estimated as from about –10% to +30% in 2100.

particles are expected to fall substantially because of the spread of policies to abate the damaging consequences of air pollution and 'acid rain' deposition to both humans and ecosystems.[5] The influence of sulphate particles in tending to reduce the warming due to increased greenhouse gases is therefore now projected to be much less than for projections made in the mid 1990s (see the IS 92a scenario for sulphur dioxide in Figure 6.1).[6] In fact it is likely that sulphate particles will be reduced to well below their 1990 levels during the twenty-first century.[7] The other anthropogenic sources of particles in the atmosphere included in Figure 3.11 will also contribute small amounts of positive or negative radiative forcing during the twenty-first century.[8] Table 6.1 includes a 2005 estimate of future total radiative forcing from all aerosol sources.

Model projections

Results that come from the most sophisticated coupled atmosphere–ocean models of the kind described in the last chapter provide fundamental information on which to base climate projections. However, because they are so

Table 6.1 Radiative forcing (W m^{-2}) globally averaged, for greenhouse gases and aerosols from the year 1750 to 2005 and from SRES scenarios to 2050 and 2100

Greenhouse gas	Year	Radioactive forcing (W m^{-2})	A1B	A1T	A1FI	A2	B1	B2	IS 92a
CO_2	2005	1.66							
	2050		3.36	3.08	3.70	3.36	2.92	2.83	3.12
	2100		4.94	3.85	6.61	5.88	3.52	4.19	4.94
CH_4	2005	0.48							
	2050		0.70	0.73	0.78	0.75	0.52	0.68	0.73
	2100		0.56	0.62	0.99	1.07	0.41	0.87	0.91
N_2O	2005	0.16							
	2050		0.25	0.23	0.33	0.32	0.27	0.23	0.29
	2100		0.31	0.26	0.55	0.51	0.32	0.29	0.40
O_3(trop)	2005	0.35							
	2050		0.59	0.72	1.01	0.78	0.39	0.63	0.67
	2100		0.50	0.46	1.24	1.22	0.19	0.78	0.90
Halocarbons	2005	0.34							
Total aerosols	2005	−1.2[a]							

[a] Including both direct and indirect effects.

demanding on computer time only a limited number of results from such models are available. Many studies have also therefore been carried out with simpler models. Some of these, while possessing a full description of atmospheric processes, only include a simplified description of the ocean; these can be useful in exploring regional change. Others, sometimes called energy balance models (see box on page 144), drastically simplify the dynamics and physics of both atmosphere and ocean and are useful in exploring changes in the global average response with widely different emission scenarios. Results from simplified models need to be carefully compared with those from the best coupled atmosphere–ocean models and the simplified models 'tuned' so that, for the particular parameters for which they are being employed, agreement with the more complete models is as close as possible. The projections presented in the next sections depend on results from all these kinds of models.

In order to assist comparison between models, experiments with many models have been run with the atmospheric concentration of carbon dioxide doubled from its pre-industrial level of 280 ppm. The global average temperature rise under steady conditions of doubled carbon dioxide concentration is known as the climate sensitivity.[9] The Intergovernmental Panel on Climate Change (IPCC) in its 1990 Report gave a range of 1.5 to 4.5 °C for the climate sensitivity with a 'best estimate' of 2.5 °C; the IPCC 1995 and 2001 Reports confirmed these values. The 2007 Report stated; 'it is likely to be in the range 2 to 4.5 °C with a best estimate of 3 °C, and is very unlikely to be less than 1.5 °C. Cloud feedbacks [see Chapter 5] remain the largest source of uncertainty.'[10] The projections presented in this chapter follow the IPCC 2007 Assessment.[11]

An estimate of climate sensitivity can also be obtained from paleoclimate information over the last million years (see Chapter 4) that connects variations of global average temperature with variations of climate forcings arising from changes in ice cover, vegetation and greenhouse gas concentrations (Chapter 4, Figures 4.6 and 4.7). The estimate of 3 ± 0.5 °C obtained in this way reported by James Hansen[12] agrees very well with the model estimates mentioned above.

Projections of global average temperature

When information of the kind illustrated in Figures 6.1 and 6.2 is incorporated into simple or more complex models, projections of climate change can be made. As we have seen in earlier chapters, a useful proxy for climate change that has been widely used is the change in global average temperature.

The projected increases in global average near surface temperature over the twenty-first century due to increase in greenhouse gases and aerosols as assumed by the six marker SRES scenarios is illustrated in Figure 6.4a. It shows increases for the different scenarios with best estimates for the year 2100 ranging from about 2 to 4 °C. When uncertainties are added, the overall likely range is from just over 1 to over 6 °C – that wide range resulting from the large uncertainty regarding future emissions and also from the uncertainty that remains regarding the feedbacks associated with the climate response to the changing atmospheric composition (as described in Chapter 5).

Compared with the temperature changes normally experienced from day to day and throughout the year, changes of between 1 and 6 °C may not seem very large. But, as was pointed out in Chapter 1, it is in fact a large amount when considering *globally averaged* temperature. Compare it with the 5 or 6 °C change in global average temperature that occurs between the middle of an ice age and the warm period in between ice ages (Figure 4.6). The changes projected for the

Simple climate models

In Chapter 5 a detailed description was given of general circulation models (GCMs) of the atmosphere and the ocean and of the way in which they are coupled together (in AOGCMs) to provide simulations of the current climate and of climate perturbed by anthropogenic emissions of greenhouse gases. These models provide the basis of our projections of the detail of future climate. However, because they are so elaborate, they take a great deal of computer time so that only a few simulations can be run with these large coupled models.

To carry out more simulations under different future emission profiles of greenhouse gases or of aerosols or to explore the sensitivity of future change to different parameters (for instance, parameters describing the feedbacks in the atmosphere which largely define the climate sensitivity), extensive use has been made of simple climate models.[13] These simpler models are 'tuned' so as to agree closely with the results of the more complex AOGCMs in cases where they can be compared. The most radical simplification in the simpler models is to remove one or more of the dimensions so that the quantities of interest are averaged over latitude circles (in two-dimensional models) or over the whole globe (in one-dimensional models). Such models can, of course, only simulate latitudinal or global averages – they can provide no regional information.

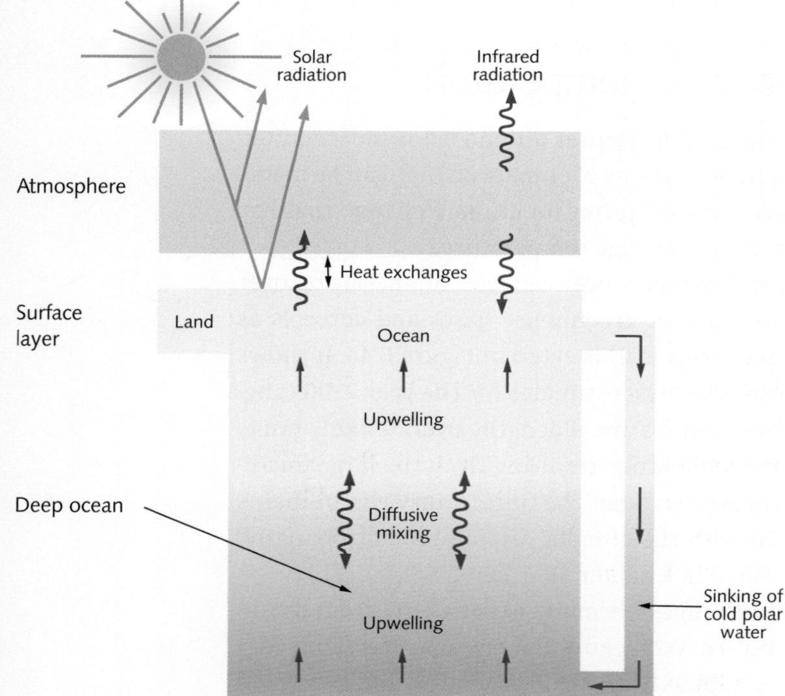

Figure 6.3 The components of a simple 'upwelling–diffusion' climate model.

Figure 6.3 illustrates the components of such a model in which the atmosphere is contained within a 'box' with appropriate radiative inputs and outputs. Exchange of heat occurs at the land surface (another 'box') and the ocean surface. Within the ocean allowance is made for vertical diffusion and vertical circulation. Such a model is appropriate for simulating changes in global average surface temperature with increasing greenhouse gases or aerosols. When exchanges of carbon dioxide across the interfaces between the atmosphere, the land and the ocean are also included, the model can be employed to simulate the carbon cycle.

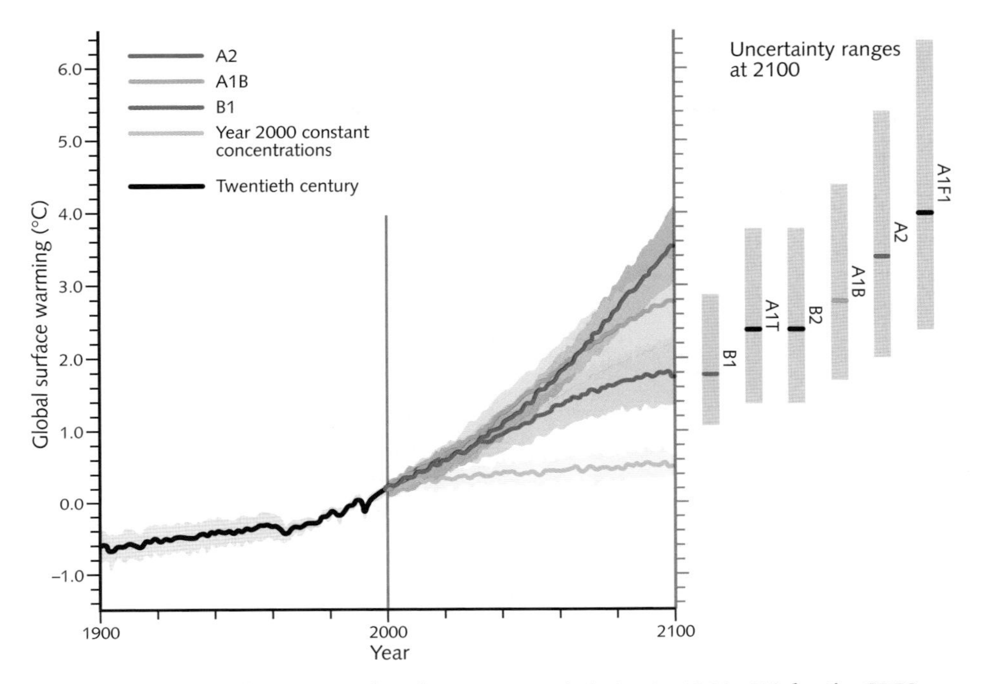

Figure 6.4 (a) Global averages of surface warming (relative to 1980–99) for the SRES scenarios A2, A1B and B1, shown as continuations of twentieth-century simulations. Each curve is a multi-model average from a number (typically around 20) of AOGCMs; shading denotes the one standard deviation range of individual model annual means. A curve is also shown for a scenario in which greenhouse gas concentrations were held constant at year 2000 values. The grey bars at the right indicate for year 2100 the best estimate and *likely* range for the six SRES marker scenarios taking into account both the spread of AOGCM results and uncertainties associated with representations of feedbacks (see Chapter 5). To obtain temperature increases from pre-industrial times, add 0.6 °C.

twenty-first century are from one-third to a whole ice age in terms of the degree of climate change!

Figure 6.4b compares the observed global mean warming from 1990 to 2006 with model projections and their ranges from 1990 to 2025 as presented by the IPCC in its first three assessment reports. Figure 6.4c illustrates the results from 21 different models (as used in constructing the average in Figure 6.4a) for the temperature increase under SRES scenario A1B.

Beginning with the first IPCC report in 1990, the IPCC has consistently projected forecasts of global average temperature increase in the range 0.15 to 0.3 °C per decade from 1990 to 2005. This can now be compared with observed values of about 0.2 °C per decade and projections for all SRES scenarios of about

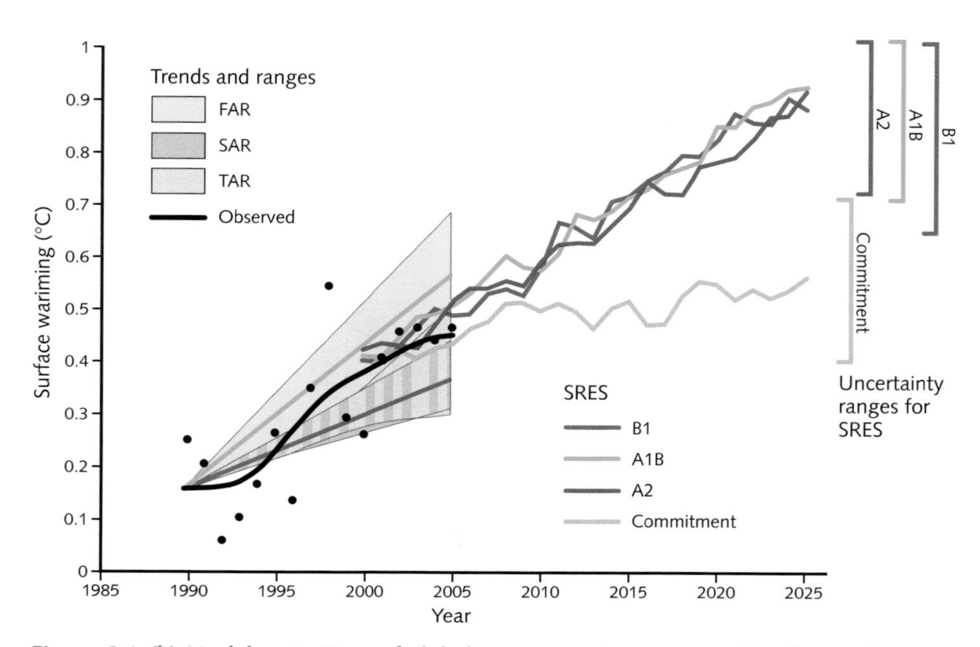

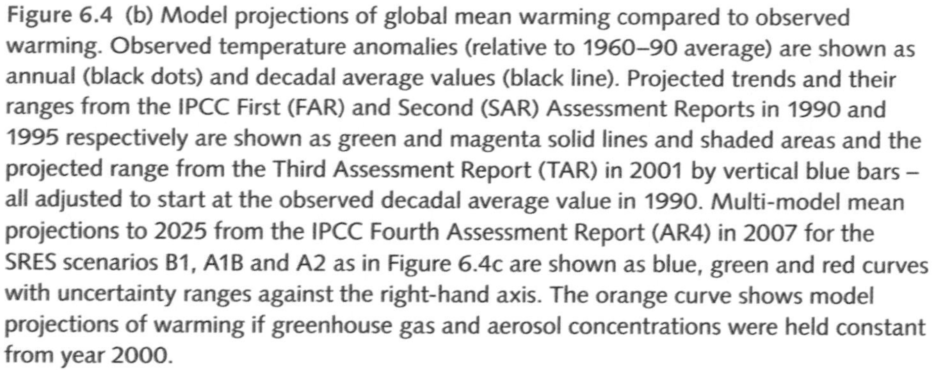
Figure 6.4 (b) Model projections of global mean warming compared to observed warming. Observed temperature anomalies (relative to 1960–90 average) are shown as annual (black dots) and decadal average values (black line). Projected trends and their ranges from the IPCC First (FAR) and Second (SAR) Assessment Reports in 1990 and 1995 respectively are shown as green and magenta solid lines and shaded areas and the projected range from the Third Assessment Report (TAR) in 2001 by vertical blue bars – all adjusted to start at the observed decadal average value in 1990. Multi-model mean projections to 2025 from the IPCC Fourth Assessment Report (AR4) in 2007 for the SRES scenarios B1, A1B and A2 as in Figure 6.4c are shown as blue, green and red curves with uncertainty ranges against the right-hand axis. The orange curve shows model projections of warming if greenhouse gas and aerosol concentrations were held constant from year 2000.

this value (largely independent of which scenario) over the next two or three decades (Figure 6.4b). Again, these might seem small rates of change; most people would find it hard to detect a change in temperature of a fraction of a degree. But remembering again that these are global averages, such rates of change become very large. Indeed, they are much larger than any rates of change the global climate has experienced for at least the past 10 000 years as inferred from palaeoclimate data. As we shall see in the next chapter, the ability of both humans and ecosystems to adapt to climate change depends critically on the rate of change.

The changes in global average temperature shown in Figure 6.4 from the IPCC 2007 Report and similar ones from the 2001 Report are substantially greater than those shown in the IPCC 1995 Report. The main reason for the difference is the much smaller aerosol emissions in the SRES scenarios compared with the IS 92 scenarios. For instance, the global average temperature in 2100 relating to the IS 92a scenario is similar to that for the SRES B2 scenario even though the carbon dioxide emissions at that date for IS 92a are 50% greater than those for B2.

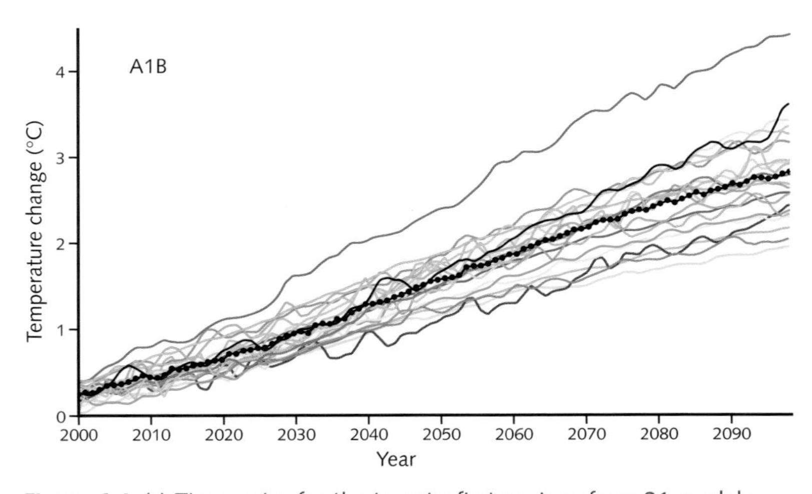

Figure 6.4 (c) Time series for the twenty-first century from 21 models run by climate modelling centres around the world, of annual means of globally averaged surface temperature change (relative to 1980–99 average) under SRES scenario A1B. Multi-model mean is marked with black dots.

To complete this section on likely temperature changes in the twenty-first century, Figure 6.5 shows multi-model mean temperature changes for the A1B scenario from 1990 to 2065 and 2099 at levels in the troposphere and different depths in the ocean. It shows cooling in the stratosphere, substantial warming in the troposphere especially in the tropics and gradual penetration of warming into the ocean from the surface downwards.

Equivalent carbon dioxide (CO_2e)

In many of the modelling studies of climate change, the situation of doubled pre-industrial atmospheric carbon dioxide has often been introduced as a benchmark especially to assist in comparisons between different model projections and their possible impacts. Since the pre-industrial concentration was about 280 ppm, doubled carbon dioxide is about 560 ppm. From the curves in Figure 6.2 this is likely to occur sometime in the second half of the twenty-first century, depending on the scenario. But other greenhouse gases are also increasing and contributing to the radiative forcing. To achieve an overall picture, it is convenient to convert other greenhouse gases to equivalent amounts of carbon dioxide, in other words to amounts of carbon dioxide that would

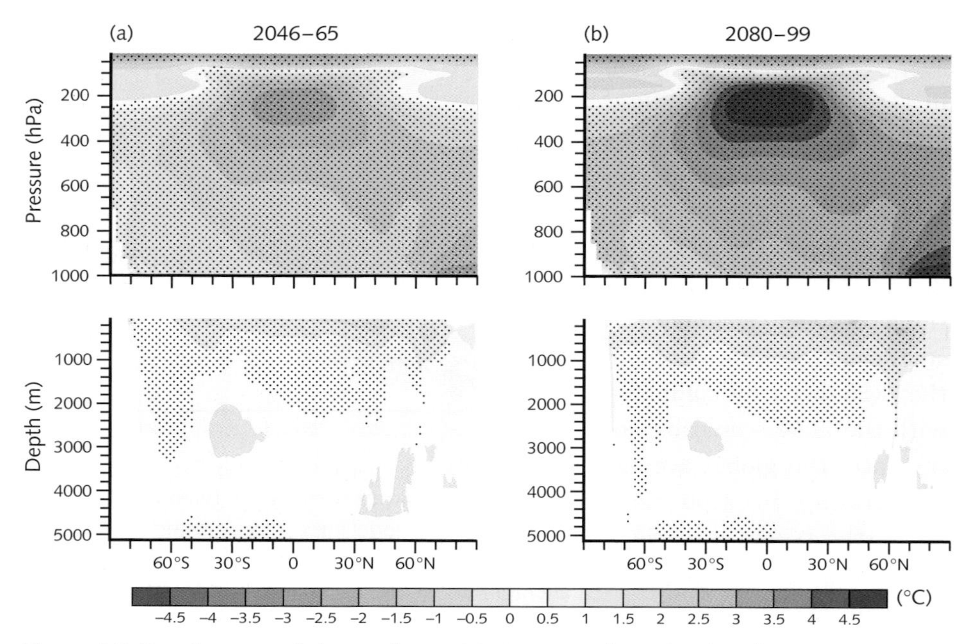

Figure 6.5 Zonal means of change from 1990 in atmospheric (top) and oceanic (bottom) temperatures (°C) shown as cross-sections. Values are multi-model means for the A1B scenario for 2046–65 (a) and 2080–99 (b). Stippling denotes regions where the multi-model ensemble mean divided by the multi-model standard deviation exceeds 1.0 (in magnitude).

give the same radiative forcing.[14] Such equivalent carbon dioxide amounts are denoted by CO_2e. The information in Table 6.1 enables the conversion to be carried out.[15] For instance, the increases in greenhouse gases (including ozone) other than carbon dioxide to date produce a radiative forcing equivalent to about three-quarters of that from carbon dioxide to date (see Figure 3.11). This proportion will drop substantially during the next few decades as the growth in carbon dioxide becomes more dominant in nearly all scenarios.

Calculations of equivalent carbon dioxide (CO_2e) have often been made including only the contributions from other greenhouse gases, sometimes only the long-lived greenhouse gases (i.e. not including ozone). However, as Figure 3.11 and Table 6.1 show, there are substantial contributions to radiative forcing from aerosols in the atmosphere that are predominantly negative. Unless otherwise stated, calculations of CO_2e in this book include the aerosol contributions.

Noting that doubled carbon dioxide produces a radiative forcing of about 3.7 W m^{-2}, it can be seen from Table 6.1 that doubling of the CO_2e amount from pre-industrial times will occur for the SRES marker scenarios around 2050 or before. For scenario A1B, assuming halocarbons and the cooling effect of aerosols in 2050 remain as in 2005 (Table 6.1), radiative forcing in 2050 is approximately equivalent to that from doubled CO_2e (i.e. 3.7 W m^{-2}). Referring now to Figure 6.4a, note that in 2050 for scenario A1B the temperature rise from pre-industrial times is about 2.2 °C. This is only about 75% of the 3 °C (the best value for climate sensitivity for the models employed to provide the results presented in Figure 6.4) that would be expected for doubled CO_2e under steady conditions. As was shown in Chapter 5, this difference occurs because of the slowing effect of the oceans on the temperature rise. This means that, as the CO_2e concentration continues to increase, at any given time there exists a commitment to further significant temperature rise that has not been realised at that time. This is illustrated by the temperature profile (see Figure 6.4a and b) for a scenario for which the concentration of all greenhouse gases and aerosols is kept constant at year 2000 levels. For this profile, warming continues throughout the century beginning with about 0.1 °C per decade for the first few decades.

What about the value of CO_2e now? If the contributions to radiative forcing for 2005 (Figure 3.11) from all greenhouse gases and aerosols are summed and turned into CO_2e,[16] a value of around 375 ppm results (note that without the aerosol and ozone contributions the value would be about 455 ppm). That is not far from the present concentration of carbon dioxide itself, because the negative forcing of aerosol in global average terms has approximately offset the positive forcing of the increased contributions from gases other than carbon dioxide. Note that this calculation is only approximate as there is substantial uncertainty surrounding the magnitude of aerosol forcing (Figure 3.11). Will this aerosol offset continue to apply in the future? All scenarios continue to include substantial if reducing aerosol contributions during the twenty-first century.[17] These considerations will surface again in Chapter 10 when we are looking at possibilities for the stabilisation of CO_2e.

Regional patterns of climate change

So far we have been presenting global climate change in terms of likely increases in global average surface temperature that provide a useful overall indicator of the magnitude of climate change. In terms of regional implications, however, a global average conveys rather little information. What is required is spatial detail. It is in the regional or local changes that the effects and impacts of global climate change will be felt.

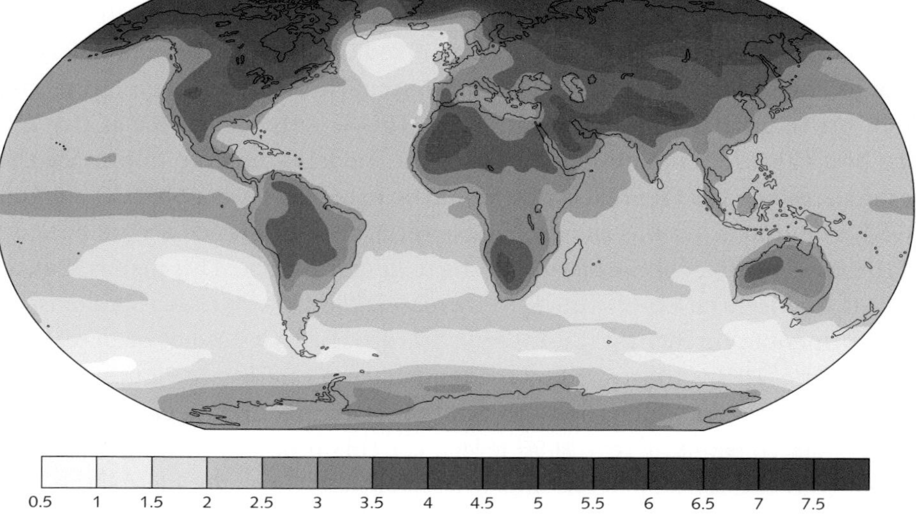

Figure 6.6 Projected pattern average of surface temperature changes in °C for the twenty-first century – the period 2090–99 compared with 1980–99 – for the SRES scenario A1B, from multi-model AOGCM averages.

With respect to regional change, it is important to realise that, because of the way the atmospheric circulation operates and the interactions that govern the behaviour of the whole climate system, climate change over the globe will not be at all uniform. We can, for instance, expect substantial differences between the changes over large land masses and over the ocean; land possesses a much smaller thermal capacity and so can respond more quickly. Listed below are some of the broad features on the continental scale that characterise the projected temperature changes; more detailed patterns are illustrated in Figure 6.6. Reference to Chapter 4 indicates that many of these characteristics are already being found in the observed record of the last few decades.

- Generally greater surface warming of land areas than of the oceans typically by about 40% compared with the global average, greater than this in northern high latitudes in winter (associated with reduced sea-ice and snow cover) and southern Europe in summer; less than 40% in south and southeast Asia in summer and in southern South America in winter.

Across the world torrential rain and flooding has increased, and sights such as these in Canada have been more commonplace in recent years.

- Minimum warming around Antarctica and in the northern North Atlantic which is associated with deep oceanic mixing in those areas.
- Little seasonal variation of the warming in low latitudes or over the southern circumpolar ocean.
- A reduction in diurnal temperature range over land in most seasons and most regions; night-time lows increase more than daytime highs.

So far we have been presenting results solely for atmospheric temperature change. An even more important indicator of climate change is precipitation. With warming at the Earth's surface there is increased evaporation from the oceans and also from many land areas leading on average to increased atmospheric water vapour content and therefore also on average to increased precipitation. Since the water-holding capacity of the atmosphere increases by about 6.5% per degree Celsius,[18] the increases in precipitation as surface temperature rises can be expected to be substantial. In fact, model projections indicate increases in precipitation broadly related to surface temperature increases of about 3% per degree Celsius.[19] Further, since the largest component of the energy input to the atmospheric circulation comes from the release of latent heat as water vapour condenses, the energy available to the atmosphere's circulation will increase in proportion to the atmospheric water content. A characteristic therefore of anthropogenic climate change due to the increase of greenhouse gases will be a more intense hydrological cycle. The likely effect of this on precipitation extremes will be discussed in the next section.

In Figure 6.7 are shown projected changes in the distribution of precipitation as global warming increases. Three broad characteristics of precipitation changes are as follows.[20]

- In addition to overall global average precipitation increase, there are large regional variations, areas with decreases in average precipitation, changes in its seasonal distribution and a general increase in the spatial variability of precipitation, contributing for instance to a reduction of rainfall in the sub-tropics and an increase at high latitudes and parts of the tropics.
- The poleward expansion of the sub-tropical high pressure regions, combined with the general tendency towards reduction in sub-tropical precipitation, creates robust projections of a reduction in precipitation on the poleward edges of the sub-tropics. Most of the regional projections of reductions in precipitation in the twenty-first century are associated with areas adjacent to these sub-tropical highs. For instance, southern Europe, Central America, southern Africa and Australia are likely to have drier summers with increasing risk of drought.
- A tendency for monsoonal circulations to result in increased precipitation due to enhanced moisture convergence, despite a tendency towards weakening of the monsoonal flows themselves. However, many aspects of tropical climatic responses remain uncertain.

Much natural climate variability occurs because of changes in, or oscillations between, persistent climatic patterns or regimes. The Pacific–North Atlantic Anomaly (PNA – which is dominated by high pressure over the eastern Pacific and western North America and which tends to lead to very cold winters in the

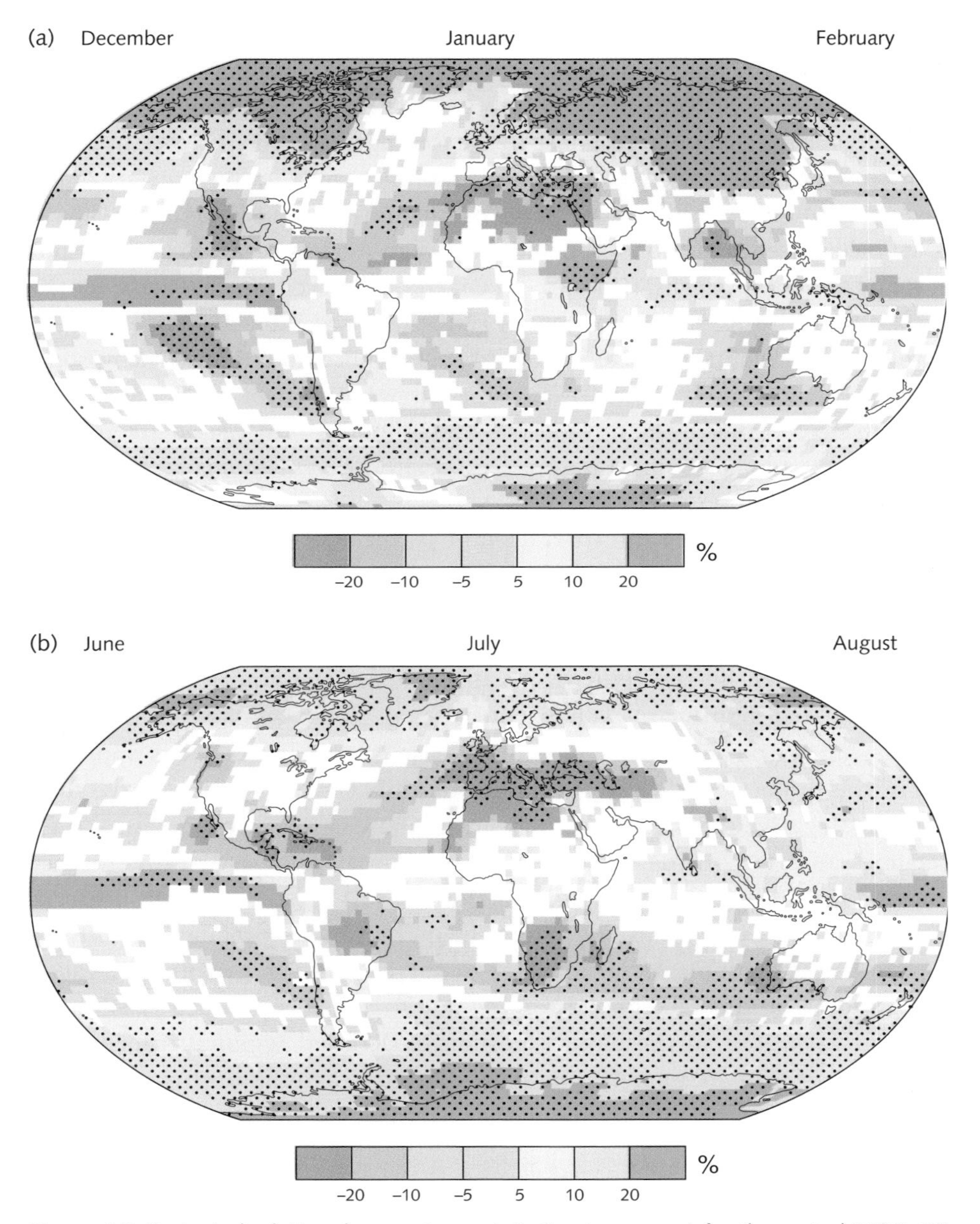

Figure 6.7 Projected relative changes in precipitation in per cent for the period 2090–99 relative to 1980–99, for the SRES scenario A1B, from multi-model AOGCM averages, for (a) December to February and (b) June to August. White areas are where fewer than 66% of the models agree in the sign of the change and stippled areas are where more than 90% of the models agree in the sign of the change.

eastern United States), the North Atlantic Oscillation (NAO – which has a strong influence on the character of the winters in northwest Europe) and the El Niño events mentioned in Chapter 5 are examples of such regimes. Important components of climate change in response to the forcing due to the increase in greenhouse gases can be expected to be in the form of changes in the intensity or the frequency of established climate patterns illustrated by these regimes.[21] There is little consistency at the present time between models regarding projections of many of these patterns. However, recent trends in the tropical Pacific for the surface temperature to become more El Niño-like (see Table 4.1 on page 73–5), with the warming in the eastern tropical Pacific more than that in the western tropical Pacific and with a corresponding eastward shift of precipitation, are projected to continue by many models.[22] The influence of increased greenhouse gases on these major climate regimes, especially the El Niño, is an important and urgent area of research.

A complication in the interpretation of patterns of climate change arises because of the differing influence of atmospheric aerosols as compared with that of greenhouse gases. Although in the projections based on SRES scenarios the influence of aerosols is less than in those based on the IS 92 scenarios published by the IPCC in its 1995 Report their projected radiative forcing is still significant. When considering global average temperature and its impact on, for instance, sea-level rise (see Chapter 7) it is appropriate in the projections to use the values of globally averaged radiative forcing. The negative radiative forcing from sulphate aerosol, for instance, then becomes an offset to the positive forcing from the increase in greenhouse gases. However, because the effects of aerosol forcing are far from uniform over the globe (Figure 3.7), the effects of increasing aerosol cannot only be considered as a simple offset to those of the increase in greenhouse gases. The large variations in regional forcing due to aerosols produce substantial regional variations in the climate response. Detailed regional information from the best climate models is being employed to assess the climate change under different assumptions about the increases in both greenhouse gases and aerosols.

Changes in climate extremes

The last section looked at the likely regional patterns of climate change. Can anything be said about likely changes in the frequency or intensity of climate extremes in the future? It is, after all, not the changes in average climate that are generally noticeable, but the extremes of climate – droughts, floods, storms and extremes of temperature in very cold or very warm periods – which provide the largest impact on our lives (see Chapter 1).[23]

The most obvious change we can expect in extremes is a large increase in the number of extremely warm days and heatwaves (Figure 6.8) coupled with a decrease in the number of extremely cold days. Many continental land areas are experiencing substantial increases in maximum temperature and more heatwaves. An outstanding example is the heatwave in central Europe in 2003 (see box on page 215). Model projections indicate, as shown in Figure 6.8c, much increased frequency and intensity of such events as the twenty-first century progresses.

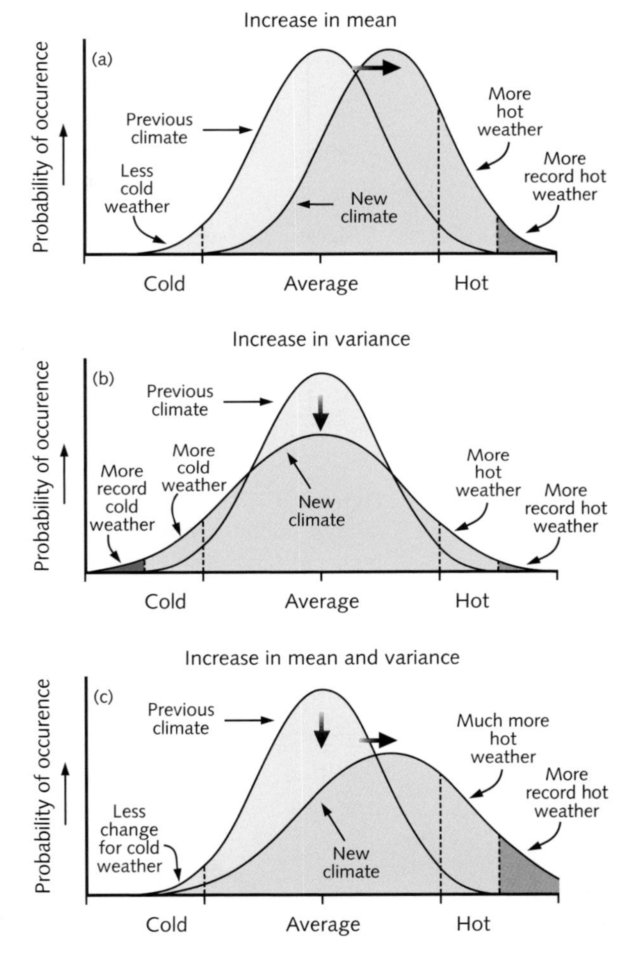

Figure 6.8 Schematic diagrams showing the effects on extreme temperatures when (a) the mean increases leading to more record hot weather, (b) the variance increases and (c) when both the mean and the variance increase, leading to much more record hot weather.

Of even more impact are changes in extremes connected with the hydrological cycle. In the last section it was explained that in a warmer world with increased greenhouse gases, average precipitation increases and the hydrological cycle becomes more intense.[24] Consider what might occur in regions of increased rainfall. Under the more intense hydrological cycle the larger amounts of rainfall will come from increased convective activity: more really heavy showers and more intense thunderstorms. This is well illustrated by Figure 6.9 which shows how, on doubling the carbon dioxide concentration, the number of days with large rainfall amounts (greater than 25 mm day^{-1}) doubled. Although from a climate model of some years ago, it illustrates a robust result from all climate models that more intense precipitation events and more dry days are to be expected during the twenty-first century as global warming increases. The substantial degree of model agreement is illustrated in Figure 6.10 which shows an analysis of results from nine different models, for precipitation intensity. Similar information for other indices related to extremes, for instance heatwaves, frost days, dry days, etc. is provided in the article from which Figure 6.10 is taken.[25]

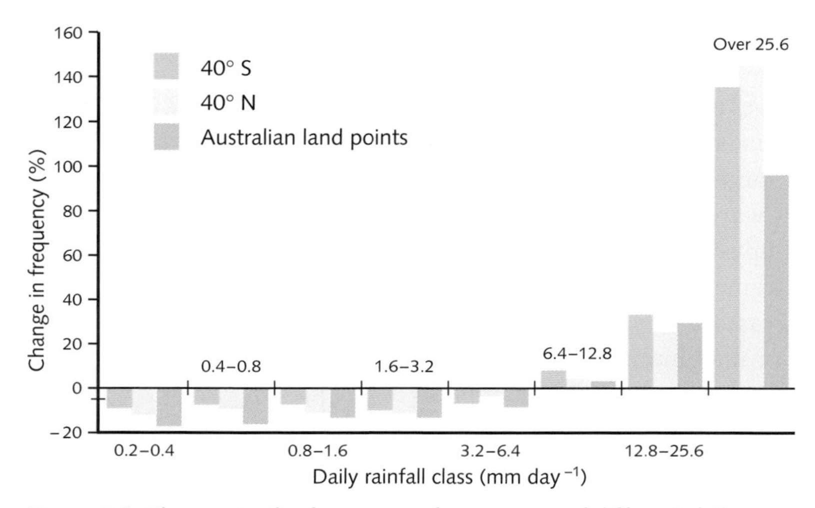

Figure 6.9 Changes in the frequency of occurrence of different daily rainfall amounts with doubled carbon dioxide as estimated by a CSIRO model in Australia.

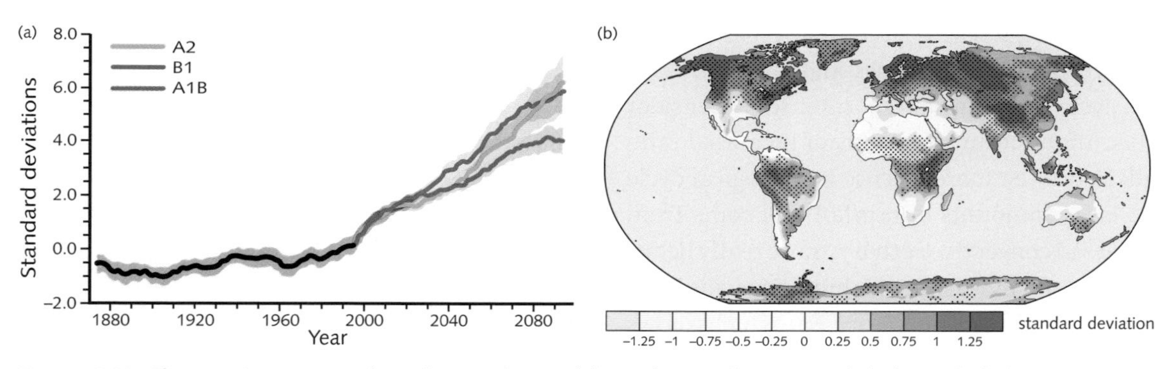

Figure 6.10 Changes in extremes based on multi-model simulations from nine global coupled climate models, adapted from Tebaldi *et al.* (2002). (a) Globally averaged changes in precipitation intensity (defined as the annual total precipitation divided by the number of wet days) for low (SRES B1), middle (SRES A1B) and high (SRES A2) scenarios. (b) Changes of spatial pattern of precipitation intensity based on simulations between two 20-year means (2080–99 minus 1980–99) for the A1B scenario. Solid lines in (a) are the 10-year smoothed multi-model ensemble means, the envelope indicates the ensemble mean standard deviation. Stippling in (b) denotes areas where at least five of the nine models concur in determining that the change is statistically significant. Extreme indices are calculated only over land and are calculated following Frich *et al.* 2002. Because the study focused on analysing the direction and significance of the changes and the degree of inter-model agreement, the indices plotted are shown in units of standard deviation rather than absolute magnitude. Each model's time series was centred around its 1980–99 average and normalised (rescaled) by its standard deviation computed (after detrending) over the period 1960–2099, then the models were aggregated into an ensemble average, both at the global average and the grid-box level.

Drought.

What does this mean in terms of floods and droughts? More intense precipitation means more likelihood of floods. Illustrated in Figure 6.11 is a modelling study showing that if atmospheric carbon dioxide concentration is doubled from its pre-industrial value, the probability of extreme seasonal precipitation in winter is likely to increase substantially over large areas of central and northern Europe. The increase in parts of central Europe is such that the return period of extreme rainfall events would decrease by about a factor of five (e.g. from 50 years to 10 years). Similar results have been obtained in a study of major river basins around the world.[26]

Note also from Figure 6.9 that the number of days with lighter rainfall events (less than 6 mm day^{-1}) is expected to decrease in the globally warmed world. This is because, with the more intense hydrological cycle, a greater proportion of the rainfall will fall in the more intense events and, furthermore, in regions of convection the areas of downdraught become drier as the areas of updraught become more moist and intense. In many areas with relatively low rainfall, the rainfall will tend to become less. Take, for instance, the

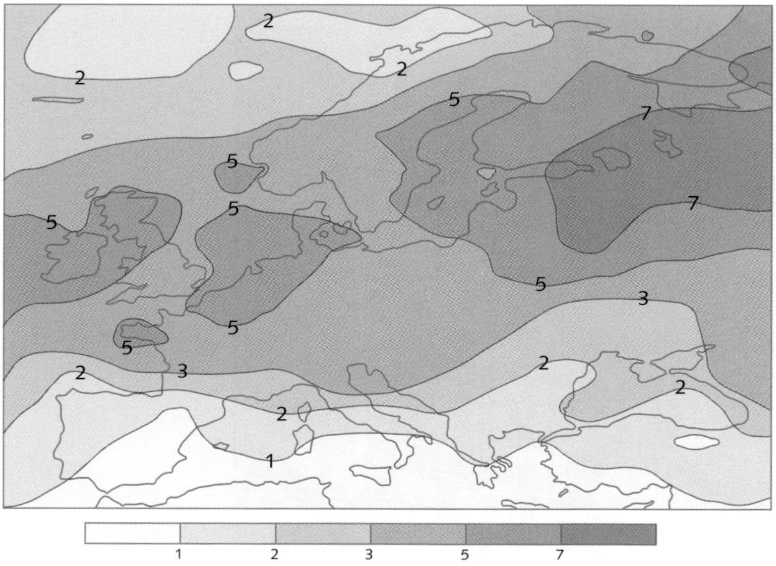

Figure 6.11 The changing probability of extreme seasonal precipitation in Europe in winter as estimated from an ensemble of 19 runs with a climate model starting from slightly different initial conditions. The figure shows the ratio of probabilities of extreme precipitation events in the years 61 to 80 of 80-year runs that assumed an increase of carbon dioxide concentration of 1% per year (hence doubling in about 70 years) compared with control runs with no change in carbon dioxide.

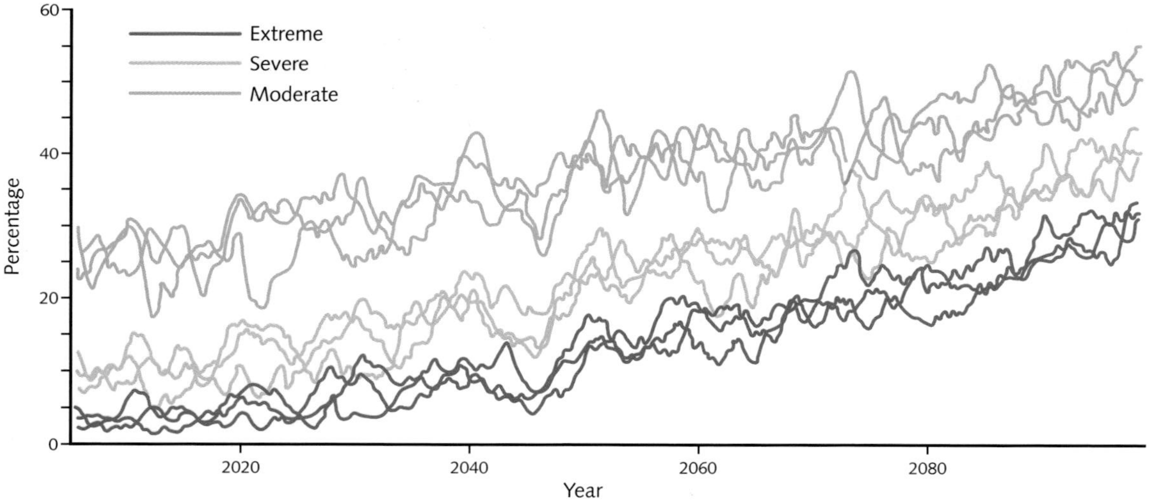

Figure 6.12 Proportion of the world's land surface in drought (extreme, severe and moderate) each month as projected for the twenty-first century by the Hadley Centre climate model. In each case results from three simulations with the A2 emissions scenario are shown.

situation in regions where the average summer rainfall falls substantially – as is likely to occur, for instance, in southern Europe (Figure 6.7). The likely result of such a drop in rainfall is not that the number of rainy days will remain the same, with less rain falling each time; it is more likely that there

The sea surface temperature in August 2005. Orange and red depict regions where the conditions are suitable for hurricanes to form (at 28°C or higher). Hurricane winds are sustained by the heat energy of the ocean and Hurricane Katrina in 2005 caused catastrophic damage along the Gulf coast from Florida to Texas, and in particular in New Orleans, Louisiana

will be substantially fewer rainy days and considerably more chance of prolonged periods of no rainfall at all. Further, the higher temperatures will lead to increased evaporation reducing the amount of moisture available at the surface – thus adding to the drought conditions. The proportional increase in the likelihood of drought is much greater than the proportional decrease in average rainfall.

A recent study of the incidence of drought[27] has employed the Hadley Centre climate model to simulate droughts over all continents, first during the second half of the twentieth century so that confidence in the model could be established by comparing simulations with observed droughts. Droughts are divided into three categories: extreme, severe and moderate.[28] Averaged over the period 1952–98, the percentages of the world's land area at any one time under extreme, severe and moderate drought were 1%, 5% and 20%. By the beginning of the twenty-first century these proportions had risen to 3%, 10% and 28%. Projections for the twenty-first century under the SRES A2 scenario (Figure 6.12) show the proportions of land area under extreme drought rising to over 10% by 2050 and 30% by 2100, the increases occuring not because droughts are much more frequent but much longer in duration. Their study indicates the areas most vulnerable to drought, broadly in agreement with the areas of reduced rainfall indicated in Figure 6.7.

In the warmer world of increased greenhouse gases, therefore, different places will experience more frequent droughts and floods – we noted in Chapter 1 that these are the climate extremes which cause the greatest impacts and will be considered in more detail in the next chapter.

What about other climate extremes, intense storms, for instance? How about hurricanes and typhoons, the violent rotating cyclones that are found over the tropical oceans and which cause such devastation when they hit land? The energy for such storms largely comes from the latent heat of the water which has been evaporated from the warm ocean surface and which condenses in the clouds within the storm, releasing energy. It might be expected that warmer sea temperatures would mean more energy release, leading to more frequent and intense storms. However, ocean temperature is not the only parameter controlling the genesis of tropical storms; the nature of the overall atmospheric flow is also important. Further, although based on limited data, observed variations in the intensity and frequency of tropical cyclones show no clear trends in the last half of the twentieth century. AOGCMs can take all the relevant factors into account but, because of the relatively large size of their grid, they are unable to simulate reliably the detail of relatively small disturbances such as tropical cyclones. From projections with these models there is no consistent evidence of changes in the frequency of tropical cyclones or their areas of formation. However, during the last few years a number of studies with regional models and more adequate resolution with large-scale variables taken from AOGCMs (see next section on regional modelling) project some consistent increases in peak wind intensities and mean and peak precipitation intensities. An indication of the size of the increases is provided from one study that projected an increase in 6% in peak surface wind intensities and 20% increase in precipitation.[29]

Regarding storms at mid latitudes, the various factors that control their incidence are complex. Two factors tend to an increased intensity of storms. The first, as with tropical storms, is that higher temperatures, especially of the ocean surface, tend to lead to more energy being available. The second factor is that the larger temperature contrast between land and sea, especially in the northern hemisphere, tends to generate steeper temperature gradients, which in turn generate stronger flow and greater likelihood of instability. The region around the Atlantic seaboard of Europe is one area where such increased storminess might be expected, a result indicated by some model projections. However, such a picture may be too simple; other models suggest changes in storm tracks that bring very different changes in some regions and there is little overall consistency between model projections.

For other extremes such as very small-scale phenomena (e.g. tornadoes, thunderstorms, hail and lightning) that cannot be simulated in global models, although they may have important impacts, there is currently insufficient information to assess recent trends, and understanding is inadequate to make firm projections.

Table 6.2 summarises the state of knowledge regarding the likely future incidence of extreme events. Although many of the trends are clear, many more studies are required that provide *quantitative* assessments on regional scales of likely changes in the frequency or intensity of extreme events or in climate variability.

Regional climate models

Most of the likely changes that we have presented have been on the scale of continents. Can more specific information be provided about change for smaller regions? In Chapter 5 we referred to the limitation of global circulation models (GCMs) in the simulation of changes on the regional scale arising from the coarse size of their horizontal grid – typically 300 km or more.[30] Also in Chapter 5 we introduced the regional climate model (RCM) which typically possesses a resolution of 50 km and can be 'nested' in a global circulation model. Examples are shown in Figures 6.13 and 7.9 of the improvement achieved by RCMs in the simulation of extremes and in providing regional detail that in many cases (especially for precipitation) shows substantial disagreement with the averages provided by a GCM.

Regional models are providing a powerful tool for the investigation of detail in patterns of climate change. In the next chapter the importance of such detail will be very apparent in studies that assess the impacts of climate change.

Table 6.2 Estimates of confidence in observed and projected changes in extreme weather and climate events

Confidence in observed changes (latter half of the twentieth century)	Changes in phenomenon	Confidence in projected changes (during the twenty-first century)
Very likely[a]	Higher maximum temperatures and more hot days over nearly all land areas	*Very likely*
Very likely	Higher minimum temperatures, fewer cold days and frost days over nearly all land areas	*Very likely*
Very likely	Reduced diurnal temperature range over most land areas	*Very likely*
Likely, over most areas	Increase of heat index[b] over land areas	*Very likely*, over most areas
Likely, over many northern hemisphere mid- to high-latitude land areas	More intense precipitation events	*Very likely*, over most areas
Likely, increases in total area affected over many land areas	Increased summer continental drying and associated risk of drought	*Likely*, in most sub-tropical areas & many mid-latitude continental areas (*very likely*, over Mediterranean, south Australia, New Zealand)
Likely, trends towards greater storm intensity, no trend in frequency	Tropical cyclones	*Likely*, increase in peak wind and precipitation intensities
Likely, net increase in intensity and poleward shift in track over many northern hemisphere land areas	Extra-tropical cyclones	*Likely*, increase in intensity over many areas (e.g. North Atlantic, Central Europe and southern New Zealand)

[a] See Note 1 of Chapter 4 for explanation *of likely, very likely*, etc.

[b] Heat index is a combination of temperature and humidity that measures effects on human comfort.

However, it is important to realise that, even if the models were perfect, because much greater natural variability is apparent in local climate than in climate averaged over continental or larger scales, projections on the local and regional scale are bound to be more uncertain than those on larger scales.

Longer-term climate change

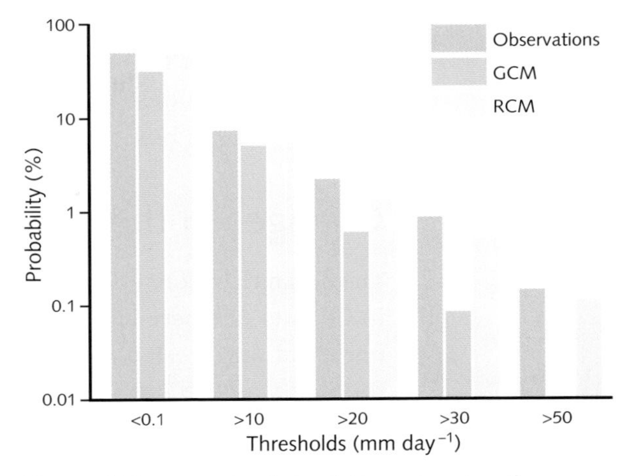

Figure 6.13 Example of simulations showing the probability of winter days over the Alps with different daily rainfall thresholds, as observed, simulated by a 300-km resolution GCM and by a 50-km resolution RCM. Red bars observed, green bars, simulated by GCM; blue bars, simulated by the RCM. The RCM shows much better agreement with observations especially for higher thresholds.

Most of the projections of future climate that have been published cover the twenty-first century. For instance, the curves plotted in Figures 6.1 to 6.6 extend to the year 2100. They illustrate what is likely to occur if fossil fuels continue to provide most of the world's energy needs during that period.

From the beginning of the industrial revolution until 2000 the burning of fossil fuels released approximately 300 Gt of carbon in the form of carbon dioxide into the atmosphere. Under the SRES A1B scenario it is projected that a further 1500 Gt will be released by the year 2100. As Chapter 11 will show, the reserves of fossil fuels in total are sufficient to enable their rate of use to continue to grow well beyond the year 2100. If that were to happen the global average temperature would continue to rise and could, in the twenty-second century, reach very high levels, perhaps up to 10 °C higher than today (see Chapter 9). The associated changes in climate would be correspondingly large and would almost certainly be irreversible.

A further longer-term effect that may become important during this century is that of positive feedbacks on the carbon cycle due to climate change. This was mentioned in Chapter 3 (see box on page 48–9) and the +30% uncertainty in 2100 in the atmospheric concentrations of carbon dioxide shown in Figure 6.2 was introduced to allow for it. Similarly the upper part of the uncertainty bars on the right-hand side of Figure 6.4a for the various SRES marker scenarios makes allowance for this uncertainty – for the A2 scenario in 2100 it amounts to about one degree Celsius. Some of the further implications of this feedback will be considered in Chapter 10 on page 311.

Especially when considering the longer term, there is also the possibility of surprises – changes in the climate system that are unexpected. The discovery of the 'ozone hole' is an example of a change in the atmosphere due to human activities, which was a scientific 'surprise'. By their very nature such 'surprises' cannot, of course, be foreseen. However, there are various parts of the system which are, as yet, not well understood, where such possibilities might be looked for;[31] for instance, in the deep ocean circulation (see box in Chapter 5, page 120) or in the stability of the major ice sheets (see paragraph on climate response, Chapter 5, page 132). In the next section we shall look in more detail at the first of these possibilities; the second will be addressed in the section of the next chapter entitled 'How much will sea level rise?'.

Changes in the ocean thermohaline circulation

The ocean thermohaline circulation (THC) was introduced in the box on page 120 in Chapter 5, where Figure 5.18 illustrates the deep ocean currents that transport heat and fresh water between all the world's oceans. Also mentioned in the box was the influence on the THC in past epochs of the input of large amounts of fresh water from ice melt to the region in the North Atlantic between Greenland and Scandinavia where the main source region for the THC is located.

With climate change due to increasing greenhouse gases we have seen that both temperature and precipitation will increase substantially especially at high latitudes (Figures 6.6 and 6.7), leading to warmer surface water and additional fresh water input to the oceans. Some increased melting of the Greenland ice cap would add further fresh water. The cold dense salty water in the North Atlantic source region for the THC will become less cold and less salty and therefore less dense. As a result the THC will weaken and less heat will flow northward from tropical regions to the north Atlantic. All coupled ocean–atmosphere GCMs show this occurring during the twenty-first century though to a varying extent from a small amount to over 50% change; a typical example is shown in Figure 6.14, which indicates a weakening of about 20% by 2100. Although there is disagreement between the models as to the extent of weakening, all model projections of the pattern of temperature change under increasing greenhouse gases show less warming in the region of the north Atlantic (Figure 6.6) – but none shows actual cooling occurring in this region during the twenty-first century.

The question is often raised as to whether an abrupt transition could occur in the THC or whether it might actually be cut off as has occurred in the past (see Chapter 4, page 89). From model projections so far it is considered very unlikely that an abrupt transition will occur during the twenty-first century. However, the stability of the THC further into the future is bound to be of concern especially if global warming continues largely unchecked and if the rate of meltdown of the Greenland ice cap accelerates. Intense research is being pursued – both observations and modelling – to elucidate further likely changes in the THC and their possible impact.

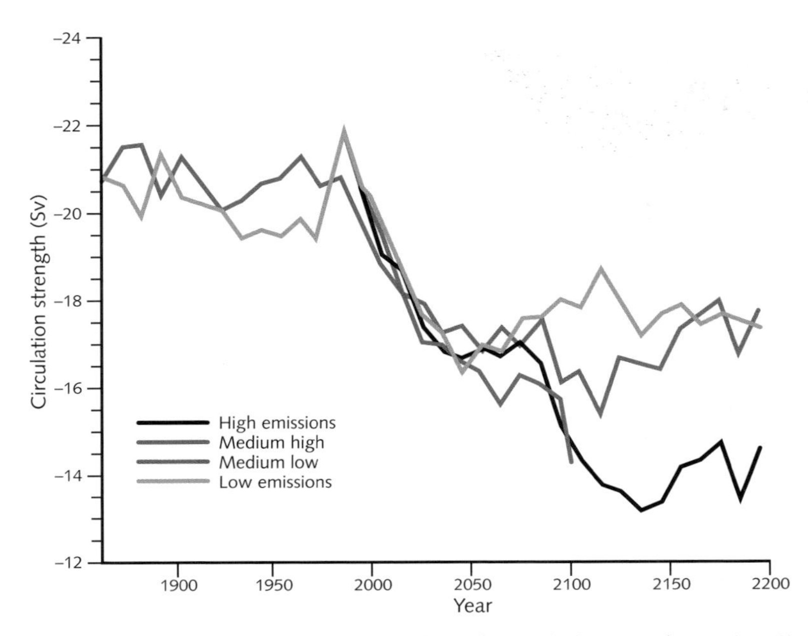

Figure 6.14 Change in the strength of the thermohaline circulation (THC) in the north Atlantic as simulated by the Hadley Centre climate model for four SRES scenarios (A1FI, A2, B2 and B1). The unit of circulation is the Sverdrup, 10^6 m^3 s^{-1}.

Other factors that might influence climate change

So far climate change due to human activities has been considered. Are there other factors, external to the climate system, which might induce change? Chapter 4 showed that it was variations in the incoming solar energy as a result of changes in the Earth's orbit that triggered the ice ages and the major climate changes of the past. These variations are, of course, still going on: what influence are they having now?

Over the past 10 000 years, because of these orbital changes, the solar radiation incident at 60° N in July has decreased by about 35 W m^{-2}, which is quite a large amount. But over 100 years the change is only at most a few tenths of a watt per square metre, which is much less than the changes due to the increases in greenhouse gases (remember that doubling carbon dioxide alters the thermal radiation, globally averaged, by about 4 W m^{-2} – see Chapter 2). Looking to the future and the effect of the Earth's orbital variations, over at least the next 50 000 years the solar radiation incident in summer on the polar regions will be unusually constant so that the present interglacial is expected

Does the Sun's output change?

Some scientists have suggested that all climate variations, even short-term ones, might be the result of changes in the Sun's energy output. Such suggestions are bound to be somewhat speculative because the only direct measurements of solar output that are available are those since 1978, from satellites outside the disturbing effects of the Earth's atmosphere. These measurements indicate a very constant solar output, changing by about 0.1% between a maximum and a minimum in the cycle of solar magnetic activity indicated by the number of sunspots.

It is known from astronomical records and from measurements of radioactive carbon in the atmosphere that solar sunspot activity has, from time to time over the past few thousand years, shown large variations. Of particular interest is the period known as the Maunder Minimum in the seventeenth century when very few sunspots were recorded.[33] At the time of the IPCC TAR in 2001, studies of recent measurements of solar output correlated with other indicators of solar activity, when extrapolated to this earlier period, suggested that the Sun was a little less bright in the seventeenth century, perhaps by about 0.4% in the average solar energy incident on the Earth's surface and that this reduction in solar energy may have been a cause of the cooler period at that time known as the 'Little Ice Age'.[34] More recently some of the assumptions in this work have been questioned and estimates made that over the past two centuries variations in the solar energy incident on the Earth's surface are unlikely to be greater than about 0.1% (Figure 6.15).[35] This is about the same as the change in the energy regime at the Earth's surface due to two or three years' increase in greenhouse gases at the current rate.

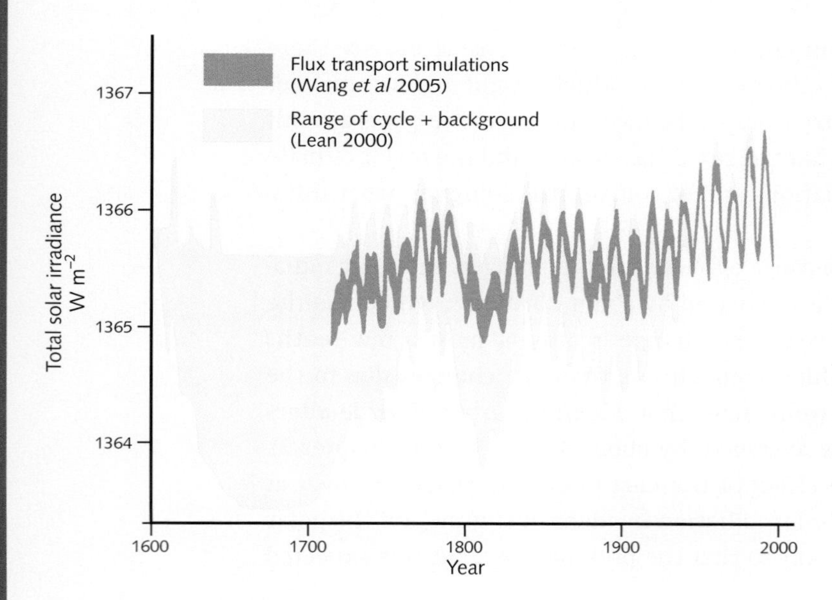

Figure 6.15 Reconstructions of the total solar irradiance from 1600 to the present showing estimates (blue) of the range of the irradiance variations arising from the 11-year solar activity cycle and the period in the seventeenth century when no sunspots were recorded. The lower envelope is the reconstruction by J. Lean, in which the long-term trend was inferred from brightness changes in sun-like stars. The recent reconstruction by Y. Wang *et al.* (purple) is based on solar considerations alone.

to last for an exceptionally long period.[32] Suggestions therefore that the current increase of greenhouse gases might delay the onset of the next ice age are unfounded.

These orbital changes only alter the *distribution* of incoming solar energy over the Earth's surface; the total amount of energy reaching the Earth is hardly affected by them. Of more immediate interest are suggestions that the actual energy output of the Sun might change with time. As I mentioned in Chapter 3 (see Figure 3.11) and as is described in the box, such changes, if they occur, are estimated to be much smaller than changes in the energy regime at the Earth's surface due to the increase in greenhouse gases.

There have also been suggestions of indirect mechanisms whereby effects on the Sun might influence climate on Earth. Changes in solar ultraviolet radiation will influence atmospheric ozone and hence could have some influence on climate. It has been suggested that the galactic cosmic ray flux, modified by the varying Sun's magnetic field, could influence cloudiness and hence climate. Although studies have been pursued on such connections, their influence remains speculative. So far as the last few decades are concerned, there is firm evidence from both observational and theoretical studies that none of these mechanisms could have contributed significantly to the rapid global temperature rise that has been observed.[36]

Another influence on climate comes from volcanic eruptions. Their effects, lasting typically a few years, are relatively short term compared with the much longer-term effects of the increase of greenhouse gases. The large volcanic eruption of Mount Pinatubo in the Philippines which occurred in June 1991 has already been mentioned (Figure 5.21). Estimates of the change in the net amount of radiation (solar and thermal) at the top of the atmosphere resulting from this eruption are of about 0.5 W m^{-2}. This perturbation lasts for about two or three years while the major part of the dust settles out of the atmosphere; the longer-term change in radiation forcing, due to the minute particles of dust that last for longer in the stratosphere, is much smaller.

SUMMARY

- Increase in greenhouse gases is by far the largest of the factors that can lead to climate change during the twenty-first century.
- Likely climate changes for a range of scenarios of greenhouse gas emissions have been described in terms of global average temperature and in terms

of regional change of temperature and precipitation and the occurrence of extremes.
- The rate of change is likely to be larger than any the Earth has seen at any time during the past 10000 years.
- The changes that are likely to have the greatest impact will be changes in the frequencies, intensities and locations of climate extremes, especially droughts and floods.
- Sufficient fossil fuel reserves are available to provide for continuing growth in fossil fuel emissions of carbon dioxide well into the twenty-second century. If this occurred the climate change could be very large indeed and lead to unpredictable features or 'surprises'.

The next chapter will look at the impact of such changes on sea level, on water, on food supplies and on human health. Later chapters of the book will then suggest what action might be taken to slow down and eventually to terminate the rate of change.

QUESTIONS

1 Suggest, for Figure 6.8, an appropriate temperature scale for a place you know. Define what might be meant by a very hot day and estimate the percentage increase in the probability of such days if the average temperature increases by 1, 2 and 4 °C.

2 It is stated in the text describing extremes that in convective regions, with global warming, as the updraughts become more moist the downdraughts tend to be drier. Why is this?

3 Look at the assumptions underlying the full range of SRES emission scenarios in the IPCC 2007 Report. Would you want to argue that some of the scenarios are more likely to occur than the others? Which (if any) would you designate as the most likely scenario?

4 It is sometimes suggested that northwest Europe could become colder in the future while most of the rest of the world becomes warmer. What could cause this and how likely do you think it is to occur?

5 How important do you consider it is to emphasise the possibility of 'surprises' when presenting projections of likely future climate change?

6 Estimate the effect, on the projected carbon dioxide concentrations for 2100 shown in Figure 6.2, the projected radiative forcing for 2100 shown in Table 6.1 and the projected temperatures for 2100 shown in Figure 6.4a, of

assuming the climate feedback on the carbon cycle illustrated in Figure 3.5 (note: first turn the accumulated atmospheric carbon in Figure 3.5 into an atmospheric concentration).

7 From newspapers or websites look up articles purporting to say that there is no human-induced climate change or that what there is does not matter. Assess them in the light of this and other chapters in this book. Do you think any of their arguments are credible?

FURTHER READING AND REFERENCE

Solomon, S., Qin, D., Manning, M., Chen, Z., Marquis, M., Averyt, K. B., Tignor, M., Miller, H.L. (eds.) 2007. *Climate Change 2007: The Physical Science Basis. Contribution of Working Group I to the Fourth Assessment Report of the Intergovernmental Panel on Climate Change*. Cambridge: Cambridge University Press.

Technical Summary (summarises basic information about future climate projections)

Chapter 10 Global Climate Projections

Chapter 11 Regional Climate Projections

Nakicenovic, N. *et al.* (eds.) 2000. *IPCC Special Report on Emissions Scenarios* Cambridge: Cambrige University Press.

WMO/UNEP, 2007. *Climate Change 2007, IPCC Synthesis Report*, www.ipcc.ch

McGuffie, K., Henderson-Sellers, A. 2005. *A Climate Modeling Primer*, third edition. New York: Wiley.

Palmer, T. Hagedorn, R. (eds.) 2006. *Predictability of Weather and Climate*. Cambridge: Cambridge University Press.

Schnellnhuber, H.J. *et al.* (eds.) 2006. *Avoiding Dangerous Climate Change*. Cambridge: Cambridge University Press.

NOTES FOR CHAPTER 6

1 Nakicenovic, N. *et al.* (eds.) 2000. *Special Report on Emission Scenarios (SRES): A Special Report of the IPCC*. Cambridge: Cambridge University Press.

2 For details of IS 92a see Leggett, J., Pepper, W. J., Swart, R. J. 1992. Emission scenarios for the IPCC: an update. In Houghton, J. T., Callender, B. A., Varney, S. K. (eds.) *Climate Change 1992: The Supplementary Report to the IPCC Assessments*. Cambridge: Cambridge University Press, pp. 69–95. Small modifications have been made to the IS 92a scenario to take into account developments in the Montreal Protocol.

3 The +30% amounting to an addition of between 200 and 300 ppm to the carbon dioxide concentration in 2100 (see box on carbon feedbacks on page 48).

4 This summary is based on the Summary of SRES in the Summary for policymakers. In Houghton *et al.* (eds.), *Climate Change 2001: The Scientific Basis*, p. 18. Also Summary for policymakers, in Solomon *et al.* (eds.) *Climate Change 2007: The Physical Science Basis*.

5 The World Energy Council Report. 1995 *Global Energy Perspectives to 2050 and Beyond*. London: World Energy Council projects at 2050 global sulphur emissions

that are little more than half the 1990 levels. Also see United Nations Environmental Programme. 2007. *Global Environmental Outlook GEO4*, Nairobi, Kenya: UNEP, Chapter 9, pp. 435–445.

6 See Figure 3.11.

7 Metz, B., David, O., Bosch, P., Dave, R., Meyer, L. (eds.) 2007. *Climate Change 2007: Mitigation of Climate Change. Contribution of Working Group III to the Fourth Assessment Report of the Intergovernmental Panel on Climate Change.* Cambridge: Cambridge University Press, Figure 3.12, Chapter 3.

8 Detailed values listed in Ramaswamy, V. *et al.* 2001. Chapter 6, in Houghton *et al.* (eds.), *Climate Change 2001: The Science Basis*; also see Meehl, G.A., Stocker, T.F. *et al.* 2007. Chapter 10, in Solomon *et al.* (eds.) *Climate Change 2007: The Physical Science Basis.*

9 Note that because the response of global average temperature to the increase of carbon dioxide is logarithmic in the carbon dioxide concentration, the increase of global average temperature for doubling of carbon dioxide concentration is the same whatever the concentration that forms the base for the doubling, e.g. doubling from 280 ppm or from 360 ppm produces the same rise in global temperature. For a discussion of 'climate sensitivity' see Cubasch, U., Meehl, G. A. *et al.* 2001. Chapter 9, in Houghton *et al.* (eds.) *Climate Change 2001: The Scientific Basis*, also in Meehl, *et al.*, Chapter 10, in Solomon *et al.* (eds.) *Climate Change 2007: The Physical Science Basis.*

10 *Summary for Policy Makers, ibid.*, p. 9.

11 *Ibid.*

12 James Hansen, Bjerknes Lecture at American Geophysical Union, 17 December 2008 at www.columbia.edu/~jeh1/2008/AGUBjerknes_20081217.pdf.

13 See Harvey, D. D. 1997. An introduction to simple climate models used in the IPCC Second Assessment Report. In *IPCC Technical Paper 2*. Geneva: IPCC.

14 The assumption that greenhouse gases may be treated as equivalent to each other is a good one for many purposes. However, because of the differences in their radiative properties, accurate modelling of their effect should treat them separately. More details of this problem are given in Gates, W. L. *et al.* 1992. Climate modelling, climate prediction and model validation. In Houghton *et al.* (eds.) *Climate Change 1992: The Supplementary Report*, pp. 171–5. Also see Forster, P., Ramaswamy, V. *et al.*, Chapter 2, Section 2.9, in Solomon *et al.* (eds.) *Climate Change 2007: The Physical Science Basis.*

15 Alternatively gases other than carbon dioxide can be converted to equivalent amounts of carbon dioxide by using their Global Warming Potentials (see Chapter 3 and Table 10.2).

16 The relationship between radiative forcing R and CO_2 concentration C is $R = 5.3\ln(C/C_0)$ where C_0 = the pre-industrial concentration of 280 ppm.

17 See for instance, Metz *et al.* (eds.) *Climate Change 2007: Mitigation*, Chapter 3, Fig 3.12. For more information about aerosol assumptions for the twenty-first century and how aerosol forcing is treated in models see Johns, T.C. *et al.* 2003. *Climate Dynamics*, **20**, 583–612.

18 Related through the Clausius–Clapeyron equation, $e^{-1}\, de/dT = L/RT^2$, where e is the saturation vapour pressure at temperature T, L the latent heat of evaporation and R the gas constant.

19 Allen, M. R., Ingram, W. J. 2002. *Nature*, **419**, 224–32.

20 Christensen, J.H., Hewitson, B. *et al.* 2007. Regional climate projections. Chapter 11, Executive Summary, in Solomon *et al.* (eds.) *Climate Change 2007: The Physical Science Basis.*

21 For more on this see Palmer, T. N. 1993. *Weather*, **48**, 314–25; and Palmer, T. N. 1999. *Journal of Climate*, **12**, 575–91.

22 See Figure 10.16 in Meehl, *et al.*, Chapter 10, in Solomon *et al.* (eds.) *Climate Change 2007: The Physical Science Basis.*

23 For a review of the science of extreme events see Mitchell, J.F.B., *et al*, 2006, *Philosophical Transactions of the Royal Society A*, **364**, 2117–33; also see Meehl, *et al.*, Chapter 10 in Solomon *et al.* (eds.) *Climate Change 2007: The Physical Science Basis.*

24 For a more detailed discussion of the effect of global warming on the hydrological cycle, see Allen, M. R., Ingram, W. J. 2002.

25 Tebaldi, C. *et al.* 2002. *Climatic Change*, **79**, 185–211.

26 Milly, P. C. D. *et al.* 2002. Increasing risk of great floods in a changing climate. *Nature*, **415**, 514–17;

see also Meehl *et al.*, in Solomon *et al.* (eds.) *Climate Change 2007: The Physical Science Basis.*

27 Burke, E. J., Brown, S. J. Christidis, N. 2006. Modeling the recent evolution of global drought and projections for the 21st century with the Hadley Centre climate model. *Journal of Hydrometrology*, **7**, 1113–25.

28 Defined with relation to the Palmer Drought Severity Index.

29 Knutson, T.R., Tuleya, R.E. 2004. *Journal of Climate*, **17**, 3477–95. See also Meehl, *et al.*, in Solomon *et al.* (eds.) *Climate Change 2007: The Physical Science Basis.*

30 For definition of continental and regional scales see Note 28 in Chapter 5.

31 See also Table 7.4.

32 Berger, A., Loutre, M. F. 2002. *Science*, **297**, 1287–8.

33 Studies of other stars are providing further information; see Nesme-Ribes, E. *et al.* 1996. *Scientific American*, **August**, 31–6.

34 Lean, J. 2000 Evolution of the Sun's spectral irradiance since the Maunder Minimum. *Geophysics Research Letters*, **27**, 2425–8.

35 Wang Y. *et al* 2005. Modeling the Sun's magnetic field and irradiance since 1713, *Astrophysical Journal*, **625**, 522–38.

36 Lockwood M., Frohlich, C. 2007. Recent oppositely directed trends in solar climate forcings and the global mean surface air temperature. *Proceedings of the Royal Society A* doi:10.1098/rspa.2007.1880.

7 The impacts of climate change

Droughts in the Masai region of Africa.

THE LAST two chapters have detailed the climate change in terms of temperature and rainfall that we can expect during the twenty-first century because of human activities. To be useful to human communities, these details need to be turned into descriptions of the impact of climate change on human resources and activities. The questions to which we want answers are: how much will sea level rise and what effect will that have?; how much will water resources be affected?; what will be the impact on agriculture and food supply?; will natural ecosystems suffer damage?; how will human health be affected? and can the cost of the likely damage be estimated? This chapter considers these questions.[1]

A complex network of changes

In outlining the character of the likely climate change in different regions of the world, the last chapter showed that it is likely to vary a great deal from place to place. For instance, in some regions precipitation will increase, in other regions it will decrease. Not only is there a large amount of variability in the character of the likely change, there is also variability in the sensitivity (for definition see box below) of different systems to climate change. Different ecosystems, for instance, will respond very differently to changes in temperature or precipitation.

There will be a few impacts of the likely climate change that will be positive so far as humans are concerned. For instance, in parts of Siberia, Scandinavia or northern Canada increased temperature will tend to lengthen the growing season with the possibility in these regions of growing a greater variety of crops. Also, in winter there will be lower mortality and heating requirements. Further, in some places, increased carbon dioxide will aid the growth of some types of plants leading to increased crop yields.

However, because, over centuries, human communities have adapted their lives and activities to the present climate, most changes in climate will tend to produce an adverse impact. If the changes occur rapidly, quick and possibly costly adaptation to a new climate will be required by the affected

Sensitivity, adaptive capacity and vulnerability: some definitions[2]

Sensitivity is the degree to which a system is affected, either adversely or beneficially, by climate-related stimuli. These encompass all the elements of climate change, including mean climate characteristics, climate variability, and the frequency and magnitude of extremes. This may be direct (e.g. a change in crop yield in response to a change in the mean, range or variability of temperature) or indirect (e.g. damage caused by an increase in the frequency of coastal flooding due to sea level rise).

Adaptive capacity is the ability of a system to adjust to climate change (including climate variability and extremes), to moderate potential damage, to take advantage of opportunities or to cope with the consequences.

Vulnerability is the degree to which a system is susceptible to, or unable to cope with, adverse effects of climate change, including climate variability and extremes. Vulnerability is a function of the character, magnitude and rate of climate change and also the extent to which a system is exposed, its sensitivity and its adaptive capacity.

Both the magnitude and the rate of climate change are important in determining the sensitivity, adaptability and vulnerability of a system.

community. An alternative might be for the affected community to migrate to a region where less adaptation would be needed – a solution that has become increasingly difficult or, in some cases, impossible in the modern crowded world.

As we consider the questions posed at the start of this chapter, it will become clear that the answers are far from simple. It is relatively easy to consider the effects of a particular change (in say, sea level or water resources) supposing nothing else changes. But other factors will change. Some adaptation, for both ecosystems and human communities, may be relatively easy to achieve; in other cases, adaptation may be difficult, very costly or even impossible. In assessing the effects of global warming and how serious they are, allowance must be made for response and adaptation. The likely costs of adaptation also need to be put alongside the costs of the losses or impacts connected with global warming.

Sensitivity, adaptive capacity and vulnerability (see box above) vary a great deal from place to place and from country to country. In particular, developing countries, especially the least developed countries, have less capacity to adapt than developed countries, which contributes to the relative high vulnerability to damaging effects of climate change in developing countries.

The assessment of the impacts of global warming is also made more complex because global warming is not the only human-induced environmental problem. For instance, the loss of soil and its impoverishment (through poor agricultural practice), the over-extraction of groundwater and the damage due to acid rain are examples of environmental degradations on local or regional scales that are having a substantial impact now.[3] If they are not corrected they will tend to exacerbate the negative impacts likely to arise from global warming. For these reasons, the various effects of climate change so far as they concern human communities and their activities will be put in the context of other factors that might alleviate or exacerbate their impact.

The assessment of climate change impacts, adaptations and vulnerability draws on a wide range of physical, biological and social science disciplines and consequently employs a large variety of methods and tools. It is therefore necessary to integrate information and knowledge from these diverse disciplines; the process is called Integrated Assessment (see box in Chapter 9 on page 280).

Table 7.1 summarises some expected impacts for different increases in global average temperature that might occur during the twenty-first century. The following paragraphs consider detail of the various impacts in turn and then bring them together in a consideration of the overall impact.

Table 7.1 Examples of global impacts projected for changes in climate associated with different increases in global average surface temperature in the twenty-first century. Add 0.6 °C to obtain temperature increases from pre-industrial times. Also shown are the projections of temperature increases associated with SRES scenarios as in Figure 6.4. Adaptation to climate change is not included in these estimates.

SRES: AR4 WG I multiple sources	2020s, 2050s, 2080s, 2090s
B1	
B2	
A1B	
A2	5.4

	Impacts
WATER	Increased water availability in moist tropics and high latitudes Decreasing water availability and increasing drought in mid-latitudes and semi-arid low latitudes 0.4 to 1.7 billion → 1.0 to 2.0 billion → 1.1 to 3.2 billion → Additional people with increased water stress
ECOSYSTEMS	Increasing amphibian extinction → About 20 to 30% species at increasingly high risk of extinction → Major extinctions around the globe Increased coral bleaching → Most coral bleached → Wildspread coral mortality Increasing species range shifts and wildfire risk → Terrestrial biosphere tends toward a net carbon source, as: −15% → −40% of ecosystems affected
FOOD	Crop productivity: Low latitudes: Decreases for some cereals → All cereals decrease Mid to high latitudes: Increases for some cereals → Decreases in some regions
COAST	Increased damage from floods and storms About 30% loss of coastal wetlands Additional people at risk of coastal flooding each year: 0 to 3 million → 2 to 15 million
HEALTH	Increasing burden from malnutrition, diarrhoeal, cardio-respiratory and infectious diseases Increased morbidity and mortality from heatwaves, floods and droughts Changed distribution of some disease vectors → Substantial burden on health services
SINGULAR EVENTS	Local retreat of ice in Greenland and West Antarctic → Long term commitment to several metres of sea-level rise due to ice sheet loss → Leading to reconfiguration of coastlines world wide and inundation of low-lying areas Ecosystem changes due to weakening of the meridional overturning circulation

0 1 2 3 4 5°

Global mean annual temperature change relative to 1980–99 (°C)

How much will sea level rise?

There is plenty of evidence for large changes in sea level during the Earth's history. For instance, during the warm period before the onset of the last ice age, about 120 000 years ago, the global average temperature was a little warmer than today (Figure 4.6). Average sea level then was about 5 or 6 m higher than it is today. When ice cover was at its maximum towards the end of the ice age, some 18 000 years ago, sea level was over 100 m lower than today, sufficient, for instance, for Britain to be joined to the continent of Europe.

The main cause of the large sea level changes was the melting or growth of the large ice-sheets that cover the polar regions. The low sea level before 18 000 years ago was due to the amount of water locked up in the large extension of the polar ice-sheets. In the northern hemisphere these extended in Europe as far south as southern England and in North America to south of the Great Lakes. Also the 5 or 6 m higher sea level during the last warm interglacial period resulted from a reduction in the Antarctic and Greenland ice-sheets. But changes over shorter periods are largely governed by other factors that combine to produce a significant effect on the average sea level.

During the twentieth century observations show that the average sea level rose by about 20 cm.[4] The largest contribution to this rise is from thermal expansion of ocean water; as the oceans warm the water expands and the sea level rises (see box below). Other significant contributions come from the melting of glaciers and as a result of long-term adjustments that are still occurring because of the removal of the major ice-sheets after the end of the last ice age. Contributions from the ice caps of Greenland and Antarctica are relatively small. A further small contribution to sea level change arises from changes in terrestrial storage of water, for instance from the growth of reservoirs or irrigation.

Since around 1990 much improved observations of changes of sea level with global coverage have become possible through satellite-borne altimeters that can measure with great accuracy the height of the sea surface at any location. In Figure 7.1 are shown the largest contributions to sea level rise for the periods 1961–2003 and 1993–2003 as estimated from climate models and indicates that these contributions when summed show good agreement with observations, providing some confidence in the modelling methods. The figure also illustrates the substantial increase in the rate of sea level rise, especially due to thermal expansion, experienced during the most recent decade, 1993–2003.

The same methods and models used to estimate twentieth-century sea level trends have been applied to provide estimates of the sea level rise during the twenty-first century. An example for SRES scenario A1B is shown in Figure 7.2 with an uncertainty range (5% to 95% probability) for the decade 2090–99

Thermal expansion of the oceans

A large component of sea level rise is due to thermal expansion of the oceans. Calculation of the precise amount of expansion is complex because it depends critically on the water temperature. For cold water the expansion for a given change of temperature is small. The maximum density of sea water occurs at temperatures close to 0 °C; for a small temperature rise at a temperature close to 0 °C, therefore, the expansion is negligible. At 5 °C (a typical temperature at high latitudes), a rise of 1 °C causes an increase of water volume of about 1 part in 10000 and at 25 °C (typical of tropical latitudes) the same temperature rise of 1 °C increases the volume by about 3 parts in 10000. For instance, if the top 100 m of ocean (which is approximately the depth of what is called the mixed layer) was at 25 °C, a rise to 26 °C would increase its depth by about 3 cm.

A further complication is that not all the ocean changes temperature at the same rate. The mixed layer fairly rapidly comes into equilibrium with changes induced by changes in the atmosphere. The rest of the ocean changes comparatively slowly (the whole of the top kilometre will, for instance, take many decades to warm); some parts may not change at all. Therefore, to calculate the sea level rise due to thermal expansion – its global average and its regional variations – it is necessary to employ the results of an ocean climate model, of the kind described in Chapter 5.[5]

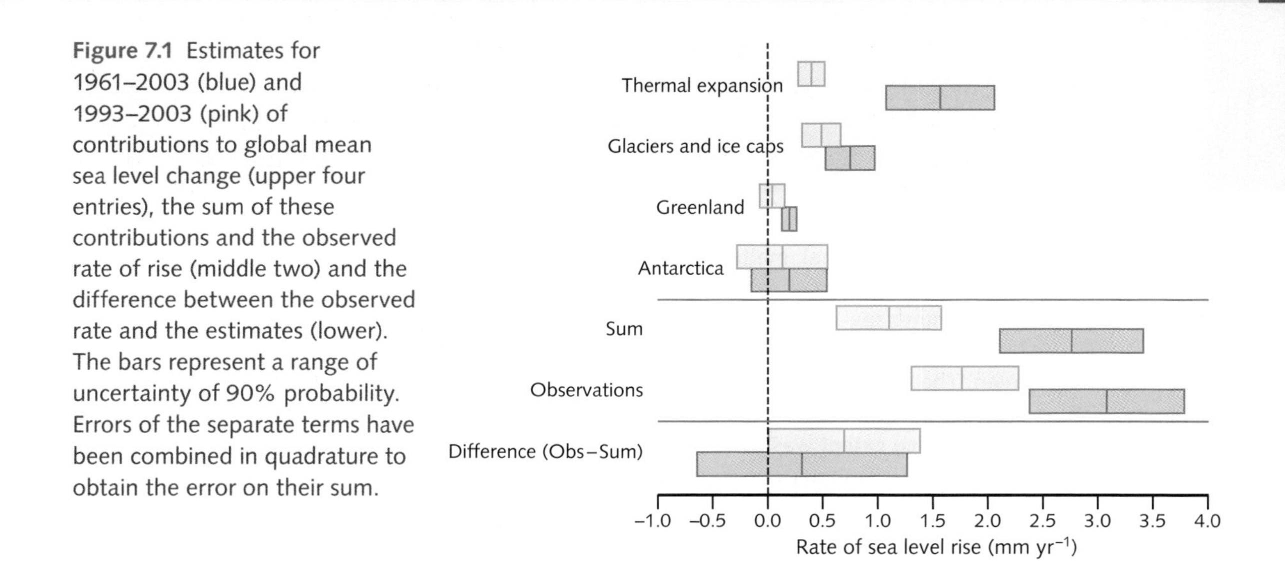

Figure 7.1 Estimates for 1961–2003 (blue) and 1993–2003 (pink) of contributions to global mean sea level change (upper four entries), the sum of these contributions and the observed rate of rise (middle two) and the difference between the observed rate and the estimates (lower). The bars represent a range of uncertainty of 90% probability. Errors of the separate terms have been combined in quadrature to obtain the error on their sum.

relative to 1980–99 of 0.21–0.48 m. Apart from a smaller uncertainty range this shows little change from the 0.12–0.7 m range for the A1B scenario given in the IPCC 2001 Report. The overall range covering the six SRES marker scenarios is from 0.18 to 0.59 m. The largest contribution to sea level rise (0.13–0.32 m for A1B) is expected to continue to be from thermal expansion of ocean water

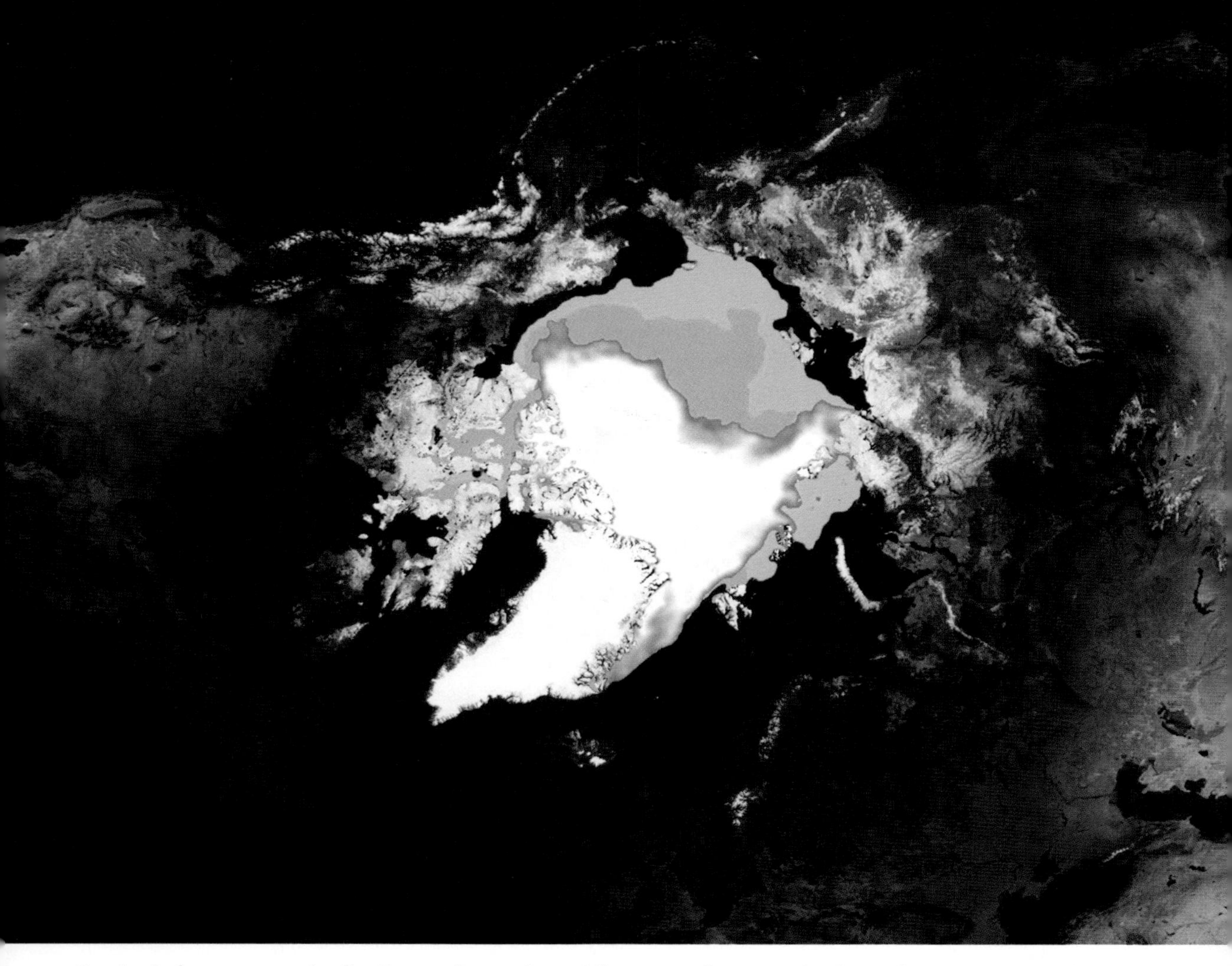

Sea-ice is frozen sea water floating on the surface of the ocean. Some sea-ice is semi-permanent, persisting from year to year, and some is seasonal, melting and refreezing from season to season. The sea-ice cover reaches its minimum extent at the end of each summer and the remaining ice is called the perennial ice cover. The 2007 Arctic summer sea-ice (white) reached the lowest extent of perennial ice cover on record – nearly 25% less than the previous lowest in 2005 (orange). The average minimum sea-ice from 1979 to 2007 is shown in green. The area of perennial ice has been steadily decreasing since the satellite record began in 1979, at a rate of about 10% per decade. Such a dramatic loss has implications for ecology, climate and industry. The 2008 sea-ice extent was even less than that of 2007. With this increasingly rapid rate of change, it is possible that Arctic summer sea-ice could reduce to zero by 2020.

calculated in detail from the ocean component of climate models. The second largest (0.08–0.16 m for A1B) is expected from the melting of glaciers and ice caps outside Greenland and Antarctica. It is derived from estimates of their mass balance – the difference between the amount of snowfall on them (mainly in winter) and melting (mainly in summer); both winter snowfall and average summer temperature are critical and have to be carefully estimated using climate models.

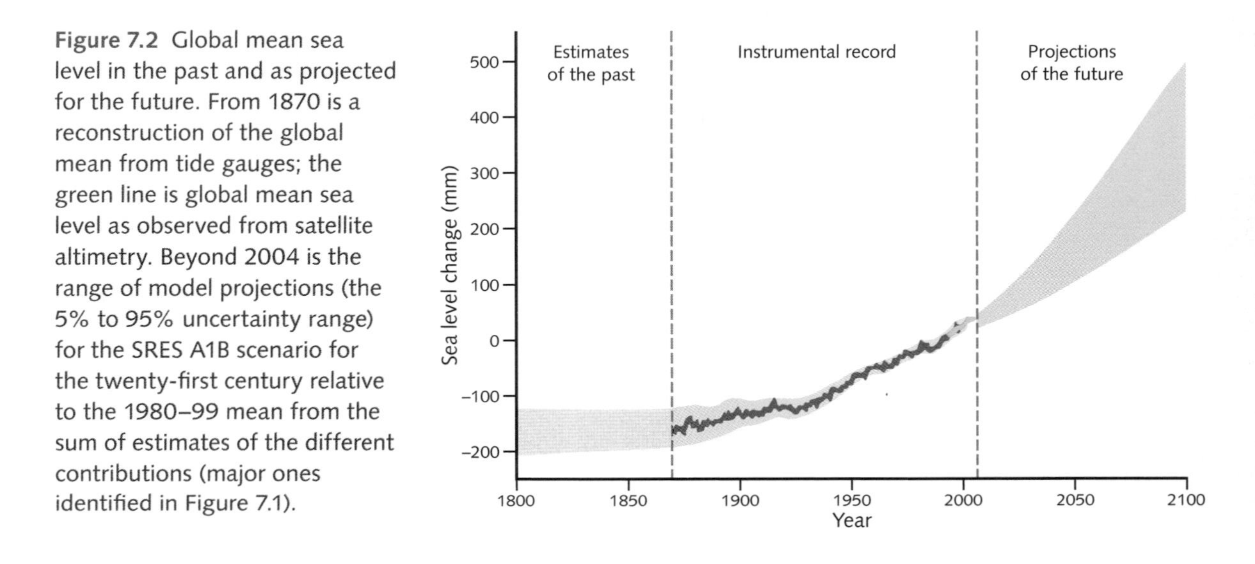

Figure 7.2 Global mean sea level in the past and as projected for the future. From 1870 is a reconstruction of the global mean from tide gauges; the green line is global mean sea level as observed from satellite altimetry. Beyond 2004 is the range of model projections (the 5% to 95% uncertainty range) for the SRES A1B scenario for the twenty-first century relative to the 1980–99 mean from the sum of estimates of the different contributions (major ones identified in Figure 7.1).

The third largest contribution is expected from changes in the Antarctic and Greenland ice-sheets. It is perhaps surprising that the net contribution from them over the twentieth century is small and is still not large (Figure 7.1). For both ice-sheets there are two competing effects. In a warmer world, there is more water vapour in the atmosphere which leads to more snowfall. But as surface temperature increases, especially at high latitudes, there is also more ablation (erosion by melting) of the ice around the boundaries of the ice-sheets where melting of the ice and calving of icebergs occur during the summer months. During the last few decades, both ice-sheets have been close to balance (Figure 7.1). For the twenty-first century, the IPCC AR4 2007 projects that Antarctica will continue close to balance but that for Greenland, ablation will be greater than accumulation leading to a net loss amounting to less than 0.1 m by the end of the century.

During the last year, many pictures have been published of rapid ice melting and discharge from the Greenland ice cap and concern expressed that the rate of melting could accelerate substantially as the warming progresses. This possibility was recognised in IPCC AR4 where it was pointed out that if ice flow from Greenland increased linearly with temperature rise the upper range of sea level rise in 2100 (Figure 7.2) would increase by 0.1 to 0.2 m. However, the rate of increase may accelerate more rapidly. Four indications of this, the first two suggested by P. Christoffersen and M.J. Hambrey[6] and the last two by J. Hansen and his co-authors,[7] are:

(1) Observations from satellite radar altimeters show that the total ice-mass loss from the Greenland ice cap rose from 90 km^3 in 1996 to 140 km^3 in 2000 and 220 km^3 in 2005. Similar losses have been observed from the West

Antarctic ice-sheet. A loss of 400 km^3 per year transfers into a global sea level rise of about 1 mm per year or 0.1 m per century.

(2) Observations of acceleration in the movement of coastal outlet glaciers to now more than 10 km per year. Similar losses and movement are occurring with the West Antarctic ice-sheet.

(3) Observations of increased melt water, from the operation of the ice-albedo feedback (see Chapter 5, page 114), penetrating to the bed of the ice-sheet and through its lubrication enhancing ice motion and instabilities near the ice-sheet base.

(4) Palaeo evidence of periods of rapid melting with associated global sea level rise of up to several metres per century occurring for instance during the recovery from the last ice age about 14 000 years ago.

The possibility of acceleration due to these non-linear processes was acknowledged by IPCC, AR4 although they did not feel able to provide any quantification of their likely size. A careful assessment published in *Science*[8] as this book is going to press concludes with a best estimate, including accelerated conditions of ice melt, of 0.8 m total sea level rise by 2100. It also concludes that a rise of up to 2 m by 2100 'could occur under physically possible glaciological conditions but only if all variables are quickly accelerated to extremely high limits'. Improved quantification should be possible as new satellite data (e.g. from the GRACE gravity mission) is obtained and interpreted.[9]

If we look into the future beyond the twenty-first century, as temperatures around Greenland rise more than 3 °C above their pre-industrial level, model studies show that meltdown of the ice cap will begin; its complete melting will cause a sea level rise of about 7 m. The time taken for meltdown to occur will depend on the amount of temperature rise; estimates for the time to 50% meltdown vary from a few centuries to more than a millennium.

The portion of the Antarctic ice-sheet that is of most concern is that in the west of Antarctica (around 90° W longitude); its disintegration would result in about a 5-m sea level rise. Because a large portion of it is grounded well below sea level it has been suggested that rapid ice discharge could occur if the surrounding ice shelves are weakened. From studies so far of ice dynamics and flow there is no agreement that rapid disintegration is likely although, as with Greenland, it is recognised that the possibility exists.

The projections in Figure 7.2 apply to the next 100 years. During that period, because of the slow mixing that occurs throughout much of the oceans, only a small part of the ocean will have warmed significantly. Sea level rise resulting from global warming will therefore lag behind temperature change at the surface. During the following centuries, as the rest of the oceans gradually warm, sea level will continue to rise at about the same rate, even if the average temperature at the surface were to be stabilised.

The estimates of average sea level rise in Figure 7.2 provide a general guide as to what can be expected during the twenty-first century. Sea level rise, however, will not be uniform over the globe.[10] The effects of thermal expansion in the oceans varies considerably with location. Further, movements of the land occurring for natural reasons due, for instance, to tectonic movements or because of human activities (for instance, the removal of groundwater) can have comparable effects to the rate of sea level rise arising from global warming. At any given place, all these factors have to be taken into account in determining the likely value of future sea level rise.

Impacts in coastal areas

A rise in average sea level of 10 to 20 cm by 2030 and about up to 1 metre by the end of the next century may not seem a great deal. Many people live sufficiently above the level of high water not to be directly affected. However, half of humanity inhabits the coastal zones around the world.[11] Within these, the lowest lying are some of the most fertile and densely populated. To people living in these areas, even half a metre increase in sea level can add enormously to their problems. Their vulnerability is increased by the likelihood of storm surges either due to more intense tropical cyclones or mid latitude storms and by other problems such as local land subsidence and the increased intrusion of salt into groundwater.

Especially vulnerable are large river deltas; in the largest 40 of these in the world over 300 million people live who are increasingly affected by the rate of sea level rise that is occurring even in the absence of climate change (Figure 7.3). Since 1980 over a quarter of a million lives have been lost due to tropical cyclones or storm surges – as I write this paragraph in May 2008, over 100 000 have been lost in a cyclone and storm surge in the Irrawaddy Delta in Myanmar. By 2080, even with only half a metre of sea level rise and no further flood defence, over 100 million in these deltas will be liable to flooding.[12] As an example of a delta area I shall first consider Bangladesh; I shall then consider the Netherlands as an example of an area very close to sea level where sea defences are already in place. Thirdly, I shall look at the plight of small low-lying islands in the Pacific and other oceans.

Bangladesh is a densely populated country of about 150 million people located in the complex delta region of the Ganges, Brahmaputra and Meghna Rivers.[13] About

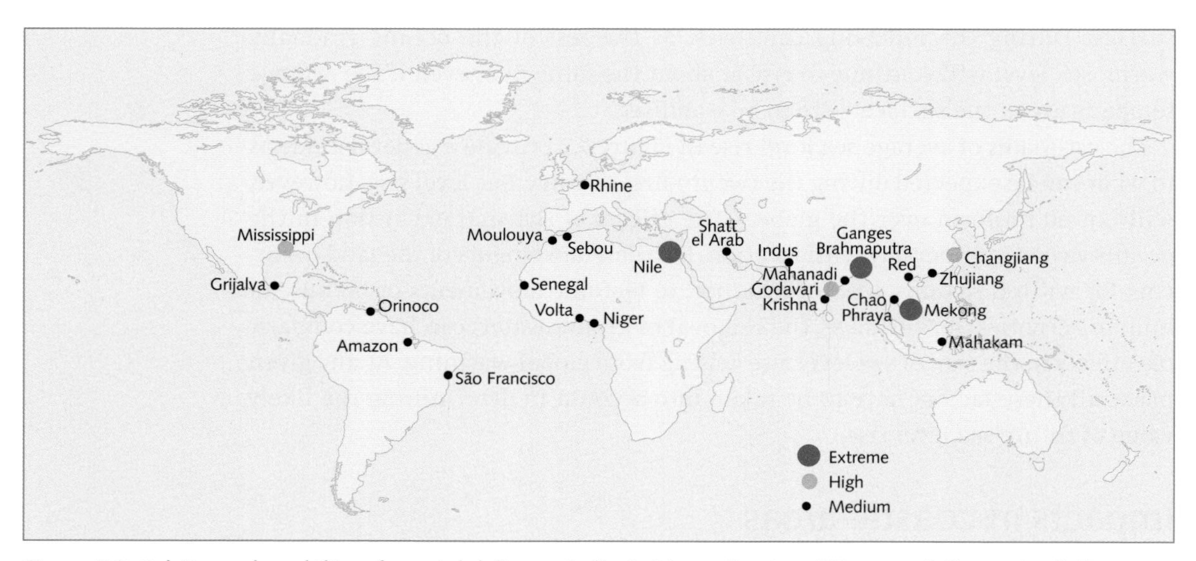

Figure 7.3 Relative vulnerability of coastal deltas as indicated by estimates of the population potentially displaced by current sea level trends to 2050 (extreme, >1 million; high, 1 million to 50000; medium, 50000 to 5000). Climate change would exacerbate these impacts.

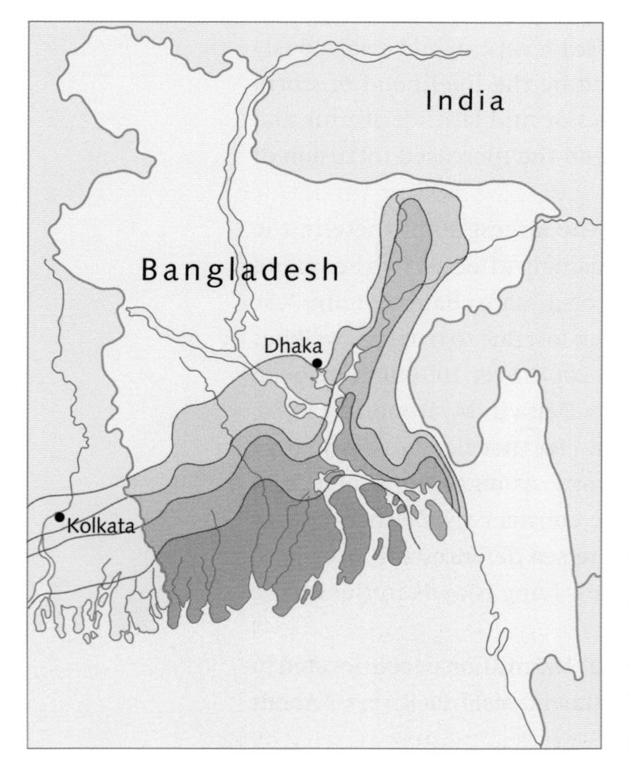

Figure 7.4 Land affected in Bangladesh by various amounts of sea level rise. The 1, 2, 3 and 5 m contours are shown.

10% of the country's habitable land (with about 6 million population) would be lost with half a metre of sea level rise and about 20% (with about 15 million population) would be lost with a 1-m rise (Figure 7.4).[14] Estimates of the sea level rise are of about 1 m by 2050 (compounded of 70 cm due to subsidence because of land movements and removal of groundwater and 30 cm from the effects of global warming) and nearly 2 m by 2100 (1.2 m due to subsidence and 70 cm from global warming)[15] – although there is significant uncertainty in these estimates.

It is quite impractical to consider full protection of the long and complicated coastline of Bangladesh from sea level rise. Its most obvious effect, therefore, is that substantial amounts

Floods help make the cultivable land in Bangladesh fertile and this helps the agriculture sector of the country. But excessive flooding is considered a calamity. The rise in sea water levels, the narrow north tip to the Bay of Bengal, tropical storms that whip up wind speeds of up to 140 mph (225 km h^{-1}) send waves up to 26 feet (8 m tall) crashing into the coast, the shallow sea bed , the fact that water coming down from the Rivers Ganges and Brahmaputra can not escape when the water level rises and the monsoons, all contribute to the severe flooding of the Bangladesh coastline.

of good agricultural land will be lost. This is serious: half the country's economy comes from agriculture and 83% of the nation's population depends on agriculture for its livelihood. Many of these people are at the very edge of subsistence.

But the loss of land is not the only effect of sea level rise. Bangladesh is extremely prone to damage from storm surges. Every year, on average, at least one major cyclone attacks Bangladesh. During the past 25 years there have been two very large disasters with extensive flooding and loss of life. The storm surge in November 1970 is probably the largest of the world's natural disasters in recent times; it is estimated to have claimed the lives of over a quarter of a million people. Well over 100 000 are thought to have lost their lives in a similar storm in April 1991. Even small rises in sea level add to the vulnerability of the region to such storms.

There is a further effect of sea level rise on the productivity of agricultural land; that is, the intrusion of salt water into fresh groundwater resources. At the present time, it is estimated that in some parts of Bangladesh salt water extends seasonally inland over 150 km. With a 1-m rise in sea level, the area affected by saline intrusion could increase substantially although, since it is also likely that climate change will bring increased monsoon rainfall, some of the intrusion of salt water could be alleviated.[16]

What possible responses can Bangladesh make to these likely future problems? Over the timescale of change that is currently envisaged it can be supposed that the fishing industry can relocate and respond with flexibility to changing fishing areas and changing conditions. It is less easy to see what the population of the affected agricultural areas can do to relocate or to adapt. No significant areas of agricultural land are available elsewhere in Bangladesh to replace that lost to the sea, nor is there anywhere else in Bangladesh where the population of the delta region can easily be located. It is clear that very careful study and management of all aspects of the problem is required. The sediment brought down by the rivers into the delta region is of particular importance. The amount of sediment and how it is used can have a large effect on the level of the land affected by sea level rise. Careful management is therefore required upstream as well as in the delta itself; groundwater and sea defences must also be managed carefully if some alleviation of the effects of sea level rise is to be achieved.

A similar situation exists in the Nile Delta region of Egypt. The likely rise in sea level this century is made up from local subsidence and global warming in much the same way as for Bangladesh – approximately 1 m by 2050 and 2 m by 2100. About 12% of the country's arable land with a population of over 7 million people would be affected by a 1-m rise of sea level.[17] Some protection from the sea is afforded by the extensive sand dunes but only up to half a metre or so of sea level rise.[18]

Many other examples of vulnerable delta regions, especially in Southeast Asia and Africa, can be given where the problems would be similar to those in Bangladesh and in Egypt. For instance, several large and low-lying alluvial plains are distributed along the eastern coastline of China. A sea level rise of just half a metre would inundate an area of about 40 000 km^2 (about the area of the Netherlands)[19] where over 30 million people currently live. A particular delta that has been extensively studied is that of the Mississippi in North America. These studies underline the point that human activities and industry are already exacerbating the potential problems of sea level rise due to global warming. Because of river management little sediment is delivered by the river to the delta to counter the subsidence occurring because of long-term movements of the Earth's crust. Also, the building of canals and dykes has inhibited the input of sediments from the ocean.[20] Studies of this kind emphasise the importance of careful management of all activities influencing such regions, and the necessity of making maximum use of natural processes in ensuring their continued viability.

In September 2008 a report by the government-appointed Delta Commission concluded that the Netherlands must spend billions of euros on dyke upgrades and coastal expansion to avoid the ravages of rising sea levels due to global warming over the coming decades

We now turn to the Netherlands, a country more than half of which consists of coastal lowlands, mainly below present sea level. It is one of the most densely populated areas in the world; 8 million of the 14 million inhabitants of the region live in large cities such as Rotterdam, The Hague and Amsterdam. An elaborate system of about 400 km of dykes and coastal dunes, built up over many years, protects it from the sea. Recent methods of protection, rather than creating solid bulwarks, make use of the effects of various forces (tides, currents, waves, wind and gravity) on the sands and sediments so as to create a stable barrier against the sea – similar policies are advocated for the protection of the Norfolk coast in eastern England.[21] Protection against sea level rise next century will require no new technology. Dykes and sand dunes will need to be raised; additional pumping will also be necessary to combat the incursion of salt water into freshwater aquifers. It is estimated[22] that an expenditure of about $US12 000 million would be required for protection against a sea level rise of 1 m.

The third type of area of especial vulnerability is the low-lying small island.[23] Half a million people live in archipelagos of small islands and coral atolls, such as the Maldives in the Indian Ocean, consisting of 1190 individual islands, and the Marshall Islands in the Pacific, which lie almost entirely within 3 m of sea level. Half a metre or more of sea level rise would reduce their areas substantially – some would have to be abandoned – and remove up to 50% of their groundwater. The cost of protection from the sea is far beyond the resources of these islands' populations. For coral atolls, rise in sea level at a rate of up to about half a metre per century can be matched by coral growth, providing that growth is not disturbed by human interference and providing also that the growth is not inhibited by a rise in the maximum sea temperature exceeding about 1–2 °C.[23]

These are some examples of areas particularly vulnerable to sea level rise. Many other areas in the world will be affected in similar, although perhaps less dramatic, ways. Many of the world's cities are close to sea level and are being increasingly affected by subsidence because of the withdrawal of groundwater. The rise of sea level due to global warming will add to this problem. There is no technical difficulty for most cities in taking care of these problems, but the cost of doing so must be included when calculating the overall impact of global warming.

So far, in considering the impact of sea level rise, places of dense population where there is a large effect on people have been considered. There are also areas of importance where few people live. The world's wetlands and mangrove swamps currently occupy an area of about 1 million square kilometres (the figure is not known very precisely), equal approximately to twice the area of France. They contain much biodiversity and their biological productivity equals or exceeds that of any other natural or agricultural system. Over two-thirds of the fish caught for human consumption, as well as many birds and animals, depend on coastal marshes and swamps for part of their life cycles, so they are vital to the total world ecology. Such areas can adjust to slow levels of sea level rise, but there is no evidence that they could keep pace with a rate of rise of greater than about 2 mm per year – 20 cm per century. What will tend to occur, therefore, is that the area of wetlands will extend inland, sometimes with a loss of good agricultural land. However, because in many places such extension will be inhibited by the presence of flood embankments and other human constructions, erosion of the seaward boundaries of the wetlands will lead more usually to a loss of wetland area. Because of a variety of human activities (such as shoreline protection, blocking of sediment sources, land reclamation, aquaculture development and oil, gas and water extraction), coastal wetlands are currently being lost at a rate of 0.5–1.5% per year. Sea level rise because of climate change would further exacerbate this loss.[25]

To summarise the impact of the half metre or more of sea level rise due to global warming which could occur during the twenty-first century: global warming is not the only reason for sea level rise but it is likely to exacerbate the impacts of other environmental problems. Careful management of human activities in the affected areas can do a lot to alleviate the likely effects, but substantial adverse impacts will remain. In delta regions, which are particularly vulnerable, sea level rise will lead to substantial loss of agricultural land and salt intrusion into freshwater resources. In Bangladesh, for instance, over 10 million people are likely to be affected by such loss. A further problem in Bangladesh and other low-lying tropical areas will be the increased intensity and frequency of disasters because of storm surges. Each year, the number of people worldwide experiencing flooding because of storm surges is estimated now at about 40 million. With a 40-cm sea level rise by the 2080s this number is estimated to quadruple – a number that might be reduced by half if coastal protection is enhanced in proportion to gross domestic product (GDP) growth.[26] Low-lying small islands will also suffer loss of land and freshwater supplies. Countries like the Netherlands and many cities in coastal regions will have to spend substantial sums on protection against the sea. Significant amounts of land will also be lost near the important wetland areas of the world. Attempts to put costs against these impacts, in both money and human terms, will be considered later in the chapter.

In this section we have considered the impacts of sea level rise for the twenty-first century. Because, as we have seen, the ocean takes centuries to adjust to an increase in surface temperature, the longer-term impacts of sea level rise also need to be emphasised. Even if the concentrations of greenhouse gases in the atmosphere were stabilised so that anthropogenic climate change is halted, the sea level will continue to rise for many centuries as the whole ocean adjusts to the new climate.

Increasing human use of fresh water resources

The global water cycle is a fundamental component of the climate system. Water is cycled between the oceans, the atmosphere and the land surface (Figure 7.5). Through evaporation and condensation it provides the main means whereby energy is transferred to the atmosphere and within it. Water is essential to all forms of life; the main reason for the wide range of life forms, both plant and animal, on the Earth is the extremely wide range of variation in the availability of water. In wet tropical forests, the jungle teems with life of enormous variety. In drier regions sparse vegetation exists, of a kind that can survive for long periods with the minimum of water; animals there are also well adapted to dry conditions.

Water is also a key substance for humankind; we need to drink it, we need it for the production of food, for health and hygiene, for industry and transport. Humans have learnt that the ways of providing for livelihood can be adapted to a wide variety of circumstances regarding water supply except, perhaps, for the completely dry desert. Water availability for domestic, industrial and agricultural use averaged per capita in different countries varies from less than 100 m^3 (22 000 imperial gallons) per year to over 100 000 m^3 (22 million imperial gallons)[27] – although quoting average numbers of that kind hides the enormous disparity between those in very poor areas who may walk many hours each day to fetch a few gallons and many in the developed world who have access to virtually unlimited supplies at the turn of a tap.

Increases in freshwater use are driven by changes in population, lifestyle, economy, technology and most particularly by demand for food which drives irrigated agriculture. During the last 50 years water use worldwide has grown over threefold (Figure 7.6); it now amounts to about 10% of the estimated global total of the river and groundwater flow from land to sea (Figure 7.5). Two-thirds of human water use is currently for agriculture, much of it for irrigation; about a quarter is used by industry; only 10% or so is used domestically. The demand is so great in some river basins, for instance the Rio Grande and the Colorado in North America, that almost no water from them reaches the sea. Increasingly, water stored over hundreds or thousands of years in underground aquifers is being tapped for current use and there are now many places in the world where groundwater is being used much faster than it is being replenished; every year the water has to be extracted from deeper levels. For instance, over more than half the land area of the United States over a quarter of the groundwater withdrawn is not replenished and around Beijing in China the water table is falling by 2 m a year as groundwater is pumped out. These are just examples of greatly increased vulnerability regarding water supplies that arise from rapid growth of demand.

The extent to which a country is *water stressed* is related to the proportion of the available freshwater supply that is withdrawn for use. In global scale assessments, basins with water stress are defined either as having per capita water availability below 1000 m^3 per year (based on long-term average run-off) or as having a ratio of withdrawals to long-term average annual run-off above 0.4.[28] Under this definition, some 1.5 to 2 billion people, one-quarter to one-third of the world's population, live in water-stressed countries – in parts of Africa, the

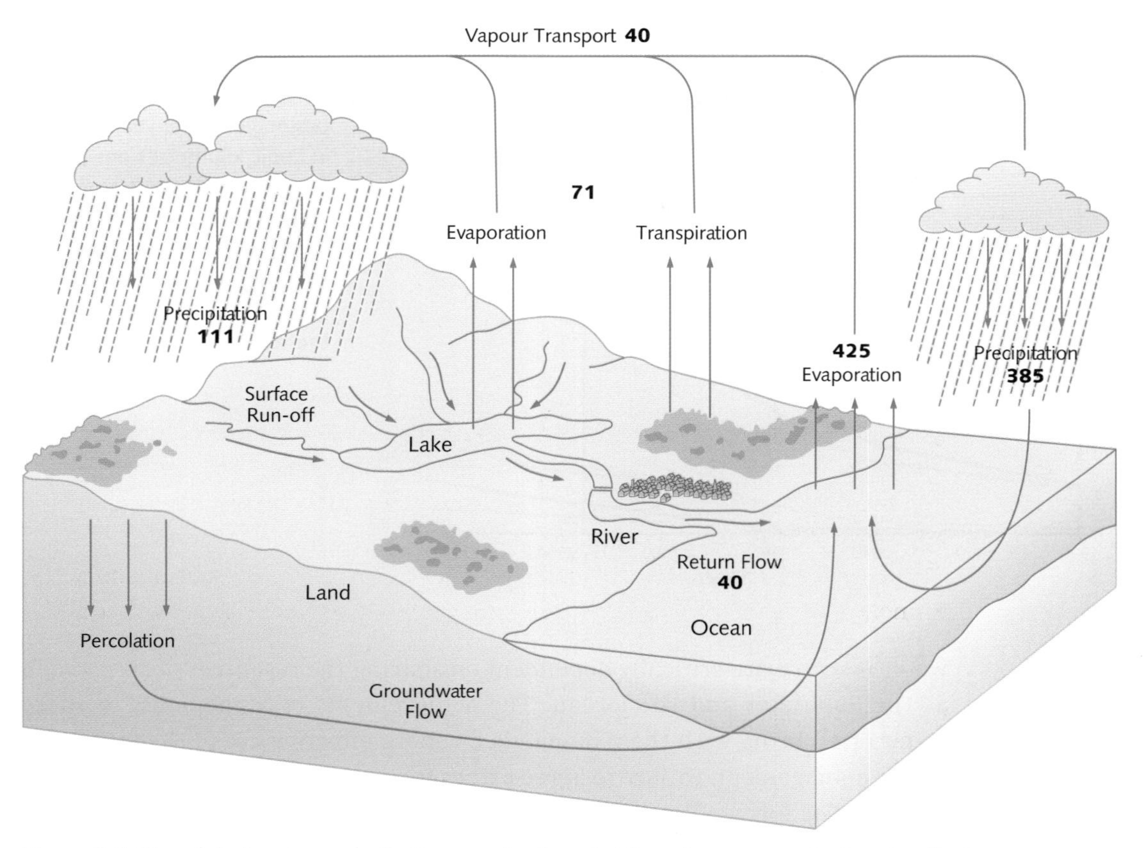

Figure 7.5 The global water cycle (in thousands of cubic kilometres per year), showing the key processes of evaporation, precipitation, transport as vapour by atmospheric movements and transport from the land to the oceans by run-off or groundwater flow.

Mediterranean region, the Near East, South Asia, northern China, Australia, USA, Mexico, northeast Brazil and the western coast of South America. This is illustrated in Figure 7.7 where examples are given of some of the current vulnerabilities of fresh water resources and their management. The number living in water-stressed countries is projected to rise steeply in coming decades even without taking into account any effect on water supplies due to climate change.

A further vulnerability arises because many of the world's major sources of water are shared. About half the land area of the world is within water basins that fall between two or more countries. There are 44 countries for which at least 80% of their land area falls within such international basins. The Danube, for instance, passes through 12 countries that use its water, the Nile water through nine, the Ganges–Brahmaputra through five. Other countries where

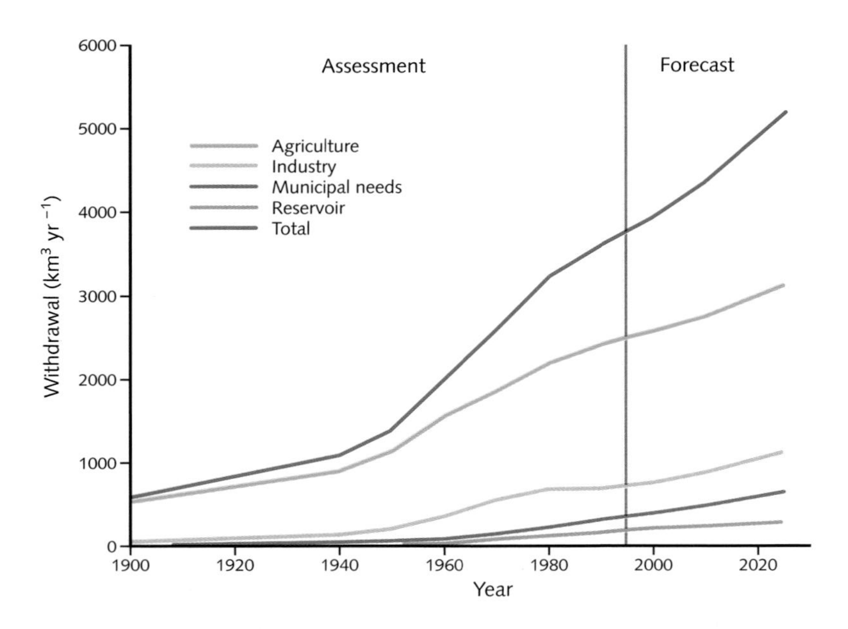

Figure 7.6 Global water withdrawal for different purposes, 1900–95, and projected to the year 2025 in cubic kilometres per year. Losses from reservoirs are also included. As some water withdrawn is reused, the total water consumption amounts to about 60% of the total water withdrawal.

water is scarce are critically dependent on sharing the resources of rivers such as the Euphrates and the Jordan. The achievements of agreements to share water often bring with them demands for more effective use of the water and better management. Failure to agree brings increased possibility of tension and conflict. The former United Nations Secretary General, Boutros Boutros-Ghali, has said that 'The next war in the Middle East will be fought over water, not politics.'[29]

The impact of climate change on fresh water resources

The availability of fresh water will be substantially changed in a world affected by global warming. Although uncertainties remain for projections of precipitation particularly on the regional or even river-basin scale, it is possible to identify many areas where substantial increases or decreases are likely (Figure 6.7). For instance, precipitation is expected to increase in high latitudes and in parts of the tropics and decrease in many mid latitude and sub-tropical regions especially in summer. Further, increase of temperature will mean that a higher proportion of the water falling on the Earth's surface will evaporate. In regions with increased precipitation, some or all of the loss due to evaporation may be made up. However, in regions with unchanged or less precipitation, there will be substantially less water available at the surface. The combined effect of

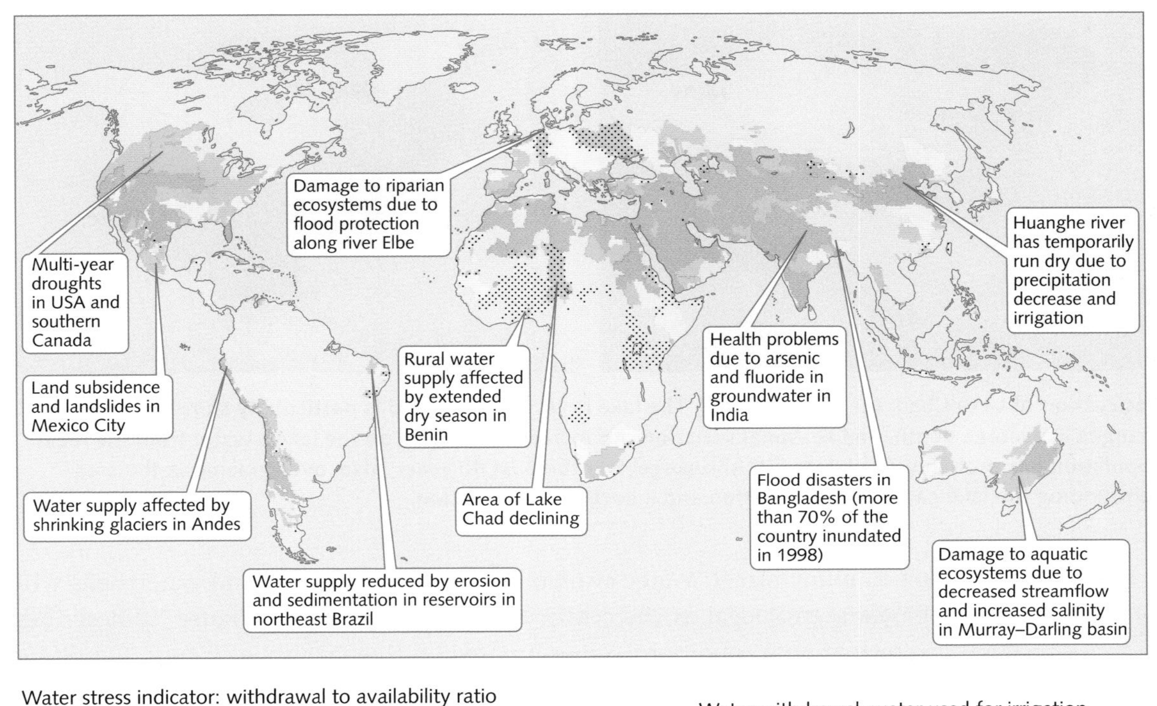

Figure 7.7 Examples of current vulnerabilities of fresh water resources and their management; in the background, a water stress map based on Alcamo *et al.* (2003a). See text for relation to climate change.

less rainfall and more evaporation means less soil moisture available for crop growth and also less run-off – in regions with marginal rainfall this loss of soil moisture can be critical. Although, increased carbon dioxide also tends to reduce plant transpiration and less water use by plants, a factor that requires further investigation.[30]

The run-off in rivers and streams is what is left from the precipitation that falls on the land after some has been taken by evaporation and by transpiration from plants; it is the major part of what is available for human use. The amount of run-off is highly sensitive to changes in climate; even small changes in the amount of precipitation or in the temperature (affecting the amount of evaporation) can have a big influence on it. To illustrate this, Figure 7.8 shows estimates of the mean change in annual run-off between 1980–2000 and 2081–2100 under the SRES A1B scenario. There are changes of up to plus 50% or minus

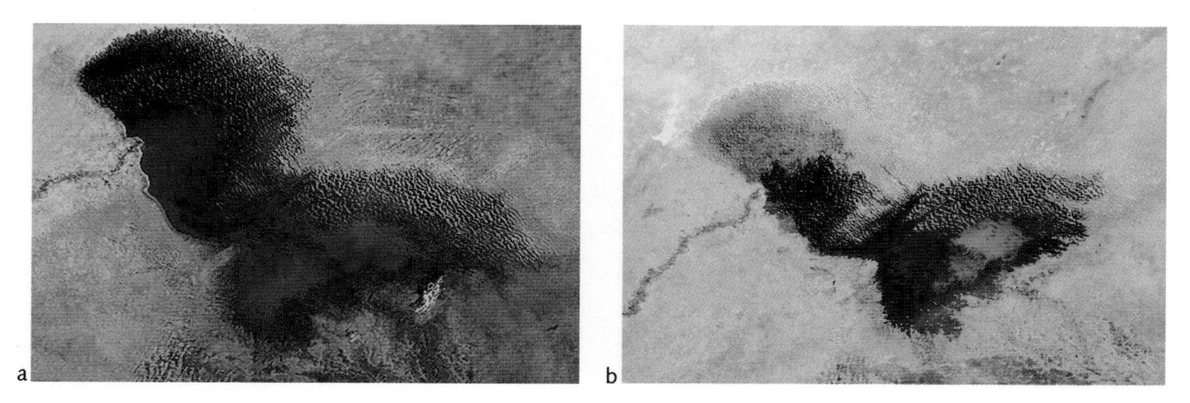

Desiccation of Lake Chad: (a) 1973 (b) 2007. The lake is very shallow and is particularly sensitive to small changes in average depth, and seasonal variation. An increased demand on the lake's water from the local population has probably accelerated its shrinkage over the past 40 years; also, over-grazing in the area surrounding the lake causes desertification and a decline in vegetation.

50% in many places. Water availability in many locations and watersheds will change a great deal as the century progresses. Note that Figure 7.8 describes average annual conditions. Superimposed on the changes of Figure 7.7 will be the variability of climate, in particular the likely increase in the incidence and intensity of climate extremes. The boxes in Figure 7.8 illustrate some particular impacts of expected changes. Eight of the changes of particular concern are the following.[31]

(1) Mentioned earlier in the chapter was that as the rate of warming increases, up to one-half of the mass of mountain glaciers and small ice caps outside the polar regions may melt away over the next hundred years. In fact if current warming rates are maintained it is projected that Himalayan glaciers could decay by 80% of their area by the 2030s. Snow melt is an important source of run-off and watersheds will be severely affected by glacier and snow cover decline. As temperatures rise, winter run-off will initially increase while spring high water, summer and autumn flows will be reduced. Particularly seriously affected will be river basins dependent on the Hindu-Kush–Himalaya glaciated region in Asia (e.g. the Indus, Ganges–Brahmaputra and Yangtze rivers) where more than one-sixth of the world's population currently lives, and those dependent on glaciers in the South American Andes. In many areas, there could be large changes in the seasonal distribution of river flow and water supply for hydroelectric generation and agriculture. For instance, in Europe a decrease in hydropower potential of about 10% has been projected by the 2070s.

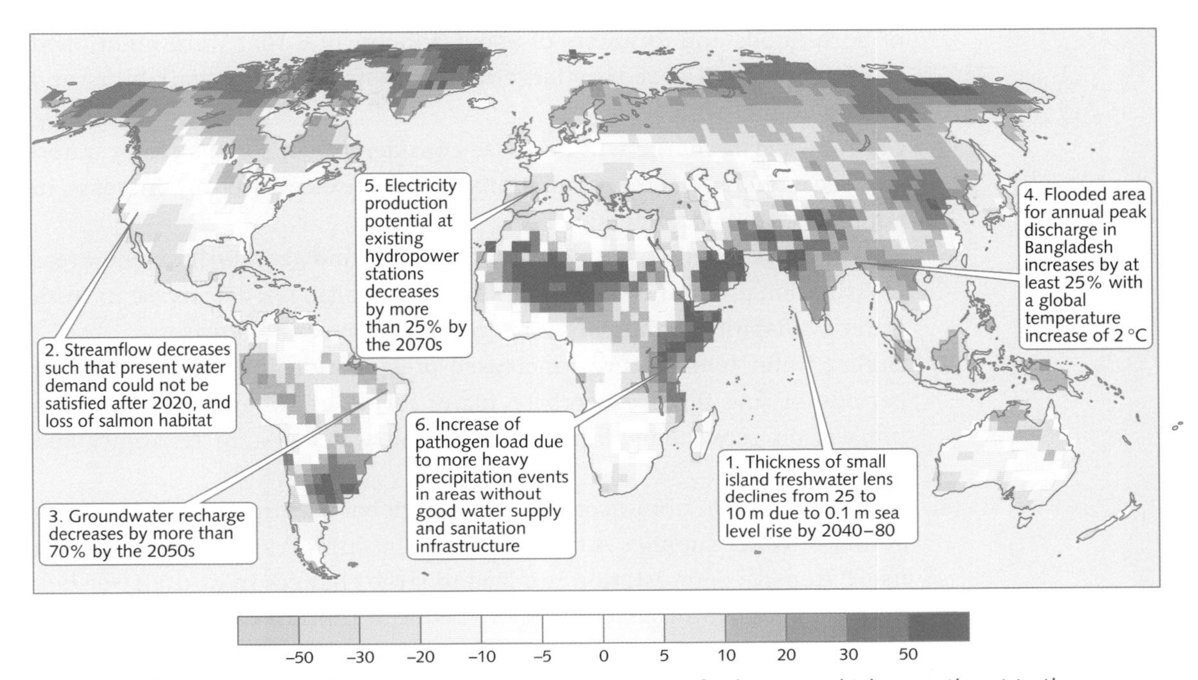

Figure 7.8 Illustrative map of future climate change impacts on fresh water which are a threat to the sustainable development of the affected regions. The background map is of the ensemble mean change in annual run-off in per cent between (1981–2000) and (2081– 2100) for the SRES A1B emissions scenario.

(2) Many semi-arid areas (e.g. Mediterranean basin, western USA, southern Africa, northeastern Brazil and parts of Australia) will suffer serious decreases in water resources due to climate change. These problems will be particularly acute in semi-arid or arid low-income countries, where precipitation and streamflow are concentrated over a few months and where the variability of precipitation is likely to increase as climate changes.

(3) Due to increases in population in addition to climate change, the number of people living in severely stressed river basins is projected to increase from about 1.5 billion in 1995 to 3 to 5 billion in 2050 for the SRES B2 scenario.

(4) As was explained in Chapter 6, the more intense hydrological cycle associated with global warming will lead to increased frequency and intensity of both floods and droughts. As was mentioned in Chapter 1, floods and droughts affect more people across the globe than all other natural disasters and their impact has been increasingly severe in recent decades. Increases by 2050 in many parts of the world in the frequency and severity

of both floods and droughts of about a factor of 5 that were mentioned in Chapter 6 will have very large implications for water availability and management.

(5) Groundwater recharge will decrease considerably in some already water-stressed regions where vulnerability may be exacerbated by increase in population and water demand.

(6) Sea level rise together with greater use of groundwater will extend areas of salination of groundwater and estuaries, resulting in a decrease in fresh water availability for humans and ecosystems in coastal areas.

(7) Higher water temperatures, increased precipitation intensity and longer periods of low flows exacerbate many forms of water pollution, with impacts on ecosystems, human health and water system reliability and operating costs.

(8) A further reason, not unconnected with global warming, for the vulnerability of water supplies is the link between rainfall and changes in land use. Extensive deforestation can lead to large changes in rainfall (see box on page 208). A similar tendency to reduced rainfall can be expected if there is a reduction in vegetation over large areas of semi-arid regions. Such changes can have a devastating and widespread effect and assist in the process of desertification. This is a potential threat to the drylands covering about one-quarter of the land area of the world (see box on page 197).

The monsoon regions of Southeast Asia are an example of an area that may be particularly vulnerable to both floods and droughts. Figure 7.9 shows the predicted change in summer precipitation over the Indian sub-continent as simulated by a regional climate model (RCM) for 2050 under a scenario similar to SRES A1B. Note the improvement in detail of the precipitation pattern that results from the use of the increased resolution of the regional model compared with the global model (GCM), for instance over the Western Ghats (the mountains that rise steeply from India's west coast) there are large increases not present in the global model simulation. The most serious reductions in water availability simulated by the regional model are in the arid regions of northwest India and Pakistan where average precipitation is reduced to less than 1 mm day^{-1} – that coupled with higher temperatures leads to a 60% decline in soil moisture. Substantial increases in average precipitation are projected for areas in eastern India and in flood-prone Bangladesh where the projected increase is about 20%. What is urgently required for this part of the world and elsewhere is much better information linking changes in

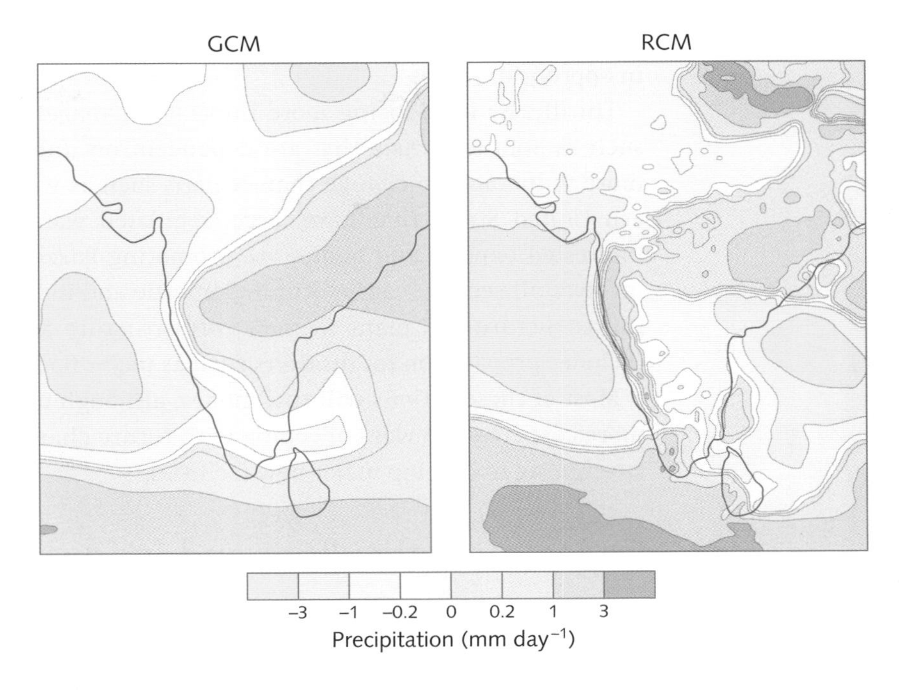

Figure 7.9 Predicted changes in monsoon rainfall (mm day^{-1}) over India between the present day and the middle of the twenty-first century from a 300-km resolution GCM and from a 50-km resolution RCM. The RCM pattern is very different in some respects from the coarser resolution pattern of the GCM.

average parameters with likely changes in frequency, intensity and location of extreme events.

It will be clear from the likely impacts as listed above that strong actions will have to be taken to lessen the impact of changes in water availability or supply and to adapt to the changes. Some of the actions that can be taken are the following.[32]

Firstly, increasing the efficiency of water use. For instance, irrigation applied to about one-sixth of the world's farmland producing about one-third of the world's crops and accounting for about two-thirds of world water use, can be made much more efficient. Most irrigation is through open ditches, which is very wasteful of water; over 60% is lost through evaporation and seepage. Microirrigation techniques, in which perforated pipes deliver water directly to the plants, provide large opportunities for water conservation, making it possible to expand irrigated fields without building new dams. Many other efficiency measures are also available, for instance, to recycle water where possible, to promote indigenous practices for sustainable water use (e.g. local rainwater storage), to conserve water (and also soil) by avoiding deforestation or increasing forested areas and to use economic incentives to encourage water conservation.

Secondly, looking for new water supplies. For instance, by increasing water storage in reservoirs or dams, by desalination of sea water, by transferring water from areas of greater abundance or by prospecting and extracting groundwater in appropriate areas.

Thirdly, by introducing more informed management. For instance, regions such as Southeast Asia that are dependent on unregulated river systems are more vulnerable to change than regions such as western Russia and the western United States that have large, regulated water resource systems. Many interested experts and bodies are promoting *Integrated water management* that involves all sectors – agricultural, domestic and industrial – relates to existing infrastructure and plans for new infrastructure and also, most importantly, includes preparation for disasters such as major floods and droughts.

Most of these actions will cost money, although many of them may be much more cost-effective ways of coping with future change in water resources than attempting to develop major new facilities.[33]

Impact on agriculture and food supply

Every farmer understands the need to grow crops or rear animals that are suited to the local climate. The distribution of temperature and rainfall during the year are key factors in making decisions regarding what crops to grow. These will change in the world influenced by global warming. The patterns of what crops are grown where will therefore also change. But these changes will be complex; economic and other factors will take their place alongside climate change in the decision-making process.

There is enormous capacity for adaptation in the growth of crops for food – as is illustrated by what was called the Green Revolution of the 1960s, when the development of new strains of many species of crops resulted in large increases in productivity. Between the mid 1960s and the mid 1980s global food production rose by an average annual rate of 2.4% – faster than global population – more than doubling over that 30-year period. Grain production grew even faster, at an annual rate of 2.9%.[36] Since the mid 1980s growth in production has continued at about 2% per year. There are concerns that factors such as the degradation of many of the world's soils largely through erosion and the slowed rate of expansion of irrigation because less fresh water is available will tend to reduce the potential for increased agricultural production in the future. However, with declining rates of population growth (SRES scenarios A1, B1, B2) and continued economic development, there remains optimism that, in the absence of major climate change, growth in world food supply is likely to continue at least to match growth in demand and that the numbers of undernourished in the world will substantially decline[37].

Desertification

Drylands (defined as those areas where precipitation is low and where rainfall typically consists of small, erratic, short, high-intensity storms) cover about 40% of the total land area of the world and support over one-fifth of the world's population. Figure 7.10 shows how these arid areas are distributed over the continents.

Desertification in these drylands is the degradation of land brought about by climate variations or human activities that have led to decreased vegetation, reduction of available water, reduction of crop yields and erosion of soil. The United Nations Convention to Combat Desertification (UNCCD) set up in 1996 estimates that over 70% of these drylands, covering over 25% of the world's land area, are degraded[34] and therefore affected by desertification. The degradation can be exacerbated by excessive land use or increased human needs (generally because of increased population), or political or economic pressures (for instance, the need to grow cash crops to raise foreign currency). It is often triggered or intensified by a naturally occurring drought.

The progress of desertification in some of the drylands will be increased by the more frequent or more intense droughts that are likely to result from climate change during the twenty-first century.

Recent research has demonstrated the complex nature of the effects of climate change on dryland ecosystems, on the interactions between the species they contain and with the local human communities who live in the dryland areas. Much more understanding is required to assess what is likely to occur and how adverse effects can be minimised.[35]

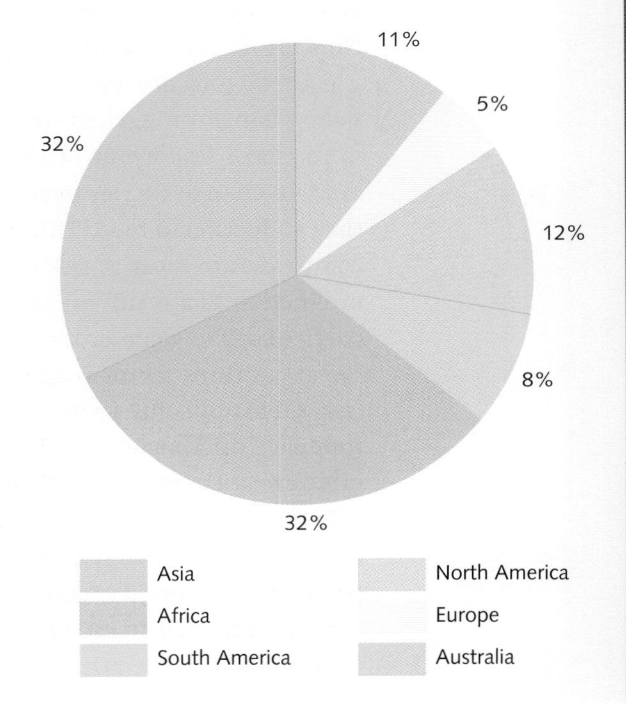

Figure 7.10 The world's drylands, by continent. The total area of drylands is about 60 million square kilometres (about 40% of the total land area), of which 10 million are hyper-arid deserts.

What will be the effect of climate change on agriculture and food supply? With the detailed knowledge of the conditions required by different species and the expertise in breeding techniques and genetic manipulation available today, there should be little difficulty in matching crops to new climatic conditions over large parts of the world. At least, that is the case for crops that mature over a year or two. Forests reach maturity over much longer periods, from decades up to a century or even more. The projected rate of climate change is such that, during this time, trees may find themselves in a climate to which they are far from suited. The temperature regime or the rainfall may be substantially changed, resulting in stunted growth or a greater susceptibility to disease, pests and fires. The impact of climate change on forests is considered in more detail in the next section.

An example of adaptation to changing climate is the way in which farmers in Peru adjust the crops they grow depending on the climate forecast for the year.[38] Peru is a country whose climate is strongly influenced by the cycle of El Niño events described in Chapters 1 and 5. Two of the primary crops grown in Peru, rice and cotton, are very sensitive to the amount and the timing of rainfall. Rice requires large amounts of water; cotton has deeper roots and is capable of yielding greater production during years of low rainfall. In 1983, following the 1982–3 El Niño event, agricultural production dropped by 14%. By 1987 forecasts of the onset of El Niño events had become sufficiently good for Peruvian farmers to take them into account in their planning. In 1987, following the 1986–7 El Niño, production actually increased by 3%, thanks to a useful forecast.

Four factors are particularly important in considering the effect of climate change on agriculture and food production. The availability of water is the most important of the factors. The vulnerability of water supplies to climate change carries over into a vulnerability in the growing of crops and the production of food. Thus the arid or semi-arid areas, mostly in developing countries, are most at risk. A second factor, which tends to lead to increased production as a result of climate change, is the boost to growth that is given, particularly to some crops, by increased atmospheric carbon dioxide (see box below). A third factor is the effect of temperature changes; as temperatures rise, yields of some crops are substantially reduced.[39] A fourth factor is the influence of climate extremes, heatwaves, floods and droughts that seriously interfere with food production.

Detailed studies have been carried out of the sensitivity to climate change during the twenty-first century of the major crops which make up a large proportion of the world's food supply (see box below). They have used the results of climate models to estimate changes in temperature and precipitation. Many of them include the effect of carbon dioxide fertilisation and some also model

The carbon dioxide 'fertilisation' effect

An important positive effect of increased carbon dioxide concentrations in the atmosphere is the boost to growth in plants given by the additional carbon dioxide. Higher carbon dioxide concentrations stimulate photosynthesis, enabling plants to fix carbon at a higher rate. This is why in glasshouses additional carbon dioxide may be introduced artificially to increase productivity. The effect is particularly applicable to what are called C3 plants (such as wheat, rice and soya bean), but less so to C4 plants (for example, maize, sorghum, sugar cane, millet and many pasture and forage grasses). Under ideal conditions it can be a large effect; for C3 crops under doubled carbon dioxide, an average of +30%,[40] although grain and forage quality tends to decline with carbon dioxide enrichment and higher temperatures. However, under real conditions on the large scale where water and nutrient availability are also important factors influencing plant growth, experiments show increases under unstressed conditions in the range 10–25% for C3 crops and 0–10% for C4 crops. Enhanced growth has been observed for young tree stands but no significant response has been measured for mature forest stands. Ozone exposure limits carbon dioxide response in both crops and forests.[41] More research is required especially for many tropical crop species and for crops grown under suboptimal conditions (low nutrients, weeds, pests and diseases). More information is also needed about possible effects on the nutritional value of the crops with carbon dioxide fertilisation.[42]

the effects of climate variability as well as changes in the means. Some also include the possible effects of economic factors and of modest levels of adaptation. These studies in general indicate that the benefit of increased carbon dioxide concentration on crop growth and yield does not always overcome the effects of excessive heat and drought. For cereal crops in mid latitudes, potential yields are projected to increase for small increases in temperature (2–3 °C) but decrease for larger temperature rises (Figure 7.11).[43] In most tropical and subtropical regions, potential yields are projected to decrease for most increases in temperature; this is because such crops are near their maximum temperature tolerance. Where there is a large decrease in rainfall, tropical crop yields would be even more adversely affected.

Taking the supply of food for the world as a whole, studies tend to show that, with appropriate adaptation, the effect of climate change on total global food supply is not likely to be large. However, none of them has adequately taken into account the likely effect on food production of climate extremes (especially of the incidence of drought), of increasingly limited water availability or of other factors such as the integrity of the world's soils, which are currently being degraded at an alarming rate.[44] A serious issue exposed by the studies is that climate change is likely to affect countries very differently. Production in developed countries with relatively stable populations may increase, whereas

Modelling the impact of climate change on world food supply

An example illustrating the key elements of a detailed study of the impact of climate change on world food supply is shown in Figure 7.12.[45]

A climate change scenario is first set up with a climate model of the kind described in Chapter 5. Models of different crops that include the effects of temperature, precipitation and carbon dioxide are applied to 124 different locations in 18 countries to produce projected crop yields that can be compared with projected yields in the absence of climate change. Included also are farm-level adaptations, e.g. planting date shifts, more climatically adapted varieties, irrigation and fertiliser application. These estimates of yield are then aggregated to provide yield-change estimates by crop and country or region.

These yield changes are then employed as inputs to a world food trade model that includes assumptions about global parameters such as population growth and economic change and links together national and regional economic models for the agricultural sector through trade, world market prices and financial flows. The world food trade model can explore the effects of adjustments such as increased agricultural investment, reallocation of agricultural resources according to economic returns (including crop switching) and reclamation of additional arable land as a response to higher cereal prices. The outputs from the total process provide information projected up to the 2080s on food production, food prices and the number of people at risk of hunger (defined as the population with an income insufficient either to produce or to procure their food requirements).

The main results with models of this kind for the 2080s regarding the impact of climate change, for SRES scenarios A1, B1 or B2, are that yields at mid to high latitudes are expected to increase, and at low latitudes (especially the arid and sub-humid tropics) to decrease. This pattern becomes more pronounced as time progresses. The African continent is particularly likely to experience marked reductions in yield, decreases in production and more people at risk of hunger as a result of climate change.

The authors emphasise that, although the models and the methods they have employed are comparatively complex, there are many factors that have not been taken into account. For instance, they have not adequately considered the impact of changes in climate extremes, the availability of water supplies for irrigation or the effects of future technological change on agricultural productivity. Further (see Chapter 6), scientists have as yet limited confidence in the regional detail of climate change. The results, therefore, although giving a general indication of the changes that could occur, should not be treated as a detailed prediction. They highlight the importance of studies of this kind as a guide to future action.

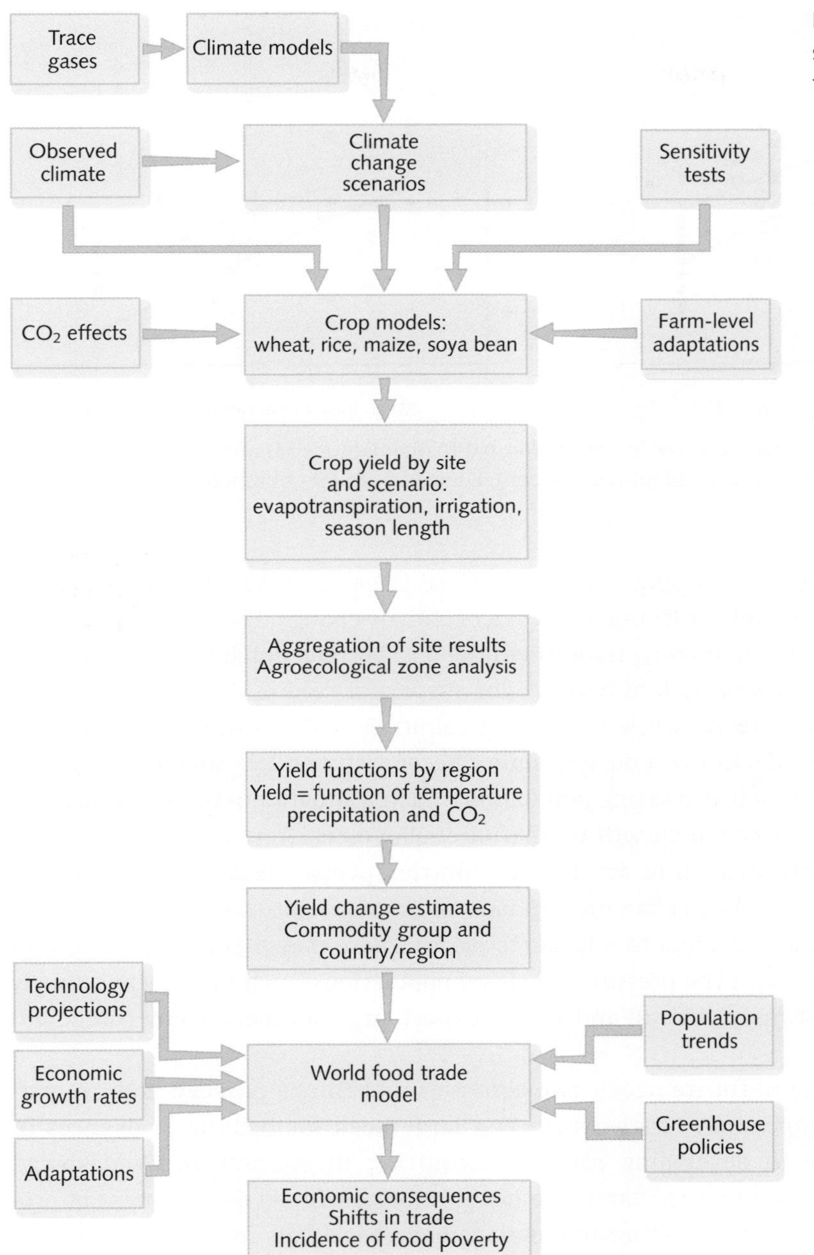

Figure 7.12 Key elements of a study of crop yield and food trade under a changed climate.

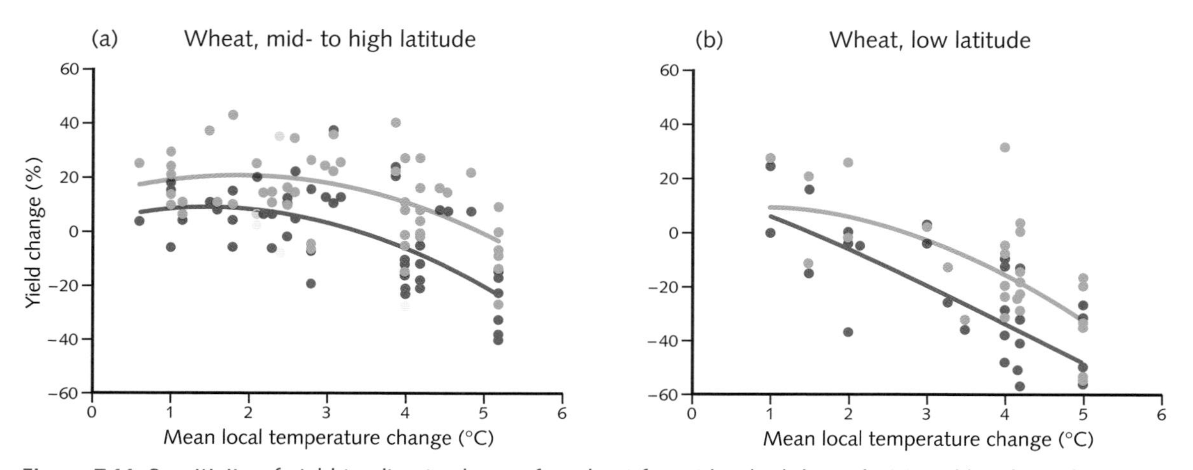

Figure 7.11 Sensitivity of yield to climate change for wheat for mid to high latitude (a) and low latitude (b). Cases without adaptation (red) and with adaptation (green). Derived from 69 published studies at multiple simulation sites.

that in many developing countries (where large increases in population are occurring) is likely to decline as a result of climate change. The disparity between developed and developing nations will tend to become much larger, as will the number of those at risk of hunger. The surplus of food in developed countries is likely to increase, while developing countries will face increasing deprivation as their declining food availability becomes much less able to provide for the needs of their increasing populations. Such a situation will raise enormous problems, one of which will be that of employment. Agriculture is the main source of employment in developing countries; people need employment to be able to buy food. With changing climate, as some agricultural regions shift, people will tend to attempt to migrate to places where they might be employed in agriculture. With the pressures of rising populations, such movement is likely to be increasingly difficult and we can expect large numbers of environmental refugees.

In looking to future needs, two activities that can be pursued now are particularly important. Firstly, there is a large need for technical advances in agriculture in developing countries requiring investment and widespread local training. In particular, there needs to be continued development of programmes for crop breeding and management, especially in conditions of heat and drought. These can be immediately useful in the improvement of productivity in marginal environments today. Secondly, as was seen earlier when considering fresh water supplies, improvements need to be made in the availability and management of water for irrigation, especially in arid or semi-arid areas of the world.

The impact on ecosystems

A little over 10% of the world's land area is under cultivation – that was the area addressed in the last section. The rest is to a greater or lesser extent unmanaged by humans. In Figure 7.13 are illustrated the world's major ecosystems (or biomes) with their global areal extent showing how they have been transformed by land use.

Ecosystems are of great importance to human communities. They provide supplies for human communities in the provision of food, water, fuel, wood and biodiversity. They also provide important regulation especially for components of the hydrological cycle. Further they possess a wide range of important cultural value. All these together are commonly called ecosystem services.

The variety of plants and animals that constitute a local ecosystem is sensitive to the climate, the type of soil and the availability of water. Ecologists divide the world into regions characterised by their distinctive vegetation. This is well illustrated by information about the distribution of vegetation over the world during past climates (e.g. for the part of North America shown in Figure 7.14), which indicates the ecosystems most likely to flourish under different climatic regimes. Changes in climate alter the suitability of a region for different species (Figure 7.15), and change their competitiveness within an ecosystem, so that even relatively small changes in climate will lead, over time, to large changes in the composition of an ecosystem.

However, changes of the kind illustrated in Figure 7.14 took place over thousands of years. With global warming similar changes in climate occur over a few decades. Most ecosystems cannot respond or migrate that fast. Fossil records indicate that the maximum rate at which most plant species have migrated in the past is less than 1 km per year. Known constraints imposed by the dispersal process (e.g. the mean period between germination and the production of seeds and the mean distance that an individual seed can travel) suggest that, without human intervention, many species would not be able to keep up with the rate of movement of their preferred climate niche projected for the twenty-first century, even if there were no barriers to their movement imposed by land use.[46] Natural ecosystems will therefore become increasingly unmatched to their environment. How much this matters will vary from species to species: some are more vulnerable to changes in average climate or climate extremes than others. But all will become more prone to disease and attack by pests. Any positive effect from added 'fertilisation' due to increased carbon dioxide is likely to be more than outweighed by negative effects from other factors.

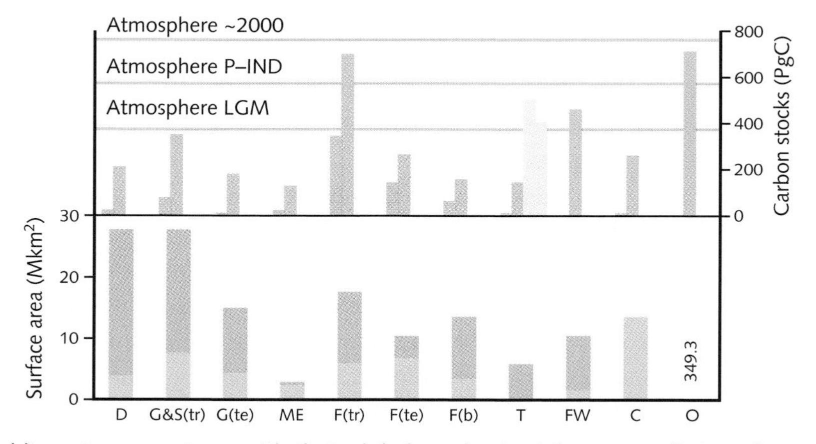

Figure 7.13 The world's major ecosystems with their global areal extent (lower panel), transformed by land use in yellow, untransformed in purple and total carbon stores (upper panel) in plant biomass (green), soil (brown) and yedoma/permafrost (light blue). D, deserts; G & S(tr), tropical grasslands and savannas; G(te), temperate grasslands; ME, Mediterranean ecosystems; F(tr), tropical forests; F(te), temperate forests; F(b), boreal forests; T, tundra; FW, freshwater lakes and wetlands; C, croplands; O, oceans. Approximate carbon content of the atmosphere is indicated for last glacial maximum (LGM), pre-industrial (P-IND) and current (about 2000).

Forests cover about 30% of the world's land area and are among the most productive of terrestrial ecosystems. They represent a large store of carbon. Of terrestrial carbon, 80% of above-ground and 40% of below-ground is in forests, together storing about twice as much as is in the atmosphere (Figure 3.1). They are particularly important in the context of climate change. Current levels of deforestation are responsible for around 20% of the additional carbon dioxide emitted to the atmosphere each year due to human activities. What effect will climate change have on the world's forests and how in turn might that effect the climate?

Trees are long-lived and take a long time to reproduce, so cannot respond quickly to climate change. Further, many trees are surprisingly sensitive to the average climate in which they develop. The environmental conditions (e.g. temperature and precipitation) under which a species can exist and reproduce are known as its niche. Climate niches for some typical tree species are illustrated in Figure 7.16; under some conditions a change as small as 1 °C in annual average temperature can make a substantial difference to a tree's productivity. For the likely changes in climate in the twenty-first century, a substantial proportion of existing trees will be subject to unsuitable climate conditions. This will be particularly the case in the boreal forests of the northern hemisphere where, as trees become less healthy, they will be more prone

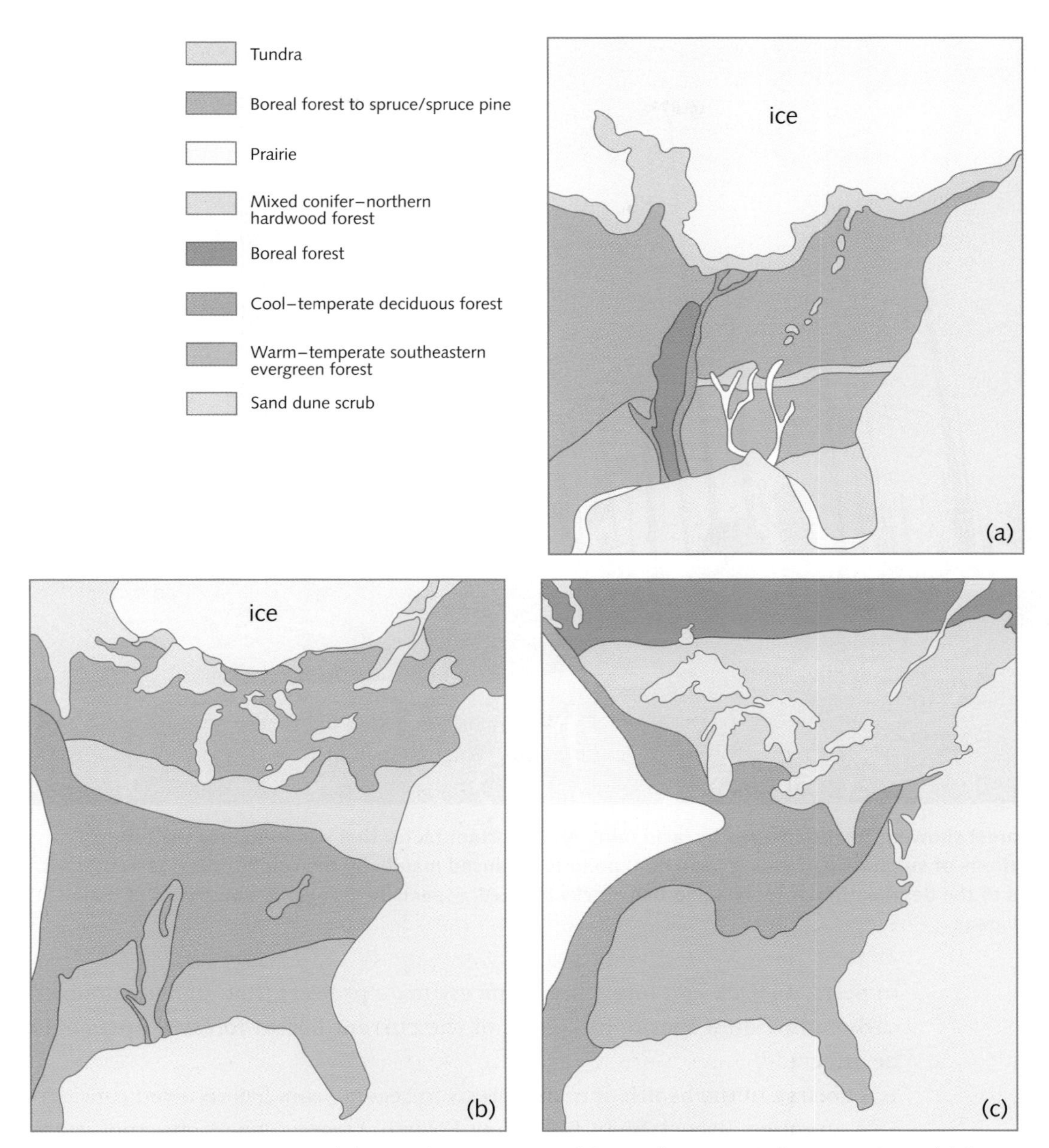

Figure 7.14 Vegetation maps of the southeastern United States during past climate regimes: (a) for 18 000 years ago at the maximum extent of the last ice age, (b) for 10 000 years ago, (c) for 5000 years ago when conditions were similar to present. A vegetation map for 200 years ago is similar to that in (c).

Conifer forest showing trees damaged by 'acid rain'. An important factor that will influence the future concentrations of sulphate particles is 'acid rain' pollution, caused mainly by the sulphur dioxide emissions. This leads to the degradation of forests and fish stocks in lakes, especially in regions downwind of major industrial areas.

to pests, dieback and forest fires. One estimate projects that, under a doubled carbon dioxide scenario, up to 65% of the current boreal forested area could be affected.[47]

A decline in the health of many forests in recent years has received considerable attention, especially in Europe and North America where much of it has been attributed to acid rain and other pollution originating from heavy industry, power stations and motor cars. Not all damage to trees, however, is thought to have this origin. Studies in several regions of Canada, for instance, indicate that the dieback of trees there is related to changes in climatic conditions, especially to successions of warmer winters and drier summers.[48] In some cases it may be the double effect of pollution and climate stress causing the problem;

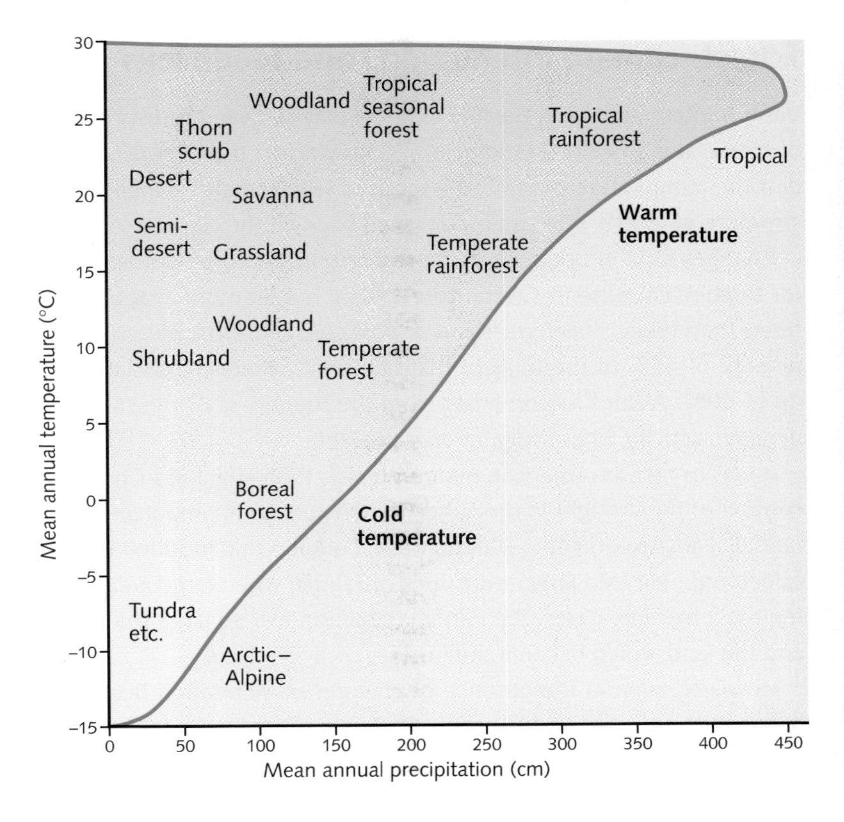

Figure 7.15 The pattern of world biome types related to mean annual temperature and precipitation. Other factors, especially the seasonal variations of these quantities, affect the detailed distribution patterns (after Gates).

trees already weakened by the effects of pollution fail to cope with climate stress when it comes. Stresses on the world's forests due to climate change (see box on page 208) will be concurrent with other problems associated with forests, in particular those of continuing tropical deforestation and of increasing demand for wood and wood products resulting from rapidly increasing populations especially in developing countries.

If a stable climate is eventually re-established, given adequate time (which could be centuries), different trees will be able to find again at some location their particular climatic niche. It is during the period of rapid change that most trees will find themselves unsatisfactorily located from the climate point of view. If, because of the rate of climate change, substantial stress and die-back occur in boreal and tropical forests (see box below) a release of carbon will occur. This positive feedback was mentioned in Chapter 3 (see the box on page 48–9). Just how large this will be is uncertain but estimates as high as 240 Gt over the twenty-first century for the above-ground component alone have been quoted.[50]

Forest–climate interactions and feedbacks

Various interactions and feedbacks are in play between forests and climate. Extensive changes in the area of forests due to deforestation can seriously affect the climate in the region of change. Changes in carbon dioxide, temperature or rainfall associated with climate change can have a major impact on the health or structure of forests that can in turn feed back on the climate. We consider some of these effects in turn.

Changes in land use such as those brought about by deforestation can affect the amount of rainfall, for three main reasons. Over a forest there is a lot more evaporation of water (through the leaves of the trees) than there is over grassland or bare soil, hence the air will contain more water vapour. Also, a forest reflects 12–15% of the sunlight that falls on it, whereas grassland will reflect about 20% and desert sand up to 40%. A third reason arises from the roughness of the surface so stimulating convection and other dynamic activity where vegetation is present.

It was in fact an American meteorologist, Professor Jules Charney, who suggested in 1975 that, in the context of the drought in the Sahel, there could be an important link between changes of vegetation and rainfall. Early experiments with numerical models that included these physical processes demonstrated the effect and indicated large reductions of rainfall when large areas of forest were replaced by grassland. In the most extreme cases, the rainfall reduction was so large that grassland would no longer be supported and the land would become semi-arid.

However, even in the absence of changes of vegetation because of human action, interactions occur between the climate and the forest that can effect large changes. Three important feedbacks that lead to reduced precipitation are:

- increased carbon dioxide causes stomatal closure within the leaves of the trees so reducing evaporation;
- increased temperature tends to cause forest dieback, again leading to reduced evaporation;
- increased temperature causes increased respiration of carbon dioxide from the soil so leading to further global warming – the climate/carbon-cycle feedback mentioned in Chapter 3.

For Amazonia, when these three feedbacks are added to the effect on the forest of the local climate change that is likely to occur because of global circulation changes, under a scenario of continuing increase in carbon dioxide emissions, simulations suggest major losses of forest cover in the Amazon basin during the twenty-first century. Large areas would be replaced by shrubs and grasses and part of Amazonia could become semi-desert.[49] Such results are still subject to considerable uncertainties (for instance, those associated with the model simulations of El Niño events under climate change conditions and the connections between these events and the climate over Amazonia), but they illustrate the type of impacts that might occur and emphasise the importance of understanding the interactions between climate and vegetation.

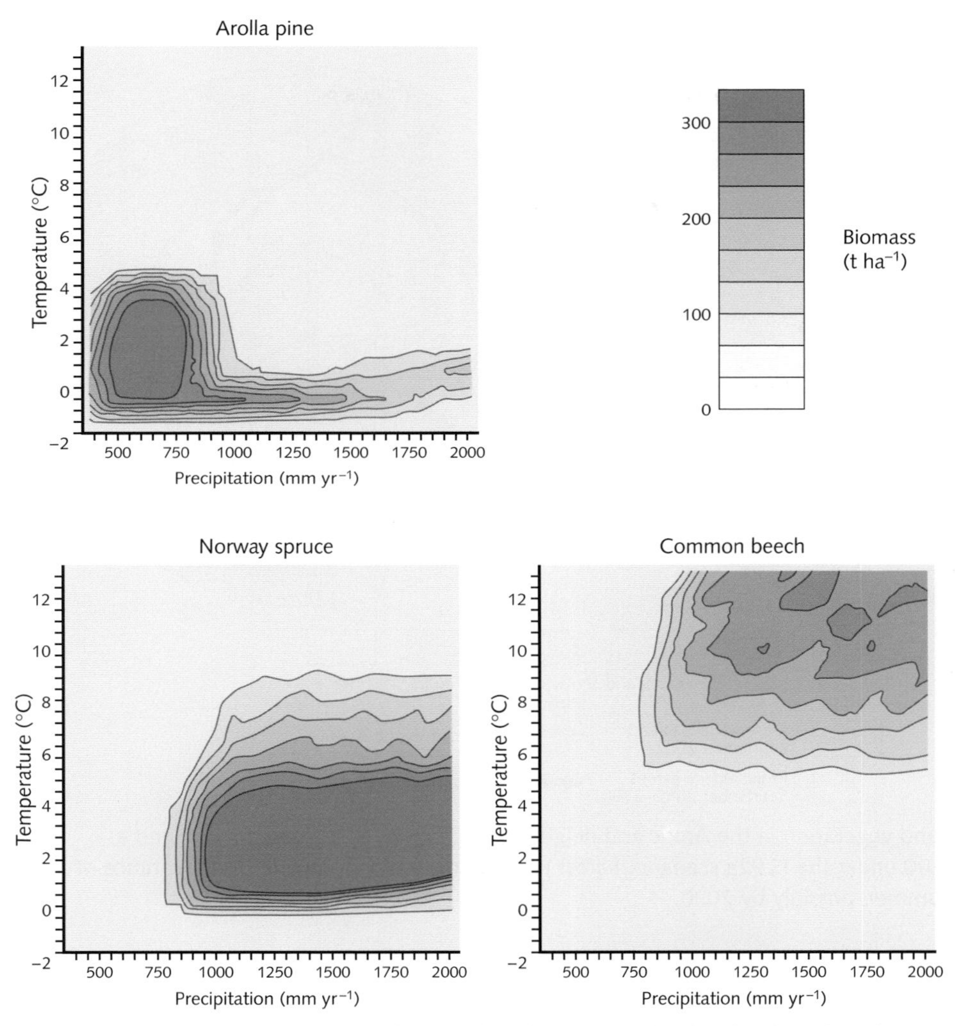

Figure 7.16 Simulated environmental realised niches (the realised niche describes the conditions under which the species is actually found) for three tree species, Arolla Pine, Norway Spruce and Common Beech. Plots are of biomass generated per year against annual means of temperature and precipitation. Arolla Pine is a species with a particularly narrow niche. The narrower the niche, the greater the potential sensitivity to climate change.

The above discussion has related to the impact of climate change on natural forests where the likely impacts are largely negative. Studies of the impacts on managed forests are more positive.[51] They suggest that with appropriate adaptation and land and product management, even without forestry projects that increase the capture and storage of carbon (see Chapter 10), a small amount of

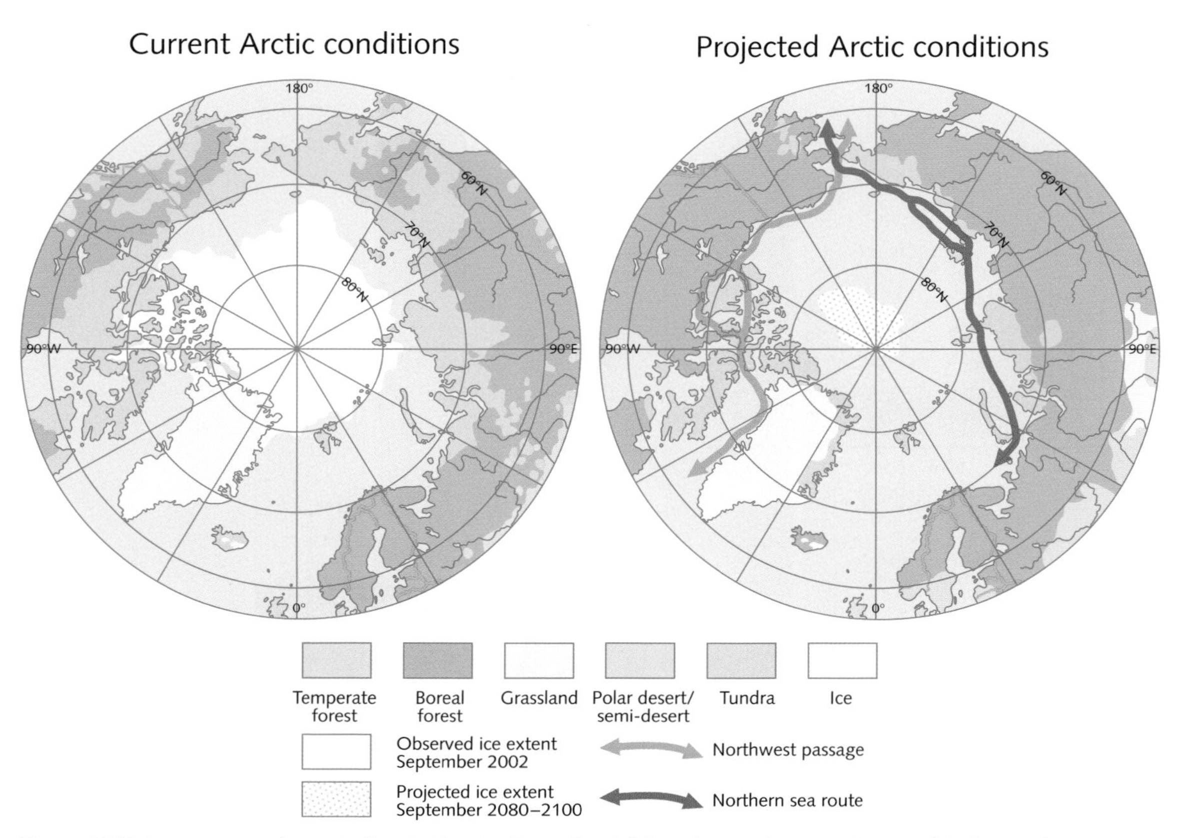

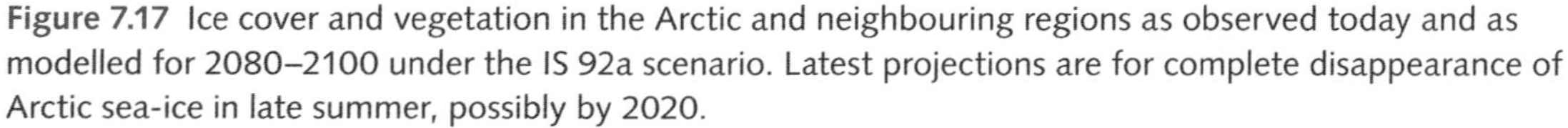

Figure 7.17 Ice cover and vegetation in the Arctic and neighbouring regions as observed today and as modelled for 2080–2100 under the IS 92a scenario. Latest projections are for complete disappearance of Arctic sea-ice in late summer, possibly by 2020.

climate change could increase global timber supply and enhance existing market trends towards rising market share in developing countries.

Large changes are also expected in the polar regions (Figure 7.17) both in the amount of sea-ice cover and in vegetation with large implications for managed and unmanaged ecosystems both on land and marine areas. Changes in drylands, in their extent or the nature of their vegetation are also of great concern (see box on page 197).

A further concern about natural ecosystems relates to the diversity of species that they contain and the loss of species and hence of biodiversity due to the impact of climate change. Significant disruptions of ecosystems from disturbances such as fire, drought, pest infestation, invasion of species, storms and coral bleaching events are expected to increase. The stresses caused by climate change, added to other stresses on ecological systems (e.g. land conversion, land degradation, deforestation, harvesting and pollution) threaten substantial damage to or complete loss of some unique ecosystems, and the extinction of some endangered species. Coral reefs and atolls, mangroves, boreal and tropical forests, polar and alpine ecosystems, prairie wetlands and remnant native grasslands are examples of systems threatened by climate change. In some cases the threatened ecosystems are those that could mitigate against some climate change impacts (e.g. coastal systems that buffer the impact of storms). Possible adaptation methods to reduce the loss of biodiversity include the establishment of refuges, parks and reserves with corridors to allow migration of species, and the use of captive breeding and translocation of species.[52]

So far we have been considering ecosystems on land. What about those in the oceans; how will they be affected by climate change? Although we know much less about ocean ecosystems, there is considerable evidence that biological activity in the oceans has varied during the cycle of ice ages. Chapter 3 noted (see box on page 43) the likelihood that it was these variations in marine biological activity which provided the main control on atmospheric carbon dioxide concentrations during the past million years (see Figure 4.6). Changes in ocean water temperature and possible changes in some of the patterns of ocean circulation will result in changes in the regions where upwelling occurs and where fish congregate. Some fisheries could collapse and others expand. At the moment the fishing industry is not well adapted to address major change.[53]

Some of the most important marine ecosystems are found within coral reefs that occur in many locations throughout the tropical and sub-tropical world. They are especially rich in biodiversity and are particularly threatened by global warming. Within them the species diversity contains more phyla than rainforests and they harbour more than 25% of all known marine fish.[54] They represent a significant source of food for many coastal communities. Corals are particularly sensitive to sea surface temperature and even 1 °C of persistent warming can cause bleaching (paling in colour) and extensive mortality accompanies persistent temperature anomalies of 3 °C or more. Much recent bleaching, for instance that in 1998, has been associated with El Niño events.[55]

Added to the stresses caused by climate change will be those that arise from increased ocean acidification that results from the increase of carbon dioxide in ocean water (Figure 7.18). These increased stresses will be most serious for a wide range of planktonic and shallow benthic marine organisms that use aragonite to make their shells or skeletons, such as corals and marine snails (pteropods). Research on how serious these stresses will be is being actively pursued.[56]

Coral bleaching is a vivid sign of corals responding to stress which can be induced by increased or reduced water temperatures, increased solar irradiance, changes in water chemistry, starvation caused by a decline in zooplankton levels as a result of over-fishing, and increased sedimentation.

Summarising the impacts of climate change on ecosystems with a warming of global average temperature of 2 °C or more from its pre-industrial value, there are five areas of particular concern.[57]

(1) The resilience of many ecosystems (their ability to adapt) is likely to be exceeded by an unprecedented combination of change in climate, associated disturbances (e.g. flooding, drought, wildfire, insects, ocean acidification) and in other drivers such as land-use change, pollution and over exploitation of resources.

(2) The terrestrial biosphere is currently a net carbon sink (see Table 3.1). As was mentioned in Chapter 3, during the twenty-first century, it is likely to become a net carbon source thus amplifying climate change.

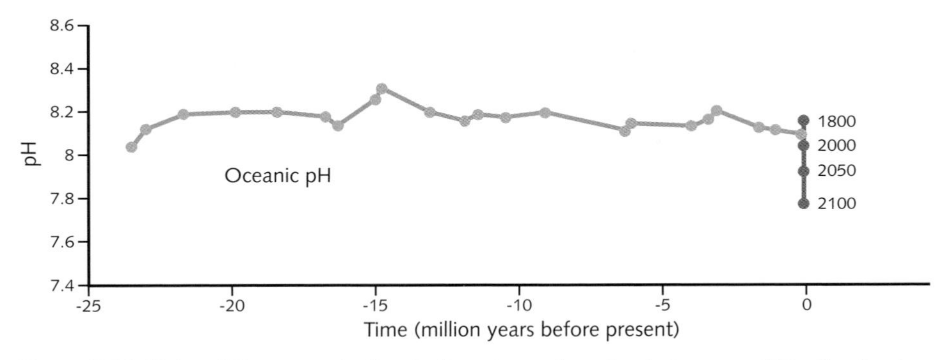

Figure 7.18 Rising CO_2 concentration in the atmosphere leads to move CO_2 dissolved in the ocean with rapid increase in ocean acidity (lower pH) to levels not encountered for millions of years. Past (blue spots, data from Pearson and Palmer 2000) and contemporary variability of marine pH (purple spots, with dates). Future predictions are model derived values assuming atmospheric carbon dioxide concentration of 500 ppm in 2050 and 700 ppm in 2100.

(3) Approximately 20–30% of plant and animal species so far assessed (in an unbiased sample) are likely to be at increasingly high risk of extinction.
(4) Substantial changes in structure and functioning of terrestrial ecosystems are very likely to occur with some positive impacts due to the carbon dioxide fertilisation effect but with extensive forest and woodland decline in mid to high latitudes and the tropics associated particularly with changing disturbance regimes (e.g. through wildfire and insects).
(5) Substantial changes in structure and functioning of marine and other aquatic ecosystems are very likely to occur. In particular the combination of climate change and ocean acidification will have a severe impact on corals.

The impact on human health

Human health is dependent on a good environment. Many of the factors that lead to a deteriorated environment also lead to poor health. Pollution of the atmosphere, polluted or inadequate water supplies and poor soil (leading to poor crops and inadequate nutrition) all present dangers to human health and well-being and assist the spread of disease. As has been seen in considering the impacts of global warming, many of these factors will be exacerbated through the climate change occurring in the warmer world. The greater likelihood of extremes of climate, such as droughts and floods, will also bring greater risks to health from increased malnutrition and from a prevalence of conditions more likely to lead to the spread of diseases from a variety of causes.

How about direct effects of the climate change itself on human health? Humans can adapt themselves and their buildings so as to live satisfactorily in very varying conditions and have great ability to adapt to a wide range of climates. The main difficulty in assessing the impact of climate change on health is that of unravelling the influences of climate from the large number of other factors (including other environmental factors) that affect health.

The main direct effect on humans will be that of heat stress in the extreme high temperatures that will become more frequent and more widespread especially in urban populations (see box and Figure 6.6). In large cities where heatwaves commonly occur death rates can be doubled or tripled during days of unusually high temperatures.[58] Although such episodes may be followed by periods with fewer deaths showing that some of the deaths would in any case have occurred about that time, most of the increased mortality seems to be directly associated with the excessive temperatures with which old people in particular find it hard to cope. On the positive side, mortality due to periods of severe cold in winter will be reduced. The results of studies are equivocal regarding whether the reduction in winter mortality will be greater or less than the increase in summer mortality. These studies have largely been confined to populations in developed countries, precluding a more general comparison between changes in summer and winter mortality.

A further likely impact of climate change on health is the increased spreading of diseases in a warmer world. Many insect carriers of disease thrive better in warmer and wetter conditions. For instance, epidemics of diseases such as viral encephalitides carried by mosquitoes are known to be associated with the unusually wet conditions that occur in the Australian, American and African continents associated with different phases of the El Niño cycle.[60] Some diseases, currently largely confined to tropical regions, with warmer conditions could spread into mid latitudes. Malaria is an example of such a disease that is spread by mosquitoes under conditions that are optimum in the temperature range of 15–32 °C with humidities of 50–60%. It currently represents a huge global public health problem, causing annually around 300 million new infections and over 1 million deaths. Under climate change scenarios, most predictive model studies indicate a net increase in the geographic range (and in the populations at risk) of potential transmission of malaria and dengue infections, each of which currently impinge on 40–50% of the world's population. Other diseases that are likely to spread for the same reason are yellow fever and some viral encephalitides. In all cases, however, actual disease occurrence will be strongly influenced by local environmental conditions,

Heatwaves in Europe and India, 2003

Record extreme temperatures were experienced in Europe during June, July and August 2003. At many locations temperature rose to over 40 °C. In France, Italy, the Netherlands, Portugal and Spain, over 20,000 (possibly as many as 35 000) additional deaths were attributed to the unrelenting heat. Spain, Portugal, France and countries in Central and Eastern Europe suffered from intense forest fires.[59] Figure 7.19 illustrates the rarity of this event showing that it is well outside normal climate variability. Studies indicate that most of the risk of this event arose from increase in greenhouse gases due to human activities. They also indicate that it will represent a normal year by 2050 and a cool year by 2100.

Extreme heat was also experienced in 2003 in other parts of the world; for instance in Andhra Pradesh in India over 1000 people died through extreme temperatures above 45 °C that occurred most unusually for 27 consecutive days.

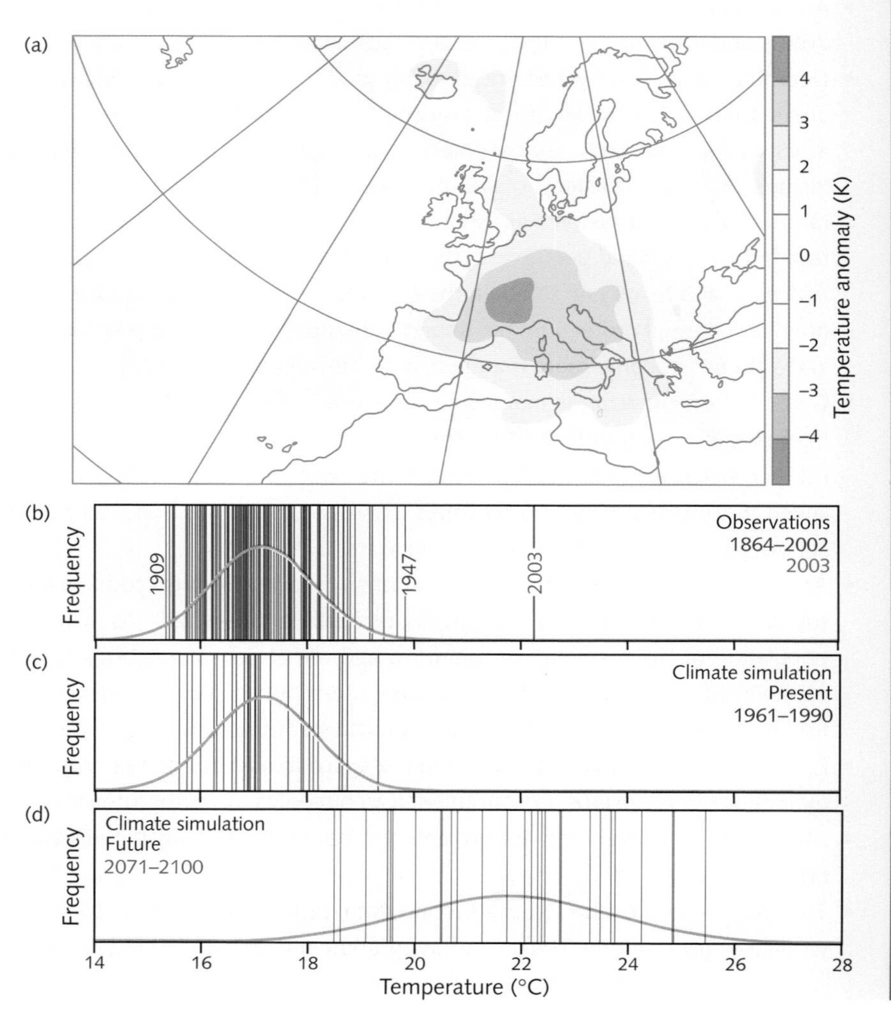

Figure 7.19 Characteristics of the summer 2003 heatwave in Europe. (a) June, July, August (JJA) temperature anomaly with respect to 1961–90; (b) to (d) JJA temperatures for Switzerland; (b) observed during 1864–2003; (c) simulated with a regional model for the period 1961–90; (d) simulated for 2071–2100 under the SRES A2 scenario. The vertical bars in (b) to (d) represent mean summer surface temperature for each year of the time period considered; the fitted Gaussian distribution is indicated in black.

Impacts on Africa

Africa is one of the most vulnerable continents to climate change and variability, a situation that is exacerbated by existing developmental challenges such as endemic poverty; complex governance and institutional dimensions; limited access to capital, including markets, infrastructure and technology; ecosystem degradation; and complex disasters and conflicts – all of which contribute to Africa's weak adaptive capacity to climate change.[61] Some projected climate impacts for Africa are summarised as follows:[62]

- The impacts of climate change in Africa are likely to be greatest where they co-occur with a range of other stresses (e.g., unequal access to resources, enhanced food insecurity, poor health management systems. These stresses, enhanced by climate variability and change, further enhance the vulnerabilities of many people in Africa.**
- An increase of 5% to 8% (60 to 90 million ha) of arid and semi-arid land in Africa is projected by the 2080s under a range of climate-change scenarios.**
- Declining agricultural yields are probably due to drought and land degradation, especially in marginal areas. Changes in the length of growing period have been noted under various scenarios. In the A1F1 SRES scenario, which has an emphasis on globally integrated ecconomic growth, areas of major change include the coastal systems of southern and eastern Africa. Under both the A1 and B1 scenarios, mixed rain-fed, semi-arid systems are shown to be heavily affected by changes in climate in the Sahel. Mixed rain-fed and highland perennial systems in the Great Lakes region in East Africa and in other parts of East Africa are also heavily affected. In the B1 SRES scenario, which assumes development within a framework of environmental protection, the impacts are, however, generally less, but marginal areas (e.g., the semi-arid systems) become more marginal, with the impacts on coastal systems becoming moderate.**
- Current stress on water in many areas of Africa is likely to be enhanced by climate variability and change. Increases in run-off in East Africa (possibly floods) and decreases in run-off and likely increased drought risk in other areas (e.g. southern Africa) are projected by the 2050s. Current water stresses are not only linked to climate variations, and issues of water governance and water-basin management must also be considered in any future assessments of water in Africa.**
- Any changes in the primary production of large lakes are likely to have important impacts on local food supplies. For example, Lake Tanganyika currently provides 25% to 40% of animal protein intake for the population of the surrounding countries, and climate change is likely to reduce primary production and possible fish yields by roughly 30%. The interaction of human management decisions, including overfishing, is likely to further compound fish offtakes from lakes.**
- Ecosystems in Africa are likely to experience major shifts and changes in species range and possible extinctions (e.g., fynbos and succulent Karco biomes in southern Africa).*
- Mangroves and coral reefs are projected to be further degraded, with additional consequences for fisheries and tourism.**
- Towards the end of the twenty-first century, projected sea-level rise will affect low-lying coastal areas with large populations. The cost of adaptation will exceed 5% to 10% of GDP.**

socio-economic circumstances, advances in treatment or prevention and public health infrastructure.

The potential impact of climate change on human health could be large. However, the factors involved are complex and quantitative conclusions require careful study of the direct effects of climate on humans and of the epidemiology of the diseases particularly affected. Some remarks about how the health impacts of extremes and disasters might be reduced are given in the next section.

Adaptation to climate change

As we have seen, some of the impacts of climate change are already apparent. A degree of adaptation[63] therefore has already become a necessity. Numerous possible adaptation options for responding to climate change have already been identified – examples are given in Table 7.2. Because it takes some decades for the oceans to warm, there also exists a substantial commitment to further climate change even if carbon dioxide emissions were to be halted. Urgent action is therefore necessary to consider the further substantial adaptation which will be necessary.

These can reduce adverse impacts and enhance beneficial effects of climate change and can also produce immediate ancillary benefits, but they cannot prevent all damages. Many of the options listed are presently employed to cope with current climate variability and extremes; their expanded use can enhance both current and future capacity to cope. But such actions may not be as effective in the future as the amount and rate of climate change increase. To make a list of possible adaptation options is relatively easy. If they are to be applied effectively, much more information needs to be generated regarding the detail and the cost of their application over the wide range of circumstances where they will be required.

Of particular importance is the requirement for adaptation to extreme events and disasters such as floods, droughts and severe storms.[64] Vulnerability to such events can be substantially reduced by much more adequate preparation.[65] For instance, following Hurricanes George and Mitch, the Pan American Health Organisation (PAHO) identified a range of policies to reduce the impact of such events:[66]

- undertaking vulnerability studies of existing water supply and sanitation systems and ensuring that new systems are built to reduce vulnerability;
- developing and improving training programmes and information systems for national programmes and international cooperation on emergency management;

- developing and testing early warning systems that should be coordinated by a single national agency and involve vulnerable communities. Provision is also required for providing and evaluating mental care, particularly for those who may be especially vulnerable to the adverse psychosocial effects of disasters (e.g. children, the elderly and the bereaved).

Table 7.2 Selected examples of planned adaptation by sector

Sector	Adaptation option/ strategy	Underlying policy framework	Key constraints and opportunities for implementation (Normal font=constraints; *italics=opportunities*)
Water	Expanded rainwater harvesting; water storage and conservation techniques; water reuse; desalination; water use and irrigation efficiency	National water policies and integrated water resources management; water-related hazards management	Financial, human resources and physical barriers; *integrated water resources management; synergies with other sectors*
Agriculture	Adjustment of planting dates and crop variety; crop relocation; improved land management, e.g. erosion control and soil protection through tree planting	R&D policies; institutional reform; land tenure and land reform; training; capacity building; crop insurance; financial incentives, e.g. subsidies and tax credits	Technological and financial constraints; access to new varieties; markets; *longer growing season in higher latitudes; revenues from 'new' products*
Infrastructure/ settlement (including coastal zones)	Relocation; seawalls and storm surge barriers; dune reinforcement; land acquisition and creation of marshlands/wetlands as buffer against sea level rise and flooding; protection of existing natural barriers	Standards and regulations that integrate climate change considerations into design; land-use policies; building codes; insurance	Financial and technological barriers; availability of relocation space; *integrated policies and management; synergies with sustainable development goals*
Human health	Heat-health action plans; emergency medical services; improved climate-sensitive disease surveillance and control; safe water and improved sanitation	Public health policies that recognise climate risk; strengthened health services; regional and international cooperation	Limits to human tolerance (vulnerable groups); knowledge limitations; financial capacity; *upgraded health services; improved quality of life*

Table 7.2 Cont.

Sector	Adaptation option/ strategy	Underlying policy framework	Key constraints and opportunities for implementation (Normal font = constraints; *italics = opportunities)*
Tourism	Diversification of tourism attractions and revenues; shifting ski slopes to higher altitudes and glaciers; artificial snow-making	Integrated planning (e.g. carrying capacity; linkages with other sectors); financial incentives, e.g. subsidies and tax credits	Appeal/ marketing of new attractions; financial and logistical challenges; potential adverse impact on other sectors (e.g. artificial snow-making may increase energy use); *revenues from 'new' attractions; involvement of wider group of stakeholders*
Transport	Realignment/relocation; design standards and planning for roads, rail and other infrastructure to cope with warming and drainage	Integrating climate change considerations into national transport policy; investment in R&D for special situations, e.g. permafrost areas	Financial and technological barriers; availability of less vulnerable routes; *Improved technologies and integration with key sectors (e.g. energy)*
Energy	Strengthening of overhead transmission and distribution infrastructure; underground cabling for utilities; energy efficiency; use of renewable sources; reduced dependence on single sources of energy	National energy policies, regulation, and fiscal and financial incentives to encourage use of alternative sources; incorporating climate change in design standards	Access to viable alternatives; financial and technological barriers; acceptance of new technologies; *stimulation of new technologies; use of local resources*

Note: Other examples from many sectors would include early warning systems.

Costing the impacts: extreme events

In the previous paragraphs the impacts of climate change have been described in terms of a variety of measures; for instance, the number of people affected (e.g. by mortality, disease or by being displaced), the gain or loss of agricultural or forest productivity, the loss of biodiversity, the increase in desertification, etc. However, the most widespread measure, looked for by many policymakers, is monetary cost or benefit. But before describing what has been done so far to estimate the overall costs of impacts, we need to consider what is known about the cost of damage due to extreme events (such as floods, droughts, windstorms or tropical cyclones). As has been constantly emphasised in this chapter these probably constitute the most important element in climate change impacts.

Because the incidence of such extreme events has increased significantly in recent decades, information about the cost of the damage due to them has been tracked by insurance companies. They have catalogued both the insured losses and, so far as they have been able to estimate, the total economic losses – these latter have shown an approximately tenfold increase from the 1950s to the 1990s (see Figure 1.2 and box below). Although factors other than climate change have contributed to this increase, climate change is probably the factor of most significance. The estimates for the 1990s of annual economic losses from weather-related disasters amount to approximately 0.2% of global world product (GWP) and vary from about 0.3% of aggregate GDPs for the North and Central American and the Asian regions to less than 0.1% for Africa (Table 7.3). These average figures hide big regional and temporal variations. For instance, the annual loss in China from natural disasters from 1989 to 1996 is estimated to range from 3% to 6% of GDP, averaging nearly 4%[67] – over ten times the world average. The reason why the percentage for Africa is so low is not because there are no disasters there – Africa on the whole has more than its fair share (see box on page 216) – but because most of the damage in African disasters is not realised in economic terms, nor does it appear in economic statistics. Further such averaged numbers hide the severe impact of disasters on individual countries or regions which, as we mention below with the example of Hurricane Mitch, can prove to be very large indeed. Even for the United States, the total economic cost of Hurricane Katrina which struck New Orleans in 2005 has been estimated at around 1% of the United States GDP.

The percentages we have quoted are conservative in that they do not represent all relevant costs. They relate to direct economic costs only and do not include associated or knock-on costs of disasters. This means, for instance, that the damage due to droughts is seriously underestimated. Droughts tend to happen slowly and many of the losses may not be recorded or borne by those not directly affected.

Another reason for treating the information in the box with caution is because of the large disparities between different parts of the world and countries regarding per capita wealth, standard of living and degree of insurance cover. For instance, probably the most damaging hurricane ever, Hurricane

Table 7.3 Fatalities, economic losses and insured losses (both in 1999 US dollars) for disasters in different regions as estimated by the insurance industry for the period 1985–99. The percentage from weather-related disasters (including windstorms, floods, droughts, wildfire, landslides, land subsidence, avalanches, extreme temperature events, lightning, frost and ice/snow damages) is indicated in each case. Total losses are higher than those summarised in Figure 1.2 because of the restriction of Figure 1.2 to losses from large catastrophic events

	Africa	America: South	America: North, Central, Caribbean	Asia	Australia	Europe	World
Number of events	810	610	2260	2730	600	1810	8820
Weather-related	91%	79%	87%	78%	87%	90%	85%
Fatalities	22990	56080	37910	429920	4400	8210	559510
Weather-related	88%	50%	72%	70%	95%	96%	70%
Economic losses ($US billion)	7	16	345	433	16	130	947
Weather-related	81%	73%	84%	63%	84%	89%	75%
Insured losses ($US billion)	0.8	0.8	119	22	5	40	187
Weather-related	100%	69%	86%	78%	74%	98%	87%

Mitch, which hit Central America in 1998 does not appear in Table 7.4 as the total insured losses were less than $US1 billion. In that storm, 600 mm of rainfall fell in 48 hours, there were 9000 deaths and economic losses estimated at over $US6 billion. The losses in Honduras and Nicaragua amounted to about 70% and 45% respectively of their annual gross national product (GNP). Another example that does not appear in Table 7.4 for the same reason is the floods in central Europe in 1997 which caused the evacuation of 162 000 people and over $US5 billion of economic damage.

How about the likely costs of extreme events in the future? To estimate those we need much more quantitative information about their likely future incidence and intensity. In Chapter 6, an estimate of a factor of 5 was quoted for the likely increase of the risk of floods by mid twenty-first century (Figure 6.11). A speculative but probably conservative calculation of a global average figure for the future might be obtained as follows. Beginning with the 0.2% or 0.3% of GDP from the insurance companies' estimate of the current average costs due to

The insurance industry and climate change

The impact of climate on the insurance industry is mainly through extreme weather events. In developing countries there may be very high mortality from extreme weather but relatively small costs to the industry because of low insurance penetration. In developed countries the loss of life may be much less but the costs to the insurance industry can be very large. Figure 1.2 illustrates the large growth in weather-related disasters and the associated economic and insured losses since the 1950s and Table 7.3 the distribution of the disasters, fatalities and economic losses from 1985 to 1999 around the continents. Some idea of the types of disaster that cause the largest economic loss can be gleaned from Table 7.4.

Part of the observed upward trend in historical disaster losses is linked to socio-economic factors such as population growth, increased wealth and urbanisation in vulnerable areas; part is linked to climatic factors such as changes in precipitation, flooding and drought events. There are differences in balance between the causes by region and type of event. Because of the complexities involved in delineating both the socio-economic and the climatic factors, the proportion of the contribution from human-induced

Table 7.4 Individual events included in the aggregates in Table 7.3 that incurred over $US5 billion of economic loss and over $US1 billion of insured loss

Year	Event	Area	Economic losses (billion $US)	Ratio: insured/ economic losses
1995	Earthquake	Japan	112.1	0.03
1994	Northridge Earthquake	USA	50.6	0.35
1992	Hurricane Andrew	USA	36.6	0.57
1998	Floods	China	30.9	0.03
1993	Floods	USA	18.6	0.06
1991	Typhoon Mireille	Japan	12.7	0.54
1989	Hurricane Hugo	Caribbean, USA	12.7	0.50
1999	Winterstorm Lothar	Europe	11.1	0.53
1998	Hurricane Georges	Caribbean, USA	10.3	0.34
1990	Winterstorm Daria	Europe	9.1	0.75
1993	Blizzard	USA	5.8	0.34
1996	Hurricane Fran	USA	5.7	0.32
1987	Winterstorm	W. Europe	5.6	0.84
1999	Typhoon Bart	Japan	5.0	0.60

climate change cannot be defined with any certainty – although it is interesting to note that the growth rate in damage cost of weather-related events was three times that of non-weather-related events for the period 1960–99.

Recent history has shown that weather-related losses can stress insurers to the point of bankruptcy. Hurricane Andrew in 1992 broke the $US20 billion barrier for insured loss and served as a wake-up call to the industry. Hurricane Katrina in 2005 that had been a Category 5 storm but weakened to a Category 3 before landfall caused a storm surge, supplemented by waves that reached around 5 m above sea level in the city of New Orleans, overtopping and breaching sections of the city's sea defences, flooding 70–80% of New Orleans. Well over 1000 people died, private insurance claims in excess of $US40 billion were made and total economic losses are estimated to be over $US100 billion or around 1% of US annual GDP.[69] Katrina was the costliest hurricane ever in terms of economic damage. Other records regarding Atlantic hurricanes were also broken in 2005,[70] the most hurricanes (13), strongest ever (Wilma) and costliest in total ($US200 billion +).

As a result of such events, in many flood-prone areas insurance premiums have risen dramatically and for many properties insurance flood cover has been withdrawn. In order to formulate their future business policy, the insurance industry is actively studying likely future trends in the incidence of disasters due to climate change along with related socio-economic trends in both the developed and the developing world.

weather-related extreme events, then multiplying by two to allow for the factors mentioned above (e.g. associated or knock-on costs) and further multiplying by four to allow for the likely increase in extreme events, say by the middle of the twenty-first century, we end up with a figure of about 2% of GDP. This can be compared with an estimate of 0.5–1% in the Stern Review on the economics of climate change of 2006,[68] where a much smaller increase was assumed in the risk of extreme events. However the Stern Review emphasises that a steeper rise is likely later in the century because of the steep rise in the incidence of extremes as global average temperature increases even more. Further, my estimate here and those in the Stern review are 'money' estimates made in the context of developed countries' economies. As the Stern Review also points out, the real total costs of extreme events taking into account all damages (including those that cannot be expressed in money terms) are likely to be very much larger especially in many developing countries.

Costing the total impacts

We now turn to consider all the impacts of anthropogenic climate change, attempts that have been made to express their cost in monetary terms and the validity of the methods employed. The IPCC 1995 Report contained a review of four cost studies[71] of the impacts of climate change in a world where the atmospheric carbon dioxide concentration had doubled from its pre-industrial level,[72] the most detailed studies being carried out for the United States. For those impacts against which some value of damage can be placed, estimates fell in the range of 1.0% to 1.5% of the US GDP in 1990. For other countries in the developed world, estimates of the cost of impacts in terms of percentage of GDP were similar. For the developing world, estimates of annual cost were typically around 5% of GDP (with a range of from 2% to 9% of GDP). These studies provided the first indication of the scale of the problem in economic terms. However, as the authors of these economic studies explain, their estimates were crude, were based on very broad assumptions, were mostly calculated in terms of the impact on today's economies rather than future ones and should not be considered as precise values.

Modelling the monetary impacts of climate change requires quantitative analysis connecting environmental, economic and social issues. The main tool for such studies is the Integrated Assessment Model (IAM) (see box in Chapter 9 on page 280) which includes all the elements illustrated in Figure 1.5. The Stern Review[73] has reviewed recent work employing such models pointing out the elements that need to be included in making cost estimates, in particular adaptation (which provides large potential for damage reduction but the cost of adaptation must be added), damage from extreme events (omitted in most studies to date) and non-market impacts (e.g. mortality from diseases, heat and cold stress, etc. omitted in many studies).

Adaptation is especially important in the agricultural sector.[74] In that sector, under changes in global average temperature of less than about 3 °C, when adaptation is taken into account, estimates of global aggregate economic impact cost vary from the slightly negative (i.e. slightly beneficial) to the moderately positive depending on underlying assumptions (see also Section 'Impact on agriculture and food supply', pages 196–202). However, these studies have largely ignored the increasing influence of climate extremes and as yet inadequately considered important factors such as water availability – largely because of the lack of detailed information regarding these.

Further, the aggregate hides large regional differences. Beneficial effects are expected predominantly at mid to high latitudes in the developed world especially where increased temperatures may bring longer growing seasons. Strongly negative effects are expected for populations at lower latitudes where any increase in average temperature or in dryness brings lower crop yields, where there is less capacity to adapt (e.g. because of lack of infrastructure, capital or education) or where there are poorer connections to regional and global

Only the rectangular support pillars that once held up the Highway 90 bridge spanning Biloxi Bay remained standing after Hurricane Katrina battered the Mississippi shoreline. The image shows widespread destruction along the shoreline, with only white foundations where buildings and homes once stood in some places. The Ikonos satellite captured this image on 2 September 2005, four days after Katrina came ashore.

trading systems. Overall, climate change will tend to tip agriculture production in favour of rich and well-fed regions at the expense of poorer and less well-fed regions.

For global average temperature increases of 2–3 °C from pre-industrial levels (i.e. up to a situation of doubled carbon dioxide concentration) which are expected to occur by early in the second half of the twenty-first century, the Stern Review has reviewed recent estimates and concluded that the cost of climate change could be equivalent to a loss of 0 to 3% in global GDP from what could have been achieved in a world without climate change.[76] The Stern Review goes on to point out that poor countries will suffer higher costs. Further, as

A US Coast Guardsman searches for survivors in New Orleans in the aftermath of Hurricane Katrina.

we saw in the last section I consider that the Stern Review has underestimated the likely cost of extreme events perhaps by a factor of two. Taking these into account would result in estimates of, say 1% to 4% loss of global GDP in developed countries and much greater loss of 5% or even 10% or more in many developing countries.

Two important factors have been omitted from the figures just cited. The first is that impacts cannot be quantified in terms of monetary cost alone. For instance, the loss of life (see Question 7), human amenity, natural amenity or the loss of species cannot be easily expressed in money terms. This can be illustrated by focusing on those who are likely to be particularly disadvantaged by global warming. Most of them will be in the developing world at around the subsistence level. They will find their land is no longer able to sustain them because it has been lost either to sea level rise or to extended drought. They will therefore wish to migrate and will become environmental refugees. It has been estimated that, under a business-as-usual scenario, the total number of persons displaced by the impacts of global warming could total of the order of 150 million by the year 2050 (or about 3 million per year on average) – about

Estimates of impacts costs under business-as-usual (BAU) from the Stern Review

Using the PAGE 2002 Integrated Assessment Model (IAM),[75] the Stern Review considers estimates of cost to the world's economies over the next two centuries if emissions of greenhouse gases continue on a 'business-as-usual' (BAU) path taking global average temperature increases possibly to 4 °C by 2100 (cf. Figure 6.4) and 8 °C by 2200. It is pointed out that modelling over many decades, regions and possible outcomes demands that distributional and ethical judgements are made systematically and explicitly, and that model results have to be treated with appropriate caution. Stern expresses the expected loss of future welfare due to climate change in terms of the future consumption that is forecast to occur with climate change compared with what would occur in the absence of climate change. It is explained that 'costs measured in this way are like a tax levied on consumption now and for ever, the proceeds of which are simply poured away'.

Stern presents the results from his modelling work as follows.[80]

- Under a basic calculation, the total cost of BAU climate change is estimated to equate to an average reduction in global per capita consumption of 5%, at a minimum, now and for ever. However, this calculation omitted three important factors as follows.
- Firstly, when direct impacts on the environment and human health (non-market impacts) are included, increases in the total cost of BAU climate change rise from 5% to 11%, although valuations here raise difficult ethical and measurement issues. But this does not fully include 'socially contingent' impacts such as social and political instability, which are very difficult to measure in monetary terms.
- Secondly, if climate feedbacks are taken into account, the projected increases in global average temperature would tend to the higher end of the range (see Chapter 6, page 143–6) and the estimated costs for BAU climate change could increase from 5% to 7% or from 11% to 14% if non-market impacts are included.
- Thirdly, a disproportionate burden of climate change impacts falls on poor regions of the world. Giving this burden stronger relative weight could increase the cost of BAU climate change by more than one-quarter.

Putting all these factors together increases the total cost of BAU climate change to the equivalent of around a 20% reduction in current per capita consumption now and for ever. Distributional judgements, a concern with living standards beyond those elements reflected in GDP and modern approaches to uncertainty all suggest that the appropriate estimate of damages may well lie in the upper part of the range 5–20%. Much but not all of that loss could be avoided through a strong mitigation policy. It is argued in later chapters of the Stern Review that this can be achieved at lower cost.

100 million due to sea level rise and coastal flooding and about 50 million due to the dislocation of agricultural production mainly due to the incidence and location of areas of drought.[78] The cost of resettling 3 million displaced persons per year (assuming that is possible) has been estimated at between $US1000 and $US5000 per person, giving a total of about $US10 000 million per year.[79] What the estimated cost for resettlement does not include, however (as the authors of the study themselves emphasise), is the human cost associated with displacement. Nor does it include the social and political instabilities that ensue when substantial populations are seriously disrupted because their means of livelihood has disappeared. The effects of these could be very large.

The second factor not being taken into account in the above estimates of total cost is the influence of the longer term – they only concern climate change impacts up to about the middle of the twenty-first century under the possibility of a doubling of equivalent carbon dioxide concentration. Soon after the end of the twenty-first century, under the scenarios with higher carbon dioxide emissions (in other words, if strong action is not taken to curb emissions), a further doubling of the equivalent carbon dioxide concentration will have occurred and it will be continuing to rise. The impacts of the additional climate change that would occur with a second effective doubling of carbon dioxide will be substantially more severe than those of the first doubling.[80] The Stern Review has considered these (see box) and estimated that the total cost of business-as-usual (BAU) climate change to be equivalent to around a 5–20% reduction in current global per capita consumption now and for ever, with a strong likelihood that it will be in the upper part of that range and with disproportionate losses tending to fall on poorer countries.

Impacts that may be a century away may not easily claim our attention. However, because of the long lifetime of some greenhouse gases, because of the long memory of the climate system, because some of the impacts may turn out to be irreversible and also because of the time taken for human activities and ecosystems to respond and change course, it is important to have an eye on the longer term. Looking at the longer term also raises for consideration the possibility of what are often called 'singular events' or irreversible events of large or unknown impact. Some of these have been mentioned earlier in this chapter or in previous chapters. Examples are given in Table 7.5. It is clearly difficult to provide quantitative estimates of the probability of such events. Nevertheless it is important that they are not ignored. One recent study[81] has allocated a potential damage cost to these of about 1% of GWP for a warming of 2.5 °C and about 7% of GWP for a warming of 6 °C. Such calculations are necessarily based on highly speculative assumptions, but in that particular study these singular events represent the largest single contributor to the total overall cost.

Table 7.5 Examples of singular non-linear events and their impacts[a]

Singularity	Causal process	Impacts
Non-linear response of thermohaline circulation (THC)	Changes in thermal and freshwater forcing could result in complete shutdown of North Atlantic THC or regional shutdown in the Labrador and Greenland Seas. In the Southern Ocean, formation of Antarctic bottom water could shut down. Such events are simulated by models and also found in the palaeoclimatic record.	Consequences for marine ecosystems and fisheries could be severe. Complete shutdown would lead to a stagnant deep ocean, with reducing deepwater oxygen levels and carbon uptake, affecting marine ecosystems. It would also represent a major change in heat budget and climate of northwest Europe.
Disintegration of West Antarctic Ice Sheet (WAIS)	WAIS may be vulnerable to climate change because it is grounded below sea level. Its disintegration could raise global sea level by 4 to 6 m. Large sea level rise from this cause is unlikely during the twenty-first century.	Considerable and rapid sea level rise would widely exceed adaptive capacity for most coastal structures and ecosystems.
Positive feedbacks in the carbon cycle	Climate change could reduce the efficiency of current oceanic and biospheric carbon sinks. Under some conditions the biosphere could become a source.[b] Gas hydrate reservoirs also may be destabilised, releasing large amounts of methane to the atmosphere.	Rapid, largely uncontrollable increases in atmospheric carbon concentrations and subsequent climate change would increase all impact levels and strongly limit adaptation possibilities.
Destabilisation of international order by environmental refugees and emergence of conflicts as a result of multiple climate change impacts	Climate change – alone or in combination with other environmental pressures – may exacerbate resource scarcities in developing countries. These effects are thought to be highly non-linear, with potential to exceed critical thresholds along each branch of the causal chain.	This could have severe social effects, which, in turn, may cause several types of conflict, including scarcity disputes between countries, clashes between ethnic groups and civil strife and insurgency, each with potentially serious repercussions for the security interests of the developed world.

[a] For recent comment see Frequently Asked Questions 10.2 in Solomon *et al.* (eds.) *Climate Change 2007: the Physical Science Basis*. pp. 818–19.

[b] See box on climate carbon/cycle feedbacks in Chapter 3, pages 48–9.

Table 7.6 Examples of impacts due to changes in extreme weather and climate events[b]

Phenomenon and direction of trend	Likelihood of future trends based on projections for twenty-first century using SRES scenarios	Examples of major projected impacts by sector
		Agriculture, forestry and ecosystems
Over most land areas, warmer and fewer cold days and nights, warmer and more frequent hot days and nights	*Virtually certain*[a]	Increased yields in colder environments; decreased yields in warmer environments; increased insect outbreaks
Warm spells/heatwaves. Frequency increases over most land areas	*Very likely*	Reduced yields in warmer regions due to heat stress; increased danger of wildfire
Heavy precipitation events. Frequency increases over most areas	*Very likely*	Damage to crops; soil erosion, inability to cultivate land due to waterlogging of soils
Area affected by drought increases	*Likely*	Land degradation; lower yields/ crop damage and failure; increased livestock deaths; increased risk of wildfire
Intense tropical cyclone activity increases	*Likely*	Damage to crops; windthrow (uprooting) of trees; damage to coral reefs
Increased incidence of extreme high sea level (excludes tsunamis)[c]	*Likely*[d]	Salinisation of irrigation water, estuaries and fresh-water systems

[a] See Note 1, Chapter 4 for further details regarding definitions.

[b] These examples do not take into account developments in adaptive capacity.

Table 7.6

Water resources	Human health	Industry, settlement and society
Effects on water resources relying on snow melt; effects on some water supplies	Reduced human mortality from decreased cold exposure	Reduced energy demand for heating; increased demand for cooling; declining air quality in cities; reduced disruption to transport due to snow, ice; effects on winter tourism
Increased water demand; water quality problems, e.g. algal blooms	Increased risk of heat-related mortality, especially for the elderly, chronically sick, very young and socially isolated	Reduction in quality of life for people in warm areas without appropriate housing; impacts on the elderly, very young and poor
Adverse effects on quality of surface and groundwater; contamination of water supply; water scarcity may be relieved	Increased risk of deaths, injuries and infectious, respiratory and skin diseases	Disruption of settlements, commerce, transport and societies due to flooding; pressures on urban and rural infrastructures; loss of property
More widespread water stress	Increased risk of food and water shortage; increased risk of malnutrition; increased risk of water- and food borne diseases	Water shortage for settlements, industry and societies; reduced hydropower generation potentials; potential for population migration
Power outages causing disruption of public water supply	Increased risk of deaths, injuries, water- and food- borne diseases; post-traumatic stress disorders	Disruption by flood and high winds; withdrawal of risk coverage in vulnerable areas by private insurers; potential for population migrations; loss of property
Decreased fresh-water availability due to salt water intrusion	Increased risk of deaths and injuries by drowning in floods; migration-related health effects	Costs of coastal protection versus costs of land-use relocation; potential for movement of populations and infrastructure; also see tropical cyclones above

[c] Extreme high sea level depends on average sea level and on regional weather systems. It is defined as the highest 1% of hourly values of observed sea level at a station for a given reference period.

[d] In all scenarios, the projected global average sea level at 2100 is higher than in the reference period. The effect of changes in regional weather systems on sea level extremes has not been assessed.

SUMMARY

- The main impacts of climate change will be due to sea level rise, increases in temperature and heat waves and a more intense hydrological cycle leading on average to more frequent and intense floods, droughts and storms (see Table 7.6 for a summary of impacts of extreme events).
- There are many ways in which the environment is being degraded due to human activities, for instance, through over-withdrawal of groundwater, loss of soil or deforestation. Global warming will exacerbate these degradations.
- To respond to climate change impacts, it will be necessary to adapt. In many cases this will involve changes in infrastructure, for instance new sea defences or water supplies. Many of the impacts of climate change will be adverse, but even when the impacts in the long term turn out to be beneficial, in the short term the process of adaptation will mostly have a negative impact and involve cost.
- Through adaptation to different crops and practices, first indications are that the total of world food production may not be seriously affected by climate change – although studies have not yet taken into account the likely occurrence of climate extremes. However, the combination of population growth and climate change will mean that disparity in per capita food supplies between the developed and the developing world will become much larger.
- Because of the likely rate of climate change, there will also be a serious impact on natural ecosystems, especially at mid to high latitudes. Forests especially will be affected by increased climate stress causing substantial dieback and loss of production, associated with which there is the positive feedback of additional carbon dioxide emissions. In a warmer world longer periods of heat stress will have an effect on human health; warmer temperatures will also encourage the spread of certain tropical diseases, such as malaria, to new areas.
- Economists have attempted to estimate the average annual cost in monetary terms of the impacts that would arise under the climate change due to a doubling of pre-industrial atmospheric carbon dioxide concentration. If allowance is added for the impact of extreme events, the estimates are typically around 1% to 4% of GDP for developed countries and 5–10 % or more for many developing countries. Later chapters will compare them with the cost of taking action to slow the onset of global warming or reduce

its overall magnitude. However, these attempts at monetary costing only represent a part of the overall impact story that must include the cost in human terms, for instance, the large social and political disruption some of the impacts will bring. In particular, it is estimated that there could be up to 3 million new environmental refugees each year or over 150 million by the middle of the twenty-first century. Refinements of all these estimates and the assumptions on which they are based are urgently required.

- Estimates of overall impact need to take the longer term into account. The cost of continuing with business-as-usual (BAU) has been estimated by the Stern Review as the equivalent of 5–20% reduction in per capita consumption now and for ever with a strong likelihood that it will be in the upper part of that range and with disproportionate losses falling on poorer countries.

However, many will ask why we should be concerned about the state of the Earth so far ahead in the future. Can we not leave it to be looked after by future generations? The next chapter will give something of my personal motivation for caring about what happens to the Earth in the future as well as now.

QUESTIONS

1. For your local region, find out about its water supply and how the water is used (e.g. by domestic users, agriculture, industry, etc.). What are likely to be the trends in its use over the next 50 years due, for instance, to population changes or changes in agriculture or industry? What are the possibilities for increased supply and how might these be affected by climate change?
2. For your local area, find out about current environmental problems such as sea level rise due to subsidence, over-use of groundwater and air pollution affecting forests. Which of these are likely to be exacerbated by climate change? Try to estimate by how much.
3. For your local region, identify the possible impacts of climate change over the next 100 years and quantify them as far as you can. Attempt to make an estimate of the cost of the damage for each impact. How far could adaptation reduce each type of damage?
4. From the information in Chapter 6, make estimates of possible climate change by the middle of the next century for typical regions of boreal forest. Then estimate from Figure 7.16 for each of the three tree species what loss of productivity might occur in each case.

5 Make an estimate of the total volume of ice in the Greenland and Antarctic ice caps. What proportion would have to melt to increase the sea level by the 6 metres or so which occurred during the last interglacial period?
6 In the past, human communities have adapted to changes of many kinds including some changes in climate. It is sometimes argued that, because the adaptability of human beings is not fully allowed for, the likely damage from the impacts of climate change in the future tends to be overestimated. Do you agree?
7 In economic cost–benefit analyses, it is often necessary to attach a value to a 'statistical life'.[81] It is not human life itself that is being valued but a change in the risk of death averaged over a population of human beings. One way of attempting this valuation is to consider a person as an economic agent capable of producing economic output. However, the preferred approach is to value a statistical life on the basis of what individuals are willing to pay or accept for changes in the risk of death. This approach tends to produce very different money values between developed countries and developing countries. Do you think this is defensible? Give up to five examples of the analysis of particular environmental problems for which you think it would be useful to include the valuation of a statistical human life. Look for values that have been attributed in different circumstances. Do questions of equity have any relevance in your examples?
8 Increasing demand for biofuels substituting for gasoline or diesel for transport is leading to increasing use of land for biofuel crops. Find out how much land might be needed and the degree of competition with other crops (e.g. food or forests) and suggest how sustainable land use can be promoted.

FURTHER READING AND REFERENCE

Solomon, S., Qin, D., Manning, M., Chen, Z., Marquis, M., Averyt, K. B., Tignor, M., Miller, H. L. (eds.) 2007: *Climate Change 2007, The Physical Science Basis, Contribution of Working Group I to the Fourth Assessment Report of the Intergovernmental Panel on Climate Change*. Cambridge: Cambridge University Press.

Technical Summary (summarises the basic science and climate projections)

Chapter 10 Global climate projections (including temperature, precipitation and sea level)

Chapter 11 Regional climate projections

Parry, M., Canziani, O., Palutikof, J., van der Linden, P., Hanson, C. (eds.) 2007. *Climate Change 2007: Impacts, Adaptation and Vulnerability. Contribution of Working Group II to the Fourth Assessment Report of the Intergovernmental Panel on Climate Change*. Cambridge: Cambridge University Press.

Technical Summary

Chapter 3 Fresh water resources
Chapter 4 Ecosystems
Chapter 5 Food, fibre and forest products
Chapter 6 Coastal systems and low-lying areas
Chapter 7 Industry, settlement and society
Chapter 8 Human health
Chapters 9 to 16 Impacts on different world regions
Chapter 17 Adaptation practices
Chapter 19 Key vulnerabilities and risk
Chapter 20 Climate change and sustainability
Cross-chapter case studies: C1 The impact of the European 2003 heatwave; C2 Impacts of climate change on coral reefs; C3 Megadeltas: their vulnerabilities to climate change; C4 Indigenous knowledge for adaptation to climate change.

Schellnhuber, H. J., Cranmer, W., Nakicenovic, N., Wigley, T. Yohe, G. 2006. *Avoiding Dangerous Climate Change*. Cambridge: Cambridge University Press. Contributions to a conference on impacts, vulnerabilities, adaptations and solutions.

World Resources Institute www.wri.org. Valuable for its catalogue of climate data (e.g. greenhouse gas emissions)

UNEP 2007, *Global Environmental Outlook* (*GEO-4*) – a comprehensive assessment and catalogue of environmental degradation.

NOTES FOR CHAPTER 7

1 For comprehensive detail about climate change impacts, see Parry *et al.* (eds.) *Climate Change 2007: Impacts*.

2 From Summary for policymakers in, McCarthy, J.J., Canziani, O., Leavy, N.A., Dotten, D.J., White, K.S. (eds.) 2001. *Climate Change 2001: Impacts, Adaptation and Vulnerability. Contribution of Working Group II to the Third Assessment Report of the Intergovernmental Panel on Climate Change*. Cambridge: Cambridge University Press; see also Parry *et al.* (eds.) *Climate Change 2007: Impacts*.

3 See, for instance, *Global Environmental Outlook GE04 (UNEP Report)*. 2007. Nairobi, Kenya; UNEP. See also Goudie, A. 2000. *The Human Impact on the Natural Environment*, fifth edition. Cambridge, Mass: MIT Press.

4 Bindoff, N., Willebrand, J. *et al.* Observations: Oceanic climate change and sea level, Chapter 5, in Parry *et al.* (eds.) *Climate Change 2007: Impacts*.

5 Lowe, J.A. Gregory, J.M. 2006. *Journal of Geophysical Research*, **111**, C11014, doi:10.1029/2005JC003421.

6 Christoffersen, P., Hambrey, M.J. 2006. Is the Greenland ice sheet in a state of collapse? *Geology Today*, **22**, 98–103.

7 Hansen, J. *et al.* 2007. Climate change and trace gases. *Philosophical Transactions of the Royal Society A*, **365**, 1925–54.

8 Pfeffer, W.T., Harper, J.T., O'Neel, S. 2008. *Science*, **321**, 1340–3.

9 Witze, A. 2008. *Nature*, **452**, 798–802.

10 Lowe and Gregory 2006.

11 More detailed information is available in Nicholls, R.J., Wong, P.P., *et al.* Coastal systems and low-lying areas, Chapter 6, in Parry *et al.* (eds.) *Climate Change 2007: Impacts*.

12 *Ibid.*, Cross-Chapter case studies, C3 Megadeltas.

13 For a comprehensive account of the impact of climate change on Bangladesh see Warrick, R.A., Ahmad, Q.K. (eds.) 1996. *The Implications of Climate and Sea Level Change for Bangladesh*. Dordrecht: Kluwer.

14 Nicholls, R.J., Mimura, N. 1998. Regional issues raised by sea level rise and their policy implications. *Climate Research*, **11**, 5–18.

15 Broadus, J.M. 1993. Possible impacts of, and adjustments to, sea-level rise: the case of Bangladesh and Egypt. In Warrick, R.A.,

Barrow, E.M., Wigley, T.M.L. (eds.) 1993. *Climate and Sea -Level Change: Observations, Projections and Implications*. Cambridge: Cambridge University Press, pp. 263–75. Note that, because of variations in the ocean structure, sea-level rise would not be the same everywhere. In Bangladesh it would be somewhat above average (see Gregory, J.M. 1993. Sea level changes under increasing CO_2 in a transient coupled ocean–atmosphere experiment. *Journal of Climate*, **6**, 2247–62).

16 See Chapter 4 in Warrick and Ahmad (eds.) *The Implications of Climate and Sea Level Change*.

17 Broadus, in Warrick *et al.* (eds.) *Climate and Sea-Level Change*, pp. 263–75.

18 Milliman, J.D. 1989. Environmental and economic implications of rising sea level and subsiding deltas: the Nile and Bangladeshi examples. *Ambio*, **18**, 340–5.

19 From a report entitled *Climate Change due to the Greenhouse Effect and its Implications for China*. 1992. Gland, Switzerland: Worldwide Fund for Nature.

20 Day, J.W. *et al.* 1993. Impacts of sea-level rise on coastal systems with special emphasis on the Mississippi river deltaic plain. In Warrick *et al.* (eds.) *Climate and Sea-Level Change*, pp. 276–96.

21 Clayton, K.M. 1993. Adjustment to greenhouse gas induced sea-level rise on the Norfolk coast: a case study. In Warrick *et al.* (eds.) *Climate and Sea-Level Change*, pp. 310–21.

22 Nicholls, R.J., Mimura, N. 1998. Regional issues raised by sea-level rise and their policy implications. *Climate Research*, **11**, 5–18. See also de Ronde, J.G. 1993. What will happen to the Netherlands if sea-level rise accelerates? In Warrick *et al.* (eds.) *Climate and Sea-Level Change*, pp. 322–35.

23 See Nurse, L., Sem, G. *et al.* 2001. Small island states. Chapter 17, in McCarthy *et al.*, (eds.) *Climate Change 2001: Impacts*.

24 Bijlsma, L. 1996. Coastal zones and small islands. In Watson, R.T., Zinyowera, M.C., Moss, R.H. (eds.) 1996. *Climate Change 1995: Impacts, Adaptations and Mitigation of Climate Change: Scientific–Technical Analyses. Contribution of Working Group II to the Second Assessment Report of the Intergovernmental Panel on Climate Change*. Cambridge: Cambridge University Press, Chapter 9.

25 McLean, R.F., Tsyban, A. *et al.* 2001. Coastal zones and marine ecosystems. In McCarthy *et al.*, (eds.) *Climate Change 2001: Impacts*, Chapter 6.

26 From Figure 3.6 in Watson, R. and the Core Writing Team (eds.) 2001. *Climate Change 2001: Synthesis Report. Contribution of Working Groups I, II and III to the Third Assessment Report of the Intergovernmental Panel on Climate Change*. Cambridge: Cambridge University Press.

27 See Table 11.8 from Shiklomanov, I.A., Rodda, J.C. (eds.) 2003. *World Water Resources at the Beginning of the Twenty-first Century*. Cambridge: Cambridge University Press.

28 Kundzewicz, Z.W. Mata, L.J. *et al.*, Fresh water resources and their management. Chapter 3, in Parry *et al.* (eds.) *Climate Change 2007: Impacts*.

29 Quoted by Geoffrey Lean in 'Troubled waters', in the colour supplement to the *Observer* newspaper, 4 July 1993.

30 Betts, R.A. *et al.* 2007, *Nature*, **448**, 1037–41.

31 All but the last are items of high confidence or very high confidence from a list in Box TS.5, p. 44 in the Technical Summary of Parry *et al.* (eds.) *Climate Change 2007: Impacts*.

32 *Ibid*., Table 3.5, in Chapter 3.

33 Waggoner, P.E. 1990. *Climate Change and US Water Resources*. New York: Wiley.

34 See UNCCD website: www.unccd.int/.

35 Solé, R. 2007. *Nature*, **449**, 151–3.

36 Crosson, P.R., Rosenberg, N.J. 1989. Strategies for agriculture. *Scientific American*, **261**, September, pp. 78–85.

37 Easterling, W., Aggarwal, P. *et al.*, Executive summary, Chapter 5, in Parry *et al.* (eds.) *Climate Change 2007: Impacts*.

38 Information in proposal for an International Research Institute for Climate Prediction. Report by Moura, A.D. (ed.) 1992. Prepared for the International Board for the TOGA project. Geneva: World Meteorological Organization.

39 See recent study by Battisti D. S. and R. L. Naylor, 2009, *Science*, **323**, 240–4.

40 Reilly, J. *et al.* 1996. Agriculture in a changing climate. Chapter 13, in Watson *et al.* (eds.) *Climate Change 1995: Impacts*.

41 Easterling, W., Aggarwal, P. *et al.* Chapter 5, in Parry *et al.* (eds.) *Climate Change 2007: Impacts*.

42 Stafford, N. 2007. *Nature*, **448**, 526–8.

43 Easterling, W., Aggarwal, P. *et al.*, Chapter 5, in Parry *et al.* (eds.) *Climate Change 2007: Impacts.*

44 See *Global Environmental Outlook 3 (UNEP Report).* 2002. London: Earthscan, pp. 63–5; also *Global Environmental Outlook 4 (GEO 4).* 2007. Nairobi, Kenya: UNEP, p. 95

45 Parry, M. *et al.* 1999. Climate change and world food security: a new assessment. *Global Environmental Change*, **9**, S51–S67.

46 From Watson *et al.* (eds.) *Climate Change 2001: Synthesis Report*, paragraph 5.17.

47 Miko, U.F. *et al.* 1996. Climate change impacts on forests. Chapter 1, in Watson *et al.*, (eds.) *Climate Change 1995: Impacts.* See also Gitay, H. *et al.* 2001. Ecosystems and their goods and services. Chapter 5, Section 5.6.3, in McCarthy *et al.* (eds.) *Climate Change 2001: Impacts.*

48 Gates, D.M. 1993. *Climate Change and Its Biological Consequences.* Sunderland, Mass.: Sinauer Associates Inc., p. 77.

49 Cox, P.M. *et al.* 2004. Amazon dieback under climate-carbon cycle projections for the twenty-first century. *Theoretical and Applied Climatology*, **78**, 137–56.

50 Melillo, J.M. *et al.* 1996. Terrestrial biotic responses to environmental change and feedbacks to climate. Chapter 9, in Houghton, J.T., Meira Filho, L.G., Callander, B.A., Harris, N., Kattenberg, A., Maskell, K. (eds.) *Climate Change 1995: The Science of Climate Change.* Cambridge: Cambridge University Press. See also Miko, U.F. *et al.* 1996. Climate change impacts on forests. Chapter 1, in Watson *et al.* (eds.) *Climate Change 1995: Impacts.*

51 Gitay, H. *et al.* 2001. Ecosystems and their goods and services. Technical Summary, Section 4.3, in McCarthy *et al.* (eds.) *Climate Change 2001: Impacts.*

52 Detail in Summary for policymakers, in Watson *et al.* (eds.) *Climate Change 2001: Synthesis Report*, pp. 68–69, paragraph 3.18. Myers, N. *et al.* 2000. *Nature*, **403**, 853–8 has proposed concentrating conservation effort in selected places with exceptional concentrations of biodiversity. For the problems of estimating the effects of global warming on biodiversity see Botkin, D.B. *et al.* 2007. *BioScience*, **57**, 227–36.

53 From Tegart, W.J., McG. Sheldon, G.W., Griffiths, D.C. (eds.) 1990. *Climate Change: The IPCC Impacts Assessment.* Canberra: Australian Government Publishing Service, pp. 6–20. Although made in 1990 this statement remains true in 2007.

54 Sale, P.F. 1999. *Nature*, **397**, 25–7. More information regarding diversity in corals available on World Resources Institute website: www.wri.org/wri/marine.

55 More information about impact on corals in special section on corals, pp. 850–7, in Parry *et al.* (eds.) *Climate Change 2007: Impacts.*

56 Turley, C. *et al.* 2006. Reviewing the impact of increased atmospheric CO_2 on oceanic pH and the marine ecosystem. In Schellnhuber, H.J. (ed.) *Avoiding Dangerous Climate Change.* pp. 65–70.

57 Fischlin, A., Midgley, G.F. *et al.* Chapter 4, p. 213 in Parry *et al.* (eds.) *Climate Change 2007: Impacts.*

58 Kalkstein, I.S. 1993. Direct impact in cities. *Lancet*, **342**, 1397–9.

59 Information on India from Dr Rajendra K. Pachauri, Tata Energy Research Institute. Information regarding Europe from World Meterological Organization, Geneva.

60 Nicholls, N. 1993. El Niño–Southern Oscillation and vector-borne disease. *Lancet*, **342**, 1284–5. The El Niño cycle is described in Chapter 5.

61 This statement taken from the summary for policymakers, Chapter 9, in Parry *et al.* (eds.) *Climate Change 2007: Impacts.*

62 Bulleted list from Box TS6, *ibid.*, Technical Summary. Meaning of asterisks: *** very high confidence (over 90% chance of being correct), ** high confidence (about 80% chance), * medium confidence (about 50% chance).

63 For a fuller discussion on adaptation see section 20.8, *ibid.*

64 See *Global Environmental Outlook 3 (UNEP Report).* 2002. London: Earthscan, pp. 274–5.

65 As an example of progress with respect to disaster preparedness, the International Red Cross has formed a Climate Change Unit based in the Netherlands.

66 PAHO Report 1999 Conclusions and Recommendations: Meeting on Revaluation of Preparedness and Response to Hurricanes George and Mitch, quoted in McMichael, A. *et al.* 2001. Human health. Chapter 9, in McCarthy *et al.* (eds.) *Climate Change 2001: Impacts.*

67 *Global Environmental Outlook 3*, p. 272.
68 Stern Review, Chapter 5.
69 Box 7.4, in Parry *et al.* (eds.) *Climate Change 2007: Impacts*.
70 Dlugolecki, A. 2006. Thoughts about the impact of climate change on insurance claims. In *Climate Change and Disaster Workshop*, Hohenkammer, Germany, www.eetd.lbl.gov/insurance/documents/060525.hohenkammer.pdf
71 Studies by Cline, Fankhauser, Nordhaus and Tol presented in Pearce, D.W. *et al.* 1996. The social costs of climate change. Chapter 6, in Bruce, J., Hoesung Lee, Haites, E. (eds.) 1996. *Climate Change 1995: Economic and Social Dimensions of Climate Change*. Cambridge: Cambridge University Press.
72 For equivalent CO_2 this is likely to occur around the middle of the twenty-first century; see Chapter 6.
73 Stern Review, Chapter 6.
74 Smith, J.B. *et al.* Vulnerability to climate change and reasons for concern: a synthesis. Chapter 19, Box 19, in McCarthy *et al.*, (eds.) *Climate Change 2001: Impacts*.
75 Details in Hope, C. 2005. Integrated Assessment Models. In Helm, D. (ed.), *Climate-Change Policy*. Oxford: Oxford University Press, pp. 77–98.
76 Stern Review, Chapter 6, pp. 161–2.
77 *Ibid.*, Chapter 6, p. 161.
78 Myers, N., Kent, J. 1995. *Environmental Exodus: An Emergent Crisis in the Global Arena*. Washington, DC: Climate Institute; also Adger, N., Fankhauser, S. 1993. Economic analysis of the greenhouse effect: optimal abatement level and strategies for mitigation. *International Journal of Environment and Pollution*, **3**, 104–19.
79 Adger and Fankhauser 1993.
80 Cline, W.R. 1992. *The Economics of Global Warming*. Washington, DC: Institute for International Economics, Chapter 2.
81 Nordhaus, W.D., Boyer, J. 2000. *Warming the World: Economic Models of Global Warming*. Cambridge, Mass: MIT Press, pp. 87–91.
82 See, for instance, Pearce, D.W. *et al.*, in Bruce *et al.* (eds.) *Climate Change 1995: Economic and Social Dimensions*.

8 Why should we be concerned?

I HAVE been describing the large changes in climate that are beginning to occur as a result of human activities, and their impact in different parts of the world. But large and devastating changes are likely to be up to a generation away. So why should we be concerned? What responsibility, if any, do we have for the planet as a whole and the great variety of other forms of life that inhabit it, or for future generations of human beings? And does our scientific knowledge in any way match up with other insights, for instance ethical and religious ones, regarding our relationship with our environment? In this chapter I want to digress from the detailed consideration of global warming (to which I shall return) in order briefly to explore these fundamental questions and to present something of my personal viewpoint on them.

Earth in the balance

Al Gore, Vice-President of the United States in the Clinton Administration, entitled his book on the environment *Earth in the Balance*,[1] implying that there are balances in the environment that need to be maintained. A small area of a tropical forest possesses an ecosystem that contains some thousands of plant and animal species, each thriving in its own ecological niche in close balance with the others. Balances are also important for larger regions and for the Earth as a whole. These balances can be highly precarious, especially where humans are concerned.

One of the first to point this out was Rachel Carson in her book *Silent Spring*,[2] first published in 1962, which described the damaging effects of pesticides on the environment. Humans are an important part of the global ecosystem; as the size and scale of human activities continue to escalate, so can the seriousness of the disturbances caused to the overall balances of nature. Some examples of this were given in the last chapter.

It is important that we recognise these balances, in particular the careful relationship between humans and the world around us. It needs to be a balanced and harmonious relationship in which each generation of humans should leave the Earth in a better state, or at least in as good a state as they found it. The word that is often used for this is sustainability – politicians talk of sustainable development, a concept defined in Chapter 9 (box on page 272) and further analysed in Chapter 12 (page 393). This principle, and its link with the harmonious relationship between humans and nature, was given prominent place by the United Nations Conference on Environment and Development held at Rio de Janeiro in Brazil in June 1992. The first principle in a list of 27 in the Rio Declaration adopted by the Conference is 'Human beings are at the centre of concerns for sustainable development. They are entitled to a healthy and productive life in harmony with nature.'[3]

However, despite such statements of principle from a body such as the United Nations, many of the attitudes that we commonly have towards the Earth are not balanced, harmonious or sustainable. Some of these are briefly outlined in the following paragraphs.

Exploitation

Humankind has over many centuries been exploiting the Earth and its resources. It was at the beginning of the Industrial Revolution some 200 years ago that the potential of the Earth's minerals began to be realised. Coal, the result of the decay of primaeval forests and laid down over many millions of years, was the main source of energy for the new industrial developments. Iron ore to make steel was mined in vastly increased quantities. The search for other metals such as zinc, copper and lead was intensified until today many millions of tonnes are mined each year. Around 1960, oil took over from coal as the dominant world source of energy; oil and gas between them now supply over twice the energy supplied by coal.

We have not only been exploiting the Earth's mineral resources. The Earth's biological resources have also been attacked. Forests have been cut down on a large scale to make room for agriculture and for human habitation. Tropical forests are a particularly valuable resource, important for maintaining the climate of tropical regions. They have also been estimated to contain perhaps half of all the Earth's biological species. Yet only about half of the mature tropical forests that existed a few hundred years ago still stand.[4] At the present rate of destruction virtually all will be gone by the end of the twenty-first century.

Great benefits have come to humankind through the use of fossil fuels, minerals and other resources. Yet, much of this exploitation has been carried out with little or no thought as to whether this use of natural resources has been a responsible one. Early in the Industrial Revolution it seemed that resources were essentially limitless. Later on, as one source ran out others became available to more than take its place. Even now, for most minerals new sources are being found faster than present sources are being used. But the growth of use is such that this situation cannot continue. In many cases known reserves or even likely reserves will begin to run out during the next hundred or few hundred years. These resources have been laid down over many millions if not billions of years. Nature took about a million years to lay down the amount of fossil fuel that we now burn worldwide every year – and in doing so it seems that we are causing rapid change of the Earth's climate. Such a level of exploitation is clearly not in balance, not harmonious and not sustainable.

'Back to nature'

Almost the reverse of this attitude is the suggestion that we all adopt a much more primitive lifestyle and give up a large part of industry and intensive farming – that we effectively put the clock back two or three hundred years to before the Industrial Revolution. That sounds very seductive and some individuals can clearly begin to live that way. But there are two main problems.

The first is that it is just not practical. The world population is now some six times what it was 200 years ago and about three times that of 50 years ago. The world cannot be adequately fed without farming on a reasonably intensive scale and without modern methods of food distribution. Further,

The golden toad is an amphibian which was indigenous to only a 5-km^2 region of Costa Rica, and is now believed to be extinct. It is considered by some as one of the first creatures whose extinction can be definitively blamed on global warming. These toads only mate during a few weeks in April and May and depend upon seasonal pools of rainwater in which to lay their eggs. Warming sea surface temperatures in the adjacent oceans are blamed for decreased rainfall and drier conditions in the cloud forest where the golden toad made its home.

most people that have them would not be prepared to be without the technical aids – electricity, central heating, refrigerator, washing machine, television and so on – that give the freedom, the interest and the entertainment that is so much taken for granted. Moreover, increasing numbers of people in the developing world are also taking advantage of and enjoy these aids to a life of less drudgery and more freedom.

The second problem is that it fails to take account of human creativity. Human scientific and technical development cannot be frozen at a given point in history, insisting that no further ideas can be developed. A proper balance between humans and the environment must leave room for humans to exercise their creative skills.

Again, therefore, a 'back to nature' viewpoint is neither balanced nor sustainable.

The technical fix

A third common attitude to the Earth is to invoke the 'technical fix'. As a senior environmental official from the United States said to me some years ago, 'We cannot change our lifestyle because of the possibility of climate change, we just need to fix the biosphere.' It was not clear just what he supposed the technical fixes would turn out to be. The point that he was making is that, in the past, humans have been so effective at developing new technology to meet the problems as they arise, can it not be assumed that this will continue? Concern about the future then turns into finding the 'fixes' as they are required.

On the surface the 'technical fix' route may sound a good way to proceed; it demands little effort and no foresight. It implies that damage can be corrected when it has been created rather than avoided in the first place. But damage already done to the environment by human activities is causing problems now. It is as if, in looking after my home, I decided not to carry out any routine maintenance but 'fixed' the failures as they occurred. For my home that would be a high-risk route to follow: failure to rewire when necessary could easily lead to a disastrous fire. A similar attitude to the Earth is both arrogant and irresponsible. It fails to recognise the vulnerability of nature to the large changes that human activities are now able to generate.

Science and technology possess enormous potential to assist in caring for the Earth, but they must be employed in a careful, balanced and responsible way. The 'technical fix' approach is neither balanced nor sustainable.

The unity of the Earth

Having described attitudes that are not balanced or harmonious in their relationship to the Earth and that fail to contribute to sustainability, I now turn to describe attitudes that adequately address the problems that I have been presenting in this book, namely the damage to the Earth's ecosystems destroying species at an alarming rate and the damage to large numbers of the world's peoples especially those who are already poor and disadvantaged. These are bound to represent the responsibilities that we all have not just for each other but also for the larger world of all living things. We are, after all, part of that larger world. There is good scientific justification for this. We are becoming increasingly aware of our dependence on the rest of nature and of the interdependencies that exist between different forms of life, between living systems and the physical and chemical environment that surrounds life on the Earth – and indeed between ourselves and the rest of the universe.

The scientific theory named Gaia after the Greek Earth goddess and publicised particularly by James Lovelock emphasises these interdependencies.[5] Lovelock points out that the chemical composition of the Earth's atmosphere is very different from that of our nearest planetary neighbours, Mars and Venus. Their atmospheres, apart from some water vapour, are almost pure carbon dioxide. The Earth's atmosphere, by contrast, is 78% nitrogen, 21% oxygen and only 0.03% carbon dioxide. So far as the major constituents are concerned, this composition has remained substantially unchanged over many millions of years – a fact that is very surprising when it is realised that it is a composition that is very far from chemical equilibrium.

This very different atmosphere on the Earth has come about because of the emergence of life. Early in the history of life, plants appeared which photosynthesise, taking in carbon dioxide and giving out oxygen. There followed other living systems that 'breathe', taking in oxygen and giving out carbon dioxide. The presence of life therefore influences and effectively controls the environment to which living systems in turn adapt. It is the close match of the environment to the needs of life and its development that seems so remarkable and which Lovelock has brought to our notice. He gives many examples; I will quote one concerned with oxygen in the atmosphere. There is a critical connection between the oxygen concentration and the frequency of forest fires.[6] Below an oxygen concentration of 15%, fires cannot be started even in dry twigs. At concentrations above 25% fires burn extremely fiercely even in the damp wood of a tropical rainforest. Some species are dependent on fires for their survival; for instance, some conifers require the heat of fire to release their seeds from the seed pods. Above 25% concentration of oxygen there would be no forests; below 15%, the regeneration that fires provide in the world's forests would be absent. The oxygen concentration of 21% is ideal.

It is this sort of connection that has driven Lovelock to propose that there is tight coupling between the organisms that make up the world of living systems and their environment. He has suggested a simple model of an imaginary world called Daisyworld (see box below), which illustrates the type of feedback mechanisms that can lead to this coupling and exert control. This model is similar to one he proposed for the biological and chemical history of the Earth during the first 1000 million years after primitive life first appeared on the Earth some 3500 million years ago.

The real world is, of course, enormously more complex than Daisyworld, which is why the Gaia hypothesis has led to so much debate. Lovelock's first statement in 1972 of the hypothesis was that 'Life, or the biosphere, regulates or maintains the climate and the atmospheric composition at an optimum for itself.'[7] In his later writings he introduced the analogy between the Earth and a living organism, introducing a new science which he calls geophysiology[8] – a more recent book is entitled *Gaia: The Practical Science of Planetary Medicine*.

An advanced organism such as a human being has many built-in mechanisms for controlling the interactions between different parts of the organism and for self-regulation. In a similar way, Lovelock argues, the ecosystems on the Earth are so tightly coupled to their physical and chemical environments that the ecosystems and their environment could be considered as one organism with an integrated 'physiology'. In this sense he believes that the Earth is 'alive'.

That elaborate feedback mechanisms exist in nature for control and for adaptation to the environment is not in dispute. But many scientists feel that Lovelock has gone too far in suggesting that ecosystems and their environment can be considered as a single organism. Although Gaia has stimulated much scientific comment and research, it remains a hypothesis.[9] What the debate has done, however, is to emphasise the interdependencies that connect all living systems to their environment – the biosphere is a system in which is incorporated a large measure of self-control.

There is the hint of a suggestion in the Gaia hypothesis that the Earth's feedbacks and self-regulation are so strong that we humans need not be concerned about the pollution we produce – Gaia has enough control to take care of anything we might do. Such a view fails to recognise the effect on the Earth's system of substantial disturbances, in particular vulnerability of the environment with respect to its suitability for humans. To quote Lovelock:[10]

> Gaia, as I see her, is no doting mother tolerant of misdemeanours, nor is she some fragile and delicate damsel in danger from brutal mankind. She is stern and tough, always keeping the world warm and comfortable for those who obey the rules, but ruthless in her destruction of those who transgress. Her unconscious goal is a planet fit for life. If humans stand in the way of this, we shall be eliminated with as little pity as would be shown by the micro-brain of an intercontinental ballistic nuclear missile in full flight to its target.

The Gaia scientific hypothesis can help to bring us back to recognise two things: firstly, the inherent value of all parts of nature, and secondly our dependence, as human beings, on the Earth and on our environment. Michael Northcott has pointed out, for instance, that Gaian theory 'suggests all human beings, all creatures, are relationally interconnected by carbon cycle of the planet'.[12] Gaia remains a scientific theory – although some have seen it as a religious idea, supporting ancient religious beliefs. Many of the world's religions have drawn attention to the close relationship between humans and the Earth.

The Native American tribes of North America lived close to the Earth. One of their chiefs when asked to sell his land expressed his dismay at the idea and said, 'The Earth does not belong to man, man belongs to the Earth. All things are connected like the blood that unites us all.'[13] An ancient Hindu saying, 'The Earth is our mother, and we are all her children'[14] also emphasises a feeling of closeness to the Earth. Those who have worked closely with indigenous peoples have given many examples of the care with which, in a balanced way, they look after the trees, plants and animals in their local ecosystem.[15]

The Islamic religion teaches the value of the whole environment, for instance in a saying of the prophet Mohammed: 'He who revives a dead land will be rewarded accordingly, and that which is eaten by birds, insects and animals out of that land will be charity provided by God' – so pointing both to our duty to care for the natural environment and our obligation to allow all living creatures their rightful place within it.[16]

Judaism and Christianity share the stories of creation in the early chapters of the Bible that emphasise the responsibility of humans to care for the Earth – we shall refer to these stories again later on in the chapter. Further on in the

Daisyworld and life on the early Earth

Daisyworld is an imaginary planet spinning on its axis and orbiting a sun rather like our own. Only daisies live in Daisyworld; they are of two hues, black and white. The daisies are sensitive to temperature. They grow best at 20 °C; below 5 °C they will not grow and above 40 °C they wilt and die. The daisies influence their own temperature by the way they absorb and emit radiation: black ones absorb more sunlight and therefore keep warmer than white ones.

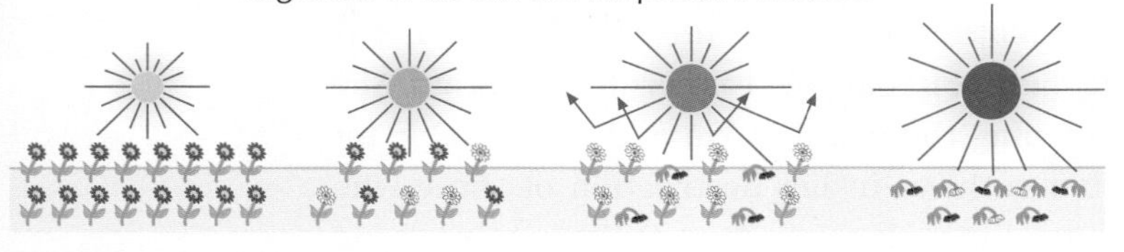

Figure 8.1 Daisyworld.

In the early period of Daisyworld's history (Figure 8.1), the sun is relatively cool and the black daisies are favoured because, by absorbing sunlight, they can keep their temperature closest to 20 °C. Most of their white cousins die because they reflect sunlight and fail to keep above the critical 5 °C. However, later in the planet's history, the sun becomes hotter. Now the white daisies can also flourish; both sorts of daisies are present in abundance. Later still as the sun becomes even hotter the white daisies become dominant as conditions become too warm for the black ones. Eventually, if the sun continues to increase its temperature even the white ones cannot keep below the critical 40 °C and all the daisies die.

Daisyworld is a simple model employed by Lovelock to illustrate the sort of feedbacks and self-regulation that occur in very much more complex forms within the living systems on the Earth.[11]

Lovelock proposes a similar simple model as a possible description of the early history of life on the Earth (Figure 8.2). The dashed line shows the temperature that would be expected on a planet possessing no life but with an atmosphere consisting, like our present atmosphere, mostly of nitrogen with about 10% carbon dioxide. The rise in temperature occurs because the sun gradually became hotter during this period. About 3500 million years ago primitive life appeared. Lovelock, in this model, assumes just two forms of life, bacteria that are anaerobic photosynthesisers – using carbon dioxide to build up their bodies but not giving out oxygen – and bacteria that are decomposers, converting organic matter back to carbon dioxide and methane. As life appears the temperature decreases as the concentration of the greenhouse gas, carbon dioxide, decreases. At the end of the period about 2300 million years ago, more complicated life appears; there is an excess of free oxygen and the methane abundance falls to low values, leading to another fall in temperature, methane also being a greenhouse gas. The overall influence of these biological processes has been to maintain a stable and favourable temperature for life on the Earth.

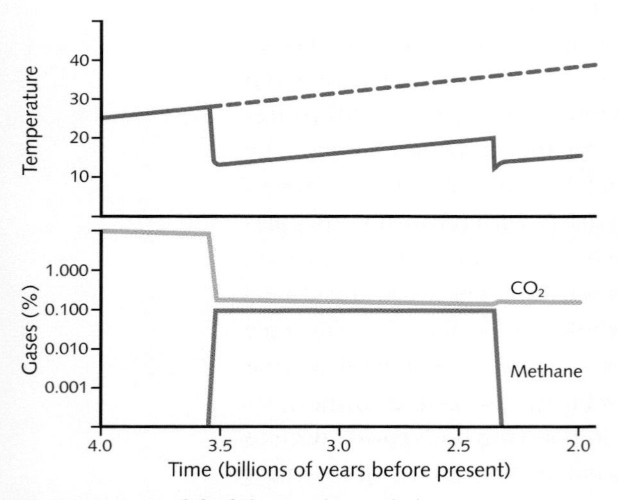

Figure 8.2 Model of the Earth's early history, as proposed by Lovelock.

Old Testament detailed instructions are given regarding care for the land and the environment.[17] Christianity was described by William Temple, Archbishop of Canterbury 60 years ago, as 'the most materialistic of the great religions'. Because of its central belief that God became human in Jesus (an event Christians call the Incarnation), Temple goes on to say 'by the very nature of its central doctrine Christianity is committed to a belief ... in the reality of matter and its place in the divine scheme'.[18] For the Christian, the twin doctrines of Creation and Incarnation demonstrate God's interest in and concern for the Earth and the life it contains.

In looking for themes that emphasise the unity between humans and their environment, we need not confine ourselves to the Earth. There is a very much larger sphere in which a similar perspective of unity is becoming apparent. Some astronomers and cosmologists, overwhelmed by the size, scale, complexity, intricacy and precision of the Universe, have begun to realise that their quest for an understanding of the evolution of the Universe right from the Big Bang some 14 000 million years ago is not just a scientific project but a search for meaning.[19] Why else has Stephen Hawking's book *A Brief History of Time*[20] become one of the bestsellers of our time?

In this new search for meaning, the perspective has arisen that the Universe was made with humans in mind – an idea expressed in some formulations of the 'anthropic principle'.[21] Two particular pointers emphasise this. Firstly, we have already seen that the Earth itself is fitted in a remarkable way for advanced forms of life. Cosmology is telling us that, in order for life on our planet to be possible, the Universe itself at the time of the Big Bang and in its early history needed to be 'fine-tuned' to an incredible degree.[22] Secondly, there is the remarkable fact that human minds, themselves dependent on the whole Universe for their existence, are able to appreciate and understand to some extent the fundamental mathematical structure of the Universe's design.[23] As Albert Einstein commented, 'The most incomprehensible thing about the universe is that it is comprehensible.' In the theory of Gaia, the Earth itself is central and humans are just one part of life on Earth; the insights of cosmology suggest that humans have a particular place in the whole scheme of things.

This section has recognised the intrinsic unity and interdependencies that exist not only on our Earth but also within the whole Universe, and the particular place that we humans have in the Universe. Being aware of these has large implications for our attitude to our environment.

Environmental values

What do we value in the environment and how do we decide what we need to preserve, to foster or improve? At the basis of our discussion so far have been several assumptions regarding the value or importance of different fundamental attitudes or actions, some of which I have associated with ideas that come from the underlying environmental science. Is it legitimate, however, to make connections of this kind between science and values? It is often argued that science itself is value free. But science is not an activity in isolation. As Michael Polanyi has pointed out, the facts of science cannot sensibly be considered apart from the participation and the commitment of those who discover those facts or incorporate them into wider knowledge.[24]

In the methodology and the practice of science are many assumptions of value. For instance, that there is an objective world of value out there to discover, that there is value in the qualities of elegance and economy in scientific theory, that complete honesty and cooperation between scientists are essential to the scientific enterprise. Further, progress in science demands a balanced view of all the data relevant to the area of investigation, not distorted by vested interests or personal or political agendas.

Values can also be suggested from the perspective of the underlying science as we have shown earlier in the chapter.[25] For instance, we have described the Earth in terms of balance, interdependency and unity. Since all of these are critical to the Earth as we know it, we can argue that they are of fundamental value and worth preserving. We have also provided some scientific evidence that humans have a particular place in the overall scheme of the natural world, that they possess special knowledge – which suggests that they also possess special responsibility.

Moving away from science, we have already referred to values related to the environment that come from our basic experiences as human beings. These are often called 'shared values' because they are common to different members of a human community – which may be a local community, a nation or ultimately the global community taking in the whole human race. An outstanding example is the conservation of the Earth and its resources, not just for our generation but for future generations. Other examples may involve how resources are used now for the benefit of the present generation of humans and how they are shared between different communities or nations. Holmes Rolston shows that in these areas of shared values, *natural* values (valuing the natural world) and *cultural* values (interpersonal, social and community values) belong together. He writes of 'a domain of hybrid values … the resultant of integrated influences from nature and culture'.[26]

When shared values are applied to real situations, however, conflicts often arise. For instance, how much should we forgo now in order to make provision for future generations, or how should resources be shared between different countries, for instance between those in the relatively rich 'North' and those in the relatively poor 'South'? How do we exercise our responsibility as humans to share the Earth with other parts of the creation? How much resource should be deployed to maintain particular ecosystems or to prevent loss of species? How do we apply principles of justice and equity in the real world? Discussion within and between human communities can assist in the definition and application of such shared values.

Many of these shared values have their origins in the cultural and religious backgrounds of human communities. Discussions about values need therefore to recognise fully the cultural and religious traditions, beliefs and assumptions that underlie many of our attitudes and reasoning about ethical concerns.

An obstacle to the recognition of religious assumptions in the attempt to establish environmental values is the view that religious belief is not consistent with a scientific outlook. Some scientists maintain that only science can provide real explanations based on provable evidence whereas the assertions of religion cannot be tested in an objective way.[27] Other scientists, however, have suggested that the seeming inconsistency between science and religion arises because of misunderstandings about the questions being addressed by the two disciplines and that there is more in common between the methodologies of science and religion than is commonly thought.[28]

Scientists are looking for descriptions of the world that fit into an overall scientific picture. They are working towards making this picture as complete as possible. For instance, scientists are looking for mechanisms to describe the 'fine-tuning' of the Universe (these are known as 'Theories of Everything'!) mentioned earlier. They are also looking for mechanisms to describe the inter-dependencies between living systems and the environment.

But the scientific picture can only depict part of what concerns us as human beings. Science deals with questions of 'how' not questions of 'why'. Most questions about values are 'why' questions. Nevertheless, scientists do not always draw clear distinctions between the two. Their motivations have often been associated with the 'why' questions. That was certainly true of the early scientists in the sixteenth and seventeenth centuries, many of whom were deeply religious and whose main driving force in pursuit of the new science was that they might 'explore the works of God'.[29]

That science and religion should be seen as complementary ways of looking at truth is a point made strongly by Al Gore in *Earth in the Balance*[30] which lucidly discusses current environmental issues such as global warming. He blames much of our lack of understanding of the environment on the modern approach, which tends to separate scientific study from religious and ethical issues. Science and technology are often pursued with a clinical detachment and without thinking about the ethical consequences. 'The new power derived from scientific knowledge could be used to dominate nature with moral impunity,' he writes.[31] He goes on to describe the modern technocrat as 'this barren spirit, precinct of the disembodied intellect, which knows the way things work but not the way they are'.[32] However, he also points out that 'there is now a powerful impulse in some parts of the scientific community to heal the breach'[33] between science and religion. In particular, as we pursue an understanding of the Earth's environment, it is essential that scientific studies and technological inventions are not divorced from their ethical and religious context.

Stewards of the Earth

The relationship between humans and the Earth that I have been advocating is often described as one of stewardship. We are on the Earth as its stewards. The word implies that we are carrying out our duty as stewards on behalf of someone else – but whom? Some environmentalists see no need to answer the question specifically, others might say we are stewards on behalf of future generations or on behalf of a generalised humanity. A religious person would want to be more specific and say that we are stewards on behalf of God. The religious person would also argue that to associate the relationship of humans to God with the relationship of humans to the environment is to place the latter relationship in a wider, more integrated, context – providing additional insights and a more complete basis for environmental stewardship.[34]

In the Judaeo-Christian tradition in the story of creation in the early chapters of the Bible is a helpful 'model' of stewardship – that of humans being 'gardeners' of the Earth. It is not only appropriate for those from those particular traditions – it is a model that can be widely applied. That story tells that humans were created to care for the rest of creation – the idea of human stewardship of creation is a very old one – and were placed in a garden, the Garden of Eden, 'to work it and take care of it'.[35] The animals, birds and other living creatures were brought to Adam in the garden for him to name them.[36] We are left with a picture of the first humans as 'gardeners' of the Earth – what does our work as 'gardeners' imply? I want to suggest four things:

- A garden provides food and water and other materials to sustain life and human industry. Part of the garden in the Genesis story contained mineral resources – 'the gold of that land is good; aromatic resin and onyx are also there'.[37] The Earth provides resources of many kinds for humans to use as they are needed.

- A garden is to be maintained as a place of beauty. The trees in the Garden of Eden were 'pleasing to the eye'.[38] Humans are to live in harmony with the rest of creation and to appreciate the value of all parts of creation. Indeed, a garden is a place where care is taken to preserve the multiplicity of species, in particular those that are most vulnerable. Millions of people each year visit gardens that have been especially designed to show off the incredible variety and beauty of nature. Gardens are meant to be enjoyed.
- A garden is a place where humans, created as described in the Genesis story in the image of God,[39] can themselves be creative. Its resources provide for great potential. The variety of species and landscape can be employed to increase the garden's beauty and its productivity. Humans have learnt to generate new plant varieties in abundance and to use their scientific and technological knowledge coupled with the enormous variety of the Earth's resources to create new possibilities for life and its enjoyment. However, the potential of this creativity is such that increasingly we need to be aware of where it can take us; it has potential for evil as well as for good. Further, good gardeners intervene in natural processes with a good deal of restraint.
- A garden is to be kept so as to be of benefit to future generations. In this context, I shall always remember Gordon Dobson, a distinguished scientist, who in the 1920s developed new means for the measurement of ozone in the atmosphere. His home outside Oxford in England possessed a large garden with many fruit trees. When he was 85, a year or so before he died, I remember finding him hard at work in his garden replacing a number of apple trees; in doing so he clearly had future generations in mind.

How well do we humans match up to the description of ourselves as gardeners caring for the Earth? Not very well, it must be said; we are more often exploiters and spoilers than cultivators. Some blame science and technology for the problems, although the fault must lie with the craftsman rather than with the tools! Others have tried to place part of the blame on attitudes[40] that they believe originate in the early chapters of Genesis, which talk of human beings having rule over Creation and subduing it.[41] Those words, however, should not be taken out of context – they are not a mandate for unrestrained exploitation. The Genesis chapters also insist that human rule over Creation is to be exercised under God, the ultimate ruler of Creation, and with the sort of care exemplified by the picture of humans as 'gardeners'. Why, therefore, do humans so often fail to get their act together?

The Garden of Eden, Jan Brueghel.

Equity – intergenerational and international

In our world community of human beings we are not all equal. Equality may be cited as an aim but before it can be pursued its terms need careful definition. Reality is full of inequities of many kinds. In the context of global warming, because it is long term and global, two equity issues are particularly important – both have already been mentioned. Firstly, there is our responsibility to future generations. A basic instinct is that we wish to see our children and grandchildren well set up in the world and wish to pass on to them some of our most treasured possessions. A similar desire would be that they inherit from us an Earth which has been well looked after and which does not pose to them more difficult problems than those we have had to face. But such an attitude is not universally held. I remember well, after a presentation in 1990 I made on global warming to Mrs Thatcher's Cabinet at Number Ten, Downing Street in London, a senior politician commented that the problem would not become serious in his lifetime and could be left for its solution to the next generation. I do not think he had appreciated that the longer we delay in taking action, the larger the problem becomes and the more difficult to solve. There is a need to face up to the problem now for the sake of the next and subsequent generations. We have no right to act as if there is no tomorrow. We also have a responsibility to give to those who follow us a pattern for their future based on the principle of sustainable development.

The second major equity issue is that of international equity where climate change creates an enormous challenge to the international community. The world's developed and richest nations have largely grown their wealth over 200 years from the cheap and bountiful energy available from coal, oil and gas without realising the damage that would do to the Earth and its climate – damage that will fall disproportionately on the poorest countries and people in the world. It is not just a problem of the past but the current disparity between the industrialised and the developing world in carbon dioxide emissions from fossil fuel burning continues to be very large (Figure 10.4). This disparity presents a strong moral imperative to the developed world, firstly to take strong action to reduce their carbon emissions and so reduce the damage they are continuing to cause, secondly to use their wealth and their skills to assist the developing world to develop their energy sources as sustainably as possible, and thirdly to find ways of providing some compensation for the damage already caused.[42] This is in fact an imperative expressed in the international Framework Convention on Climate Change (FCCC) (see Chapter 10) that states because of the benefits so far received by developed countries, they have to be the first to take action.

The will to act

Many of the principles I have been enunciating are included at least implicitly in the declarations, conventions and resolutions that came out of the United Nations Conference on Environment and Development held in Rio de Janeiro in June 1992; indeed, they form the background of many statements emanating from the United Nations or from official national sources. We are not short of statements of ideals. What tend to be lacking are the capability and resolve to carry them out. Sir Crispin Tickell, a British diplomat who has lectured widely on the policy implications of climate change, has commented 'Mostly we know what to do but we lack the will to do it.'[43]

Many recognise this lack of will to act as a 'spiritual' problem (using the word spiritual in a general sense), meaning that we are too obsessed with the 'material' and the immediate and fail to act according to generally accepted values and ideals particularly if it means some cost to ourselves or if it is concerned

The United Nations Conference on Environment and Development held in Rio de Janerio in 1992 (also known as the 'Earth Summit') resulted in several declarations, agendas and conventions, which formed the framework for the Kyoto Protocol of 1997.

with the future rather than with the present. We are only too aware of the strong temptations we experience at both the personal and the national levels to use the world's resources to gratify our selfishness and greed. Because of this, it has been proposed that at the basis of stewardship should be a principle extending what has traditionally been considered wrong[44] – or in religious parlance as sin – to include unwarranted pollution of the environment or lack of care for it.[45]

Those with religious belief tend to emphasise the importance of coupling together the relationship of humans to the environment to the relationship of humans to God.[46] It is here, religious believers would argue, that a solution for the problem of 'lack of will' can be found. That religious belief can provide an important driving force for action is often also recognised by those who look elsewhere than religion for a solution.

SUMMARY

This chapter has particularly pointed out that action addressing environmental problems is dependent not only on knowledge about them but on the values we place on the environment and our attitudes towards it. Assessments of environmental value and appropriate attitudes can be developed from the following:

- The perspectives of balance, interdependence and unity in the natural world generated by the underlying science.
- A recognition – some would argue suggested by the science – that humans have a special place in the Universe, which in turn implies that humans have special responsibilities with respect to the natural world.

- A recognition that to damage the environment or to fail to care for it is to do wrong.
- An interpretation of human responsibility in terms of stewardship of the Earth based on 'shared' values, generally recognised by human communities, that strives for equity and justice as between different human communities and different generations.
- A recognition of the importance of the cultural and religious basis for the principles of stewardship – humans as 'gardeners' of the Earth is suggested as a 'model' of such stewardship.
- The moral imperative for the sharing of wealth and skills based on recognition of the wealth created in developed countries through the availability of cheap fossil fuel energy and the damage caused by fossil fuel burning that falls disproportionately on developing countries that in their turn need to develop.
- A recognition that, just as the totality of damage to the environment is the sum of the damage done by a large number of individuals, the totality of action to address environmental problems is the sum of a large number of individual actions to which we can all contribute.[47]

The pursuit of many of these issues and their practical outworking involves the principle of sustainable development to which I shall return in later chapters (Chapter 9, page 272 and Chapter 12, page 393).

Finally, let me recall some words of Thomas Huxley, an eminent biologist from the nineteenth century, who emphasised the importance in the scientific enterprise of 'humility before the facts'. An attitude of humility is also one that lies at the heart of responsible stewardship of the Earth.

In the next chapter we shall reflect on the uncertainties associated with the science of global warming and consider how they can be taken into account in addressing the imperative for action. For instance, should action be taken now or should we wait until the uncertainties are less before deciding on the right action to take?

QUESTIONS

1 There is a debate regarding the relationship of humans to the environment. Should humans be at the centre of the environment with everything else and other life related to the human centre – in other words an anthropocentric view? Or should higher prominence be given to the non-human part of nature in our scheme of things and in our consideration of values – a more ecocentric view? If so, what form should this higher prominence take?

2 How far can science be involved in the generation and application of environmental values?
3 How far do you think environmental values can be generated through debate and discussion in a human community without reference to the cultural or religious background of that community?
4 It has been suggested that religious belief (especially strongly held belief) is a hindrance in the debate about environmental values. Do you agree?
5 Should we strive for universally accepted values with respect to the environment? Or is it acceptable for different communities to possess different values?
6 Identify and list as many values as you can that belong to the categories *natural* and *cultural* (see page 265). In what ways do items in these categories 'belong together'?
7 What principles underlie the concept of intergenerational equity? Suggest how it might be applied in practice. Suggest limits that might be set to its application.
8 A *moral imperative* was outlined in the equity section. What principles lie behind such an imperative? Are such principles universally held?
9 Equity is frequently imported into arguments about how the costs of damage due to global warming, of adaptation or mitigation, should be shared between nations in the light of their varying responsibilities for both past and present emissions. Look up the *Green Development Rights* website and the many other similar sites sponsored by NGOs concerned with this issue. Summarise the various proposals for sharing and analyse the arguments presented. What do you think are the strongest arguments and on what principles are they based?
10 An argument for religious belief that is sometimes put forward, irrespective of whether the belief is considered to have any foundation, is that such belief motivates people more strongly than other driving forces. Do you agree with this argument?
11 Explain how the cultural or religious traditions in which you have been brought up have influenced your view of environmental concern or action. How have these influences been modified because you now hold (or do not hold) definite religious beliefs?
12 Discuss the term 'stewardship', often used as a description of the relation of humans to the environment. Does it imply too anthropocentric a relationship?
13 Discuss the model of humans as 'gardeners' of the Earth. How adequate is the picture it presents of the relationship of humans to the environment?
14 Do you agree with Thomas Huxley when he spoke of the importance of humility before the scientific facts? How important do you think humility is in this context and in the wider context of the application of scientific knowledge to environmental concern?

15 Because of the formidability of the task of stewardship of the Earth, some have suggested that it is beyond the capability of the human race to tackle it adequately. Do you agree?

16 In Chapter 9 (see box on page 280) the concept of Integrated Assessment and Evaluation is introduced which involves all the natural and social science disciplines. In what ways could ethical or religious values be introduced into such evaluations? Is it appropriate and necessary that they be included?

17 It has been said that it is not easy for humans to make connection, especially of a moral kind, between taking a trip in an aircraft and a flooding disaster in Bangladesh. Can you suggest how this connection can be presented so that it appears relevant?

18 Brazil has proposed to the FCCC that nations should contribute to the solution of climate change in proportion to the damage from their historic emissions. Look up the UNFCCC website www.unfccc.int and other sources and find information about the Brazilian Proposal. What are its advantages and disadvantages? How do you think it could be modified so as to make it acceptable to all countries?

19 Climate change will impact on the world's poor more than on the world's rich people. Find out and compare how caring for the poor, especially those in regions or countries most disadvantaged by climate change, is approached by the world's major religions and by secular societies.

FURTHER READING AND REFERENCE

Gore, A. 1992. *Earth in the Balance*. Boston, Mass.: Houghton Mifflin Company.

Lovelock J. E. 1988. *The Ages of Gaia*. Oxford: Oxford University Press.

Northcott, M. 2007. *A Moral Climate: The Ethics of Global Warming*. London: Darton, Longman and Todd.

Russell C. 1994. *Earth, Humanity and God*. London: UCL Press. A review of environmental prospects for the planet from a Christian perspective.

Polkinghorne J. 1988. *Science and Creation*. London: SPCK; Polkinghorne, J. 1996. *Beyond Science*. Cambridge: Cambridge University Press. (On science, religion, values and culture).

Houghton J. 1995. *The Search for God: Can Science Help?* London: Lion Publishing, recently reprinted and available from Regent College, Vancouver book shop or the John Ray Initiative, www.jri.org.uk. Chapters on science and religion and one on global warming.

Berry R. J. (ed.) 2006. *Environmental Stewardship*. London: T & T Clark. Collection of 25 articles and essays on environmental stewardship.

Spencer, N. White, R. 2007. *Christianity, Climate Change and Sustainable Living*. London: SPCK. The challenge of achieving sustainability addressed especially to Christians.

NOTES FOR CHAPTER 8

1 Gore, A. 1992. *Earth in the Balance*. Boston, Mass.: Houghton Mifflin Company.
2 Carson, R. 1962. *Silent Spring*. Boston, Mass.: Houghton Mifflin Company.
3 See box in Chapter 9, page 272.
4 See Lean, G., Hinrichsen, D., Markham, A. 1990. *Atlas of the Environment*. London: Arrow Books.
5 Lovelock, J.E. 1979. *Gaia*. Oxford: Oxford University Press; Lovelock, J.E. 1988. *The Ages of Gaia*. Oxford: Oxford University Press.
6 Lovelock, *The Ages of Gaia*, pp. 131–3.
7 Lovelock, J.E., Margulis, L. 1974. *Tellus*, **26**, 1–10.
8 Lovelock, J.E. 1990. Hands up for the Gaia hypothesis. *Nature*, **344**, 100–12; also Lovelock, J.E. 1991. *Gaia: The Practical Science of Planetary Medicine*. London: Gaia Books.
9 Colin Russell discusses Gaia as a scientific hypothesis and also its possible religious connections in *The Earth, Humanity and God*. London: UCL Press, 1994.
10 Lovelock, *The Ages of Gaia*, p. 212.
11 For more details see Lovelock, *The Ages of Gaia*.
12 Northcott, M. *A Moral Climate*. 2007. London: Darton, Longman and Todd, p. 163.
13 Quoted by Gore, *Earth in the Balance*, p. 259.
14 Quoted by Gore, *Earth in the Balance*, p. 261.
15 Ghillean Prance, Director of Kew Gardens in the UK, provides examples from his extensive work in countries of South America in his book *The Earth under Threat*. Glasgow: Wild Goose Publications, 1996.
16 Khalil, M.H. 1993. Islam and the ethic of conservation. *Impact* (Newsletter of the Climate Network Africa), December, 8.
17 A number of injunctions were given to the Jews in the Old Testament regarding care for plants and animals and care for the land; for example, Leviticus 19:23–25, Leviticus 25:1–7, Deuteronomy 25:4.
18 Temple, W. 1964. *Nature, Man and God*. London: Macmillan (first edition 1934).
19 See for instance Davies, P. 1992. *The Mind of God*. London: Simon and Schuster. I have also addressed this theme in Houghton, J.T. 1995. *The Search for God: Can Science Help?* London: Lion Publishing – recently reprinted by the John Ray Initiative www.jri.org.uk.
20 Hawking, S. 1989. *A Brief History of Time*. London: Bantam.
21 See for instance Davies, *The Mind of God*; also Barrow, J., Tipler, F.J. 1986. *The Anthropic Cosmological Principle*. Oxford: Oxford University Press.
22 Barrow and Tipler, *The Anthropic Cosmological Principle*; Gribbin, J., Rees, M. 1991. *Cosmic Coincidences*. London: Black Swan.
23 Davies, *The Mind of God*.
24 Polanyi, M. 1962. *Personal Knowledge*. London: Routledge and Kegan Paul.
25 The relation of science to value is explored in Rolston, H. III. 1999. *Genes, Genesis and God*. Cambridge: Cambridge University Press, Chapter 4.
26 Rolston, H. III. 1988. *Environmental Ethics*. Philadelphia, Penn.: Temple University Press, p. 331.
27 See, for instance, Dawkins, R. 1986. *The Blind Watchmaker*. Harlow: Longman. Dawkins, R. 2006. *The God Delusion*, Bantam Press.
28 See, for instance, McGrath, A., McGrath, J. C. 2007. *The Dawkins Delusion*? London: SPCK. Polkinghorne, J. 1986. *One World*. London: SPCK; Polkinghorne, J. 1986. *Beyond Science*. Cambridge: Cambridge University Press; Houghton, *The Search for God*.
29 See, for instance, Russell, C. 1985. *Cross-Currents: Interactions between Science and Faith*. Leicester: Intervarsity Press.
30 Gore, *Earth in the Balance*.
31 *Ibid*., p. 252.
32 *Ibid*., p. 265.
33 *Ibid*., p. 254.
34 See Berry R.J. (ed.) 2006. *Environmental Stewardship*. London: T&T clark; also Berry R.J. (ed.) 2007. *When Enough is Enough*. Leicester: Invervarsity Press.
35 Genesis 2:15.
36 Genesis 2:19.
37 Genesis 2:12.
38 Genesis 2:9.
39 Genesis 1:27.
40 The best-known exposition of this position is, for instance, White, L. Jr. 1987. The historical roots of our ecological crisis. *Science*, **155**, 1203–7; see Russell, *The Earth, Humanity and God*, for a commentary on this thesis.
41 Genesis 1:26–8.

42 Brazil has proposed to the Framework Convention on Climate Change that nations should contribute to the solution of climate change in proportion to the damage from their historic emissions, a proposal that has been widely analysed, exposing the difficulty raised by the uncertain nature of much of the past data; see www.unfccc.int.

43 *The Doomsday Letters*, broadcast on BBC Radio 4, UK, 1996.

44 Patriarch Bartholomew of Constantinople and Pope John Paul II have both addressed this point – see Northcott, *A Moral Climate*, p. 153.

45 This was the first of the principles that came out of a symposium (called the Patmos Principles since the climax of the symposium, held in celebration of the 1900th anniversary of the writing of the Book of Revelation, was on the island of Patmos) I attended in 1995 sponsored by the Ecumenical Patriarch Bartholomew I of the Greek Orthodox Church and Prince Philip in his capacity as President of the World Wildlife Fund. An extremely eclectic group of scientists, politicians, environmentalists and theologians attended from a wide range of religious backgrounds and beliefs. John, the Metropolitan of Pergamon, who was chairman of the symposium's scientific committee, kept emphasising that we should consider pollution of the environment, or lack of care for the environment, as a sin – not only against nature but a sin against God. His message struck a strong chord with the symposium. The principle goes on to explain that this new category of sin should include activities that lead to 'species extinction, reduction in genetic diversity, pollution of the water, land and air, habitat destruction and disruption of sustainable life styles'. The symposium's report is edited by Sarah Hobson and Jane Lubchenco and published under the title *Revelation and the Environment: AD 95–1995*. Singapore: World Scientific Publishing, 1997.

46 In Judaeo-Christian teaching the coupling of these two relationships begins with the Creation stories in Genesis. These stories go on to describe how humans disobeyed God (Genesis 3) and broke the partnership. But the Bible continually explains how God offers a way back to partnership. A few chapters on in Genesis (9:8–17), the basis of the relationship between God and Noah is a covenant agreement in which 'all life on the Earth' is included as well as humans. A relationship based on covenant is also the basis of the partnership between God and the Jewish nation in the Old Testament. But, after many times when that relationship was broken, the Old Testament prophets looked forward to a new covenant based not on law but on a real change of heart (Jeremiah 31:31–34). The New Testament writers (for example Hebrews 8:10–11) see this new covenant being worked out through the life and particularly through the death and resurrection of Jesus, the Son of God. Jesus promised his followers the Holy Spirit (John 15, 16), whose influence would enable the partnership between them and God to work. Paul, in his letters, is constantly referring to the dependent relationship which forms the basis of his own partnership with God (Galatians 2:20, Philippians 4:13) and which has been the experience of millions of Christians down the centuries. Included in Paul's theology is the whole of Creation (Romans 8:19–22).

47 Edmund Burke, a nineteenth-century British politician said, 'no one made a greater mistake than he who did nothing because he could only do so little' – quoted at the end of Chapter 12.

9 Weighing the uncertainty

Pocerady power station in the Czech Republic is the backdrop to a commerical crop of sunflowers

THIS BOOK is intended to present clearly the current scientific position on global warming. A key part of this presentation concerns the uncertainty associated with all parts of the scientific description, especially with the prediction of future climate change, which forms an essential consideration when decisions regarding action are being taken. However, uncertainty is a relative term; utter certainty is not often demanded on everyday matters as a prerequisite for action. Here the issues are complex; we need to consider how uncertainty is weighed against the cost of possible action. First, we address the scientific uncertainty.

The scientific uncertainty

Before considering the 'weighing' process and the cost of action, we begin by explaining the nature of the scientific uncertainty and how it has been addressed by the scientific community.

In earlier chapters I explained in some detail the science underlying the problem of global warming and the scientific methods that are employed for the prediction of climate change due to the increases in greenhouse gases. The basic physics of the greenhouse effect is well understood. If atmospheric carbon dioxide concentration doubles and nothing else changes apart from atmospheric temperature, then the average global temperature near the surface will increase by about 1.2 °C. That figure is not disputed among scientists.

However, the situation is complicated by feedbacks and regional variations. Numerical models run on computers are the best tools available for addressing these complications because they are able effectively to add together all the non-linear interactions. Although they are highly complex, climate models are capable of giving useful information of a predictive kind. As was explained in Chapter 5, confidence in the models comes from the considerable skill with which they simulate present climate and its variations (including perturbations such as the Pinatubo volcanic eruption) and from their success in simulating past climates; these latter are limited as much by the lack of data as by inadequacies in the models.

However, model limitations remain, which give rise to uncertainty (see box below). The predictions presented in Chapter 6 reflected these uncertainties, the largest of which are due to the models' failure to deal adequately with clouds and with the effects of the ocean circulation. These uncertainties become of greatest importance when changes on the regional scale, for instance in regional patterns of rainfall, are being considered.

With uncertainty in the basic science of climate change and in the predictions of future climate, especially on the regional scale, there are bound also to be uncertainties in our assessment of the impacts of climate change. As Chapter 7 shows, however, some important statements can be made with reasonable confidence. Under nearly all scenarios of increasing carbon dioxide emissions this century, the rate of climate change is likely to be large, probably greater than the Earth has seen for many millennia. Many ecosystems (including human beings) will not be able to adapt easily to such a rate of change. The most noticeable impacts are likely to be on the availability of water (especially on the intensity of heat waves, the frequency and severity of droughts and floods), on the distribution (though possibly not on the overall size) of global food production and on sea level in low-lying areas of the world. Further, although most of our predictions have been limited in range

The reasons for scientific uncertainty

The Intergovernmental Panel on Climate Change[1] has described the scientific uncertainty as follows.

There are many uncertainties in our predictions particularly with regard to the timing, magnitude and regional patterns of climate change, due to our incomplete understanding of:

- sources and sinks of greenhouse gases, which affect predictions of future concentrations,
- clouds, which strongly influence the magnitude of climate change,
- oceans, which influence the timing and patterns of climate change,
- polar ice-sheets, which affect predictions of sea level rise.

These processes are already partially understood, and we are confident that the uncertainties can be reduced by further research. However, the complexity of the system means that we cannot rule out surprises.

to the end of the twenty-first century, it is clear that by the century beyond 2100 the magnitude of the change in climate and the impacts resulting from that change are likely to be very large indeed.

The statement in the box regarding scientific uncertainty was formulated for the IPCC 1990 Report. Eighteen years later it remains a good statement of the main factors that underlie scientific uncertainty. That this is the case does not imply little progress since 1990. On the contrary, as the subsequent IPCC Reports show, a great deal of progress has taken place in both scientific understanding and the development of models. There is now much more confidence that the signal of anthropogenic climate change is apparent in the observed climate record. Models now include much more sophistication in their scientific formulations and possess increased skill in simulating the important climate parameters. For regional scale simulation and prediction, regional climate models (RCMs) with higher resolution have been developed that are nested within global models (see Chapters 5 and 6). These RCMs are beginning to bring more confidence to regional projections of climate change. Further, over the last decade, a lot of progress has been made with studies in various regions of the sensitivity to different climates of these regions' resources, such as water and food. Coupling such studies with regional scenarios of climate change produced by climate models enables more meaningful impact assessments to be carried out[2] and also enables appropriate measures to be assessed. Particularly in some regions large uncertainties remain; it will be seen for instance from Figure 6.7 that current models perform better for some regions than for others.

Summarised in Figure 9.1 are the various components that are included in the development of projections of climate change or its impacts. All of these possess uncertainties that need to be aggregated appropriately in arriving at estimates of uncertainties in different impacts.

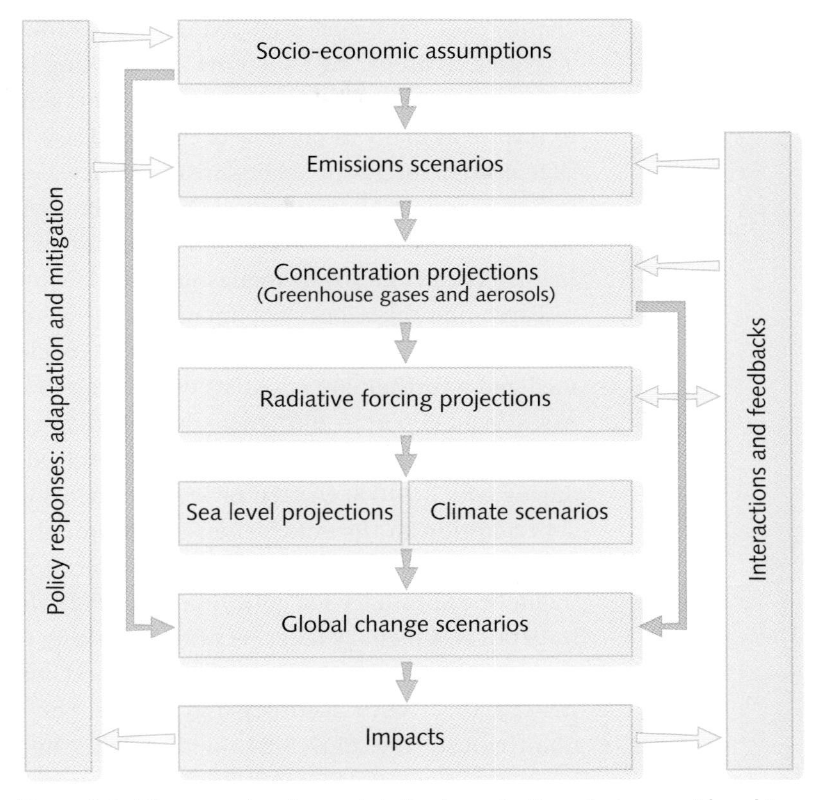

Figure 9.1 The cascade of uncertainties in projections to be considered in developing climate and related scenarios for climate change impact, adaptation and mitigation assessment.

The IPCC Assessments

Because of the scientific uncertainty, it has been necessary to make a large effort to achieve the best assessment of present knowledge and to express it as clearly as possible. For these reasons the IPCC was set up jointly by two United Nations' bodies, the World Meteorological Organization (WMO) and the United Nations Environmental Programme (UNEP).[3] The IPCC's first meeting in November 1988 was timely; it was held just as strong political interest in global climate change was beginning to develop. The Panel realised the urgency of the problem and, under the overall chairmanship of Professor Bert Bolin from Sweden, established three working groups, one to deal with the science of climate change, one with impacts and a third one to deal with policy responses. The IPCC has produced four main comprehensive Reports,[4] in 1990, 1995, 2001 and 2007, together with a number of special reports covering particular issues. Previous chapters have already referred widely to these reports.

I would like to say more about the Physical Science Assessment Working Group (of which I was chairman from 1988 until 1992 and co-chairman from 1992 until 2002).[5] Its task has been to present in the clearest possible terms our knowledge of the science of climate change together with our best estimate of the climate change over the twenty-first century that is likely to occur as a result of human activities. In preparing its reports the Working Group realised from the start that if they were to be really authoritative and taken seriously, it would be necessary to involve as many as possible of the world scientific community in their production. A small international organising team was set up at the Hadley Centre of the United Kingdom Meteorological Office at Bracknell and through meetings, workshops and a great deal of correspondence most of those scientists in the world (both in universities and government-supported laboratories) who are deeply engaged in research into the science of climate change were involved in the preparation and writing of the reports. For the first report, 170 scientists from 25 countries contributed and a further 200 scientists were involved in its peer review. For the Fourth Assessment Report in 2007, these numbers had grown to 152 lead authors and over 500 contributing authors and over 600 involved in the two-stage review process during which 30 000 written review comments were received and processed.

In addition to the comprehensive, thorough and intensively reviewed background chapters that form the basic material for each assessment, each report includes a Summary for policymakers (SPM), the wording of which is approved in detail at a plenary meeting of the Working Group, the object being to reach agreement on the science and on the best way of presenting the science to policymakers with accuracy and clarity. The plenary meeting which agreed unanimously the 2007 SPM, held in Paris in January 2007, was attended by representatives of 113 countries, a number of scientists representing the lead authors of the scientific chapters together with representatives from non-governmental organisations. There has been very lively discussion at these plenary meetings, most of which has been concerned with achieving the most informative and accurate wording rather than fundamental dispute over scientific content.

During the preparation of the reports, a considerable part of the debate amongst the scientists has centred on just how much can be said about the likely climate change in the twenty-first century. Particularly to begin with, some felt that the uncertainties were such that scientists should refrain from making any estimates or predictions for the future. However, it soon became clear that the responsibility of scientists to convey the best possible information could not be discharged without making estimates of the most likely magnitude of the change coupled with clear statements of our assumptions and the level of uncertainty in the estimates. Weather forecasters have a similar, although much more short-term, responsibility. Even though they may feel uncertain about tomorrow's weather, they cannot refuse to make a forecast. If they do refuse, they withhold from the public most of the useful information they possess. Despite the uncertainty in a weather forecast, it provides useful guidance to a wide range of people. In a similar way the climate models, although subject to uncertainty, provide useful guidance for policy.

An important feature of the Third and Fourth Science Assessments has been the presentation of uncertainty wherever possible in terms of probabilities. Words to express uncertainty have been associated with probabilities, for instance, *very likely* (more than 90% probability), *likely* (more than 67%), etc. This has substantially increased the value of future climate projections especially when considering the impacts of climate change and policy concerning adaptation to them.

I have given these details of the work of the Physical Science Assessment Group in order to demonstrate the degree of commitment of the scientific community to the understanding of global climate change and to the communication of the best scientific information to the world's politicians and policymakers. After all, the problem of global environmental change is one of the largest problems facing the world scientific community. No previous scientific assessments on this or any other subject have involved so many scientists so widely distributed both as regards their countries and their scientific disciplines. The IPCC Reports can therefore be considered as authoritative statements of the contemporary views of the international scientific community.

A further important strength of the IPCC is that, because it is an intergovernmental body, governments are involved in its work. In particular, government representatives assist in making sure that the presentation of the science is both clear and relevant from the point of view of the policymaker. Having been part of the process, governments as well as scientists are in a real sense owners of the resulting assessments – an important factor when it comes to policy negotiations.

In the presentation of the IPCC assessments to politicians and policymakers, the degree of scientific consensus achieved has been of great importance in persuading them to take seriously the problem of global warming and its impact. In the run-up to the United Nations Conference on Environment and Development (UNCED) at Rio de Janeiro in June 1992, the fact that they accepted the reality of the problem led to the formulation of the international Framework Convention on Climate Change (FCCC) which was signed by over 160 countries – President George Bush, Snr, signed for the United States and the United States Senate subsequently unanimously ratified it. It has often been commented that without the clear message that came from the world's scientists, orchestrated by the IPCC, the world's leaders would never have agreed to sign the Climate Convention.

After the publicity arising from the Earth Summit in 1992, debate concerning the scientific findings of the IPCC intensified in the world's press. Some of this

Rajendra Pachauri, chairman of IPCC, and other members of the IPCC delegation with Al Gore at the Nobel Peace Prize celebrations in Oslo, December 2007.

was honest scientific debate – argument and debate are, after all, intrinsic to the scientific process. Some, however, was stimulated by strong vested interests particularly in the United States that attempted to discredit the work of the IPCC and to persuade the public at large either of the absence of scientific evidence for global warming or even if it were occurring that it was not an issue requiring urgent attention.[6] From my standpoint from within the IPCC, these attacks tended to lead to enhanced clarity and accuracy in our reports although for the public at large the propagation of so much misleading material created a great deal of misunderstanding and confusion.[7]

Over the past 20 years as the global climate has steadily warmed and scientific effort to understand climate change has grown, evidence that climate change is bringing with it serious adverse impacts has continued to strengthen and to be widely recognised. This is well illustrated by the statement published in June 2005 by the Academies of Science of the world's 11 most important countries (the G8 plus India, China and Brazil) endorsing the conclusions of the IPCC and urging the governments meeting at the G8 summit that year in Edinburgh to take urgent action to address climate change.[8] The world's top scientists could not have spoken more strongly. A further endorsement of the IPCC and its work came in 2007 with the award of Nobel Peace Prizes to the IPCC and to Al Gore.

I have illustrated the work of the IPCC by describing in some detail the activity of the Physical Science Assessment Working Group. The IPCC has two other Working Groups that have followed similar procedures and have dealt with the Impacts of Climate Change, with Adaptation and Mitigation strategies and with the Economics and Social Dimensions of Climate Change. Contributions to their work have not only come from natural scientists; increasingly social scientists, especially economists, have become involved. In these social science areas much fresh ground has been broken as consideration has been given to questions of what, in the global context, might form the basis of appropriate political and economic response to climate change. The rest of this chapter and the following chapters will draw heavily on their work.

Narrowing the uncertainty

A key question constantly asked by policymakers is: 'How long will it be before scientists are more certain about the projections of likely climate change, in particular concerning the regional and local detail?' They were asking that question 20 years ago and then I generally replied that in 10 to 15 years we would know a lot more. As we have already seen, there is now more confidence that anthropogenic climate change has been detected and more confidence too in climate change projections than was the case a decade ago. However, some of the key uncertainties remain and their reduction is urgently needed. Not surprisingly, policymakers are still asking for more certainty. What can be done to provide it?

For the basic science of change, the main tools of progress are observations and models. Both need further development and expansion. Observations are required to detect climate change in all its aspects as it occurs and also to validate models. That means that regular, accurate and consistent monitoring of the most important climate parameters is required with good coverage in both space and time. Monitoring may not sound very exciting work, often even less exciting is the rigorous quality control that goes with it, but it is absolutely essential if climate changes are to be observed and understood. Because of this, a major international programme, the Global Climate Observing System (GCOS) has been set up to orchestrate and oversee the provision of observations on a global basis. Models are needed to integrate all the scientific processes that are involved in climate change (most of which are non-linear, which means they cannot be added together in any simple manner) so that they can assist in the

Space observations of the climate system

For forecasting the weather round the world – for airlines, for shipping, for many other applications and for the public – meteorologists rely extensively on observations from satellites. Under international agreements, five geostationary satellites are spaced around the equator for weather observation; moving pictures from them have become familiar to us on our television screens. Information from polar orbiting satellites is also available to the weather services of the world to provide input into computer models of the weather and to assist in forecasting (see for instance Figure 5.4).

Figure 9.2 The ENVISAT Earth Observation Satellite of the European Space Agency launched in 2002. Instruments included in its payload are: the Advanced Along-Track Scanning Radiometer (AATSR), the Michelson Interferometer for Passive Atmospheric Sounding (MIPAS), the MEdium Resolution Imaging Spectrometer (MERIS), the SCanning Image Absorption spectroMeter for Atmospheric CartograpHY (SCIAMACHY), the MicroWave Radiometer (MWR), the Global Ozone Monitoring by Observation of Stars (GOMOS), the Radar Altimeter – second generation (RA-2), the Advanced Synthetic Aperture Radar (ASAR), and other instruments for communication and exact tracking. DORIS stands for Doppler Orbitography and Radiopositioning Integrated by Satellite. In its 800-km Sun-synchronous orbit with the solar array deployed, it measures 26 m × 10 m × 5 m and weighs 8.1 tonnes.

These weather observations provide a basic input to climate models. But for climate prediction and research, comprehensive observations from other components of the climate system, the oceans, ice and land surface are required. ENVISAT, a satellite launched by the European Space Agency in 2002, is one example of the most recent generation of large satellites in which the latest techniques are directed to observing the Earth (Figure 9.2). The instruments are directed at the measurement of atmospheric temperature and composition including aerosols (MIPAS, SCIAMACHY and GOMOS), sea surface temperature and topography, the latter for ocean current information (AATSR and RA-2), information about ocean biology and land surface vegetation (MERIS) and sea-ice coverage and ice-sheet topography (ASAR and RA-2).

analysis of observations and provide a method of projecting climate change into the future.

Take, for instance, the example of cloud-radiation feedback that remains the source of greatest single uncertainty associated with climate sensitivity.[9] It was mentioned in Chapter 5 that progress with understanding this feedback will be made by formulating better descriptions of cloud processes for incorporation into models and also by comparing model output, especially of radiation quantities, with observations especially those made by satellites. To be really useful such measurements need to be made with extremely high accuracy – to within the order of 0.1% in the average radiation quantities – that is proving highly demanding. Associated with the better measurements of clouds is the need for all aspects of the hydrological (water) cycle to be better observed.

There is also inadequate monitoring at present of the major oceans of the world, which cover a large fraction of the Earth's surface. However, this is beginning to be remedied with the introduction of new methods of observing the ocean surface from space vehicles (see box) and new means of observing the interior of the ocean. But not only are better physical measurements required: to be able to predict the detailed increases of greenhouse gases in the atmosphere, the problems of the carbon cycle must be unravelled; for this, much more comprehensive measurements of the biosphere in the ocean as well as that on land are needed.

Stimulated by internationally organised observing programmes such as the GCOS, space agencies around the world have been very active in the development of new instruments and the deployment of advanced space platforms that are beginning to provide many new observations relevant to the problems of climate change (see box).

Alongside the increased understanding and more accurate predictions of climate change coming from the community of natural scientists, much more effort is now going into studies of human behaviour and activities, how they will influence climate through changes in emissions of greenhouse gases and how they in turn might be affected by different degrees of climate change. Much better quantification of the impacts of climate change will result from these studies. Economists and other social scientists are pursuing detailed work on possible response strategies and the economic and political measures that will be necessary to achieve them. It is also becoming increasingly realised that there is an urgent need to interconnect more strongly research in the natural sciences with that in the social sciences. The integrated framework presented in Chapter 1 (Figure 1.5) illustrates the scope of interactions and of required integration between all the intellectual disciplines involved.

Sustainable development

So much for uncertainty in the science of global warming. But how does this uncertainty map on to the world of political decision-making? A key idea is that of sustainable development.

One of the remarkable movements of the last few years is the way in which problems of the global environment have moved up the political agenda. In her speech at the opening in 1990 of the Hadley Centre at the United Kingdom Meteorological Office, Margaret Thatcher, the former British Prime Minister, explained our clear responsibility to the environment:

> We have a full repairing lease on the Earth. With the work of the IPCC, we can now say we have the surveyor's report; and it shows there are faults and that the repair work needs to start without delay. The problems do not lie in the future, they are here and now: and it is our children and grandchildren, who are already growing up, who will be affected.

Many other politicians have similarly expressed their feelings of responsibility for the global environment. Without this deeply felt and widely held concern, the UNCED conference at Rio, with environment as the number one item on its agenda, could never have taken place.

But, despite its importance, even when concentrating on the long term, the environment is only one of many considerations politicians must take into account. For developed countries, the maintenance of living standards, full employment (or something close to it) and economic growth have become dominant issues. Many developing countries are facing acute problems in the short term: basic survival and large debt repayment; others, under the pressure of large

This multitemporal radar image, from ENVISAT's Advanced Synthetic Aperture Radar (ASAR) instrument, is composed of two images – one acquired on 26 July 2007 and another on 12 April 2007 – and highlights the flooding in Bangladesh and parts of India brought on by two weeks of persistent rain. ASAR is able to peer through clouds, rain or local darkness, and is well suited for differentiating between waterlogged and dry land. Areas in black and white denote no change, while areas outlined in blue are potentially flooded spots. Areas in red may also indicate flooding, but could also be related to agricultural practices.

increases in population, are looking for rapid industrial development. However, an important characteristic of environmental problems, compared with many of the other issues faced by politicians, is that they are long term and potentially irreversible – which is why Tim Wirth, the Under Secretary of State for Global Affairs in the United States Government during the Clinton Administration, said, 'The economy is a wholly owned subsidiary of the environment.' More recently Gordon Brown, in a speech in 2005 when he was the UK's Chancellor of the Exchequer, said:[10]

> Environmental issues – including climate change – have traditionally been placed in a category separate from the economy and economic policy. But this is not longer tenable. Across a range of environmental issues – from soil erosion to the depletion of marine stocks, from water scarcity to air pollution – it is clear now not just that economic activity is their cause, but that these problems in themselves threaten future economic activity and growth.

Sustainable development: how is it defined?

A number of definitions of sustainable development have been produced. The following two well capture the idea.

According to the Bruntland Commission Report *Our Common Future* presented in 1987, sustainable development is 'meeting the needs of the present without compromising the ability of future generations to meet their own needs'.

A more detailed definition is contained in the White Paper *This Common Inheritance*, published by the United Kingdom Department of the Environment in 1990: 'sustainable development means living on the Earth's income rather than eroding its capital' and 'keeping the consumption of renewable natural resources within the limits of their replenishment'. It recognises the intrinsic value of the natural world explaining that sustainable development 'means handing down to successive generations not only man-made wealth (such as buildings, roads and railways) but also natural wealth, such as clean and adequate water supplies, good arable land, a wealth of wildlife and ample forests'.

Further discussion of sustainable development and its definition is in Chapter 12, page 393.

A balance, therefore, has to be struck between the provision of necessary resources for development and the long-term need to preserve the environment. That is why the Rio Conference was about Environment and Development. The formula that links the two is called sustainable development (see box below) – development that does not carry with it the overuse of irreplaceable resources or irreversible environmental degradation.

The idea of sustainable development echoes what was said in Chapter 8, when addressing more generally the relationship of humans to their environment and especially the need for balance and harmony. The Climate Convention signed at the Rio Conference also recognised the need for this balance. In the statement of its objective (see box on pages 291–2 in Chapter 10), it states the need for stabilisation of greenhouse gas concentrations in the atmosphere. It goes on to explain that this should be at a level and on a timescale such that ecosystems are allowed to adapt to climate change naturally, that food production is not threatened and that economic development can proceed in a sustainable manner.

It is also increasingly realised that the idea of sustainability not only applies to the environment but also to human communities – a theme addressed in Chapter 8. Sustainable development is often therefore assumed to include wider social factors as well as environmental and economic ones. The provision of social justice and equity are important components of a drive to sustainable communities. Considerations of equity include not only equity between nations but also equity between generations: we should not leave the world in a poorer state for the next generation. These and other aspects of sustainability are considered further in Chapter 12.

Why not wait and see?

The debate about climate change not only addresses how much action is required but also when it needs to be taken. In the light of scientific uncertainty, it has often been argued that the case is not strong enough for much action to be taken now. What we should do is to obtain as quickly as possible, through appropriate research programmes, much more precise information about future climate change and its impact. We would then, so the argument goes, be in a much better position to decide on relevant action. However, such a wait-and-see attitude is inadequate for a number of reasons.

In the first place, enough is already known for it to be realised that the rate of climate change due to increasing greenhouse gases will almost certainly bring substantial deleterious effects and pose a large problem to the world. It will hit some countries much more than others. Those worst hit are likely to be those in the developing world that are least able to cope with it. Some countries may actually experience a more beneficial climate. But in a world where there is increasing interdependence between nations, no nation will be immune from the effects. Further, as we saw in Chapter 6 (Figure 6.4a), because of the time taken for the oceans to warm, we are already committed to substantially more climate change than has yet been experienced.

Secondly, the timescales of both atmospheric and human responses are long. Carbon dioxide emitted into the atmosphere today will contribute to the increased concentration of this gas and the associated climate change for over 100 years. The more that is emitted now, the more difficult it will be to reduce atmospheric carbon dioxide concentration to the levels that will eventually be required. With regard to the human response, the major changes that are likely to be needed, for instance in large-scale infrastructure, will take many decades. Large power stations that will produce electricity in 30 or 40 years' time are being planned and built today. The demands that are likely to be placed on all of us because of concerns about global warming need to be brought into the planning process now. Trends from 2000 until the present strongly reinforce this argument. Although much consideration and talk has been given to ways of reducing emissions, a significant upturn in global emissions of carbon dioxide has in fact occurred since 2000 (more on this in Chapter 10) which points to the need for even more urgent action.

Thirdly, many of the required actions not only lead to substantial reductions in greenhouse gas emissions but they are good to do for other reasons which bring other direct benefits – such proposals for action are often described as 'no regrets' proposals. Many actions addressing increased efficiency lead also to net savings in cost (sometimes called 'win–win' measures). Other actions lead to improvements in performance or additional comfort.

Fourthly, there are more general beneficial reasons for some of the proposed actions. In Chapter 8 it was pointed out that humans are far too profligate in their use of the world's resources. Fossil fuels are burnt and minerals are used, forests are cut down and soil is eroded without any serious thought of the needs of future generations. The imperative of the global warming problem will help us to use the world's resources in a more sustainable way. Further, the technical innovation that will be required in the energy industry – in energy efficiency and conservation and in renewable energy development – will provide a challenge and opportunity to the world's industry to develop important new technologies – more of that in Chapter 11. And in all these matters, developed countries in taking the first actions (as demanded by the FCCC) need to show the way to developing countries as they develop their economies.

The Precautionary Principle

Some of these arguments for action are applications of what is often called the Precautionary Principle, one of the basic principles that was included in the Rio Declaration at the Earth Summit in June 1992 (see box below). A similar statement is contained in Article 3 of the Framework Convention on Climate Change (see box on pages 291–2 in Chapter 10).

We often apply the Precautionary Principle in our day-to-day living. We take out insurance policies to cover the possibility of accidents or losses; we carry out precautionary maintenance on housing or on vehicles; and we readily accept that in medicine prevention is better than cure. In all these actions we weigh up the cost of insurance or other precautions against the possible damage and conclude that the investment is worthwhile. The arguments are similar as the Precautionary Principle is applied to the problem of global warming.

In taking out an insurance policy we often have in mind the possibility of the unexpected. In fact, when selling their policies, insurance companies often trade on our fear of the unlikely or the unknown, especially of the more devastating possibilities. Although covering ourselves for the most unlikely happenings is not our main reason for taking out the insurance, our peace of mind is considerably increased if the policy includes these improbable events. In a similar way, in arguing for action concerning global warming, some have strongly emphasised the need to guard against the possibility of surprises (see examples in Table 7.4). They point out that, because of positive feedbacks that are not yet well understood (see Chapter 3), the increase of some greenhouse gases could be much larger than is currently predicted. They also point to the evidence that rapid changes of climate have occurred in the past (Figures 4.5 and 4.6) possibly because of dramatic changes in ocean circulation; they could presumably occur again.

The risk posed by such possibilities is impossible to assess. It is, however, salutary to call attention to the discovery of the ozone 'hole' over Antarctica in 1985. Scientific experts in the chemistry of the ozone layer were completely taken by surprise by that discovery. In the years since its discovery, the 'hole' has substantially increased in depth. Resulting from this knowledge, international action to ban ozone-depleting chemicals has progressed much more rapidly. Ozone levels are beginning to recover – full recovery will take about a century. The lesson for us here is that the climate system may be more vulnerable to disturbance than we have often thought it to be. When it comes to future climate change, it would not be prudent to ignore the possibility of surprises.

However, in weighing the action that needs to be taken with regard to future climate change, although the possibility of surprises should be kept in mind, that possibility must not be allowed to feature as the main argument for action. Much stronger in the argument for precautionary action is the realisation that significant anthropogenic climate change is not an unlikely possibility but a near certainty; it is no change of climate that is unlikely. The uncertainties that mainly have to be weighed lie in the magnitude of the change and the details of its regional distribution.

An argument that is sometimes advanced for doing nothing now is that by the time action is really necessary, more technical options will be available. By acting now, we might foreclose their use. Any action taken now must, of course, take into account the possibility of helpful technical developments. But the argument also works the other way. The thinking and the activity generated by considering appropriate actions now and by planning for more action later will itself be likely to stimulate the sort of technical innovation that will be required.

While speaking of technical options, I should briefly mention possible options to counteract global warming by the artificial modification of the environment (sometimes referred to as geoengineering).[11] The suggestion of iron fertilisation of the oceans was mentioned in Chapter 3. Other proposals have concentrated on techniques that might reduce the amount of sunlight absorbed by the Earth, for instance, the installation of mirrors in space to cool the Earth by reflecting sunlight away from it; the addition of dust to the upper atmosphere to provide a similar cooling effect and the alteration of cloud amount and type by adding cloud condensation nuclei to the atmosphere.[12] None of these however has been demonstrated to be either feasible or effective, nor would any of them make any difference to the problem of increasing acidity of the oceans due to increasing carbon dioxide. Further, they suffer from the serious problem that none of them would exactly counterbalance the effect of increasing greenhouse gases. As has been shown, the climate system is far from simple. The results of any attempt at large-scale climate modification could not be perfectly predicted and might not be what is desired. With the present state of knowledge, extreme caution must be exercised when considering implementation of proposals for the introduction of artificial climate modification.

The conclusion from this section – and the last one – is that to 'wait and see' would be an inadequate and irresponsible response to what we know. That was recognised over 15 years ago by the FCCC signed in Rio (see box on pages 291–2 in Chapter 10) in 1992 and has often been reiterated since. Just what the required action should be is the subject of the next chapter.

Principles for international action

From the three previous sections and from the discussion in Chapter 8, four distinct principles can be identified to form the basis of international action. They are all contained in the Rio Declaration on Environment and Development (see box below) agreed by over 160 countries at the United Nations Conference on Environment and Development (the Earth Summit) held in Rio de Janeiro in 1992. They can also be identified in one form or another in the FCCC (see box on pages 291–2 in Chapter 10). The Principles (with references to the Principles of the Rio Declaration and the Articles of the FCCC) are:

- The Precautionary Principle (Principle 15)
- The Principle of Sustainable Development (Principles 1 and 7)
- The Polluter-Pays Principle (Principle 16)
- The Principle of Equity – International and Intergenerational (Principles 3 and 5).

In the next chapter we shall consider how these principles can be applied.

Some global economics

So far our attempt to balance uncertainty against the need for action has been considered in terms of issues. Is it possible to carry out the weighing in terms of cost? In a world that tends to be dominated by economic arguments, quantification of the costs of action against the likely costs of the consequences of inaction must at least be attempted. It is also helpful to put these costs in context by comparing them with other items of global expenditure.

The costs of anthropogenic climate change fall into three parts. Firstly, there is the cost of the damage due to that change; for instance, the cost of flooding due to sea level rise or the cost of the increase in the number or intensity of disasters such as floods, droughts or windstorms, and so on. Secondly, there is the cost of adaptation that reduces the damage or the impact of the climate change. Thirdly, there is the cost of mitigating action to reduce the amount of climate change. The roles of adaptation and mitigation are illustrated in Figure 1.5. Because there is already a commitment to a significant degree of climate change, a need for significant adaptation is apparent. That need will continue to increase through the twenty-first century, an increase that will eventually be mollified as the effects of mitigation begin to bite. Mitigation is beginning now but the degree of mitigation that is eventually undertaken will depend on an assessment of the effectiveness and cost of adaptation. The costs, disadvantages and benefits of both adaptation and mitigation need therefore to be assessed and weighed against each other.

The models providing estimates of cost need to include all aspects of the climate change issue, for instance, interactions between the factors driving climate change and its impacts both on humans and ecosystems, human activities that are influencing those factors and the response to climate change both of humans and ecosystems – in fact all the elements illustrated in Figure 1.5. This process is often called Integrated Assessment (see box below) and is supported by Integrated Assessment Models (IAMs) that address all the relevant elements in as complete a manner as possible.

At the end of Chapter 7, estimates of the cost of damage from global warming were presented. Many of these estimates also included some of the costs of adaptation; in general adaptation costs have not been separately identified. Many of these cost estimates assumed a situation for which, resulting from human activities, the increase in greenhouse gases in the atmosphere was equivalent to a doubling of the carbon dioxide concentration – under business-as-usual this is likely to occur around the middle of the twenty-first century. The estimates were typically in the range 1% to 4% of GDP for developed countries. In developing countries, because of their greater vulnerability to climate change and because a greater proportion of their expenditure is dependent on activities such as agriculture and water, estimates of the cost of damage are greater, typically in the range 5% to 10% of GDP or more. It was also pointed out in Chapter 7 that the cost estimates only included those items that could be costed in money terms. Those items of damage or disturbance for which money is not an appropriate measure (e.g. the generation of large numbers of environmental

The Rio Declaration 1992

The Rio Declaration on Environment and Development was agreed by over 160 countries at the United Nations Conference on Environment and Development (the Earth Summit) held in Rio de Janeiro in 1992. Some examples of the 27 principles enumerated in the Declaration are as follows.

Principle 1 Human beings are at the centre of concerns for sustainable development. They are entitled to a healthy and productive life in harmony with nature.

Principle 3 The right to development must be fulfilled so as to equitably meet developmental and environmental needs of present and future generations.

Principle 5 All States and all people shall cooperate in the essential task of eradicating poverty as an indispensable requirement for sustainable development, in order to decrease the disparities in standards of living and better meet the needs of the majority of the people of the world.

Principle 7 States shall cooperate in a spirit of global partnership to conserve, protect and restore the health and integrity of the Earth's ecosystem. In view of the different contributions to global environmental degradation, States have common but differentiated responsibilities. The developed countries acknowledge the responsibility they bear in the international pursuit of sustainable development in view of the pressures their societies place on the global environment and of the technologies and financial resources they command.

Principle 15 In order to protect the environment, the precautionary approach shall be widely applied by States according to their capabilities. Where there are threats of serious or irreversible damage, lack of full scientific certainty shall not be used as a reason for postponing cost-effective measures to prevent environmental degradation.

Principle 16 National authorities should endeavour to promote the internalisation of environmental costs and the use of economic instruments, taking into account the approach that the polluter should, in principle, bear the cost of pollution, with due regard to the public interest and without distorting international trade and investment.

refugees) also need to be exposed and taken into account in any overall appraisal.

The longer-term damage, should greenhouse gases more than double in concentration, is likely to rise somewhat more steeply in relation to the concentration of carbon dioxide. For quadrupled equivalent carbon dioxide concentration, for instance, estimates of damage cost of the order of two to four times that for doubled carbon dioxide have been made – suggesting that the damage might follow something like a quadratic law relative to the expected temperature rise.[15] In addition the much larger degree of climate change would considerably enhance the possibilities of singular events (see Table 7.4), irreversible change and of possible surprises. The Stern Review (see box in Chapter 7 on page 227) estimates the cost of continuing with business-as-usual beyond 2100 as equivalent to a reduction of 5–20% in current per capita consumption now and for ever with a strong likelihood that it will be in the upper part of that range and with disproportionate losses tending to fall on poorer countries.

Since the main contribution to global warming arises from carbon dioxide emissions, attempts have also been made to express these costs in terms of the cost per tonne of carbon as carbon dioxide emitted from human activities. This is known as the *social cost of carbon*. A simple but crude calculation can be carried out as follows. Consider the situation when carbon dioxide concentration in the atmosphere has doubled from its pre-industrial value, i.e. when an additional amount of about 5500 Gt of CO_2 (1500 GtC) from anthropogenic sources has been emitted into the atmosphere (see Figure 3.1 and recall that about half the carbon dioxide emitted accumulates in the atmosphere). This carbon dioxide will remain in the atmosphere on average for about 100 years. Assuming a figure of 3% of global world product (GWP) – or 2000 billion per year – as the cost of the damage due to global warming in that situation, and assuming also that the damage remains over the 100 years of the lifetime of carbon dioxide in the atmosphere, the cost per tonne of CO_2 turns out to be about $US36.

Calculations of the cost per tonne of carbon can be made with much more sophistication by considering that it is the *incremental* damage cost (that is, the cost of the damage due to one extra tonne of carbon emitted now) that is really required and also by allowing through a discount rate for the fact that it is damage some time in the future that is being costed now. Estimates made by different economists produced for the IPCC 1995 Report ranged from about $US1.5 to $US35 per tonne of CO_2 ($US5–125 per tonne C).[16] For the 2007 Report, the range of estimates is even greater, the very large range being due to the different assumptions that have been made.[17]

The estimates are particularly sensitive to the discount rate that is assumed; values at the top end of the range have assumed a discount rate of less than 2%; those at the bottom end have assumed a discount rate of 5% or more.[18] The dominant effect of the discount rate will be clear when it is realised that over 50 years a 2% discount rate devalues costs by a factor of about 3 while a 5% rate discounts by a factor of 13. Over 100 years the difference is even larger – a factor of 7 for a 2% and a factor of 170 for a 5% rate. Amongst economists there has been much debate but no agreement about how to apply discount accounting to long-term

Integrated Assessment and Evaluation[13]

In the assessment and evaluation of the impacts of different aspects of global climate change with its large complexity, it is essential that all components are properly addressed. The major components are illustrated in Figure 1.5. They involve a very wide range of disciplines from natural sciences, technology, economics and the social sciences (including ethics). Take the example of sea level rise – probably the easiest impact to envisage and to quantify. From the natural sciences, estimates can be made of the amount and rate of rise and its characteristics. From various technologies, options for adaptation can be proposed. From economics and the social sciences, risks can be assessed and evaluated. The economic costs of sea level rise might be expressed, for instance, most simply as the capital cost of protection (where protection is possible) plus the economic value of the land or structures that may be lost plus the cost of rehabilitating those persons that would be displaced. But in practice the situation is more complex. For a costing to be at all realistic, especially when it is to apply to periods of decades into the future, it must account not only for direct damage and the cost of protection but also for a range of options and possibilities for adaptation other than direct protection. The likelihood of increased storm surges with the consequent damages and the possibility of substantial loss of life need also to be addressed. Further, there are other indirect consequences; for instance, the loss of fresh water because of salination, the loss of wetlands and associated ecosystems, wildlife or fisheries and the lives and jobs of people that would be affected in a variety of ways. In developed country situations rough estimates of the costs of some of these components can be made in money terms. For developing countries, however, the possible options can less easily be identified or weighed and even rough estimates of costs cannot be provided.

Integrated Assessment Models or IAMs are important tools for Integrated Assessment and Evaluation. They represent within one integrated numerical model the physical, chemical and biological processes that control the concentration of greenhouse gases in the atmosphere, the physical processes that determine the effect of changing greenhouse gas concentrations on climate and sea level, the biology and ecology of ecosystems (natural and managed), the physical and human impacts of climate change and the socio-economics of adaptation to and mitigation of climate change. Such models are highly sophisticated and complex although their components are bound to be very simplified. They provide an important means for studying the connections and interactions between the various elements of the climate change problem. Because of their complexity and because of the non-linear nature of many of the interactions, a great deal of care and skill is needed in interpreting the results from such models.

A number of the components of impact, even for the relatively simple situation of sea level rise, cannot be readily costed in money terms. For instance, the loss of ecosystems or wildlife as it impacts tourism can be expressed in money terms, but there is no agreed way of setting a money measure for the longer-term loss or the intrinsic value of unique systems. Or a further example is that, although the cost of rehabilitation for displaced people can be estimated, other social, security or political consequences of displacement (e.g. in extreme cases the loss of whole islands or even whole states) cannot be costed in terms of money. Any appraisal therefore of impacts of anthropogenic climate change will have to draw together components that are expressed in different ways or use different measures. Policy-and decision-makers need to find ways of considering alongside each other all the components that need to be aggregated in order to make appropriate judgements.[14]

problems of this sort or about what rate is most appropriate. However, as Partha Dasgupta points out, 'the disagreement is not about economics nor about social cost–benefit analysis nor even about the numeracy of fellow scientists',[19] but is in fact more fundamental. He explains that the effects of carbon emissions could bring such large negative perturbations to future economies that the basis is threatened on which discount rates for future investment are set. Further, there are the likely damages that cannot easily be valued in money terms such as the large-scale loss of land – or even of whole countries – due to sea level rise or the large-scale loss of habitats or species. For these, even if valuation is attempted, discounting seems inappropriate. The Stern Review also believes discounting to be inappropriate and argues that the welfare of future generations should be considered on the same basis as the welfare of the current generation.[20] This is an example of an ethical question raised by the discounting process. Professor John Broome of Oxford points out that discounting is not only concerned with economics but also brings into play ethical questions that cannot be avoided – although they are frequently ignored.[21] If discount rates are to be applied at all to cost estimating for climate change, there seem cogent arguments that a smaller discount rate rather than a larger one should be employed. And in any case, for any cost estimate that is made the discount rate used should be adequately exposed.

After a thorough discussion of the factors influencing the social cost of carbon, the Stern Review, working with the PAGE 2002 IAM and a business-as-usual (BAU) scenario, estimates its value as around $US85 per tonne of CO_2.[22] The Review also points out that the social cost of carbon will rise with time (as damage increases with time) and at any one time is dependent on the future emissions trajectory (Figure 9.3); for a stabilisation scenario (see Chapter 10, page 311), for instance at 550 ppm CO_2e it would be considerably less at around $US30 per tonne of CO_2e and at 450 ppm CO_2e about $US25 per tonne CO_2e. For our broad economic arguments in later chapters we shall use estimates of the social cost of carbon in the range $US25 to $US50 per tonne CO_2e.

To slow the onset of climate change and to limit the longer-term damage, mitigating action can be taken by reducing greenhouse gas emissions, in particular the emissions of carbon dioxide. The cost of mitigation depends on the amount of reduction required in greenhouse gas emissions; large reductions will cost proportionately more than small ones. It will also depend on the timescale of reduction. To reduce emissions drastically in the very near term would inevitably mean large reductions in energy availability with significant disruption to industry and large cost. However, less drastic reductions can be made with relatively small cost through actions of two kinds. Firstly, substantial efficiency

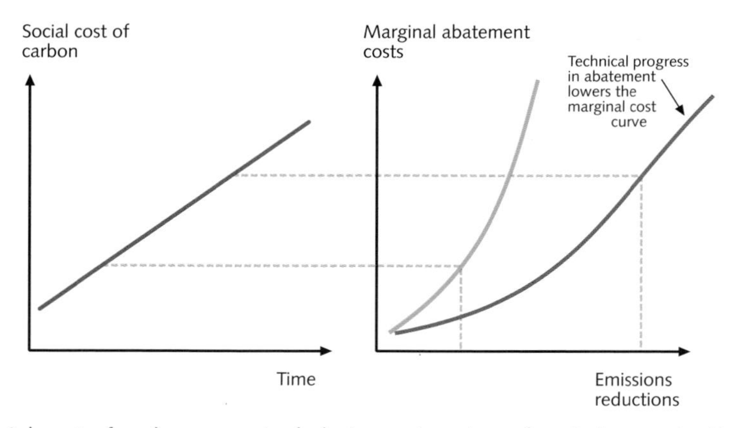

Figure 9.3 The social cost of carbon, marginal abatement costs and emissions reductions. Up to the long-term stabilisation goal, the social cost of carbon will rise over time because marginal costs do so. This is because damage costs tend to rise more rapidly than global average temperature. Abatement costs are illustrated schematically in the right-hand part of the figure. Over time, technical progress will reduce the total cost of any particular level of abatement, so that at any given price there will be more emission reductions. The dashed lines illustrate how the path for the social cost of carbon drives the extent of abatement.

gains in the use of energy can easily be achieved, many of which would lead to cost savings; these can be put into train now. Secondly, in the generation of energy, again proven technology exists for substantial efficiency improvements and also for bringing into use sources of energy generation that are not dependent on fossil fuels. These can be planned for now and changes made as energy infrastructure is replaced or new infrastructure constructed. The next two chapters will present more detail about these possible actions and how they might be achieved.

What about the cost of mitigation? Much of it will arise in the energy or transport sectors as cheap fossil fuels are replaced by other energy sources that, at least in the short term, are likely to be more expensive. Some detail is provided in the next chapter of the profile of emissions reductions required to stabilise concentrations of greenhouse gases in the atmosphere at different levels especially at 550 or 450 ppm CO_2e (for definition of equivalent carbon dioxide, i.e. CO_2e, see Chapter 6, page 147). To stabilise at 550 ppm CO_2, a reduction by 2050 of global carbon dioxide emissions back to about 1990 levels would be required (Figure 10.3). The Stern Review estimates the annual cost to the developed world economies of such reductions will be around 1% of GDP.[23] As quoted by Stern and by the IPCC, estimates from all sources of this annual cost span a range from minus 1% to around 4% with a mean of about 1%. This large range signifies the large uncertainties in the assumptions that have to be made. As might be expected, the cost is substantially dependent on the target level of carbon dioxide concentration stabilisation. For a 450 ppm stabilisation level the mitigation costs will be higher and a reduction by 2050 of global carbon dioxide emissions by about 60% from 1990 levels will be required. For levels in the range 445 to 535 ppm CO_2e the IPCC AR4 report[24] cites estimates of less than 5.5% of GDP for mitigation costs by 2050. With typical levels of economic growth being between 2% and 4% per year, the cost of achieving reductions to meet any stabilisation levels mentioned, even as low as 445 ppm CO_2e, is less than the equivalent of about two years' economic growth over 50 years.

Although the economic studies on which these estimates are based have taken into account many of the relevant factors, they are bound to be surrounded by substantial uncertainty – as is illustrated by the large range in the estimates quoted. Some of the more difficult factors to take into account are some that contribute to lower costs[25] such as the economic effects of introducing new low-emission technologies, new revenue-raising instruments, adequate inter-regional financial and technology transfers and likely future innovation. For the last of these, it is not easy to peer into the crystal ball of technical development; almost any attempt to do so is likely to underestimate its potential. For these reasons the estimates of mitigation cost are almost certainly on the high side.

How do these mitigation (or abatement) costs compare with the damage costs we listed earlier? Compared with the damage costs assuming a BAU trajectory with no significant action being taken this century, they are much less. They are more comparable with the likely damage cost for lower scenarios. However, recalling the warnings, for instance in the Stern Review, that monetary estimates only represent part of the story of the damage costs, mitigation costs appear modest compared with the likely overall damage if little or no mitigating action is taken. The right-hand part of Figure 9.3 is a schematic showing curves of marginal abatement cost (the cost of reducing by one unit of emissions at the margin). These curves rise with emissions reductions as the reductions that are cheaper to achieve will be carried out first. The figure illustrates how abatement cost might be related to the social cost of carbon. However, in practice because of the large uncertainties in future estimates both of damages and abatement costs, factors in addition to money cost estimates will tend to determine the extent of mitigation that is planned or achieved.

However, it should be noted that, even if the carbon dioxide concentration is stabilised at 450 ppm CO_2e, the world will have been committed to a very significant degree of climate change, bringing with it substantial costs and demands for adaptation. What is being mitigated is further and even more damaging climate change.

D. PARKINS

Nature's view of the key meeting to finalise the second IPCC report. Scientific integrity and political and environmental agendas met to thrash out and finally agree on the report, the results of which fed into the Kyoto Protocol. Left to right: Professor Bolin (first Chair of IPCC), myself, and Dr Gylvan Meira of Brazil, (Co-Chair of the science working group in 1995) are taking the temperature of a sick Earth with a clinical thermometer!

In considering the costs of the impacts of both global warming and adaptation or mitigation, figures of a small percentage of GDP have been mentioned. It is interesting to compare this with other items of expenditure in national or personal budgets. In a typical developed country, for example the United Kingdom, about 5% of national income is spent on the supply of primary energy (basic fuel such as coal, oil and gas, fuel for electricity supply and fuel for transport), about 9% on health and 3–4% on defence. It is, of course, clear that global warming is strongly linked to energy production – it is because of the way energy is provided that the problem exists – and this subject will be expanded in the next two chapters. But the impacts of global warming also have implications for health – such as the possible spread of disease – and for national security – for example, the possibility of wars fought over water, or the impact of large numbers of environmental refugees. Any thorough consideration of the economics of global warming needs therefore to assess the strength of these implications and to take them into account in the overall economic balance.

So far, on the global warming balance sheet we have estimates of costs and of benefits or drawbacks. What we do not have as yet is a capital account. Valuing human-made capital is commonplace, but in the overall accounting we are attempting, 'natural' capital must clearly be valued too. By 'natural' capital is meant, for instance, natural resources that may be renewable (such as a forest) or non-renewable (such as coal, oil or minerals).[26] Their value is clearly more than the cost of exploitation or extraction.

Other items, some of which were mentioned at the end of Chapter 7, such as natural amenity and the value of species, can also be considered as 'natural' capital. I have argued (Chapter 8) that there is intrinsic value in the natural world – indeed, the value and importance of such 'natural' capital is increasingly recognised. The difficulty is that it is neither possible nor appropriate to express much of this value in money. Despite this difficulty, it is now widely recognised that national and global indicators of sustainable development should be prepared that include items of 'natural' capital and ways of including such items in national balance sheets are being actively pursued.

SUMMARY

This chapter has been considering how uncertainty concerning future climate change is weighed against the costs of the likely damage due to climate change, of adaptation and of mitigation action. Its conclusions can be summarised as follows:

- The four comprehensive IPCC assessments since 1990 have provided increasingly accurate and detailed information about the climate change occurring now and likely future change – although substantial uncertainties still remain.
- Four principles for international action have been identified: the Precautionary Principle, the Principle of Sustainable Development, the Polluter-Pays Principle and the Principle of Equity.
- In assessing costs the following items in the climate change balance sheet have been identified as follows:

  (1) Due to damage from the impacts of climate change by 2050, loss in GDP in developed countries has been estimated as typically in the range 1% to 4% of GDP and of 5% to 10% or more in many developing countries.

  (2) Climate change impacts that cannot be valued in money terms; for instance, those with social consequences or that affect human amenity

or 'natural' capital or those with implications for national security. For instance, it is estimated that there could be over 150 million environmental refugees by 2050.

(3) The cost of adaptation to anthropogenic climate change. As climate change is beginning to be realised, planning for adaptation is urgently required that in some areas and sectors will lead to substantial cost.

(4) The costs of mitigation of anthropogenic climate change. Providing mitigation action is pursued urgently and properly planned, for reductions in emissions leading to stabilisation of atmospheric carbon dioxide concentration mitigation costs are typically less than one or two year's economic growth by 2050 – considerably less than estimates of climate change damage in (1) above.

(5) Refinements of all the above estimates and the assumptions on which they are based in the above list are urgently required.

The next chapter will consider some of the actions in more detail in the light of the principles we have enunciated in this chapter and in the context of the international FCCC.

QUESTIONS

1 It is sometimes argued that, in scientific enquiry, 'consensus' can never be achieved, because debate and controversy are fundamental to the search for scientific truth. Discuss what is meant by 'consensus' and whether you agree with this argument. Do you think the IPCC Reports have achieved 'consensus'?

2 How much do you think the value of IPCC Reports depends on (1) the peer review process to which they have been subjected, and (2) the involvement of governments in the presentation of scientific results?

3 Look out as many definitions of 'sustainable development' as you can find. Discuss which you think is the best.

4 Make a list of appropriate indicators that might be used to assess the degree to which a country is achieving sustainable development. Which do you think might be the most valuable?

5 Work out the value of a 'cost' today if it is 20, 50 or 100 years into the future and the assumed discount rate is 1%, 2% or 5%. Look up and summarise the arguments for discounting future costs as presented for instance in the Stern Review and various chapters of the IPCC Reports. What do you think is the most appropriate discount rate to use?

6 Construct, as far as you are able, a set of environmental accounts for your country including items of 'natural' capital. Your accounts will not necessarily be all in terms of money.
7 List, with as much detail as you can, the mitigation action that is being undertaken by your country. What are the factors that determine the extent of this mitigation action. Is enough being done by your country? If you think not, how could it be increased?
8 Because of continuing economic growth, there is an expectation that the world will be very much richer by the middle of the twenty-first century and therefore, it is sometimes argued, in a better position than now to tackle the impacts or the mitigation of climate change. Do you agree with this argument?
9 Gross National Product (GNP) is commonly employed to measure the health of a national economy. But it is commonly accepted that, in providing a relatively crude measure of economic growth, it fails to take account of many important factors such as human welfare, quality of life, use of irreplaceable resources etc. Investigate other measures that have been proposed to assess and compare national economies. Would you judge any of them to be more valuable than GNP for policymakers to use as a general measure of economic health and performance?
10 Look up information on proposals for possible geoengineering options to mitigate climate change – for instance in the volume cited in Note 12. Make an appraisal of these options particularly considering their likely effectiveness and the possibilities of unwanted or negative effects on climate or society that might also result from their introduction.

FURTHER READING AND REFERENCE

IPCC AR4 *Climate Change* 2007 Synthesis Report

Metz, B., Davidson, O., Bosch, P., Dave, R., Meyer, L. (eds.) 2007. *Climate Change 2007: Mitigation of Climate Change. Contribution of Working Group III to the Fourth Assessment Report of the Intergovernmental Panel on Climate Change*. Cambridge: Cambridge University Press.

Technical Summary

Chapter 2 Framing issues (e.g. links to sustainable development, integrated assessment)

Chapter 3 Issues relating to mitigation in the long-term context

Chapter 12 Sustainable development and mitigation

Stern, N. 2006. *The Economics of Climate Change*. Cambridge: Cambridge University Press. The Stern Review: especially Chapters 3 to 6 in Part II on the cost of climate-change impacts.
Framework Convention on Climate Change (FCCC): www.unfccc.int.
Pew Centre on Global Climate Change: www.pewclimate.org.
The Climate Group: www.climategroup.org.

NOTES FOR CHAPTER 9

1 Houghton, J.T., Jenkins, G.J., Ephraums, J.J. (eds.) 1990. *Climate Change: The IPCC Scientific Assessments*. Cambridge: Cambridge University Press, p. 365; Executive Summary, p. xii. Similar but more elaborate statements are in the 1995, 2001 and 2007 IPCC Reports.

2 For a detailed description of how the output from climate models can be combined with other information in climate studies see Mearns, L.O., Hulme, M. *et al.* 2001. Climate scenario development. In Houghton, J.T., Ding, Y., Griggs, D.J., Noguer, M., van der Linden, P.J., Dai, X., Maskell, K., Johnson, C.A. (eds.) 2001. *Climate Change 2001: The Scientific Basis. Contribution of Working Group I to the Third Assessment Report of the Intergovernmental Panel on Climate Change*. Cambridge: Cambridge University Press, Chapter 13.

3 For an overview of the history of the IPCC see Bolin, B. 2007. *A History of the Science and Politics of Climate Change*. Cambridge: Cambridge University Press.

4 Houghton *et al.* (eds.) *Climate Change: The IPCC Scientific Assessment*; Tegart, W.J. McG., Sheldon, G.W., Griffiths, D.C. (eds.) 1990. *Climate Change: The IPCC Impacts Assessment*. Canberra: Australian Government Publishing Service.

Houghton, J.T., Meira Filho, L.G., Callander, B.A., Harris, N., Kattenberg, A., Maskell, K. (eds.) 1996. *Climate Change 1995: The Science of Climate Change*. Cambridge: Cambridge University Press.

Watson, R.T., Zinyowera, M.C., Moss, R.H. (eds.) 1996. *Climate Change 1995: Impacts, Adaptations and Mitigation of Climate Change: Scientific–Technical Analyses. Contribution of Working Group II to the Second Assessment Report of the Intergovernmental Panel on Climate Change*. Cambridge: Cambridge University Press.

Bruce, J., Hoesung Lee, Haites, E. (eds.) 1996. *Climate Change 1995: Economic and Social Dimensions of Climate Change*. Cambridge: Cambridge University Press.

Houghton, J.T., Ding, Y., Griggs, D.J., Noguer, M., van der Linden, P.J., Dai, X., Maskell, K., Johnson, C.A. (eds.) 2001. *Climate Change 2001: The Scientific Basis. Contribution of Working Group I to the Third Assessment Report of the Intergovernmental Panel on Climate Change*. Cambridge: Cambridge University Press.

McCarthy, J.J., Canziani, O., Leary, N.A., Dokken, D.J., White, K.S. (eds.) 2001. *Climate Change 2001: Impacts, Adaptation and Vulnerability. Contribution of Working Group II to the Third Assessment Report of the Intergovernmental Panel on Climate Change*. Cambridge: Cambridge University Press.

Metz, B., Davidson, O., Swart, R., Pan, J. (eds.) 2001. *Climate Change 2001: Mitigation. Contribution of Working Group III to the Third Assessment Report of the Intergovernmental Panel on Climate Change*. Cambridge: Cambridge University Press.

Solomon, S., Qin, D., Manning, M., Marquis, M., Averyt, K., Tignor, M.M.B., Miller, H.L. Jr, Chen, Z. (eds.), 2007. *Climate Change 2007: The Physical Science Basis. Contribution of Working Group I to the Fourth Assessment Report of the Intergovernmental Panel on Climate Change*. Cambridge: Cambridge University Press.

Parry, M., Canziani, O., Palutikof, J., van der Linden, P., Hanson, C. (eds.) 2007. *Climate Change 2007: Impacts, Adaptation and Vulnerability. Contribution of Working Group II to the Fourth Assessment Report of the Intergovernmental Panel on Climate Change*. Cambridge: Cambridge University Press.

Metz, B., Davidson, O., Bosch, D., Dave, R., Meyer, L. (eds.) 2007. *Climate Change 2007: Mitigation of Climate Change, Contribution of Working Group III to the Fourth Assessment Report of the Intergovernmental Panel on Climate Change*. Cambridge: Cambridge University Press.

5 See also Houghton, J.T. 2002. An overview of the IPCC and its process of science assessment. In Hester, R.E., Harrison, R.M. (eds.) *Global Environmental Change*, Issues in Environmental Science and Technology, No. 17. Cambridge: Royal Society of Chemistry.

6 For an account of the 'denial industry' see Monbiot, G. 2007. *Heat: How to Stop the Planet Burning*. London: Allen Lane, Chapter 2.

7 See *Climate Change Controversies: A Simple Guide* published by the Royal Society, http://royalsociety.org/document.asp?id=6229

8 The Academies of Science 2005 statement can be found on http://royalsociety.org/document.asp?id=3222.

9 Defined on page 143 in Chapter 6.

10 Address to the Energy and Environment Ministerial Round Table, 15 March 2005.

11 See Section 11.2.2, in Metz *et al.* (eds.) *Climate Change 2007: Mitigation.*

12 For more information, see Lauder, B., Thompson, M. (eds.) 2008. *Geoscale Engineering to Avert Dangerous Climate Change*. London: Royal Society (an issue of *Philosophical Transactions of the Royal Society*).

13 Weyant, J. *et al.* Integrated assessment of climate change. In Bruce *et al.* (eds.), *Climate Change 1995: Economic and Social Dimensions*, Chapter 10; also, Chapters 2 and 3, in Parry *et al.* (eds.) *Climate Change 2007; Impacts.*

14 A full discussion of such integrated appraisal can be found in *21st Report of the UK Royal Commission on Environmental Pollution*. London: Her Majesty's Stationery Office, 1998.

15 Pearce, D.W. *et al.*, Chapter 6, in Bruce *et al.* (eds.), *Climate Change 1995: Economic and Social Dimensions.*

16 Summary for policymakers, in Bruce *et al.* (eds.), *Climate Change 1995: Economic and Social Dimensions.*

17 Chapter 3, page 232, in Metz *et al.* (eds.) *Climate Change 2007: Mitigation.*

18 Cline argues for a low rate (Cline, W.R. 1992. *The Economics of Global Warming*. Washington, DC: Institute for International Economics, Chapter 6), as does S. Fankhauser (*Valuing Climate Change*, London: Earthscan, 1995). Nordhaus (Nordhaus, W.R. 1994. *Managing the Global Commons: The Economics of Climate Change*. Cambridge, Mass.: MIT Press) has used rates in the range of 5% to 10%; see also Tol, R.S.J. 1999. The marginal costs of greenhouse gas emissions. *The Energy Journal*, **20**, 61–8.

19 Dasgupta, P. 2001. *Human Well-Being and the Natural Environment*. Oxford: Oxford University Press, p. 184; see also pp. 183–91; see also Markhndya, A., Halsnaes, K. *et al.* Costing methodologies. Chapter 7, in Metz *et al.* (eds.), *Climate Change 2001: Mitigation.*

20 Stern, N. 2007. *The Economics of Climate Change*. Cambridge: Cambridge University Press, pp. 35–7.

21 Broome, J. 2008. The ethics of climate change. *Scientific American*, **298**, 69–73.

22 Stern, *Economics of Climate Change*, p. 344.

23 Ibid., p. 239.

24 Table SPM 7 and Figure 3.25, in Metz *et al.* (eds.) *Climate Change 2007: Mitigation.*

25 Further detail in Hourcade, J.-C., Shukla, P. *et al.* 2001. Global regional and national costs and ancillary benefits of mitigation. Chapter 8; in Metz *et al.* (eds.), *Climate Change 2001: Mitigation*; see also Metz *et al.* (eds.) *Climate Change 2007: Mitigation.*

26 For a discussion of this issue see Daly, H.E. 1993. From empty-world economics to full-world economics: a historical turning point in economic development. In Ramakrishna, K., Woodwell, G.M. (eds.) *World Forests for the Future*. New Haven, Conn.: Yale University Press, pp. 79–91.

10 A strategy for action to slow and stabilise climate change

Amazon rainforest canopy.

FOLLOWING THE awareness of the problems of climate change aroused by the IPCC scientific assessments, the necessity of international action has been recognised. In particular, an Objective has been agreed to stabilise the concentrations of greenhouse gases in the atmosphere so as to eventually stabilise the climate. Nations or groups of nations are already pledging to substantial emissions reductions between now and 2050. What has yet to be agreed is the target level of stabilisation. In this chapter I discuss what target levels should be the aim and the actions that will be necessary to achieve them.

The Climate Convention

The United Nations Framework Convention on Climate Change (FCCC) signed by over 160 countries at the United Nations Conference on Environment and Development held in Rio de Janeiro in June 1992 came into force on 21 March 1994. It has set the agenda for action to slow and stabilise climate change. The signatories to the Convention (some of the detailed wording is presented in the box below) recognised the reality of global warming, recognised also the uncertainties associated with current predictions of climate change, agreed that action to mitigate the effects of climate change needs to be taken and pointed out that developed countries should take the lead in this action.

The Convention mentions one particular aim concerned with the relatively short-term and one far-reaching objective. The particular aim is that developed countries (Annex I countries in Climate Convention parlance) should take action to return greenhouse gas emissions, in particular those of carbon dioxide, to their 1990 levels by the year 2000. The long-term objective of the

Some extracts from the United Nations Framework Convention on Climate Change, signed by over 160 countries in Rio de Janeiro in June 1992

Firstly, some of the paragraphs in its preamble, where the parties to the Convention:

CONCERNED that human activities have been substantially increasing the atmospheric concentration of greenhouse gases, that these increases enhance the natural greenhouse effect, and that this will result on average in an additional warming of the Earth's surface and atmosphere and may adversely affect natural ecosystems and humankind.

NOTING that the largest share of historical and current global emissions of greenhouse gases has originated in developed countries, that per capita emissions in developing countries are still relatively low and that the share of global emissions originating in developing countries will grow to meet their social and development needs.

RECOGNISING that various actions to address climate change can be justified economically in their own right and can also help in solving other environmental problems.

RECOGNISING that low-lying and other small island countries, countries with low-lying coastal, arid and semi-arid areas or areas liable to floods, drought and desertification, and developing countries with fragile mountainous ecosystems are particularly vulnerable to the adverse effects of climate change.

AFFIRMING that responses to climate change should be coordinated with social and economic development in an integrated manner with a view to avoiding adverse impacts on the latter, taking into full account the legitimate priority needs of developing countries for the achievement of sustained economic growth and the eradication of poverty.

Continued

DETERMINED to protect the climate system for present and future generations, have AGREED as follows:

The Objective of the Convention is contained in Article 2 and reads as follows:

> The ultimate objective of this Convention and any related legal instruments that the Conference of the Parties may adopt is to achieve, in accordance with the relevant provisions of the Convention, stabilisation of greenhouse gas concentrations in the atmosphere at a level that would prevent dangerous anthropogenic interference with the climate system. Such a level should be achieved within a time-frame sufficient to allow ecosystems to adapt naturally to climate change, to ensure that food production is not threatened and to enable economic development to proceed in a sustainable manner.

Article 3 deals with principles and includes agreement that the Parties:

> take precautionary measures to anticipate, prevent or minimize the causes of climate change and mitigate its adverse effects. Where there are threats of serious or irreversible damage, lack of full scientific certainty should not be used as a reason for postponing such measures, taking into account that policies and measures to deal with climate change should be cost-effective so as to ensure global benefits at the lowest possible cost.

Article 4 is concerned with Commitments. In this article, each of the signatories to the Convention agreed:

> to adopt national policies and take corresponding measures on the mitigation of climate change, by limiting its anthropogenic emissions of greenhouse gases and protecting and enhancing its greenhouse sinks and reservoirs. These policies and measures will demonstrate that developed countries are taking the lead in modifying longer-term trends in anthropogenic emissions consistent with the objective of the Convention, recognising that the return by the end of the present decade to earlier levels of anthropogenic emissions of carbon dioxide and other greenhouse gases not controlled by the Montreal Protocol would contribute to such modification…
>
> Each signatory also agreed:
>
> in order to promote progress to this end … to communicate … detailed information on its policies and measures referred to above, as well as on its resulting projected anthropogenic emissions by sources and removals by sinks of greenhouse gases not covered by the Montreal Protocol … with the aim of returning individually or jointly to their 1990 levels these … emissions …

Convention, expressed in Article 2, is that the concentrations of greenhouse gases in the atmosphere should be stabilised 'at a level which would prevent dangerous anthropogenic interference with the climate system', the stabilisation to be achieved within a time-frame sufficient to allow ecosystems to adapt naturally to climate change, to ensure that food production is not threatened and to enable economic development to proceed in a sustainable manner. In setting this objective, the Convention has recognised that it is only by stabilising the concentration of greenhouse gases (especially carbon dioxide) in the atmosphere that the rapid climate change which is expected to occur with global warming can be halted.

Up to the end of 2008, 14 sessions of the Conference of the Parties to the Climate Convention have taken place. Those since November 1997 have largely been concerned with the Kyoto Protocol, the first formal binding legislation promulgated under the Convention. The following paragraphs will first outline the actions taken so far, then describe the Kyoto Protocol and address the further actions necessary to satisfy the Convention's objective to stabilise greenhouse gas concentrations. Options for the energy and transport sectors to achieve the reductions in emissions required will be described in Chapter 11.

Stabilisation of emissions

The target for short-term action proposed for developed countries by the Climate Convention was that, by the year 2000, greenhouse gas emissions should be brought back to no more than their 1990 levels. In the run-up to the Rio conference, before the Climate Convention was formulated, many developed countries had already announced their intention to meet such a target at least for carbon dioxide. They would do this mainly through energy-saving measures, through switching to fuels such as natural gas, which for the same energy production generates 40% less carbon dioxide than coal and 25% less than oil. In addition those countries with traditional heavy industries (e.g. the iron and steel industry) were experiencing large changes which significantly reduce fossil fuel use. More detail of these energy-saving measures are given in the next chapter, which is devoted to a discussion of future energy needs and production.

Despite the Climate Convention target, by the year 2000, compared with 1990, global emissions from fossil fuel burning had risen by about 11%. There was great variation between the emissions from different countries. In the USA they rose by 17%, in the rest of the OECD (Organization for Economic Cooperation and Development) they rose on average by 5%. Emissions in countries in the former Soviet Union (FSU – also often called Economies in Transition) fell by around 40% because of the collapse of their economies, while the total of emissions from developing countries increased by nearly 40%. Since 2000, global emissions from fossil fuel burning have continued to rise at an average of about 3% per year.

As we shall learn later in the chapter, stabilisation of carbon dioxide emissions would not lead in the foreseeable future to stabilisation of atmospheric concentrations. Stabilisation of emissions could only be a short-term aim. In the longer term much more substantial reductions of emissions are necessary.

The Montreal Protocol

The chlorofluorocarbons (CFCs) are greenhouse gases whose emissions into the atmosphere are already controlled under the Montreal Protocol on ozone-depleting substances. This control has not arisen because of their potential as greenhouse gases, but because they deplete atmospheric ozone (see Chapter 3). Emissions of CFCs have fallen sharply during the last few years and the growth in their concentrations has slowed; for some CFCs a slight decline in their concentration is now apparent. The phase-out of their manufacture in industrialised countries by 1996 and in developing countries by 2006 as required by the 1992 amendments to the Montreal Protocol will ensure that the profile of their atmospheric concentration will continue to decline. However, because of their long life in the atmosphere this decline will be slow; it will be a century or more before their contribution to global warming is reduced to a negligible amount.

The replacements for CFCs – the hydrochlorofluorocarbons (HCFCs), which are also greenhouse gases though less potent than the CFCs – are required to be phased out by 2030. It will probably be close to that date before their atmospheric concentration stops rising and begins to decline.

Because of the international agreements that now exist for control of the production of the CFCs and many of the related species that contribute to the greenhouse effect, for these gases the stabilisation of atmospheric concentration required by the Climate Convention will in due course be achieved.

Other replacements for CFCs are the hydrofluorocarbons (HFCs), which are greenhouse gases but not ozone-depleting. The controls of the Montreal Protocol do not therefore apply and, as was mentioned in Chapter 3, any substantial growth in HFCs needs to be evaluated along with the other greenhouse gases. As we shall see in the next section, they are included in the 'basket' of greenhouse gases addressed by the Kyoto Protocol.

The Kyoto Protocol

At the first meeting after its entry into force held in Berlin in 1995, the Parties to the Climate Convention (i.e. all the countries that had ratified it) decided that they needed to negotiate a more specific and quantified agreement than the Convention on its own provided. Because of the principle in the Convention that industrialised countries should take the lead, a Protocol was formulated that required commitments from these countries (known as Annex I countries) for specific quantitative reductions in emissions (listed in Table 10.1) from their level in 1990 to their average from 2008–12, called the first commitment period. The Protocol also required that a second commitment period be defined. Negotiations began at the Montreal meeting in late 2005 with the aim of completing arrangements in time for a smooth transition between the first and second periods. The Protocol carries inbuilt mechanisms that could lead to stronger action and be expanded over time to include developing countries.

The basic structure of the Protocol and the commitments required by different countries were agreed at a meeting of the Conference of the Parties in Kyoto in November 1997. But the Protocol is a highly complex agreement and over the next three years intense negotiations followed regarding the details – the range of gases covered, the basis for comparing them and the rules for monitoring, reporting and compliance. Further the Protocol incorporates a range of mechanisms (see box below) of a kind that are unprecedented in an international treaty and that enable countries to offset their domestic emission obligations against the absorption of emissions by 'sinks' (e.g. through forestation – see next section) or by investment in or trading with other countries where it might be cheaper to limit emissions.

The emissions controlled by the Protocol are from six greenhouse gases (Table 10.2 and Figure 10.1) that can be converted into an amount of carbon dioxide equivalent through the use of their global warming potentials (GWPs) which were introduced in Chapter 3, page 63.

The details of the Protocol were finally agreed at a meeting of the Conference of the Parties in Marrakesh in October–November 2001. Much of the detailed discussion related to the inclusion of carbon sinks, especially from forests and from land-use change. Because of the large uncertainties regarding the magnitude of such sinks, considerable doubts were expressed regarding their inclusion in the Protocol arrangements. However, it was agreed that they should be included in a limited way and detailed regulations were agreed concerning the inclusion of afforestation, reforestation and deforestation activities and certain kinds of land-use change. Capping arrangements were also set up that limit the extent to which removals of carbon dioxide from these activities are allowed to offset emissions elsewhere.[1]

`Before the Marrakesh meeting in 2001 the United States had announced its withdrawal from the Protocol. Despite this by the end of 2003, 120 countries had ratified the Protocol and the Annex I countries that had ratified represented 44% of Annex I country emissions. For the Protocol to come

Table 10.1 Emissions targets (1990[a]–2008/2012) for greenhouse gases under the Kyoto Protocol

Country	Target (%)
EU-15[b], Bulgaria, Czech Republic, Estonia, Latvia, Lithuania, Romania, Slovakia, Slovenia, Switzerland	−8
USA[c]	−7
Canada, Hungary, Japan, Poland	−6
Croatia	−5
New Zealand, Russian Federation, Ukraine	0
Norway	+1
Australia[c]	+8
Iceland	+10

[a] Some economies in transition (EIT) countries have a baseline other than 1990.

[b] The 15 countries of the European Union have agreed an average reduction; changes for individual countries vary from −28% for Luxembourg, −21% for Denmark and Germany to +25% for Greece and +27% for Portugal.

[c] The USA has not ratified the Protocol. Australia did not ratify until March 2008.

Table 10.2 Greenhouse gases covered by the Kyoto Protocol and their global warming potentials (GWPs) on a mass basis relative to carbon dioxide and for a time horizon of 100 years

Greenhouse gas	Global warming potential (GWP)
Carbon dioxide (CO_2)	1
Methane (CH_4)	25
Nitrous oxide (N_2O)	298
Hydrofluorocarbons (HFCs)	from 12 to 12 000[a]
Perfluorocarbons (PFCs)	from 5000 to 12 000[a]
Sulphur hexafluoride (SF_6)	22 200

[a] Range of values for different HFCs: for more information about HFCs see Moomaw, W. R., Moreira, J. R. *et al.*, in Metz *et al.* (eds.) *Climate Change 2001: Mitigation*, Chapter 3 and its appendix.

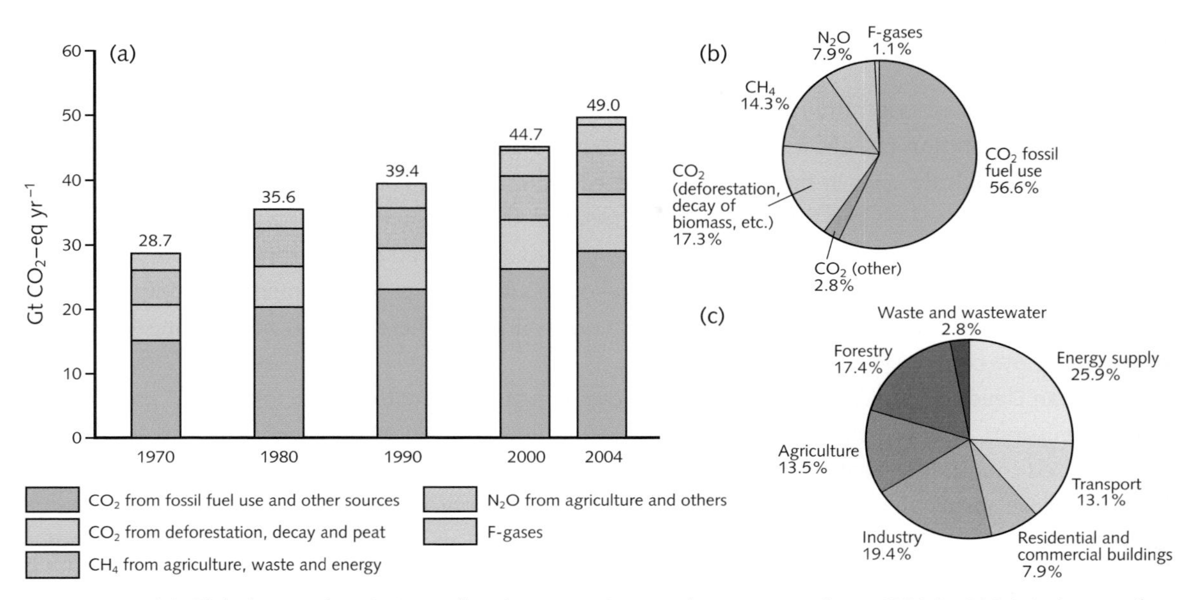

Figure 10.1 (a) Global annual emissions of anthropogenic greenhouse gases from 1970 to 2004, in terms of GtCO_2e per year (includes CO_2, CH_4, N_2O, HFCs, PFCs and SF_6 weighted by their 100-year horizon global warming potentials – see Chapter 3, page 63). (b) Share of different anthropogenic greenhouse gases in total emissions in 2004 in terms of CO_2e. (c) Share of different sectors in total anthropogenic emissions in 2004 in terms of CO_2e (forestry includes deforestation). Buildings and Industry sectors do not include electricity use which is aggregated under Energy supply. (Figure 11.2 provides proportional information under end use).

into force 55 countries had to ratify together with sufficient Annex I countries to represent 55% of Annex I country emissions. With the ratification by Russia towards the end of 2004, the Protocol finally came into force on 16 February 2005.

Concern has often been expressed about the likely cost of implementation of the Kyoto Protocol. Cost studies have been carried out using a number of international energy-economic models. For nine such studies, the range of values in impacts on the gross domestic product (GDP) of participating countries is as follows.[2] In the absence of emissions trading (see box), estimated reductions in projected GDP in the year 2010 are between 0.2% and 2% compared with a base case with no implementation of the Protocol. With emissions trading between Annex I countries, the estimated reductions in GDP are between 0.1% and 1.1%. If emissions trading with all countries is assumed through ideal CDM (see box) implementation, the estimated reductions in GDP are substantially less – between 0.01% and 0.7%. Although there are differences between countries, most of the large range in the results is due to differences in the models

The Kyoto mechanisms

The Kyoto Protocol includes three special mechanisms to assist in emissions reductions.

Joint implementation (JI) allows industrialised countries to implement projects that reduce emissions or increase removals by sinks in the territories of other industrialised countries. Emissions reduction units generated by such projects can then be used by investing Annex I countries to help meet their emissions targets. Examples of JI projects could be the replacement of a coal-fired power plant with a more efficient combined heat and power plant or the reforestation of an area of land. Joint implementation projects are expected to be mainly in EIT (economies in transition) countries where there is more scope for cutting emissions at low cost.

The Clean Development Mechanism (CDM) allows industrialised countries to implement projects that reduce emissions in developing countries. The certified emission reductions generated can be used by industrialised countries to help meet their emissions targets, while the projects also help developing countries to achieve sustainable development and contribute to the objective of the Convention. Examples of CDM projects could be a rural electrification project using solar panels or the reforestation of degraded land.

Emissions trading allows industrialised countries to purchase 'assigned amount units' of emissions from other industrialised countries that find it easier, relatively speaking, to meet their emissions targets. This enables countries to utilise lower cost opportunities to curb emissions or increase removals, irrespective of where those opportunities exist, in order to reduce the overall cost of mitigating climate change. (See the box on carbon trading on page 299 for more detail.)

The detailed regulations concerning the implementation of these mechanisms state that projects will only be approved if they lead to real, measurable and long-term benefits related to the mitigation of climate change and that they are additional to any that would have occurred without the project.

and can be considered as an expression of the large uncertainties inherent in such studies at the present stage of development.

Given that failure of any Annex I Parties to meet their emission reduction targets could undermine the efforts of the others, it is important that any Party lagging in their commitments is identified early. The Protocol has therefore established a rigorous system of reporting and verification whereby states' annual emission estimation reports are scrutinised by independent experts. A Compliance Committee encourages compliance through a combination of 'carrot' and 'stick' measures. Through its Facilitative Branch it offers Parties financial and technical assistance towards meeting their commitments whilst its Enforcement Branch is responsible for declaring cases of non-compliance and has power to penalise Parties by preventing them from making use of the flexible mechanisms. For any Party exceeding their emissions allowance at the

Carbon trading

Carbon trading is an innovative market-based solution to the problem of reducing greenhouse gas emissions[3]. Its main rationale is that, by attaching a price to carbon dioxide emissions, trading schemes will generate powerful economic incentives to cut emissions and channel investment efficiently.

Emissions trading works by setting limits on total allowable emissions that are then converted into tradable permits to be distributed amongst participants. For example, a company with commitments to reduce its emissions by 20% during a set 'commitment period' will be allocated permits equal to 80% of what it would have emitted given an agreed baseline emissions level. Participants have the option of trading a certain proportion of their allocation but must ensure that they hold enough permits to cover their emissions reduction target when the commitment period comes to an end. Those that find they can make reductions relatively cheaply have an incentive to reduce their emissions below their allocated level knowing that they can sell any excess permits to participants for whom direct reductions are too expensive. By these means, all participants will end up spending less than they would have done without the trading mechanism, helping to reduce emissions more quickly than would otherwise have been possible.

The United States successfully argued for including carbon trading as a central component of the Kyoto Protocol in 1997, citing the effectiveness of its own domestic sulphur dioxide trading scheme. International discussions subsequently provided the impetus for the European Union to develop the first international scheme. The European Trading Scheme commenced operation in 2005, covering some 11 500 industrial plants accounting for 45% of total EU carbon dioxide emissions. Since then a number of governments have announced plans for their own variants of carbon trading schemes including Australia, New Zealand, Canada and several states within the USA. Rules and procedures agreed under the Kyoto Protocol provide a framework through which future trading schemes might be linked.

Two factors are key to the success of any emissions trading scheme. First is the provision of accurate and verifiable information regarding actual emissions by different sectors and countries. Secondly, the method of allocation of permits must be transparent and fair to all participants; in practice this presents a large challenge. At the start of the European Trading Scheme, substantial over-allocations of permits occurred which led in 2006, to a collapse of the price of carbon. For the 2008–12 period, procedures have been tightened. One method of allocation, known as *grandfathering*, allocates permits in proportion to participant's current emission levels. This tends to favour the largest current emitters. Another, potentially fairer, method is to auction the permits using the auction proceeds, for instance, to assist reductions schemes that are of general value to the participants. Lessons learnt from the European Trading Scheme so far point strongly to future arrangements in which most of the permits are allocated by auction.

As an instrument of policy the market-based approach underlying emissions trading has been criticised for favouring emission reduction projects that promise low initial costs and rapid paybacks in the short term over more radical, systematic programmes that offer greater and cheaper reductions over a longer period of time.[4] It has also been criticised for its inadequate recognition of human rights especially in developing countries. Emissions trading therefore, although an important instrument for the control of emissions, must be supported by other measures mentioned later in this chapter or in Chapter 11.

end of the first commitment period, 130% of the excess is deducted from their allowance in the second period.

The Kyoto Protocol is an important start to the mitigation of climate change through reductions in greenhouse gas emissions. With its complexity and its diversity of mechanisms for implementation, it also represents a considerable achievement in international negotiation and agreement. It will stem continuing growth of emissions from many industrialised countries and achieve a reduction overall compared with 1990 from those Annex I countries that participate. A Kyoto Protocol with successful achievements will be essential for movement to be made towards a truly global system (i.e. involving both developed and developing countries) with binding targets after 2012. The much more substantial longer-term reductions that are necessary for the decades following 2012 will be discussed later in the chapter.

Forests

We now turn to the situation of the world's forests and the contribution that they can make to the mitigation of global warming. Action here can easily be taken now and is commendable for many other reasons.

Over the past few centuries many countries, especially those at mid latitudes, have removed much of their forest cover to make room for agriculture. Many of the largest and most critical remaining forested areas are in the tropics. However, during the last few decades, the additional needs of the increasing populations of developing countries for agricultural land and for fuelwood, together with the rise in demand for tropical hardwoods by developed countries, has led to a worrying rate of loss of forest in tropical regions (see box below). In many tropical countries the development of forest areas has been the only hope of subsistence for many people. Unfortunately, because the soils and other conditions were often inappropriate, some of this forest clearance has not led to sustainable agriculture but to serious land and soil degradation.[5]

Measurements on the ground and observations from orbiting satellites have been combined to provide estimates of the area of tropical forest lost. Over the decades of the 1980s and 1990s the average loss was about 1% per year (see box below) although in some areas it was considerably higher. Such rates of loss cannot be sustained if much forest is to be left in 50 or a 100 years' time. The loss of forests is damaging, not only because of the ensuing land degradation but also

The world's forests and deforestation

The total area covered by forest is almost one-third of the world's land area, of which 95% is natural forest and 5% planted forest.[9] About 47% of forests worldwide are tropical, 9% sub-tropical, 11% temperate and 33% boreal.

At the global level, the net loss in forest area during the 1990s was an estimated 940 000 km^2 (2.4% of total forest area). This was the combined effect of a deforestation rate of about 150 000 km^2 per year and a rate of forest increase of about 50 000 km^2 per year. Deforestation of tropical forests averaged about 1% per year. The rate of loss since 2000 has slightly slowed but not by enough to reduce concern.

The area under forest plantations grew by an average of about 3000 km^2 per year during the 1990s. Half of this increase was the result of afforestation on land previously under non-forest land use, whereas the other half resulted from conversion of natural forest.

In the 1990s, almost 70% of deforested areas changed to agricultural land, predominantly under permanent rather than shifting systems.

because of the contribution that loss makes to carbon emissions and therefore to global warming. There is also the dramatic loss in biodiversity (it is estimated that over half the world's species live in tropical forests) and the potential damage to regional climates (loss of forests can lead to a significant regional reduction in rainfall – see box on page 208).

For every square kilometre of a typical tropical forest there are about 25 000 tonnes of biomass (total living material) above ground, containing about 12 000 tonnes of carbon.[6] It is estimated that burning or other destruction from deforestation turns about two-thirds of this carbon into carbon dioxide. Approximately the same amount of carbon is also stored below the surface in the soil. On this basis, from the destruction of about 150 000 km^2 per year over the decades of the 1980s and 1990s (see box) about 1.2 Gt of carbon would enter the atmosphere as carbon dioxide. Although there are substantial uncertainties in the numbers, they approximately tally with the IPCC estimate, quoted in Chapter 3 (see Table 3.1), of the carbon as carbon dioxide entering the atmosphere each year from land-use change (mostly deforestation) of 1.6 ± 1.1 Gt per year – a larger fraction of the total anthropogenic emissions of carbon dioxide than results from the whole of the world's transportation sector. If all tropical forest were to be removed by 2100, between 100 and 150 ppm would be added to the CO_2 concentration at that date.[7]

Reducing tropical deforestation can therefore make a large contribution to slowing the increase of greenhouse gases in the atmosphere, as well as the

Landsat images of Bolivia taken in 1984 and 2000 show the dramatic deforestation of the Bolivian rainforest. In 1984 the rainforest had been thinned out in places, and by 2000 the rainforest had receded dramatically.

provision of other vital benefits such as guarding biodiversity, protecting water supplies, avoiding soil degradation and preserving the livelihoods of forest peoples. The Stern Review has estimated the cost of emissions savings from avoided deforestation as less than $US5 per tonne CO_2.[9]

Strong emphasis is being given internationally to reduction of deforestation as an essential contribution to mitigation of climate change. Towards the end of 2007, at the Bali conference of the UN FCCC it was agreed to work towards an agreement on deforestation in developing countries to be included as part of a post-Kyoto international Climate Change agreement.[10]

In the above discussion of deforestation, I have not mentioned the increasing interest in growing biomass for production of energy either directly or through biofuels. Land that is under forest may increasingly be taken over for such crops. This will be addressed, along with other energy issues, in the next chapter.

What about the possibilities for afforestation. For every square kilometre, a growing forest fixes between about 100 and 600 tonnes of carbon per year for a tropical forest and between about 100 and 250 tonnes for a boreal forest.[11] To illustrate the effect of afforestation on atmospheric carbon dioxide, suppose that an area of 100 000 km^2, a little more than the area of the island of Ireland, were planted each year for 40 years – starting now. By the year 2050, 4 000 000 km^2 would have been planted; that is roughly half the area of Australia. During that 40 years, the forests would continue to grow and uptake carbon for 20 to 50 years or more after planting (the actual period depending on the type of forest and site conditions) – and, assuming a mixture of tropical, temperate and boreal forest, between about 10 and 40 Gt of carbon from the atmosphere would have been sequestered or 4 $GtCO_2$ per year. This accumulation of carbon in the forests is equivalent to between about 5% and 10% of the likely emissions due to fossil fuel burning up to 2050. Add to this the emissions reductions that could arise with a near elimination of tropical deforestation and approximately 20% of

Aforestation in Burkina Faso.

anthropogenic CO_2 emissions over the period to 2050 would be accounted for.

But is such a tree-planting programme feasible and is land on the scale required available? The answer is almost certainly, yes. For instance, China is currently adding forests covering approximately 10 000 km^2 per year or one-tenth of the area we have mentioned above.[12] Studies have been carried out that have identified land which is not presently being used for croplands or settlements, much of which has supported forests in the past, totalling about the area just quoted.[13] Estimates of the cost of afforestation range from $US5 to $US15 per tonne CO_2 (the lower values in developing countries) not including the value of assocatiated local benefits (for instance, watershed protection, maintenance of biodiversity, education, tourism and recreation).[14] Compare these figures with the estimate given in Chapter 9 of between $US25 and $US50 for the cost per tonne CO_2 of the likely damage due to global warming. Such a programme therefore appears potentially attractive for alleviating the rate of change of climate due to increasing greenhouse gases in the relatively short term.

Let me insert here a note of caution. As with many environmental projects the situation, however, may not be as simple as it seems at first. One complicating factor is that introducing forest can change the albedo[15] of the Earth's surface. Dark green forests absorb more of the incoming solar radiation than arable cropland or grassland and so tend to warm the surface. This is particularly noticeable in winter months when unforested areas may possess highly reflecting snow cover. Calculations show that, particularly at high latitudes, the warming due to this 'albedo effect' can offset a significant fraction of the cooling that arises from the additional carbon sink provided by the forest.[16]

A possible afforestation programme has been presented in order to illustrate the potential for carbon sequestration. Once the trees are fully grown, of course, the sequestration ceases. What happens then depends on the use that may be made of them. They may be 'protection' forests,[17] for instance for the control of erosion or for the maintenance of biodiversity; or they may be production forests, used for biofuels or for industrial timber. If they are used as fuel for energy generation (see Chapter 11), they add to the atmospheric carbon dioxide but, unlike fossil fuels, they are a renewable resource. As with the rest of the biosphere where natural recycling takes place on a wide variety of timescales, carbon from wood fuel can be continuously recycled through the biosphere and the atmosphere.

Reduction in sources of greenhouse gases other than carbon dioxide

Methane, nitrous oxide and the halocarbons are greenhouse gases, less important than carbon dioxide, all of which show increases at the present time. In Figures 6.1 and 6.2 and Table 6.1 are shown the emissions, atmospheric concentrations and radiative forcing of these gases estimated for the twenty-first century under the various SRES scenarios, assuming no special action to reduce them. Is it possible that these further increases can be slowed or eliminated? We consider them in turn.

Methane contributes about 15% to the present level of global warming (Figure 10.1). The stabilisation of its atmospheric concentration would contribute a small but significant amount to the overall problem. Because of its much shorter lifetime in the atmosphere (about 12 years compared with 100–200 years for carbon dioxide), only a relatively small reduction in the anthropogenic emissions of this gas, about 8%, would be required to stabilise its concentration at the current level. Of the various sources of methane listed in Table 3.2, there are four sources arising from human activities that could rather easily be reduced at small cost.[18]

Firstly, methane emission from biomass burning would be cut by, say, one-third if deforestation were drastically curtailed. Secondly, methane production from landfill sites could be cut by at least a third if more waste were recycled or used for energy generation by incineration or if arrangements were made on landfill sites for the collection of methane gas (it could then be used for energy production or if the quantity were insufficient it could be flared, turning the methane into carbon dioxide which molecule-for-molecule is less effective than methane as a greenhouse gas). Waste management policies in many countries already include the encouragement of such measures.

Thirdly, the leakage from natural gas pipelines from mining and other parts of the petrochemical industry could at little cost (probably even at a saving in cost) also be reduced by, say, one-third. An illustration of the scale of the

As wastes break down, methane is produced that can be captured and burnt for power generation. The carbon dioxide then released has much less impact as a greenhouse gas than the methane that would otherwise be released.

leakage is provided by the suggestion that the closing down of some Siberian pipelines, because of the major recession in Russia, has been the cause of the fall in the growth of methane concentration in the atmosphere from about 1992. Improved management of such installations could markedly reduce leakage to the atmosphere.

Fourthly, with better management, options exist for reducing methane emissions from sources associated with agriculture, for instance, by adjustments to the diet of cattle or to the details of rice cultivation.[19]

Reductions from these four sources could reduce anthropogenic methane emissions by more than 60 million tones per year – enough to stabilise methane concentration in the atmosphere at or below the current level. Put another way, the reduction in methane emissions from these sources would be equivalent to a reduction in annual carbon dioxide emissions of about 1.4 Gt[20] or about 3% of total greenhouse gas emissions – a useful contribution towards the solution of the global warming problem.[21]

It was noted in Chapter 3 (page 50) that the growth in global atmospheric methane concentrations had largely halted since the late 1990s. However, since early 2007 there is evidence of renewed growth[22] which may be more pronounced in the northern hemisphere. Some increase from methane emissions from natural sources may be expected because of the influence of global warming on various natural methane reservoirs (see Chapter 3). Prominent among these are the very large reservoirs under the tundra at high latitudes (see box on page 48–9 and Table 7.5). As the cover of Arctic summer sea ice has reduced and as northern Siberia has warmed, more evidence of local methane emissions has emerged. Further studies are needed to ascertain whether these are connected with the recent growth in global methane concentration. As the Arctic warms further, the possibility exists of much larger releases in the longer term, especially if global average temperature rise is not halted.

Turning now to nitrous oxide which contributes about 7% to the present level of global warming and which is growing at about 0.25% per year. Much of its growth appears to rise from emissions associated with the use of nitrogen fertilisers. Careful management of the use of such fertilisers and other changes in agricultural practice could largely stem the continuing increase.

For halocarbons for many of which the manufacture is being phased out, the most important concern is that disposal of products containing these gases, for instance of foams or refrigeration equipment, is carefully controlled so as to minimise leakage to the atmosphere and to ensure that their atmospheric concentration gradually reduces during the twenty-first century.

In this section we have seen that options are available for stemming the growth and possibly reducing the concentrations of greenhouse gases other than carbon dioxide. In the following section, stabilisation of carbon dioxide concentrations will be considered together with stabilisation of the combined effect of greenhouse gases considered together. Some reductions in the concentrations of methane, nitrous oxide and halocarbons will be seen as important contributions to achieving this overall stabilisation.

Because the lifetime of methane in the atmosphere is relatively short, a small reduction in methane emissions will quickly lead to its stabilisation as required by the Climate Convention objective. The same is not true of the stabilisation of carbon dioxide concentration with its much longer and rather complicated lifetime. It is to that we shall now turn.

Stabilisation of carbon dioxide concentrations

Carbon dioxide, as we have seen, is the most important of the greenhouse gases that is increasing through human activities. As we saw in Chapter 3, emissions of carbon dioxide into the atmosphere from anthropogenic sources result from fossil fuel burning (about 80%) and from land-use changes (about 20%) – mainly deforestation. Reduction in emissions from land-use changes was considered earlier in the chapter. Reductions in emissions from fossil fuel burning will be the subject of the next chapter.

Under all the SRES scenarios, the concentration of carbon dioxide rises continuously throughout the twenty-first century and apart from scenarios B1 and A1T none comes anywhere near to stabilisation of concentration by 2100. Since the year 2000, the growth of global carbon dioxide emissions (Figure 10.1) at close to 3% per year has been faster than assumed in most of the SRES scenarios where the average growth was around 1%. During the 1980s and 1990s global carbon dioxide emissions grew at an average of just over 1% per year, the average during the 1990s being kept down because of substantial falls in emissions in the former Soviet Union and eastern Europe. Projections for the next few years indicate that global emissions are likely to continue rising at about the current rate.

Here we consider what sort of emissions scenario would lead to stabilisation of the carbon dioxide concentration. Suppose for instance that it were possible to keep global emissions for the future at the same level as now, would that be enough? Stabilising concentrations is, however, very different from stabilising emissions. With constant emissions from now, the concentration in the atmosphere would continue to rise, would reach at least 500 ppm by the year 2100 and continue to increase thereafter, although more slowly, for many centuries. Further, because of the long lifetime of carbon dioxide in the atmosphere, even if very severe action is taken to curb emissions, stabilisation of its concentration and hence stabilisation of climate will take many decades.

A large number of studies focusing on climate stabilisation and emissions profiles leading to stabilisation of greenhouse gases have been brought together in the IPCC AR4 Report, where they have been grouped together under categories leading to different stabilisation levels. For each of the chosen levels, the range of emissions profiles shown by the studies to bring about stabilisation of carbon dioxide concentration is shown in Figure 10.2 and Table 10.3. Studies that have included greenhouse gases other than carbon dioxide have been put together with carbon dioxide-only studies through the use of carbon dioxide equivalent CO_2e (for definition see Chapter 6, page 147).

Note that stabilisation at any level shown in the figure, even at an extremely high level, requires that anthropogenic carbon dioxide emissions eventually fall to a small fraction of current emissions. This highlights the fact that to maintain a constant future carbon dioxide concentration, emissions must eventually be no greater than the level of persistent natural sinks. The main known such sink is due to the dissolution of calcium carbonate from the oceans into ocean sediments which, for high levels of carbon dioxide concentration, is probably less than 0.1 GtC per year.[23] This means, for instance, that for the lowest category in Figure 10.2, anthropogenic emissions of greenhouse gases need to fall close to zero by 2100.

In the work presented in Figure 10.2, many different pathways to stabilisation could have been chosen. The particular emission profiles illustrated in Figure 10.2 begin by following the current average rate of increase of emissions and then provide a smooth transition to the time of stabilisation. To a

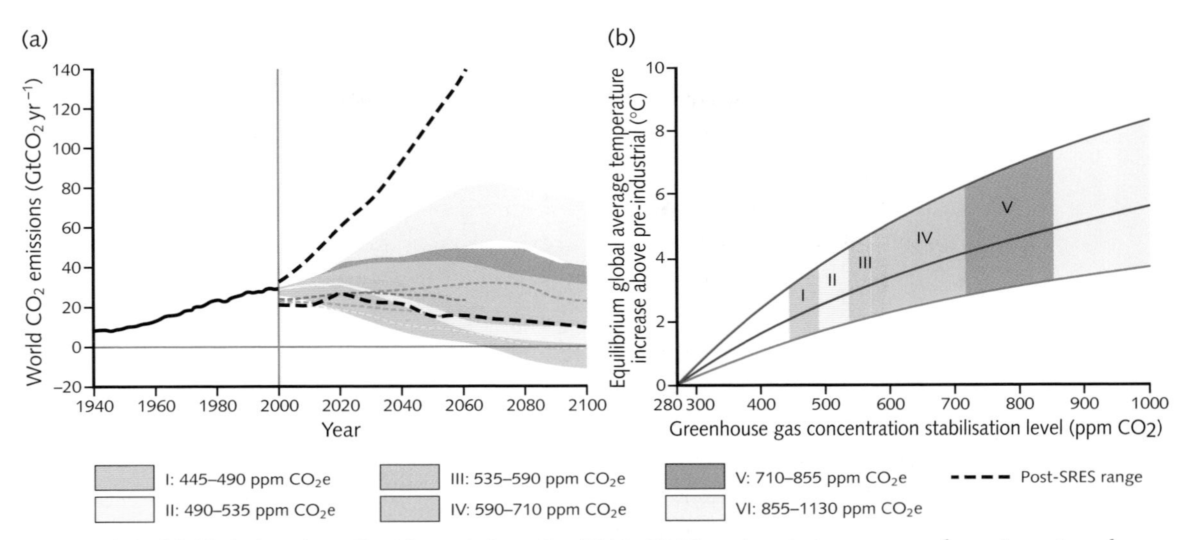

Figure 10.2 (a) Global carbon dioxide emissions for 1940–2000 and emissions ranges for categories of stabilisation scenarios from 2000 to 2100; colours show stabilisation scenarios grouped according to different levels (categories I to VI in inset and Table 10.3). The range shown covers the 10th to 90th percentiles of the full scenario distribution (numbers of scenarios included in Table 10.3). The thin dashed lines denote the lower end of the range for that category in cases where there is overlap between the categories. To convert Gt CO_2 to Gt C, divide by 3.66. The thick dashed black lines indicate the range of emissions scenarios published since 2000. (b) Relationship between stabilisation level and the likely equilibrium global average temperature increase above pre-industrial level; the dark blue line assumes a best estimate of climate sensitivity of 3 °C and the colours indicate the effect of a range in climate sensitivity from 2 to 4.5 °C. For calculating the equilibrium temperature, the simple relationship $T_{eq} = T_{2\times CO_2} \times \ln([CO_2]/280)/\ln(2)$ is employed with mean, lower and upper values of $T_{2\times CO_2}$ of 3, 2 and 4.5 °C. The relationship between radiative forcing (R in W m^{-2}) and concentration (C in ppm) is $R = 5.3\ln(C/C_0)$ where C_0 is the pre-industrial CO_2 concentration of 280 ppm.

first approximation, the stabilised concentration level depends more on the accumulated amount of carbon emitted up to the time of stabilisation than on the exact concentration path followed en route to stabilisation. This means that alternative pathways that assume higher emissions in earlier years would require steeper reductions in later years. For instance, if the atmospheric concentration of carbon dioxide is to remain below about 550 ppm, the future global annual emissions averaged over the twenty-first century cannot exceed the level of global annual emissions in the year 2000. For lower levels of stabilisation, to bring down the twenty-first-century accumulated emissions to a much lower level will require urgent and large carbon dioxide emissions reductions. Note also that for the lowest stabilisation level, Category 1, in Figure 10.2a some of the scenarios later in the century require negative emissions implying a need for substantial removal of CO_2 from the atmosphere.

Table 10.3 Characteristics of stabilisation scenarios and resulting long-term equilibrium global average[a]

Category	Anthropogenic addition to radiative forcing at stabilisation ($W m^{-2}$)	CO_2 equivalent concentration at stabilisation[b] (ppm)	Peaking year for CO_2 emissions[a, c] (year)	Change in global CO_2 emissions in 2050 (per cent of 2000 emissions) (percent)[a, c]	Global average temperature increase above pre-industrial at equilibrium, using 'best estimate' climate sensitivity[d, e] (°C)	Number of assessed scenarios
I	2.5–3.0	445–490	2000–2015	−85 to −50	2.0–2.4	6
II	3.0–3.5	490–535	2000–2020	−60 to −30	2.4–2.8	18
III	3.5–4.0	535–590	2010–2030	−30 to +5	2.8–3.2	21
IV	4.0–5.0	590–710	2020–2060	+10 to +60	3.2–4.0	118
V	5.0–6.0	710–855	2050–2080	+25 to +85	4.0–4.9	9
VI	6.0–7.5	855–1130	2060–2090	+90 to +140	4.9–6.1	5

[a] The emission reductions to meet a particular stabilisation level reported in the mitigation studies assessed here might be underestimated due to missing carbon cycle feedbacks

[b] Atmospheric CO_2 concentrations were 379 ppm in 2005. The best estimate of total CO_2 equivalent concentration in 2005 for all long-lived greenhouse gases is about 455 ppm, while the corresponding value including the net effect of all anthropogenic forcing agents e.g. aerosols is 375 ppm CO_2e.

[c] Ranges correspond to the 15th to 85th percentile of the scenario distribution. CO_2 emissions are shown so multi-gas scenarios can be compared with CO_2-only scenarios.

[d] The best estimate of climate sensitivity is 3 °C.

[e] Note that global average temperature at equilibrium is different from expected global average temperature at the time of stabilisation of greenhouse gas concentrations due to the inertia of the climate system. For the majority of scenarios assessed, stabilisation of greenhouse gas concentrations occurs between 2100 and 2150. For Categories I or II, equilibrium temperature may be reached earlier.

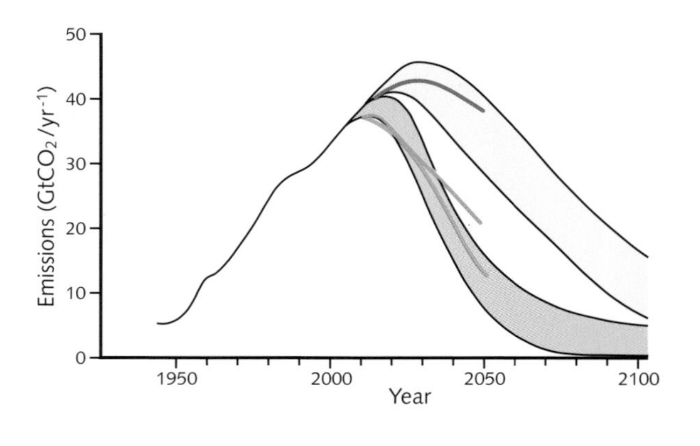

Figure 10.3 Global CO_2 emission profiles that would stabilise CO_2 at 450 ppm (pink) and 550 ppm (light blue). The shaded areas show the range of uncertainty arising because of the climate–carbon-cycle feedback (see text). Also shown are emissions profiles for the IEA scenarios ACT Map (red) and BLUE Map (blue) for fossil fuel emissions (see Chapter 11, page 332) to both of which has been added a constant 7.3 $GtCO_2$ (2GtC) per year to allow for emissions from deforestation and land use change. A further profile (green) shows the effect on the BLUE Map profile if emissions from deforestation and land use change are halted completely by 2050.

Many of the scenarios included in compiling the results shown in Figure 10.2 and Table 10.3 do not include the effect of climate feedbacks on the carbon cycle (see box in Chapter 3 on page 48–9). Two of the feedbacks are important in the context of the consideration of stabilisation scenarios; namely, increased respiration from the soil as the temperature rises and a decrease in net uptake of carbon by plants in some regions as the climate warms (e.g. in forests this can be perceived as dieback). The area over which such decrease occurs becomes larger for greater warming. As we saw in Chapter 3, the effect of these feedbacks could lead to the biosphere becoming a substantial source of carbon dioxide during the twenty-first century. The size of that source will depend on the amount of climate change. In Figure 10.3 are shown profiles for stabilisation of carbon dioxide alone at 450 and 550 ppm both without these climate change feedbacks included (the upper line) and with the range of feedbacks represented by the models studied in the IPCC 2007 Report, the largest value of the combined feedbacks being that derived by the Hadley Centre in the UK. Compared with no feedback, the effect of this largest value is to reduce the accumulated emissions allowable in the twenty-first century, for instance, for the 450 ppm and 550 ppm stabilisation scenarios, by about 600 Gt CO_2 and 900 Gt CO_2 respectively.[24] Also under this largest feedback case, emissions scenarios not allowing for feedback that are aiming at 450 ppm stabilisation, when the feedbacks are included, would in fact achieve around 500 ppm.

The choice of stabilisation level

The last few sections have addressed the main greenhouse gases and how their concentrations might be stabilised. To decide how appropriate stabilisation levels should be chosen as targets for the future we look to the guidance provided by the Climate Convention Objective (see box on pages 291–2), which states that the levels and timescales for their achievement should be such that dangerous interference with the climate system is avoided, that ecosystems are able to adapt naturally, that food production is not threatened and that economic development can proceed in a sustainable manner.

The balancing of these scientific, economic, social and political criteria presents a large challenge. In Chapter 9 (see box on page 280) the concept of Integrated Assessment and Evaluation was introduced which involves employment of the whole range of disciplines in the natural and social sciences. Taking all factors into consideration will involve different kinds of analysis, cost–benefit analysis (which was considered briefly in Chapter 9), multicriteria analysis (which takes into account factors that cannot be expressed in monetary terms) and sustainability analysis (which considers avoidance of particular thresholds of stress or of damage). Further, because uncertainty is associated both with many of the factors that have to be included and with the methods of analysis, the process of choice is bound to be an evolving one subject to continuous review – a process often described as sequential decision-making.

In Figure 10.2 were presented options for the stabilisation of carbon dioxide and other greenhouse gases in terms both of the concentration of CO_2e (a) and of the global average temperature increase since pre-industrial times (b). It is the latter measure that is closer to what actual climate change is all about. The information in Figure 10.2 and Table 10.3 covers a wide range of possible temperature increases up to over 6 °C. However, attention has recently focused on the lower end of the range, between 2 and 3 °C.

In Chapter 7 we found that many of the studies of impacts of climate change had been made under the assumption that the atmospheric CO_2e concentration had doubled from its pre-industrial value of 280 ppm to about 560 ppm and that the global average temperature had increased from its pre-industrial value by a best estimate of 3 °C – an estimate that was raised in the 2007 IPCC Report from its earlier value of 2.5 °C (see Chapter 6, page 143). We enumerated in Chapter 7 the substantial impacts that apply to this situation together with estimates of their associated costs.[25] In Chapter 9 it was pointed out, even considering only those costs that can be estimated in terms of money, that estimates of the cost of the likely damage of impacts at that level of climate change were substantially larger than the mitigation costs of stabilising CO_2e concentration at doubled pre-industrial CO_2. We also noted that damage due to anthropogenic climate change is likely to rise more rapidly as the amount of carbon dioxide in the atmosphere increases. These considerations suggest that limits set at 560 ppm CO_2e or 3 °C would be too high.

A widely publicised target for the maximum allowable rise in global average temperature is 2 °C. It was proposed by the European Union over 10 years ago[26] and has been recently reiterated by the EU,[27] by governments (e.g. by Chancellor Merkel of Germany before the 2007 G8 conference) and by many other organisations.[28]

How important is it to aim at 2 °C rather than say 3 °C? Further perspective on this can be obtained from Table 7.1 which indicates that the impacts at 2 °C above pre-industrial are substantially less severe compared with those at 3 °C, for instance in terms of water stress, extinction of species, coral mortality, decreased crop productivity, ocean acidity, increased floods, droughts and storms and risk of more rapid sea level rise. As an example, we can turn to Figure 6.12 to compare more quantitatively the risk of drought for different increases in global average temperature. Noting from Figure 6.4a, that for the A2 SRES scenario, as followed by Figure 6.12, global average temperature increases by 2 °C and 3 °C above the pre-industrial value by about 2050 and 2075 respectively. As mentioned in Chapter 6 (page 158) and shown in Figure 6.12 the proportion of land area under extreme drought increased from around 1% in 1980 to 2% or 3% now and is projected to increase to about 10% in 2050 and 20% in 2075, indicating a likelihood of a rise from today in the risk of extreme droughts of a factor of up to 8 for a rise of 3 °C in global average temperature compared with a factor of up to 4 for a 2 °C rise.

Stabilisation below 2 °C might also avoid some of the worst impacts; for instance, some of the large-scale dieback of forests and the transition of the biosphere from a sink to a source for carbon dioxide (see box in Chapter 3 on page 48–9) that would otherwise occur around the middle of the twenty-first century. It would also reduce the risk mentioned earlier in the chapter of large-scale release of methane as the Arctic ocean or tundra warm.

How achievable is 2 °C? From Figure 10.1b and Table 10.3, it can be seen that 2 °C implies an equilibrium radiative forcing of 2.5 $W m^{-2}$ and about 450 ppm stabilisation for CO_2e. Since these are best estimates, what we can say is that 450 ppm CO_2e and 2.5 $W m^{-2}$ should provide a 50% chance of achieving 2 °C.

We now need to know what 450 ppm CO_2e implies for CO_2 itself. As was explained in Chapter 6 page 149 (see also Figure 3.11), the negative radiative forcing of anthropogenic aerosols at the present time approximately balances the positive contributions from greenhouse gases other than carbon dioxide. Earlier in the chapter, it was pointed out that options are available for preventing further increases and reducing the contributions from methane, nitrous oxide and halocarbons. Also aerosol scenarios show little reduction in the magnitude of the aerosol contribution during the next few decades.[29] These considerations provide a rational basis for the time being for coupling the 2 °C temperature target with a stabilisation target of 450 ppm for CO_2 alone (Figure 10.3). A similar argument couples together a 3 °C temperature target with 550 ppm for carbon dioxide alone. It is in fact these assumptions that are made in Chapter 11 when the challenging implications of such targets are presented for the future of the energy and transport sectors.

In accepting the targets of 2 °C and 450 ppm as a basis for action, I must however inject two notes of caution. The first is that the calculations I have presented are based on best estimates only. They have not allowed for uncertainties; 450 ppm stabilisation for carbon dioxide only provides a 50% chance of achieving 2 °C. An 80% chance of achieving 2 °C would require stabilisation at about 380 ppm, the current atmospheric carbon dioxide level.[30]

The second point of caution pertains to the future of sulphate aerosols that provide the largest component towards aerosol cooling mentioned in the last paragraph. Because they lead to serious low level pollution and also to 'acid rain', there is large pressure to control and reduce the sulphur dioxide emissions that are the precursors of sulphate aerosols. Reductions can also be expected as the consumption of coal and oil is phased out. Many future aerosol scenarios therefore show large reductions of sulphur emissions especially during the second half of the twenty-first century. Now, because the life time of aerosols in the atmosphere is very short (a few days) reductions in aerosol emissions lead almost immediately to large changes in aerosol concentrations and hence also in radiative forcing. Compensating changes in forcing through changes in CO_2 emissions can only occur much more slowly because of the long life time of carbon dioxide in the atmosphere (Chapter 3 page 37). This means that, in order to maintain a 2 °C target for global average temperature rise, reductions in global sulphur dioxide emissions need to be anticipated well in advance by matching reductions in CO_2 emissions. In fact, that anticipation should begin now; it will likely mean that global CO_2 emissions should be reduced close to zero by 2050 and total greenhouse gas (CO_2e) emissions to zero before the end of the century. I return to these issues in the section entitled *A Zero Carbon Future* in Chapter 11, page 378.

Is it possible to consider targets lower than 2 °C and 450 ppm carbon dioxide? Considering the most important greenhouse gas, carbon dioxide, as we have already noted, its long life in the atmosphere provides severe constraints on the future emission profiles that lead to stabilisation at any level. The concentration of carbon dioxide in the atmosphere is already above 380 ppm which means (Figure 10.2) that stabilisation of carbon dioxide alone below 400 ppm would require an immediate drastic reduction in emissions. Such reduction could only be achieved at a large cost and with some curtailment of energy availability and would almost certainly breach the criterion that requires 'that economic development can proceed in a sustainable manner'.

Many are, however, asking the question whether a 2 °C target will be adequate to stabilise the climate against very damaging and irreversible change.

Prominent among these is Professor James Hansen at the NASA Institute for Space Studies at New York. In a recent paper,[31] Hansen argues largely from palaeoclimate evidence that a 350 ppm target is necessary to avoid the danger of rapid collapse of the Greenland and West Antarctic ice-sheets and other serious non-linear processes (see Table 7.5). Such a target could only be realised after substantial overshoot in the early years and would probably also require a large programme over many decades of sequestration of carbon dioxide already in the atmosphere. The possibility of such a programme has also recently been proposed by Professor Wally Broecker of Columbia University in the USA.[32] Targets such as that aimed at 2 °C that may be set now are bound to be reviewed and revised during the next few years and decades as more information becomes available regarding how 'dangerous' climate change can be defined and avoided.

Realising the Climate Convention Objective

Having recommended a choice of stabilisation level, a large question remains: how can the nations of the world work together to realise it in practice?

It is instructive first to look at annual emissions of greenhouse gases expressed as CO_2e and per capita. Averaged over the world in 2004 they were about 6.5 t CO_2e (~1.8 t C) per capita but they varied very much from country to country (Figure 10.4). For developed countries, including transitional economy countries, in 2000 they averaged 16 t CO_2e (ranging downwards from about 25 t for the USA) while for developing countries they averaged about 4 t. Looking ahead to the years 2050 and 2100, even if the world population rises to only 9 billion, under the profile of carbon dioxide emissions leading to stabilisation at concentrations of 450 ppm (Figure 10.3) the per capita annual emissions averaged over the world would be between 1 and 2 t CO_2e for 2050 and less than 0.4 t CO_2e for 2100[33] – much less than the current value of about 6.5 t.

The Objective of the Climate Convention is largely concerned with factors associated with the requirement for sustainable development. In Chapter 9, four principles were enunciated that should be at the basis of negotiations concerned with future emissions reductions to mitigate climate change. One of these was the Principle of Sustainable Development. The others were the Precautionary Principle, the Polluter-Pays Principle and the Principle of Equity. This last principle includes *intergenerational* equity, or weighing the needs of the present generation against those of future generations, and *international* equity, or weighing the balance of need between industrial and developed nations and the developing world. Striking this latter balance is going to be

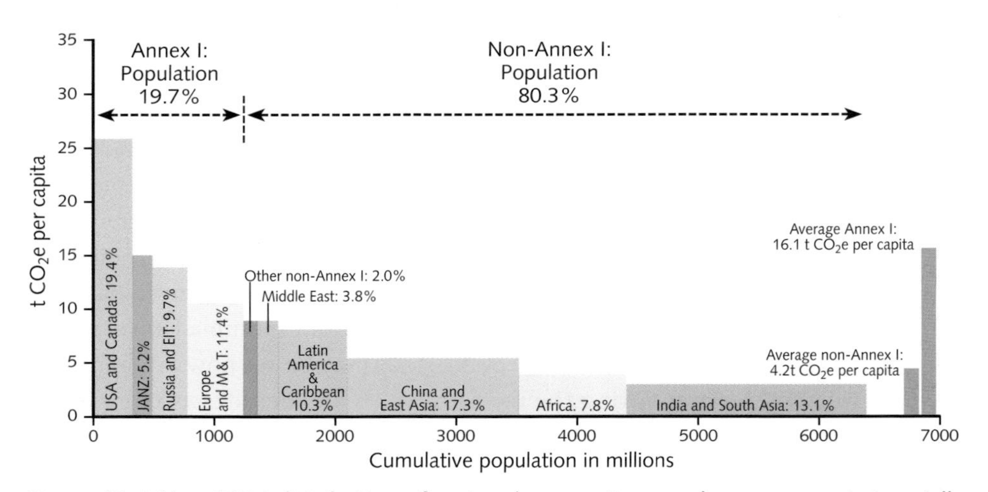

Figure 10.4 Year 2004 distribution of regional per capita greenhouse gas emissions (all gases included in the Kyoto Protocol including land-use change) expressed as carbon dioxide equivalent emissions in 2004 from different countries or groups of countries plotted against population. The percentages in the bars indicate a region's share in global greenhouse gas emissions. EIT, economies in transition; M & T, Malta and Turkey; JANZ, Japan, Australia and New Zealand. To convert tonnes CO_2 to tonnes C, divide by 3.66.

particularly difficult because of the great disparity in current carbon dioxide emissions between the world's richest nations and the poorest nations (Figure 10.4), the continuing demand for fossil fuel use in the developed world and the understandable desire of the poorer nations to escape from poverty through development and industrialisation. This last is particularly recognised in the Framework Convention on Climate Change (see box on pages 291–2) where the growing energy needs of developing nations as they achieve industrial development are clearly stated. In Chapter 8 on page 253, this current international inequity was presented as a challenging moral imperative to the developed world.

An example of how an approach to stabilisation for carbon dioxide might be achieved is illustrated in Figure 10.5. It is based on a proposal called 'Contraction and Convergence' that originates with the Global Commons Institute (GCI),[34] a non-governmental organisation based in the UK. The envelope of carbon dioxide emissions is one that leads to stabilisation at about 450 ppm (without climate feedbacks included), although the rest of the proposal does not depend on that actual choice of level. Note that, under this envelope, global fossil fuel emissions rise by about 15% to about 2025; they then fall to less than half the current level by 2100. The figure illustrates the division of emissions between major countries or groups of countries as it has been up to the present. Then

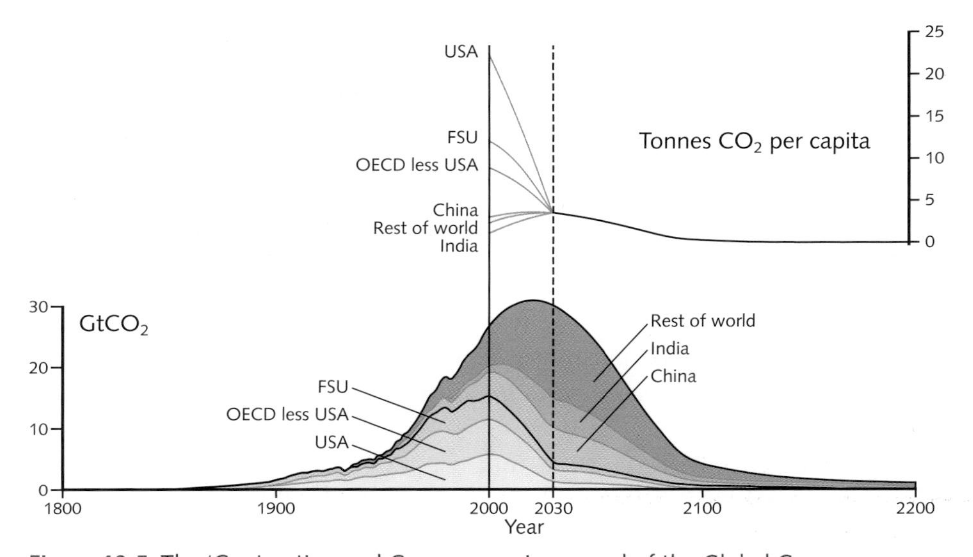

Figure 10.5 The 'Contraction and Convergence' proposal of the Global Commons Institute for achieving stabilisation of carbon dioxide concentration. The envelope of carbon dioxide emissions illustrated is one that leads to stabilisation at 450 ppm (but the effect of climate carbon-cycle feedbacks is not included). For major countries or groups of countries, up to the year 2000, historic emissions are shown. After 2030 allocations of emissions are made on the basis of equal shares per capita on the basis of population projections for that date. From now until 2030, smooth 'convergence' from the present situation to that of equal shares is assumed to occur. In the upper part of the diagram the per capita contributions that apply to different countries or groups of countries are shown. For OECD and FSU see Glossary.

the simplest possible solution is taken to the sharing of emissions between countries and proposes that, from some suitable date (in the figure, 2030 is chosen), emissions are allocated on the basis of equal shares per capita. From now until 2030 the division is allowed to converge from the present situation to that of equal per capita shares. Hence the 'Contraction and Convergence'. The further proposal is that arrangements to trade the carbon dioxide allocations are made.

The 'Contraction and Convergence' proposal addresses all of the four principles mentioned above. In particular, through its equal per capita sharing arrangements it addresses head-on the question of international equity – and the proposed trading arrangements ensure that the greatest 'polluters' pay. The value of the proposal is that it clearly suggests some of the principal ingredients of a long-term solution. However, the discussions taking place at the moment and the other proposals that have been put forward[35] demonstrate that any international agreement is bound to be more elaborate and to differentiate appropriately between countries. In particular it will have to take account, not just of very large differences between countries in their emissions but also of large differences in the threat of damage from climate change, the requirement for adaptation and in their needs and responsibilities especially within the energy sector.

Substantial complications arise in these negotiations because of international entanglement of responsibility for greenhouse gas emissions. This is well illustrated by looking at the component of emissions that may be embedded in a country's exports. For instance in 2005, 44% of China's CO_2 emissions were embedded in exports of goods and services mainly to Europe, north America and Australia.[35] Other complications arise in the allocation of responsibility for emissions from internation aviation, a sector where emissions are rising rapidly.

The Conference of Parties (COP 13) of the FCCC, meeting in Bali in late 2007, particularly addressed international action post-Kyoto. The conference set up negotiations to begin immediately and to be completed by 2009 to bring about:

> *A shared vision for long-term cooperative action, including a long-term global goal for emission reductions, to achieve the ultimate objective of the Convention, in accordance with the provisions and principles of the Convention, in particular the principle of common but differentiated responsibilities and respective capabilities, and taking into account social and economic conditions and other relevant factors.*

Significant progress was also made at the Bali meeting in addressing the important areas of adaptation, deforestation and technology transfer.

The setting of targets at the international level is, of course, only the first part of the action required. For these targets to be realised requires action at all levels from the international to the national down to the local and eventually at the level of the individual. Five essential ingredients are required. The first is an aggressive emphasis on energy saving and conservation. Much here can be achieved at zero net cost or even at a cost saving. Though much energy conservation can be shown to be economically advantageous, it is unlikely to be undertaken without significant incentives. However, it is clearly good in its own right, it can be started in earnest now and it can make a large contribution to the reduction of emissions and the slowing of global warming. The second ingredient is priority on the development of appropriate non-fossil fuel energy sources (e.g. carbon capture and storage applied to coal-fired power stations and renewable energy sources) together with very rapid growth in their implementation. The third ingredient is moving rapidly to a halting of tropical deforestation. The fourth is the transfer of technologies to developing countries that will enable them to apply the most appropriate and the most efficient technologies to their industrial development, especially in the energy sector. The fifth ingredient is to act in all these ways with the utmost urgency. Figures 10.2 and 10.3 and Table 10.3 demonstrate that to achieve the 2 °C target emissions need to peak by about 2015 and then rapidly reduce (see Figure 11.27). The required concentration of national and international effort is unprecedented.

Earlier in the chapter we noted that the Kyoto Protocol introduced various measures aimed at the stimulation of emissions reductions in efficient and cost-effective ways. Measures that include incentives, regulation, taxation and emissions trading will be part of follow-on international agreements and national policies. The challenge is to make sure not only that they achieve the necessary reductions but that they also prove beneficial in terms of their social and political implications. The next chapter will present some of these challenges as they concern the energy sector.

SUMMARY

This chapter has outlined international action to combat climate change beginning with the Framework Convention on Climate Change (FCCC) agreed by all nations in 1992. The FCCC's Objective is to achieve stabilisation of greenhouse gases and hence of climate at a level that ensures that dangerous interference with the climate system is avoided, that ecosystems can adapt naturally, that food production is not threatened and that economic development can proceed in a sustainable manner.

In 2005, the Kyoto Protocol came into force under which developed countries, except USA and Australia, agreed, by 2012 to make reductions in carbon dioxide emissions averaging about 5% below 1990 levels.

A post-Kyoto agreement is now under negotiation. Key to that agreement will be a global target limiting future climate change as required by the FCCC together with proposals for international action to achieve it. At the end of 2007, at the Bali Conference, all nations set out a timetable for an agreement by the end of 2009. The biggest challenges are to ensure fairness between developed and developing countries and to prepare financial and other measures that will ensure targets will be achieved.

Arguments have been put forward for a target, supported by many experts, governments and international bodies, that would limit global average temperature rise to no greater than 2 °C above its pre-industrial level. On the assumption of no further increases in greenhouse gases other than carbon dioxide, a 50% chance of reaching the 2 °C target implies a stabilisation level for carbon dioxide of no more than 450 ppm.

The achievement of this target will not come easily. It will require much determination and consistent political will. Urgent and aggressive actions have to begin now, many of which also bring further benefits. The necessary action is affordable and its cost much less than the cost of inaction. Further much of the action is good to do for other reasons. The most important areas of action are:

- rapid reduction in tropical deforestation and increase in afforestation;
- aggressive increase in energy saving and conservation measures,
- rapid movement to sources of energy free of carbon emissions, e.g through carbon capture and storage and renewable energy sources;
- some relatively easy-to-do reductions in emissions of greenhouse gases other than carbon dioxide, especially methane.

The next chapter presents implications for the energy and transport sectors.

Will the 2 °C target be adequate to stabilise the climate against very damaging and irreversible change? Many are asking this question. Further evidence obtained during the next few years is likely to demand a serious reappraisal with the possibility of even more severe targets.

QUESTIONS

1 From Figure 10.2, what are the rates of change of global average temperature for the profiles shown that lead to stabilisation of carbon dioxide concentration at different levels? From information in Chapter 7 or from elsewhere, can you suggest a criterion involving rate of change that might assist in the choice of a stabilisation level for carbon dioxide concentration as required by the Objective of the Climate Convention?

2 From the formula in the caption to Figure 10.2 and the information in Figure 3.11 and Table 6.1, calculate the contributions from the various components of radiative forcing (including aerosol) to the equivalent carbon dioxide concentration in 1990. How valid do you think is it to speak of equivalent carbon dioxide for components such as aerosol and tropospheric ozone?

3 From the information in Table 6.1 and the formula in the caption to Figure 10.2, calculate the equivalent carbon dioxide concentration, including

(1) the well-mixed greenhouse gases and (2) total aerosols, for SRES scenarios A1B and A2 in 2050 and 2100.

4 Associated with the choice of stabilisation level under the criteria of the Objective of the Climate Convention, different kinds of analysis were mentioned; cost–benefit analysis, multicriteria analysis and sustainability analysis. Discuss which analysis is most applicable to each of the criteria in the Objective. Suggest how the analyses might be presented together so as to assist in the overall choice.

5 From the information available in previous chapters and using the criteria laid out in the Climate Convention Objective, what stabilisation levels of greenhouse gas concentrations do you think should be chosen?

6 The arguments concerning the choice of stabilisation level and the action to be taken have concentrated on the likely costs and impacts of climate change before the year 2100. Do you think that information about continuing climate change or sea level rise (see Chapter 7) after 2100 should be included and taken into account by decision-makers, or is that too far ahead to be of importance?

7 Compare the growth of emissions since 1990 in the major countries of the world[36] and comment on the policies they appear to be following regarding future emissions.

8 Given the need for reducing emissions as quickly as possible, do you think national and international bodies are deciding and acting with sufficient urgency. If not, how might more urgent action be achieved?

9 The international response to global warming is likely to lead to decisions being taken sequentially over a number of years as knowledge regarding the science, the likely impacts and the possible responses becomes more certain. Describe how you think the international response might progress over the next 20 years. What decisions might be taken at what time?

10 Explain how the 'Contraction and Convergence' proposal meets the four principles listed in Chapter 9 and elaborated in Chapter 10. Suggest the political or economic arguments that might be used to argue against the proposal. Can you suggest other ways of sharing emissions between countries that might achieve agreement more easily?

11 Find out the details of any plans for afforestation in your country. What actions or incentives could make it more effective?

12 Assume a snow-covered area at latitude 60° with an albedo of fifty per cent is replaced by partially snow-covered forest with an albedo of twenty per cent. Make an approximate comparison between the 'cooling' effect of the carbon sink provided by the forest and the 'warming' effect of the added solar radiation absorbed, averaged over the year.

FURTHER READING AND REFERENCE

Parry, M., Canziani, O., Palutikof, J., van der Linden, P., Hanson, C. (eds.) 2007. *Climate Change 2007: Impacts, Adaptation and Vulnerability. Contribution of Working Group II to the Fourth Assessment Report of the Intergovernmental Panel on Climate Change*. Cambridge: Cambridge University Press.
Technical Summary
Chapter 5 Food, fibre and forest products
Chapter 19 Assessing key vulnerabilities and the risk from climate change
Chapter 20 Perspectives on climate change and sustainability

Metz, B., Davidson, O., Bosch, P., Dave, R., Meyer, L. (eds.) 2007. *Climate Change 2007: Mitigation of Climate Change. Contribution of Working Group III to the Fourth Assessment Report of the Intergovernmental Panel on Climate Change*. Cambridge: Cambridge University Press.
Technical Summary
Chapter 2 Framing issues (e.g. links to sustainable development, integrated assessment)
Chapter 3 Issues relating to mitigation in the long-term context
Chapter 8 Agriculture
Chapter 9 Forestry
Chapter 10 Waste management
Chapter 11 Mitigation from a cross-sectoral perspective
Chapter 12 Sustainable development and mitigation
Chapter 13 Policies, instruments and cooperative Agreements

IPCC AR4 Synthesis Report, Summary for Policymakers and Full Report (52 pages) available on www.ipcc.ch

Stern, N. 2006. *The Economics of Climate Change*. Cambridge: Cambridge University Press. The Stern Review: especially Chapters 3 to 6 in Part II on the cost of climate-change impacts.

Lynas, M. 2008. *Six Degrees*. London: HarperCollins. A readable and challenging account of the probable impacts of climate change in different parts of the world at different levels of global warming. Winner of the Royal Society's award for the best popular science book of the year.

NOTES FOR CHAPTER 10

1 More details of the Kyoto Protocol and of the detailed arrangements for the inclusion of carbon sinks can be found in Watson, R. T., Noble, I. R., Bolin, B., Ravindranath, N. H., Verardo, D. J., Dokken, D. J. (eds.) 2000. *Land Use, Land-Use Change and Forestry. A Special Report of the IPCC*. Cambridge: Cambridge University Press and on the FCCC website: www.unfccc.int/resource/convkp.html

2 More details in Watson, R. and the Core Writing Team (eds.) 2001. *Climate Change 2001: Synthesis Report. Contribution of Working Groups I, II and III to the Third Assessment Report of the Intergovernmental Panel on Climate Change*. Cambridge: Cambridge University Press, question 7, pp. 108ff. See also Hourcade, J.-C., Shukla, P. *et al.* Global, regional and national costs and ancillary benefits of mitigation. Chapter 8, in Metz, B., Davidson, O., Swart, R., Pan, J. (eds.) 2001. *Climate Change 2001: Mitigation. Contribution of Working Group III to the Third Assessment Report of the Intergovernmental Panel on*

Climate Change. Cambridge: Cambridge University Press.

3 See Stern, N. 2006 *The Economics of Climate Change*. Cambridge: Cambridge University Press. Chapter 15 for a review of practical issues concerned with carbon trading.

4 For a critical dialogue regarding carbon trading see *Carbon Trading*, Development Dialogue No. 48. Dag Hammarskjöld Foundation, Uppsala, 2006.

5 More detail in *Global Environmental Outlook GEO 3 (UNEP)*. 2002. London: Earthscan and *Global Environmental Outlook GEO 4 (UNEP)*. 2007. Nairobi, Kenya: UNEP

6 Bolin, B., Sukumar, R. *et al.* 2000. Global perspective. Chapter 1, in Watson, *et al.* (eds.) *Land Use*.

7 Information from Jonas Lowe at the Hadley Centre, UK Met. Office.

8 From *Global Environmental Outlook 3*, pp. 91–2; see also www.fao.org/forestry.

9 Stern, *Economics of Climate Change*, p. 244.

10 A programme, Reduction of Emissions from Deforestation in Developing Countries (REDD), has been initiated by the Forestry 8 countries, responsible for 80% of the world's forest cover, with the aim of attracting international funding for forest preservation.

11 Bolin and Sukumar, p 26.

12 Stern, *Economics of Climate Change*, p. 612.

13 Watson *et al.* (eds.) *Land Use*, Policymakers Summary, and also in Kauppi, P., Sedjo, R. *et al.* Technical and economic potential of options to enhance, maintain and manage biological carbon reservoirs and geo-engineering. Chapter 4, in Metz *et al.* (eds.) *Climate Change 2001: Mitigation*.

14 Stern, *Economics of Climate Change*, Chapter 9.

15 For definition see Glossary.

16 Betts, R. A. 2000. Offset of the potential carbon sink from boreal forestation by decreases in surface albedo. *Nature*, **408**, 187–90. Also Solomon, S., Qin, D., Manning, M., Chen, Z., Marquis, M., Averyt, K. B., Tignor, M., Miller, H.L. (eds.) 2007. *Climate Change 2007: The Physical Science Basis. Contribution of Working Group I to the Fourth Assessment Report of the Intergovernmental Panel on Climate Change*. Cambridge: Cambridge University Press, Chapter 2.

17 From Summary policymakers. In Houghton, J. T., Ding, Y., Griggs, D. J., Noguer, M., van der Linden, P. J., Dai, X., Maskell, K., Johnson, C. A. (eds.) *Climate Change 2001: The Scientific Basis. Contribution of Working Group I to the Third Assessment Report of the Intergovernmental Panel on Climate Change*. Cambridge: Cambridge University Press.

18 *Energy Technology Perspectives 2008*, International Energy Agency, Paris, Chapter 14. Available online at www.iea.org.

19 Chapter 8, in Metz *et al.* (eds.) *Climate Change 2007: Mitigation*.

20 This figure is calculated by multiplying the 60 million tonnes by the global warming potential for methane which, for a time horizon of 100 years, is about 23 (Table 10.2).

21 See Bousquet, P. *et al.* 2006. *Nature*, **443**, 439–43 for an analysis of methane sources and sinks since 1985 with a suggestion that emissions may soon begin to rise again.

22 Rigby, M. *et al.* 2008. *Geophys. Res. Lett.* **35**, L22805, doi: 10.1029/2008GL036037.

23 Prentice, I. C. *et al.* 2001. The carbon cycle and atmospheric carbon dioxide. In Houghton *et al.* (eds.) *Climate Change 2001: The Scientific Basis*.

24 Cox, P. M. *et al.* 2000. Acceleration of global warming due to carbon cycle feedbacks in a coupled climate model. *Nature*, **408**, 184–7; Jones, C. D. *et al.* 2003. *Tellus*, **55B**, 642–58.

25 See also Fig. 3.25, in Metz *et al.* (eds.) *Climate Change 2007: Mitigation*.

26 European Commission Communication on a Community Strategy on Climate Change; Council of Ministers Conclusion, 25–26 June 1996.

27 European Council of Ministers 2005. *Climate Strategies*. Brussels: ECM.

28 For examples of 2 °C target, see World Wildlife Fund at www.wwf.org.uk/climate/

29 Fig. 3.12, in Metz *et al.* (eds.) *Climate Change 2007: Mitigation*.

30 Table 3.9, in Metz *et al.* (eds.) *Climate Change 2007: Mitigation*.

31 Hansen, J. *et al.* 2008. Target atmospheric CO_2: where should humanity aim? *Open Journal on Atmospheric Sciences*, **2**, 217–31 and James Hansen, Bjerknes Lecture at American Geophysical

Union, 17 December 2008 at www.columbia.edu/njeh1/2008/AGUBjerknes_20081217.pdf.

32 Kunzig, R., Broecker, W. S. 2008. *Fixing Climate.* London: Profile Books. M. Meinshausen 2006. What does a 2°C target mean for greenhouse gas concentrations? *Avoiding Dangerous Climate Change.* Cambridge: Cambridge University Press. pp.268–279.

33 Additional climate carbon-cycle feedbacks have been ignored in this calculation.

34 Further details on the GCI website: www.gci.org.uk

35 See for instance Baer, P., Athanasiou, T. 2007. No. 30, *Frameworks and Proposals.* Global Issue Papers. Washington, DC: Heinrich Böll Foundation.

36 More detail in *World Energy Outlook*, International Energy Agency 2008, p. 386ff.

37 Information on emissions available from many sources, e.g. International Energy Agency and World Resources Institute.

Energy and transport for the future

11

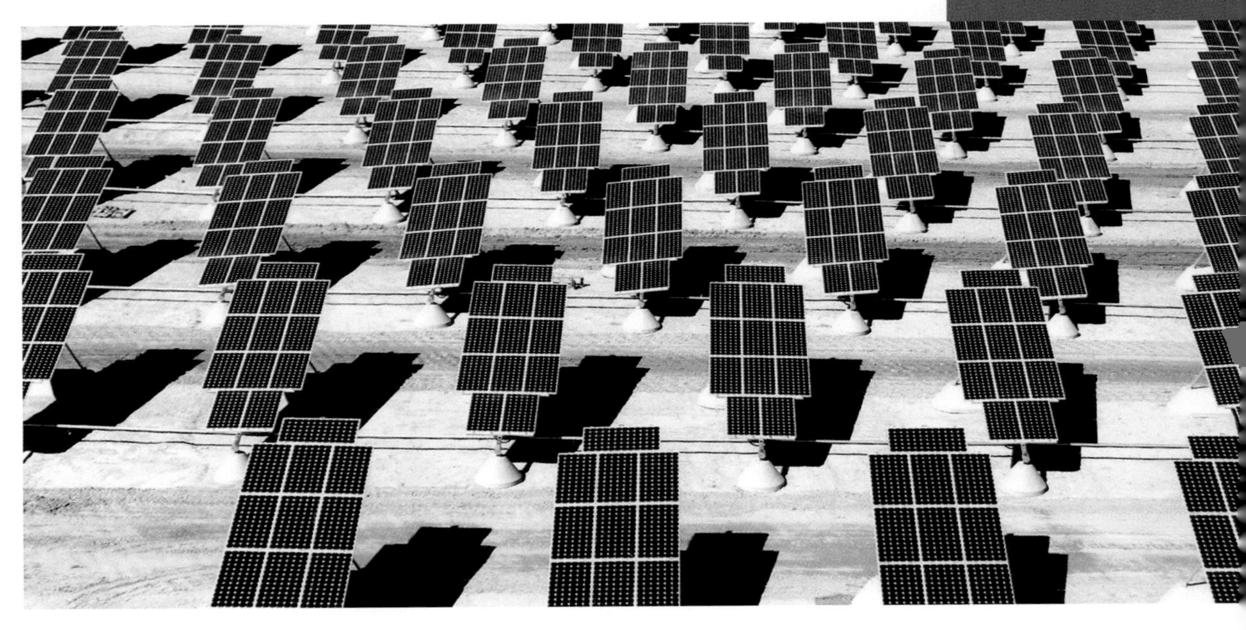

70 000 solar panels form a photovoltaic array generating 15 megawatts of solar power for the Nellis Air Force Base, Nevada.

WE FLICK a switch and energy flows. Energy is provided so easily for the developed world that thought is rarely given to where it comes from, whether it will ever run out or whether it is harming the environment. Energy is also cheap enough that little serious attention is given to conserving it. However, most of the world's energy comes from the burning of fossil fuels, which generates the major proportion of the greenhouse gas emissions into the atmosphere. If these emissions are to be reduced, a large proportion of the reduction will have to occur in the energy sector. There is a need, therefore, to concentrate the minds of policymakers and indeed of everyone on our energy requirements and usage. This chapter looks at how future energy might be provided in a sustainable manner. It also addresses how basic energy services might be made available to the more than 2 billion people in the world who as yet have no such provision.

World energy demand and supply

Most of the energy we use can be traced back to the Sun. In the case of fossil fuels (coal, oil and gas) it has been stored away over millions of years in the past. If wood (or other biomass including animal and vegetable oils), hydropower, wind or solar energy itself is used, the energy has either been converted from sunlight almost immediately or has been stored for at most a few years. These latter sources of energy are renewable; they will be considered in more detail later in the chapter. The other forms of energy that do not originate with the Sun are nuclear energy and geothermal energy, both of which result from the presence of radioactive elements in the Earth when it was formed.

Until the Industrial Revolution, energy for human society was provided from 'traditional' sources – wood and other biomass and animal power. Since 1860, as industry has developed, the rate of energy use has multiplied by about a factor of over 30 (Figure 11.1), at first mostly through the use of coal, followed, since about 1950, by rapidly increasing use of oil and then more recently by the use of natural gas. In 2005 the world consumption of primary energy was about 11 400 million tonnes of oil equivalent (toe). This can be converted into physical energy units to give an average rate of primary energy use of about 15 million million watts (or 15 terawatts = 15×10^{12} W).[1]

Great disparities exist in the amount of energy used per person in various parts of the world. The 2 billion poorest people in the world (less than $US1000 annual income per capita) each use an average of only 0.2 toe of energy annually while the billion richest in the world (more than $US22 000 annual income per capita) use nearly 25 times that amount at 5 toe per capita annually.[2] The average annual energy use per capita in the world is about 1.7 toe, an average consumption of energy of about 2.2 kilowatts (kW). The highest rates of energy consumption are in North America where the average citizen consumes an average of about 11 kW. About one-third of the world's population rely wholly on traditional fuels (wood, dung, rice husks, other forms of 'biofuels') and do not currently have access to commercial energy in any of its forms.

In Figure 11.2 is shown how the energy we consume is generated and used. Also summarised are the energy flows from source to users in the main sectors and the size of the various resources that are available using conventional technologies. Taking the world average approximately 25% of primary energy is used in transport, 35% in industry and 40% in buildings (two-thirds in residential buildings and one-third in commercial buildings). It is also interesting to know how much energy is used in the form of electricity. Rather more than one-third of primary energy goes to make electricity at an average efficiency of conversion of about one-third. Of this electrical power about half,

StatoilHydra's Sleipner T gas platform off the Norwegian coast which is sequestering one million tonnes of carbon dioxide per year.

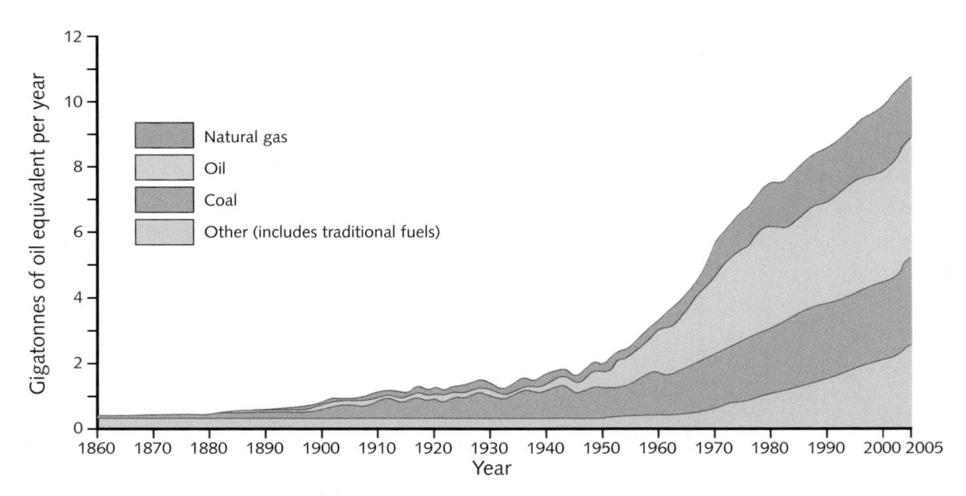

Figure 11.1 Growth in the rate of primary energy use and in the sources of energy from 1860 to 2005 in thousand millions of tonnes of oil equivalent (Gtoe) per year. In terms of primary energy units, 1 Gtoe = 41.9 exajoules. Of the 'other' in 2005, approximately 1.2 Gtoe is attributed to traditional fuels, 0.7 Gtoe to nuclear energy and 0.3 Gtoe to hydropower and other renewables (source for data up to 2000: Report of G8 Renewable Energy Task Force, July 2001; from 2000 to 2005 Fig. TS 13 in IPCC AR4 WGIII 2007).

on average, is utilised by industry and the other half in commercial activities and in homes.

How much is spent on energy? Taking the world as a whole, the amount spent per year by the average person for the 1.7 toe of energy used is about 5% of annual income. Despite the very large disparity in incomes, the proportion spent on primary energy is much the same in developed countries and developing ones.

How about energy for the future? If we continue to generate most of our energy from coal, oil and gas, do we have enough to keep us going? Current knowledge of proven recoverable reserves (Figure 11.2) indicates that at current rates of use, known reserves of fossil fuel will meet demand at least until 2050. But before then, if demand continues to expand, oil and gas production will come under increasing pressure. Further exploration will be stimulated, which will lead to the exploitation of more sources, although increased difficulty of extraction can be expected to lead to a rise in price. So far as coal is concerned, there are operating mines with resources for production for well over 100 years.

Estimates have also been made of the ultimately recoverable fossil fuel reserves, defined as those potentially recoverable assuming high but not prohibitive prices and no significant bans on exploitation. Although these are bound to be somewhat speculative, they show that, at current rates of use, reserves

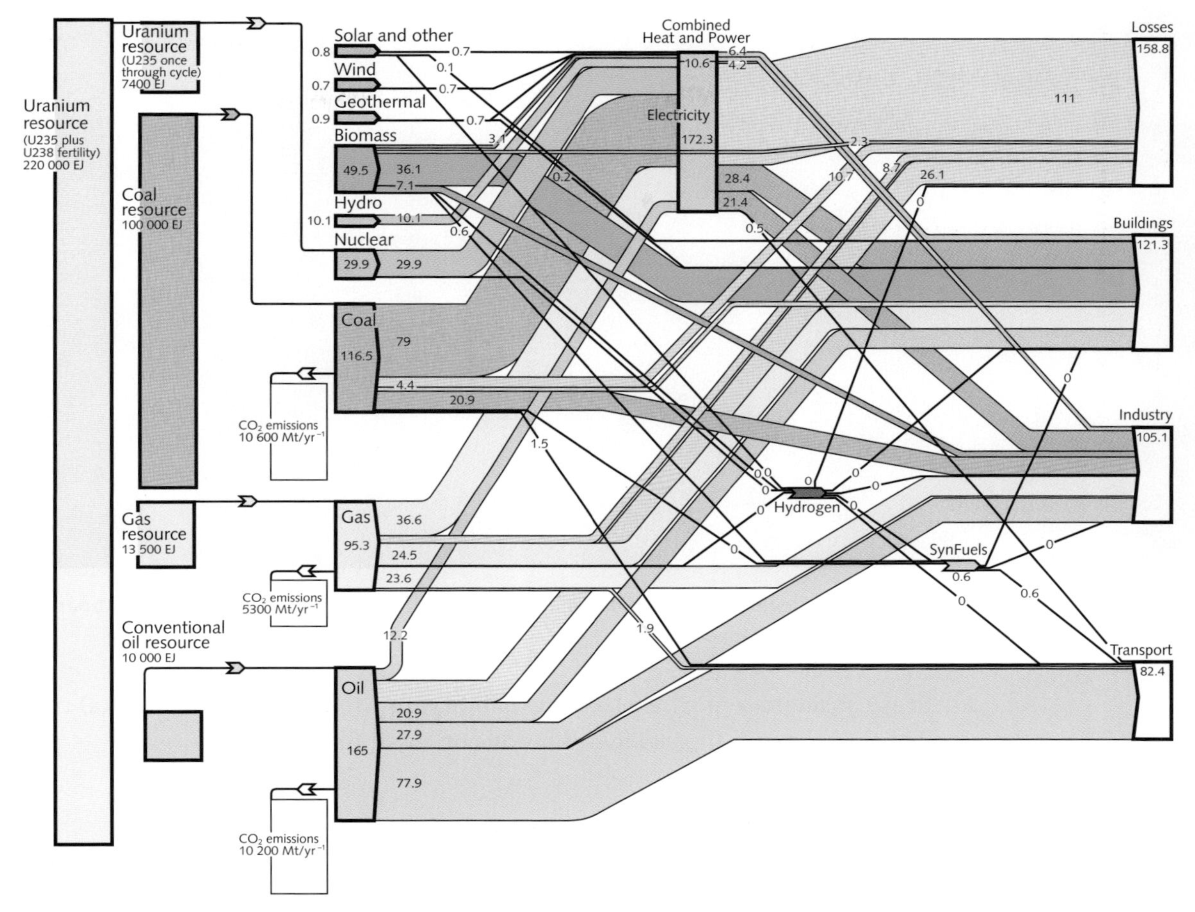

Figure 11.2 Global energy flows (EJ in 2004) from primary energy through carriers to end-users and losses in transmission, etc. Related carbon dioxide emissions from coal, oil and gas combustion are also shown as well as the size of known resources. Further energy conversions occur in the end-use sectors. Peat is included with coal, organic waste is included with biomass. The resource efficiency ratio by which fast-neutron technology increases the power-generation capability per tonne of natural uranium varies greatly in different assessments. In this diagram the ratio used is up to 240 : 1.

of oil and gas are likely to be available for 100 years and of coal for more than 1000 years. In addition to fossil fuel reserves considered now to be potentially recoverable there are reserves not included in Figure 11.2, such as the methane hydrates, which are probably very large in quantity but from which extraction would be much more difficult.

Likely reserves of uranium for nuclear power stations should also be included in this list. When converted to the same units (assuming their use in 'fast' reactors) they are believed to be substantially greater than likely fossil fuel reserves (Figure 11.2).

Human-made lights which highlight developed or populated areas of the Earth's surface, including the seaboards of Europe, the eastern United States and Japan.

It is considerations other than availability, in particular environmental considerations, that will provide limitations on fossil fuel use.

Future energy projections

In Chapter 6 were described the SRES scenarios sponsored by the IPCC that detail, for the twenty-first century, a range of possibilities regarding future energy demand (based on a range of assumptions concerning population, economic growth and social and political development), how that demand might be met and what greenhouse gas emissions might result. In that chapter were also described the implications for those scenarios regarding climate change. Chapter 10 explained the imperative set out by the Framework Convention on Climate Change (FCCC) in its Objective that greenhouse gas concentrations in the atmosphere must be stabilised so that continued anthropogenic climate change can be avoided. Scenarios of carbon dioxide equivalent (CO_2e) emissions that would be consistent with various stabilisation levels were also presented. Arguments were put forward for limiting the rise in global atmospheric temperature to 2 °C above its pre-industrial level implying a target level for atmospheric concentration of carbon dioxide equivalent of about 450 CO_2e. How the world's energy producers and consumers can meet the challenge of such a target is addressed by this chapter.

Energy intensity and carbon intensity

An index that provides an indication of a country's energy efficiency is the ratio of annual energy consumption to gross domestic product (GDP) known as the *energy intensity*. Figure 11.3 shows that from 1970 to 2005 world total GDP increased by a factor of about 3 while energy consumption increased by a factor of about 2, the result being a decrease in energy intensity of about 30% or an average of about 1% per year. There are substantial differences between countries. Within the OECD, Denmark, Italy and Japan have the lowest energy intensities and Canada and the USA the highest, with more than a factor of 2 between the lowest and the highest.

Of importance too in the context of this chapter is the *carbon intensity*, which is a measure of how much carbon is emitted for a given amount of energy. This can vary with different fuels. For instance, the carbon intensity of natural gas is 25% less than that of oil and 40% less than that of coal. For renewable sources the carbon intensity is small and depends largely on that which originates during manufacture of the equipment making up the renewable source (e.g. during manufacture of solar cells). Figure 11.3 shows that the average carbon intensity for the globe has reduced only a little since 1970.

The *Kaya Identity* expresses the level of energy-related carbon dioxide emissions as the product of four indicators, namely carbon intensity, energy intensity, gross domestic product per capita and the population, the global averages for which are all plotted in Figure 11.3. For the reductions in global carbon dioxide emissions required in the future, energy and carbon intensities have to reduce more quickly than income and population growth taken together.

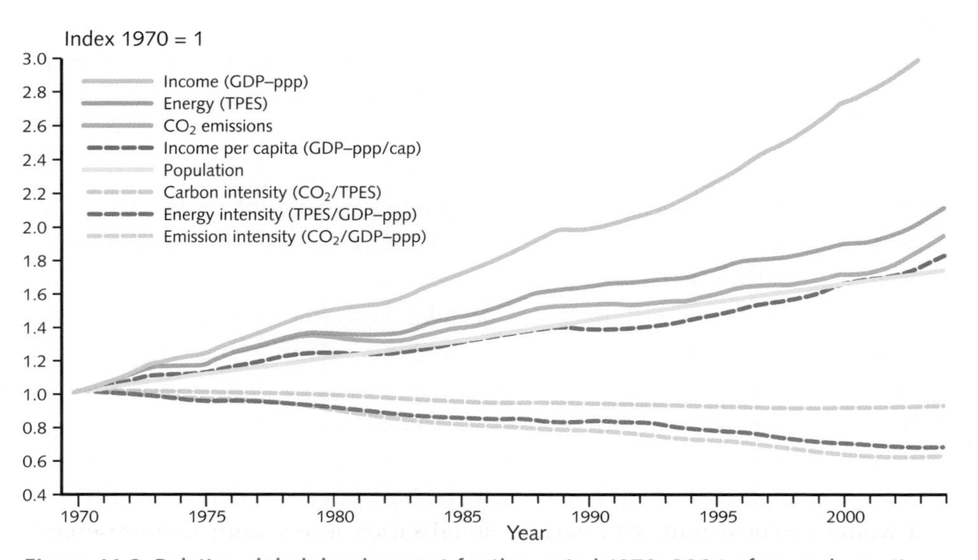

Figure 11.3 Relative global development for the period 1970–2004 of gross domestic product (GDP measured in purchasing power parity: ppp), total primary energy supply (TPES), carbon dioxide emissions (from fossil fuel burning, gas flaring and cement manufacture) and population. In dashed lines are shown income per capita, energy intensity, carbon intensity of energy supply and emission intensity.

Many national and international bodies and some energy industries have studied future energy scenarios and how they might be achieved. The most comprehensive of these are those by the International Energy Agency presented to year 2030 in its annual volume World Energy Outlook[3] and in more detail to 2050 in its Energy Technology Perspectives.[4] Three scenarios of interest are presented (Figure 11.4). The first is a reference or baseline scenario that assumes energy carbon dioxide emissions continue to rise with minimal environmental constraints throughout the period; in 2050 emissions are 2.3 times up on their level in 2005 – similar to the SRES A2 scenario (Figure 6.1). As we have already seen in Chapter 10, with this scenario global warming and climate change will continue unchecked. The second, called the ACT Map scenario, brings energy carbon dioxide emissions back to their 2005 level by 2050 and the third, the BLUE Map scenario, returns emissions to their 2005 level by 2025 and reduces them by a further factor of 2 by 2050. As shown in Chapter 10 (Figures 10.2 and 10.3), these scenarios are broadly consistent with stabilised carbon dioxide concentrations of 550 ppm CO_2e for ACT and 450 ppm CO_2e for BLUE. Also in Figure 11.4 are shown the share of emissions and emissions reductions by different sectors and illustrative options for how these reductions might be made. Reference to these options will be made later in the chapter where more detail regarding different sectors or technologies is presented.

Listed below are some key findings of the IEA that illustrate how achievement of these reducing scenarios can make the world's energy sector more sustainable.[5]

- For the Baseline scenario in 2050, OECD countries account for less than one-third of global carbon dioxide emissions. Population growth (see Chapter 12, page 393) and the need for economic development make it inevitable that developing countries will, for many decades, consume increased amounts of energy. Global emissions can only be halved if developing countries and transition economies contribute very substantially.
- There is an urgent need for aggressive and determined action in the next decade[6]. There is a danger that investments made in this period, due to the long lifespan of capital equipment such as buildings, industrial installations and power plants, could be the subject of economically wasteful early replacement or refurbishment if emission reduction targets are to be met. The BLUE scenario already envisages 350 GW of coal-fired power being replaced before the end of its lifespan.
- Deep emission cuts will require extensive application of energy efficiency measures, carbon capture and storage (CCS), renewable energy technologies

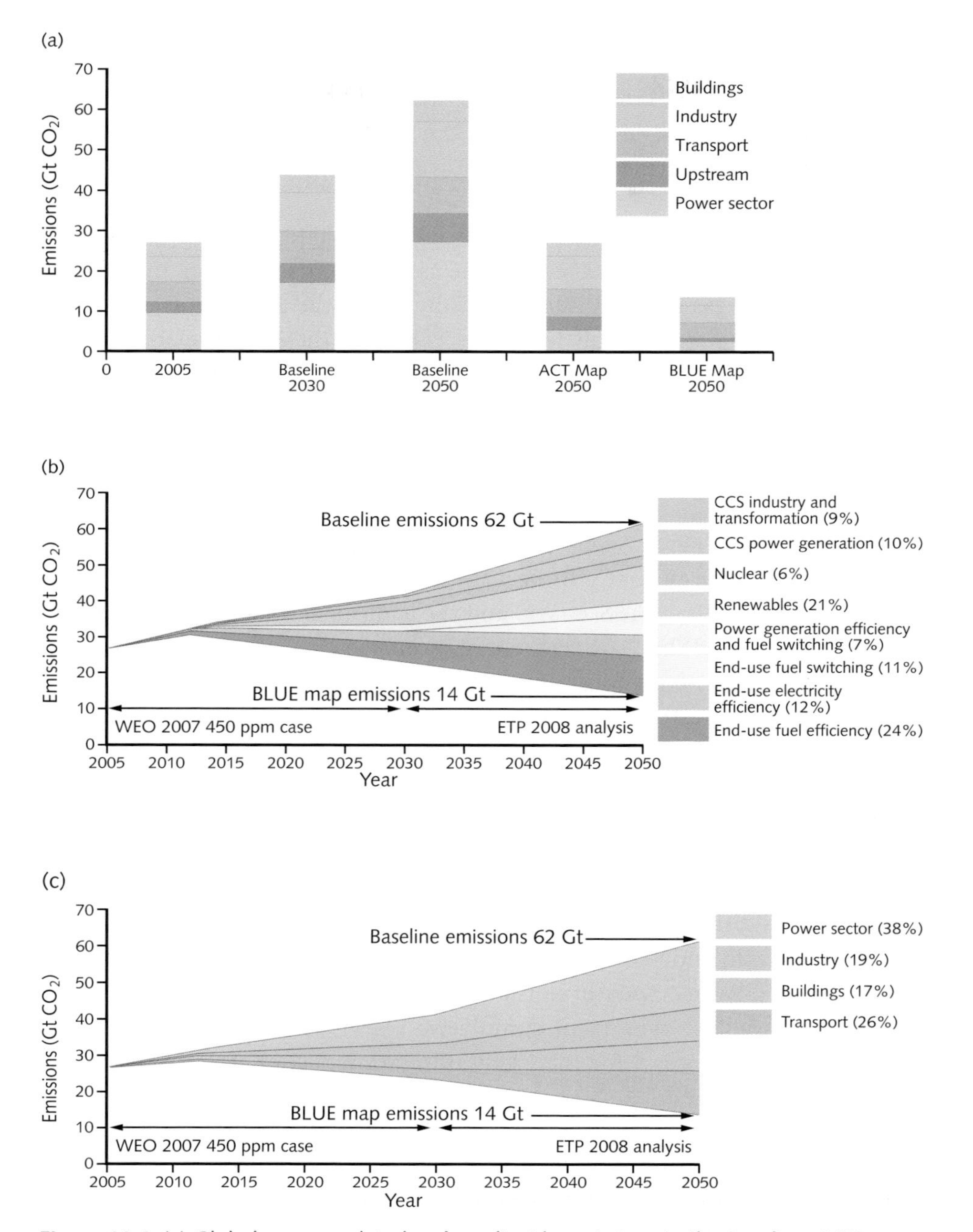

Figure 11.4 (a) Global energy-related carbon dioxide emissions in the Baseline, ACT Map and BLUE Map scenarios of the International Energy Agency (IEA) showing division into sectors. (b) and (c) Illustrative options for contributions to emissions reductions 2005–50 for the BLUE Map scenario by source (b) and by sector (c). In (c) reductions from electricity savings have been allocated to end-use sectors.

and nuclear. The transport sector especially will require new solutions with substantial cost.

- Electricity will play an increasing role as a carbon-dioxide-free energy carrier. The near elimination of emissions in the power sector is key to achieving deep emission reductions worldwide. Advances in new technologies are key to accomplishing this.
- Carbon dioxide emission reduction policies can help to avoid very significant supply challenges. This is especially the case for transportation. In both scenarios, oil and gas demand is substantially below the Baseline level in 2050. In the BLUE Map scenario, oil demand is 27% below the 2005 level. Fossil fuels, with large-scale application of carbon capture and storage, remain a key element of the world's energy supply in 2050 in all scenarios.

Projections for energy investment

The International Energy Agency has also estimated the future financial investment in global energy that will be necessary between 2005 and 2050 under their Reference or Baseline Scenario and the additional investment required to achieve the ACT and the BLUE Map energy scenarios.[7]

The total cumulative energy investment needs for the Reference scenario over this period is estimated to be about $US250 trillion (million million) or about 6% of cumulative world GDP over the period. By far the largest proportion of this relates to investments that consumers make in capital equipment that consumes energy, from vehicles to light bulbs to steel plants. In fact because of very large investment in vehicles, transport alone accounts for 84% of the total investment. For the enviromentally driven scenarios over this period, the additional investment needs are estimated as $US17 trillion for the ACT scenario and $US45 trillion for the BLUE Map scenario or increases of 7% and 18% respectively over the Reference scenario. For the BLUE Map scenario, this is just over 1% of cumulative world GDP – a figure similar to that quoted in Chapter 9 for the likely mitigation cost of stabilising CO_2e at 450 ppm.

The IEA also point out that, compared with the Reference scenario, the ACT and BLUE Map scenarios will result in significant fuel savings over the period 2005 to 2050, amounting to about $US35 trillion and $US50 trillion respectively. These are substantially larger than the additional investment needs mentioned in the last paragraph and the differences are not wiped out even if significant levels of discount are applied in working them out (see Chapter 9, page 279 for a discussion of discounting).

Socolow and Pacalas' Wedges

A simple presentation of the type of changes that will be required has been created by Professors Socolow and Pacala of Princeton University.[8] To counter the likely growth of global carbon dioxide emissions from 2005 to 2055, seven 'wedges' of reduction are proposed (Figure 11.5), each wedge amounting to 1 gigatonne of carbon per year (= 3.66 gigatonnes of carbon dioxide per year) in 2055 or 25 gigatonnes in the period 2005–55. Many combinations of technologies can be proposed to fill the wedges. Some of the possible ones are the following. They illustrate the scale of what is necessary.

- Buildings efficiency – cut electricity use by 25%
- Double fuel economy of 2 billion cars – 30 to 60 miles per gallon (~10 to 5 litres per 100 km)
- Install carbon capture and storage (CCS) at 800 large coal-fired power plants
- Install CCS at coal plants that produce hydrogen for 1.5 billion vehicles
- Wind power from 1 million 2 MWp windmills
- Solar photovoltaic power from area $(150\,km)^2$
- Nuclear power – add 700 GW = 2 × current capacity
- Biofuel production from 250 Mha of land
- Halve tropical deforestation.

Note that Socolow and Pacala proposed wedges only sufficient to counter the emissions growth to 2055. To meet the reductions to below 2005 levels in 2055 as in Figure 11.4 (b) or (c) requires 13 Gt per year of reduction in 2055 or 13 wedges.

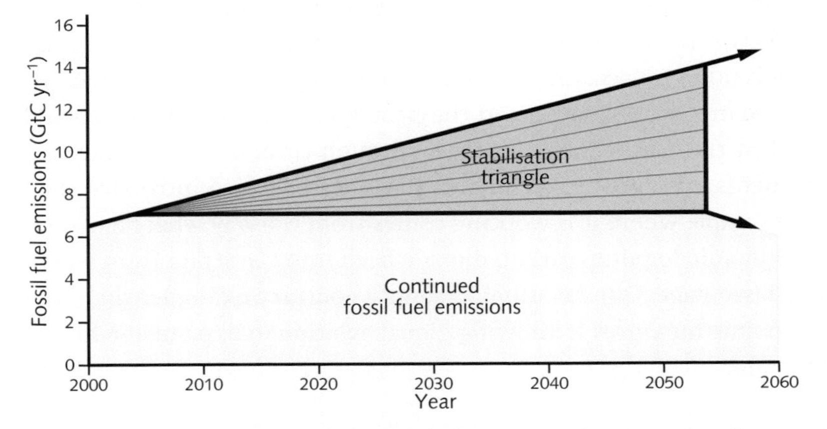

Figure 11.5 Socolow and Pacalas' Wedges illustrate the changes required to counter the likely growth of global carbon dioxide emissions from 2005 to 2055. Seven 'wedges' of reduction are proposed, each wedge amounting to a reduction of 1 gigatonne of carbon per year in 2055 (=3.66 gigatonnes of carbon dioxide per year) or 25 gigatonnes in the period 2005–55. Many combinations of technologies can be proposed, to fill the wedges.

A long-term energy strategy

Before presenting details of the implementation, I want to step back and consider how the choices between the great variety of proposed solutions and potential technologies are to be made. It is relatively easy to present paper solutions, but how do we decide between them and find the best way forward? There is no one solution to the problem and no obviously best technology; different solutions will be appropriate in different countries or regions. Simplistic answers I have often heard are: *Leave it to the market to provide*, or *The three solutions are Technology, Technology, and Technology*. The market and technology are essential and effective tools, but poor masters. Solutions need to be more carefully crafted than those tools that provide on their own.

Let me take the analogy of a motor boat with the engine representing technology, and the propeller market forces (Figure 11.6). But where is the boat heading? Without a rudder and someone steering, the course will be arbitrary or even disastrous. Every voyage needs a destination and a strategy to reach it. Some of the components of the necessary energy strategy are listed in the box on the following page.

Absolutely key is the relationship between the economy and the environment; they must be addressed together. It has been said that the economy is a wholly owned subsidiary of the environment a view echoed by Gordon Brown, then the UK 's Chancellor of the Exchequer, in a speech in 2005.[9]

Take the market. It responds overwhelmingly to price and the short term. It has been effective in reducing energy prices over the last two decades. But in its raw form it takes no account of environmental or other external factors. Economists for many years have agreed the principle that such factors should be internalised in the market, for instance, through carbon taxes or cap and trade arrangements, but most governments have been slow to introduce such measures. An example where it is working comes from Norway where a carbon tax makes it economic to pump carbon dioxide back into the strata from where gas is extracted (see page 327). Aviation presents a contrary example where the absence of economic measures is allowing global aviation to expand at a highly unsustainable rate.

Buildings: energy conservation and efficiency

If we turn lights off in our homes when we do not need them, if we turn down the thermostat by a degree or two so that we are less warm or less cool or if we add more insulation to our home, we are conserving or indeed saving energy.

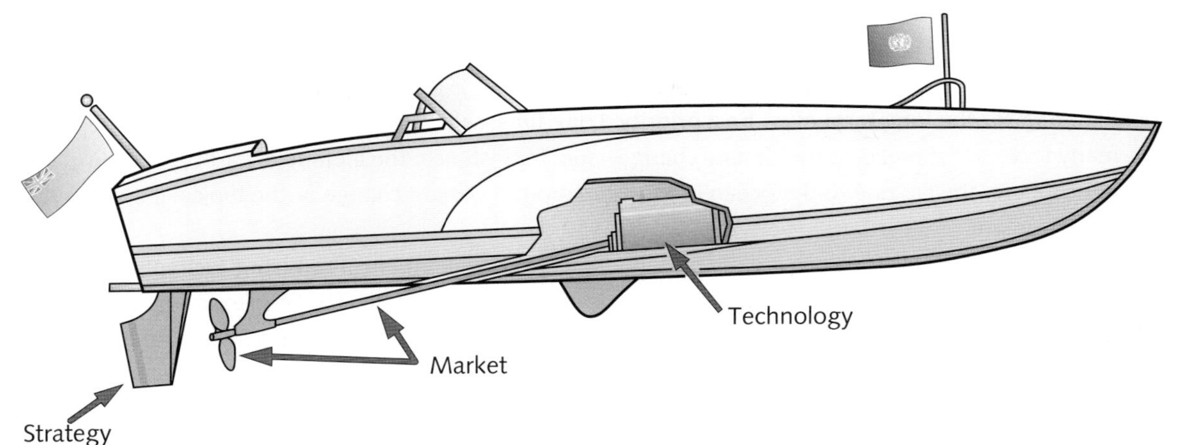

Figure 11.6 Where are we heading? – the need for an energy strategy. The boat flies national and UN flags to illustrate the need for national and international strategies.

But are such actions significant in overall energy terms? Is it realistic to plan for really worthwhile savings in our use of energy?

To illustrate what might be possible, let us consider the efficiency with which energy is currently used. The energy available in the coal, oil, gas, uranium, hydraulic or wind power is *primary energy*. It is either used directly, for instance as heat, or it is transformed into motor power or electricity that in turn provides for many uses. The process of energy conversion, transmission and transformation into its final useful form involves a proportion of the primary energy being wasted. For example, to provide one unit of electrical power at the point of use typically requires about three units of primary energy. An incandescent light bulb is about 3% efficient in converting primary energy into light energy; unnecessary use of lighting reduces the overall efficiency to perhaps no more than 1%.[12] Assessments have been carried out across all energy uses comparing actual energy use with that which would be consumed by ideal devices providing the same services. Although it is not easy to define precisely the performance of such 'ideal' devices (see box below for a discussion of thermodynamic efficiencies), assessments of this kind conclude that there are large opportunities for improvement in average energy efficiency, perhaps by a factor of 3 or more.[13] In this section we look at possibilities for energy saving in buildings; in later sections we consider possible savings in transport and in industry.[14]

To be comfortable in buildings we heat them in winter and cool them in summer. In the United States, for instance, about 36% of the total use of energy is in buildings (about two-thirds of this in electricity), including about 20% for their heating (including water heating) and about 3% for cooling.[16] Energy demand

Where are we heading? Components of energy strategy

(1) **Planning for the long term must be a priority**. Long timescales up to 50 or 100 years are involved in many factors that make up the climate change issue, for instance, the lifetime of carbon dioxide in the atmosphere, the lag due to the ocean in the realisation of climate change or the typical life of energy infrastructure.

(2) Not all potential technologies are at the same stage of development. **Promising technologies need to be brought to the starting gate** so that they can properly compete. This implies joint programmes between government and industry, the provision of adequate resources for research and development, the creation of demonstration projects, and sufficient support to see technologies through to maturity.[10] Tidal and wave energy in western Europe provides an example (see page 368).

(3) Consideration needs to be given to **social and 'quality of life' implications** arising from the way energy is provided to a community. For instance, energy provision from small local plants with community participation possesses very different social and community characteristics compared with energy from large, central installations. The best urban solutions may not be appropriate in rural locations. Addressing more than one problem at once is also part of this component of the strategy. For instance, disposal of waste and generation of energy frequently go together. It has been estimated that the potential energy value in agricultural and forestry wastes and residues could, if realised, meet at least 10% of the world's total energy requirement.[11] Local energy provision supporting the development of local industries would prevent depopulation and enable rural areas to flourish.

(4) **Energy security** is frequently mentioned and must be part of the strategy. For instance, how safe are gas pipelines crossing continents and how secure from political interference at the other end? Or how safe are nuclear power stations from terrorist attack or nuclear material from proliferation to terrorist groups? Diversity of source is clearly important. But thinking about security could be more integrated and holistic; energy security should not be disconnected from world security. As I will mention in Chapter 12, world security is dependent on solutions being found for dealing with the threats to human communities from climate change.

(5) **Partnerships of many kinds** are required as is stated and implied in the 1992 Framework Convention on Climate Change. All nations (developed and developing) need to work together with national, international and multinational industries and corporations to craft sustainable and equitable solutions. Large-scale technology transfer from developed to developing countries is vital if energy growth in developing countries is to proceed sustainably.

(6) **Attainable goals, targets and timescales must be set**, at all levels of society – international, national, local and personal. Any commercial company understands the importance of targets for successful business. Voluntary action alone will fail to bring change on the required scale.

in the buildings sector grew by about 3% per year averaged worldwide from 1970 to 1990 and, apart from countries with economies in transition, has been growing during the last decade by about 2.5% per year. How can these trends be reversed?[17]

To achieve the greatly increased energy efficiency required from the buildings sector it is essential that there be an effective programme for retrofitting existing buildings with adequate insulation so as to reduce the requirement for heating in winter (see box) and cooling in summer. Many countries, including the UK and the USA, still have relatively poor standards of building insulation compared, for instance, with Scandinavian countries. It is also essential that all new domestic and commercial buildings are designed and constructed to the highest possible standards so as to require the minimum energy input (i.e. with higher insulation standards than those listed in the box) and with maximum use of passive solar design (see box on page 362).[19] Large energy savings can also be made by improving the efficiency of appliances (see box) and through installing simple control technology (e.g. based on thermostats) to avoid energy waste. The cost of these actions would be less than the cost saved through the saving of energy. Electricity companies in some parts of the USA and elsewhere are contracting to implement some of these energy-saving measures as an alternative to the installation of new capacity – at significant profit

Thermodynamic efficiencies

When considering the efficiency of energy use, it can be important to distinguish between efficiency as defined by the First Law of Thermodynamics and efficiency as defined by the Second Law. The second particularly applies when energy is used for heating.

A furnace used to heat a building may deliver say 80% of the energy released by full combustion of the fuel, the rest being lost through the pipes, flue, etc. That 80% is a First Law efficiency. An ideal thermodynamic device delivering 100 units of energy as heat to the inside of a building at a temperature of 20 °C from the outside at a temperature of 0 °C would require only seven units of energy. So the Second Law efficiency of the furnace is less than 6%.

Heat pumps (refrigerators or air conditioners working in reverse) are devices that make use of the Second Law and deliver more energy as heat than the electrical energy they use.[15] Although typically their Second Law efficiencies are only about 30%, they are still able to deliver more heat energy than the primary energy required to generate the electricity they use. Because of their comparatively high capital and maintenance costs, however, heat pumps have not been widely used. An example of their substantial use is their contribution to district heating in the city of Uppsala in Sweden where 4 MW of electricity is employed to extract heat from the river and deliver 14 MW of heat energy.

Efficiency of appliances

There is large potential for reducing the electricity consumption from appliances used in domestic or commercial buildings. If, in replacing appliances, everyone bought the most efficient available, their total electricity consumption could easily drop by more than half.

Take lighting for instance. One-fifth of all electricity used in the United States goes directly into lighting. This can easily be reduced by the wider use of compact fluorescent light bulbs which are as bright as ordinary light bulbs, but use a fifth of the electricity and last eight times as long before they have to be replaced – with significant economic savings to the user. For instance, a 20-W compact fluorescent bulb (equivalent to a 100-W ordinary incandescent light bulb) costing £3 or less will use about £20 worth of electricity over its lifetime of 12 years. To cover the same period eight ordinary bulbs would be needed costing about £4 but using £100 worth of electricity. The net saving is therefore about £80. A further large increase in the efficiency of lighting will occur when light-emitting diodes (LEDs) giving out white light become commonplace.[18] The latest such device which is about 1 cm^2 in size and consumes only 3 W produces the same light as a 60-W incandescent bulb.

The average daily electricity use from the appliances in a home (cooker, washing machine, dishwasher, refrigerator, freezer, TVs, lighting) for typical appliances bought in the early 1990s amounts to about 10 kWh per day. If these were replaced by the most efficient available now, electricity use would fall by about two-thirds. The extra cost of the purchase of efficient appliances would soon be recovered in the savings in running cost. Similar calculations can be carried out for other appliances.

both to the companies and its customers. Similar savings would be possible in other developed countries. Major savings at least as large in percentage terms could also be made in countries with economies in transition and in developing countries if existing plant and equipment were used more efficiently.

Further large savings can be realised when buildings are being planned and designed by the employment of *integrated building design*. When buildings are designed, the systems for heating, air conditioning and ventilation are commonly developed separately from the main design. The value of integrated building design is that energy-saving opportunities can be taken associated with the synergies between many aspects of the overall design including the sizing of the systems where much of the energy use occurs. Many examples exist of buildings that take advantage of the many ways of increasing energy efficiency including integrated building design, that reduce energy use by 50% or more and that are often more acceptable and user-friendly than buildings designed in more traditional ways.[20] Some recent examples demonstrate the possibility of more radical building designs that aim at Zero Emission

Insulation of buildings

About 1500 million people live in cold climates where some heating in buildings is required. In most countries the energy demand of space heating in buildings is far greater than it need be if the buildings were better insulated (Figure 11.7).

Table 11.1 provides as an example details of two houses, showing that the provision of insulation in the roof, the walls and the windows can easily lead to the energy requirement for space heating being more than halved (from 5.8 kW to 2.65 kW). The cost of the insulation is small and is quickly recovered through the lower energy cost.

If a system for circulating air through the house is also installed, so that incoming air can exchange heat with outgoing air, the total heating requirement is further reduced. In this case it is worthwhile to add more insulation to reduce the heating requirement still further.

Figure 11.7 An image of Aberdeen (Scotland, UK) from the air taken in the infrared in the winter. Red buildings are warm as a result of poor insulation. Blue buildings are cool showing they are well insulated. Red buildings include some of the older buildings near the city centre but also some much more recent buildings in the outskirts.

Table 11.1 Two assumptions (one poorly insulated, and one moderately well insulated) regarding construction of a detached, two-storey house with ground floor of size 8 m × 8 m, and the accompanying heat losses (*U*-values express the heat conduction of different components in watts per square metre per °C)

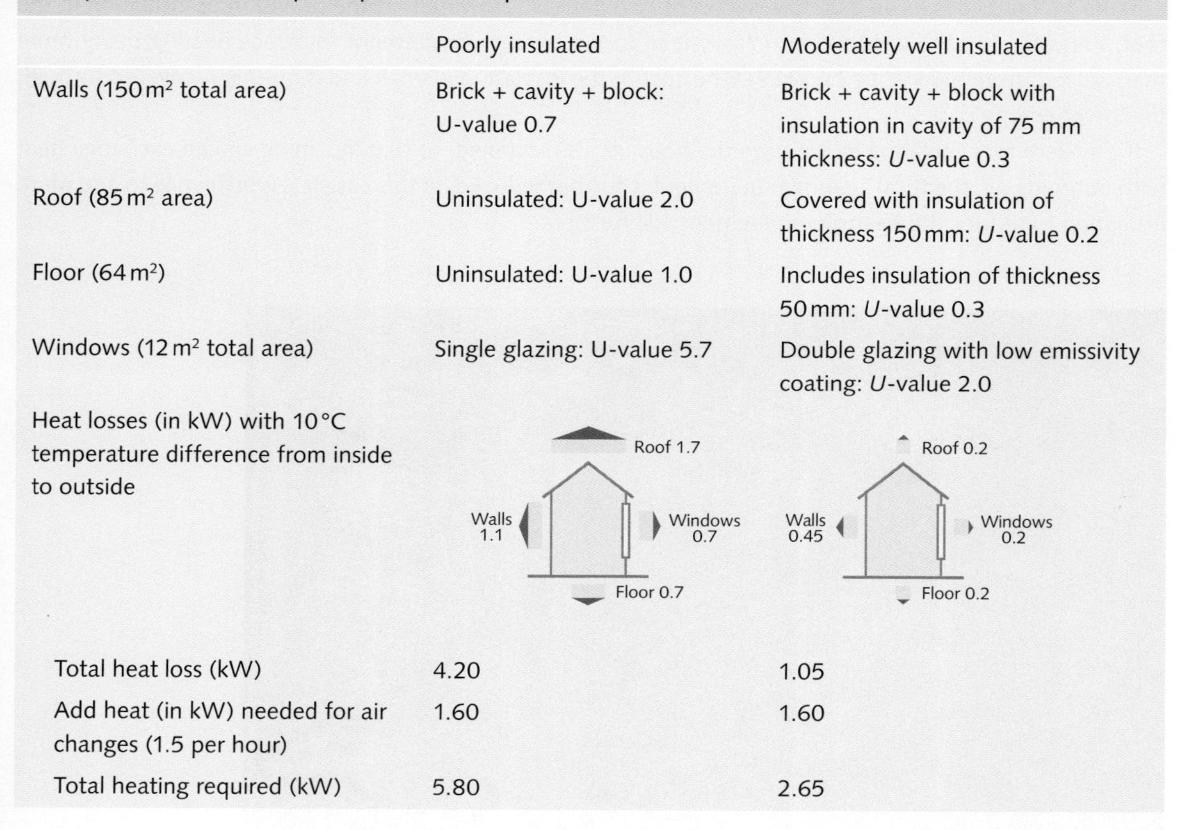

	Poorly insulated	Moderately well insulated
Walls (150 m² total area)	Brick + cavity + block: U-value 0.7	Brick + cavity + block with insulation in cavity of 75 mm thickness: *U*-value 0.3
Roof (85 m² area)	Uninsulated: U-value 2.0	Covered with insulation of thickness 150 mm: *U*-value 0.2
Floor (64 m²)	Uninsulated: U-value 1.0	Includes insulation of thickness 50 mm: *U*-value 0.3
Windows (12 m² total area)	Single glazing: U-value 5.7	Double glazing with low emissivity coating: *U*-value 2.0
Heat losses (in kW) with 10 °C temperature difference from inside to outside	Roof 1.7; Walls 1.1; Windows 0.7; Floor 0.7	Roof 0.2; Walls 0.45; Windows 0.2; Floor 0.2
Total heat loss (kW)	4.20	1.05
Add heat (in kW) needed for air changes (1.5 per hour)	1.60	1.60
Total heating required (kW)	5.80	2.65

(fossil-fuel) Developments (ZED).[21] The box illustrates a recent development in the UK along these lines.

Efficiency increases bringing cost savings sound very good in principle. In practice, however, it is frequently found that much of the energy and cost saving fails to materialise because of the increased comfort or convenience that comes from increased energy use – hence an increase in energy demand. Energy efficiency measures need therefore to be associated with adequate public education that explains the need for overall energy reductions.

Alongside the increases in energy efficiency in buildings and appliances, there need to be moves to carbon-free sources of energy supply to the buildings sector. These will be addressed in later sections.

Example of a ZED (Zero Emission (fossil-fuel) Development)

BedZED (Figure 11.8) is a mixed development urban village constructed on a brownfield wasteland in the London Borough of Sutton, providing 82 dwellings in a mixture of apartments, maisonettes and town houses together with some work/office space and community facilities.[22] The combination of super-insulation, a wind-driven ventilation system incorporating heat recovery, and passive solar gain stored within each unit in thermally massive floors and walls reduces the energy needs so that a 135 kW wood-fuelled combined heat and power (CHP) plant is sufficient to meet the village's energy requirements. A 109 kW peak photovoltaic installation provides enough solar electricity to power 40 electric cars, some pool, some taxi, some privately owned. The community has the capacity to lead a carbon-neutral lifestyle – with all energy for buildings and local transport being supplied from renewable sources.

Figure 11.8 The BEDZED development in south London, developed by the Peabody Trust and designed by Bill Dunster Architects'.

Energy and carbon dioxide savings in transport

Transport is responsible for nearly one-quarter of greenhouse gas emissions worldwide. It is also the sector where emissions are growing most rapidly (Figure 11.9). Road transport accounts for the largest proportion of this, over 70%, shipping around 20% and air transport about 10%.[23] The world population

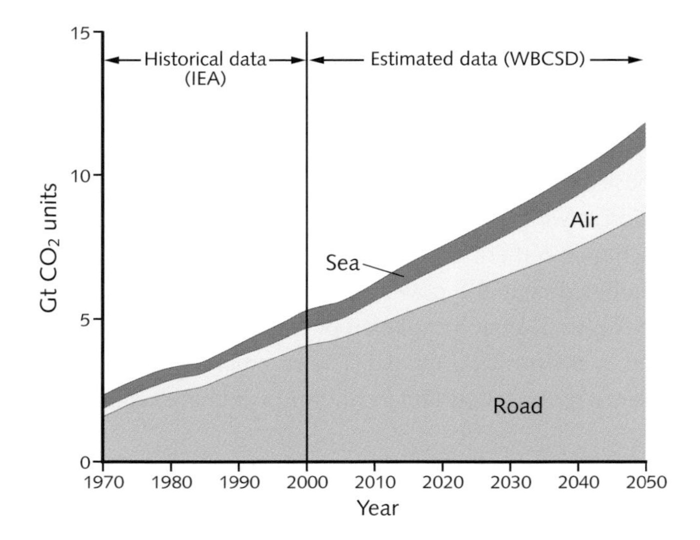

Figure 11.9 Historical and projected carbon dioxide emissions from transport by modes, 1970–2050. Projected data by the World Business Council on Sustainable Development (WBCSD) under a business-as-usual scenario; historical data from the International Energy Agency.

of light motor vehicles, currently around 750 million, is projected to rise by a factor of 2 by 2030 and a factor of 3 by 2050, most of the growth occurring in developing countries.[24] This trend seems inevitable when account is taken of the very large differences today in car ownership in different countries – in terms of persons per car, about 1.5 in the USA, 30 in China and 60 in India. Under similar assumptions, by 2050 a growth in aviation is projected by a factor of 5, again much of it in the developing world. Increased prosperity brings increased demand for personal mobility and also increased movement of freight. In the transport sector the achievement of reductions in carbon dioxide emissions will be particularly challenging.

There are three types of action that can be taken to curb the rapidly growing carbon dioxide emissions from motor transport (Figures 11.9 and 11.10). The first is to increase the efficiency of energy and fuel use and to move to non-fossil-fuel sources of energy. We cannot expect the average car to compete with the vehicle which, in 1992, set a record by covering over 12 000 km on 1 gallon of petrol – a journey which serves to illustrate how inefficiently we use energy for transport! However, it is estimated that the average fuel consumption of the current fleet of motor cars could be halved through the use of existing technology – more efficient engines, lightweight construction and low-air-resistance design (see box on p. 346) – while maintaining an adequate performance. Further, possibilities exist for the use of electric propulsion driven from larger and more efficient batteries or from fuel cells powered by hydrogen fuel supplied from non-fossil-fuel sources. The second action is to plan cities and other developments so as to lessen the need for transport and to make personalised transport less necessary – work, leisure and shopping should all

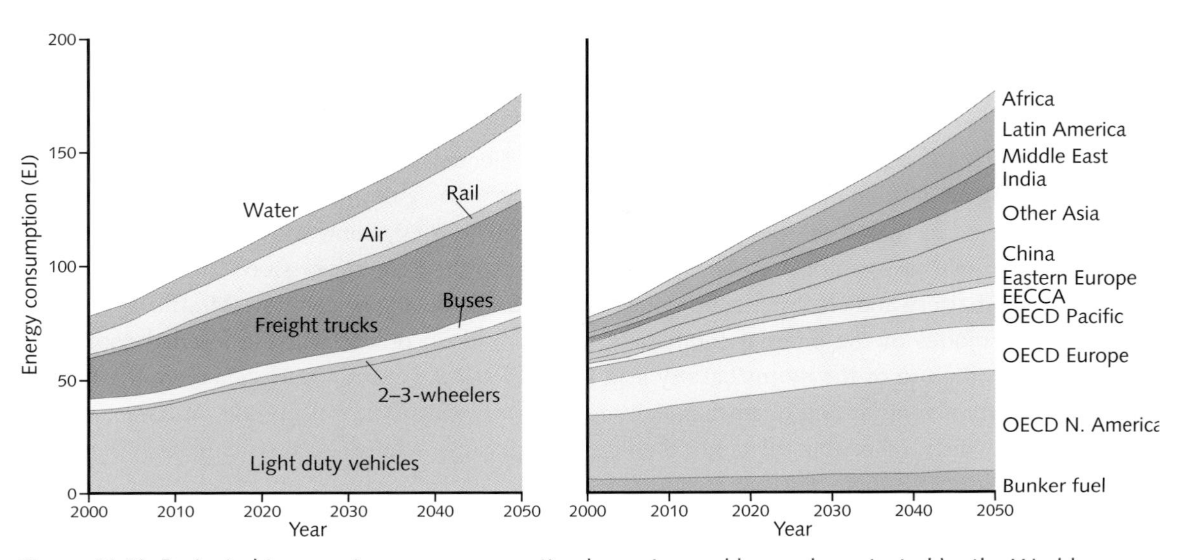

Figure 11.10 Projected transport energy consumption by region and by mode projected by the World Business Council on Sustainable Development (WBCSD) under a business-as-usual scenario.

be easily accessible by public transport, or by walking or cycling. Such planning needs also to be linked with a recognition of the importance of ensuring that public transport is reliable, convenient, affordable and safe. The third action is to increase the energy efficiency of freight transport by making maximum use of the most energy-efficient forms of freight transport, e.g. rail or water, rather than road or air and by eliminating unnecessary journeys.

Air transport is growing even faster than motor transport. Global passenger air travel, as measured in passenger-km, is projected to grow at about 5% per year over the next decade or more and total aviation fuel use – including passenger, freight and military – is projected to increase by about 3% per year, the difference being due largely to increased fuel efficiency.[26] Further increases in fuel efficiency are expected but they are unlikely to keep up with the increase in the volume of air transport. Biofuels as an alternative to kerosene are also being studied and are assumed, for instance in the IEA BLUE Map scenario, to have replaced 30% of conventional aviation fuel by 2050. Hydrogen has also been proposed as a long-term possibility but the effect on the dry upper troposphere of injection of the resulting water vapour has yet to be evaluated – it would probably lead to an unacceptable increase in cloud cover unless flight altitudes were substantially lowered.

A further problem with air transport, mentioned in Chapter 3 page 63, is that its carbon dioxide emissions are not the only contributor to global warming; increased high cloudiness due to other emissions produce an effect of similar or even greater magnitude. Operational changes to minimise this effect have

Technologies for reducing carbon dioxide emissions from motor vehicles

An important recent development is that of the hybrid electric motor car that combines an internal combustion engine with an electric drive train and battery.[25] The gains in efficiency and therefore fuel economy achieved by hybrid vehicles are typically around 50%. They mainly arise from: (1) use of regenerative braking (with the motor used as a generator and captured electricity stored in the battery), (2) running on the battery and electric traction only when in slow-moving or congested traffic, (3) avoiding low-efficiency modes of the internal combustion engine and (4) downsizing the internal combustion engine through the use of the motor/battery as a power booster. Toyota and Honda were the first two to introduce hybrid vehicles and other manufacturers are following. An imminent development is of the plug-in hybrid which will enable the larger than normal car battery to be boosted by connecting with a commercial electricity supply. For shorter journeys the plug-in hybrid could run only on the battery in which case with fossil-fuel-free electricity, its carbon dioxide emissions would be eliminated.

Other significant efficiency improvements are coming from the use of lower weight structural materials, improvements in low-air-resistance design and the availability of direct injection diesel engines, long used in heavy trucks, for automobiles and light trucks.

Developments are also occurring in battery technology that soon should enable more extensive employment of electric vehicles which will use electricity from wholly carbon-free sources. During the next few years we will begin to see the introduction of vehicles driven by fuel cells (see Figure 11.24 below) based on hydrogen fuel that can potentially be produced from renewable sources (see page 377). This new technology has the potential eventually to revolutionise much of the transport sector.

Biofuels generated from crops can be employed to fuel motor vehicles thereby avoiding fossil fuel use. For instance, ethanol has been extensively produced from sugar cane in Brazil. Biodiesel is also becoming more widely available (see later section on biomass).

been proposed but more understanding of the mechanism is needed before serious work on its reduction can be carried out. Controlling the growing influence of aviation on the climate is probably the largest challenge to be solved in the overall mitigation of climate change.

Energy and carbon dioxide savings in industry

Industry currently accounts for nearly one-third of worldwide primary energy use and about one-quarter of carbon dioxide emissions of which 30% comes from the iron and steel industry, 27% from non-metallic minerals (mainly cement) and 16% from chemicals and petrochemicals production.[27] Substantial opportunities exist for efficiency savings in all these areas. The application of appropriate control technologies, other best-available technologies (BAT) and more widespread combined heat and power (CHP) could bring 20–30% carbon dioxide emissions reduction along with substantial net saving in cost – such are known as no-regrets actions. Other potential decreases in carbon dioxide emissions can occur through the recycling of materials or waste (especially plastic waste), the use of waste as an energy source and switching to biomass feedstocks or to less carbon-intensive fuels.

Given appropriate incentives, substantial carbon dioxide savings can also be realised in the petrochemical industry with significant savings in cost. For instance, British Petroleum has set up a carbon emissions trading system within the company that encourages the elimination of waste and leaks from their operations and the application of technology to eliminate the venting of methane. In its first three years of operation, $US 600 million were saved and carbon emissions reduced to 10% below 1990 levels.[28]

Carbon capture and storage (CCS – see next section) is also an emerging option for industry. It is most suited for large sources of off-gases with high carbon dioxide concentrations such as blast furnaces (iron and steel), cement kilns, ammonia plants and also black liquor boilers or gasifiers (pulp and paper).

Industrial activity worldwide will increase substantially over the next few decades especially in developing countries where there is large demand for technology transfer that will enable the latest most efficient technologies to be employed. Because of this growth, in the absence of major actions to reduce carbon dioxide emissions, they are bound to rise. However, the opportunities for reductions are such that in the IEA BLUE Map scenario, emissions from industry in 2050 fall to 22% below the 2005 level. The necessary policy instruments and incentives to stimulate these reductions are discussed later in the chapter.

Carbon-free electricity supply

We have already noted that moving as rapidly as possible to carbon-free electricity is key to achieving the level of overall reductions of carbon dioxide emissions required by 2030 and 2050. Contributions to this movement can come in five ways: (1) increases in efficiency, (2) decreases in carbon intensity, (3) by widespread deployment of carbon capture and storage, (4) by the use of nuclear energy, (5) through the use of all possible renewable energies. From the long-term point of view, (1) and (5) are the most important. I will now briefly address each in turn.

First, regarding **energy efficiency**, the efficiency of coal-fired power stations, for instance, has improved from about 32%, a typical value of 20 or 30 years ago, to about 42% for a pressurised, fluidised-bed combustion plant of today. Gas turbine technology has also improved providing efficiency improvements such that efficiencies approaching 60% are reached by large modern gas-turbine-combined cycle plants. Large gains in overall efficiency are also available by making sure that the large quantities of low-grade heat generated by power stations is not wasted but utilised, for instance in CHP schemes. For such co-generation, the efficiencies attainable in the use of the energy from combustion of the fuel are typically around 80%. The wider deployment therefore of CHP in building schemes or in industries where both heat and power are required is an effective way of substantially increasing efficiency at the same time as producing savings in economic terms.[29]

Second, regarding **carbon intensity**, for a given production of energy, the carbon dioxide emissions from natural gas are 25% less than those from oil and 40% less than those from coal. By switching fuel to gas, therefore, substantial emissions savings can be made.

Third, an alternative to moving away from fossil fuel sources of energy is to prevent the carbon dioxide from fossil fuel burning from entering the atmosphere by the employment of **carbon capture and storage (CCS).**[30] Carbon dioxide capture is arranged either by removing it from the flue gases in a power station, or the fossil fuel feedstock can, in a gasification plant, be converted through the use of steam,[31] to carbon dioxide and hydrogen (Figure 11.11). The carbon dioxide is then relatively easy to remove and the hydrogen used as a versatile fuel. The latter option will become more attractive when the technical and logistic problems of the large-scale use of hydrogen in fuel cells to generate electricity have been overcome – this is mentioned again later in the chapter.

Various options are possible for the disposal (or sequestration) of the large amounts of carbon dioxide that result. For instance, the carbon dioxide can be pumped into spent oil or gas wells, into deep saline reservoirs or into unminable coal seams. Other suggestions have also been made such as pumping it into the deep ocean, but these are more speculative and need careful research and assessment before they can be realistically put forward. In the most favourable circumstances (for instance when power stations are close to suitable reservoirs and when the extraction cost is relatively small), the cost of removal, although significant, is only a small fraction of the total energy cost. The IPCC has estimated a range of $US15–80 per tonne CO_2 for the added cost – the cost of extraction being generally much larger than the cost of storage.

The global potential for storage in geological formations is large and has been estimated to be at least 2000 Gt carbon dioxide and possibly much larger. The likely rate of leakage is believed to be very low although more research is required into this rate and also into the risk of rapid release as a result, for instance, of seismic activity.

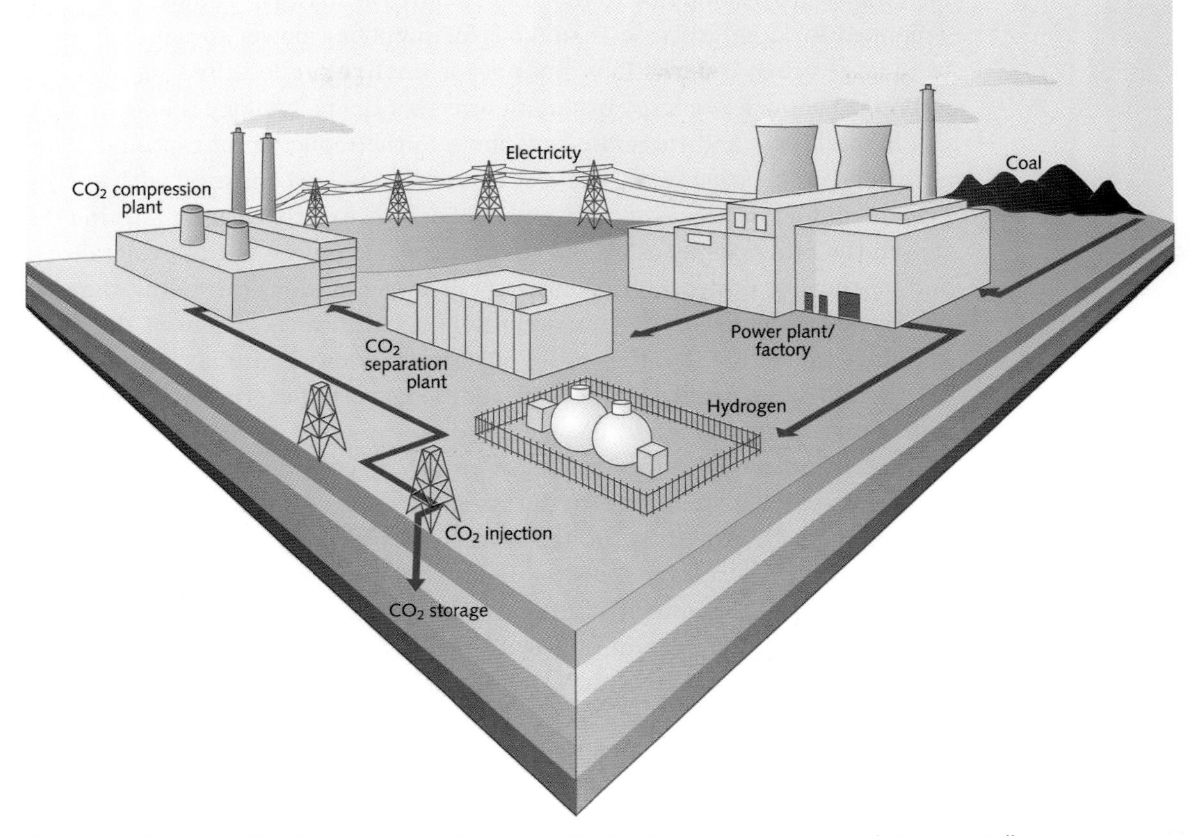

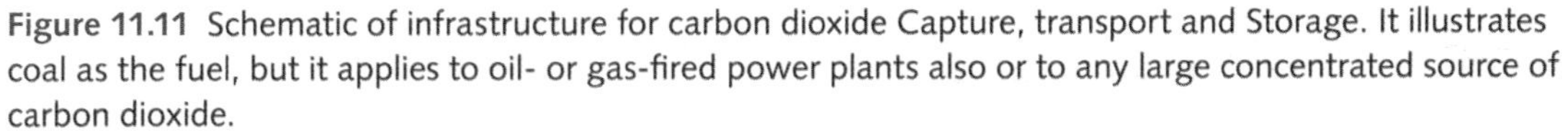

Figure 11.11 Schematic of infrastructure for carbon dioxide Capture, transport and Storage. It illustrates coal as the fuel, but it applies to oil- or gas-fired power plants also or to any large concentrated source of carbon dioxide.

Because of the rapid increase during the last few years in the number of new coal-fired power stations constructed globally (for instance, China is currently adding capacity of 2 GW per week), the need for CCS technology has become more acute. A substantial number of demonstration plants employing CCS need to be built before 2015 in the USA, Europe, China, Australia and other countries where coal remains a major source of power generation.[32] Rapid deployment of CCS to all new coal-fired power stations would enable continuing use of fossil fuels without the deleterious effects of carbon dioxide emissions.

A fourth source of carbon-free energy is **nuclear energy.**[33] It has considerable attractiveness from the point of view of sustainable development because it does not produce greenhouse gas emissions (apart from the relatively small amount associated with the materials employed in nuclear power station construction) and because the rate at which it uses up resources of radioactive material is small compared to the total resource available. It is most efficiently generated in large units, so is suitable for supplying power to national grids or to large urban conurbations, but not for small, more localised supplies. An advantage of nuclear energy installations is that the technology is known; they can be built now and therefore contribute to the reduction of carbon dioxide emissions in the short term. The cost of nuclear energy compared with energy from fossil fuel sources is often a subject of debate; exactly where it falls in relation to the others depends on the return expected on the upfront capital cost and on the cost of decommissioning spent power stations (including the cost of nuclear waste disposal), which represent a significant element of the total. Recent estimates are that the cost of nuclear electricity is similar to the cost of electricity from natural gas when the additional cost of capture and sequestration of carbon dioxide is added.

The continued importance of nuclear energy is recognised in the IEA energy scenarios, which assume growth in this energy source in the twenty-first century. How much growth is limited in the short term by the shortage of personnel with the necessary skills for design and construction of nuclear power systems and by the limited facilities available for building key components. In the longer term, the amount of growth realised will depend on how well the nuclear industry is able to satisfy the general public of the safety of its operations; in particular that the risk of accidents from new installations is negligible, that nuclear waste can be safely disposed of and that dangerous nuclear material can be effectively controlled and prevented from getting into the wrong hands. Despite the substantial safeguards that are in place internationally, this last possibility of the proliferation of dangerous nuclear material is the one that, in my view, presents the strongest argument for questioning the widespread growth of nuclear energy.[34] However, proposals are now being pursued for a fourth generation of nuclear power plants based on more advanced reactors that promise to be safer, less productive of radioactive waste and with much less danger of leading to nuclear proliferation. None of these, however, are likely to be built before 2020 and maybe 2030.

A further nuclear energy source with great potential in the more distant future depends on fusion rather than fission (see box below p. 377).

The fifth source of carbon-free energy is from the variety of **renewable energies** that have been identified and that are available. To put renewable energy in context it is relevant to realise that the energy incident on the Earth from the Sun amounts to about 180 000 million million watts (or 180 000 terawatts, 1 TW = 10^{12} W). This is about 12 000 times the world's average energy use of about 15 million million watts (15 TW). As much energy arrives at the Earth from the Sun in 40 minutes as we use in a whole year. So, providing we can harness it satisfactorily and economically, there is plenty of renewable energy coming in from the Sun to provide for all the demands human society can conceivably make.

There are many ways in which solar energy is converted into forms that we can use; it is instructive to look at the efficiencies of these conversions. If the solar energy is concentrated, by mirrors for instance, almost all of it can be made available as heat energy. Between 1% and 2% of solar energy is converted through atmospheric circulation into wind energy, which although concentrated in windy places is still distributed through the whole atmosphere. About 20% of solar energy is used in evaporating water from the Earth's surface which eventually falls as precipitation, giving the possibility of hydropower. Living material turns sunlight into energy through photosynthesis with an efficiency of around 1% for the best crops. Finally, photovoltaic (PV) cells convert sunlight into electricity with an efficiency that for the best modern cells can be over 20%.

Around the year 1900, very early in the production of commercial electricity, water power was an obvious source and from the beginning made an important contribution. Hydroelectric schemes now supply about 18% of the world's electricity. Other renewable sources of electricity, however, have been dependent on recent technology for their implementation. In 2005, only about 4% of the world's electricity came from renewable sources other than large hydro (these are often collectively known as 'new renewables').[35] Over half of this was from 'modern' biomass (called 'modern' when it contributes to commercial energy to distinguish it from traditional biomass), the rest being shared between solar, wind energy, geothermal, small hydro and marine sources.

Under the IEA BLUE Map scenario (Figure 11.12), all renewable sources will be contributing by 2050 45% of total electricity production. The main growth expected is in energy from 'modern' biomass and from solar and wind energy sources. In the following paragraphs, the main renewable sources are described in turn and their possibilities for growth considered. Most of them are employed for the production of electricity through mechanical means (for hydro and wind power), through heat engines (for biomass and solar thermal) and through direct conversion from sunlight (solar PV). In the case of biomass, liquid or gaseous fuels can also be produced.

Hydropower

Hydropower, the oldest form of renewable energy, is well established and is competitive economically with electricity generated by other means. Some hydroelectric schemes are extremely large. The world's largest, the Three Gorges

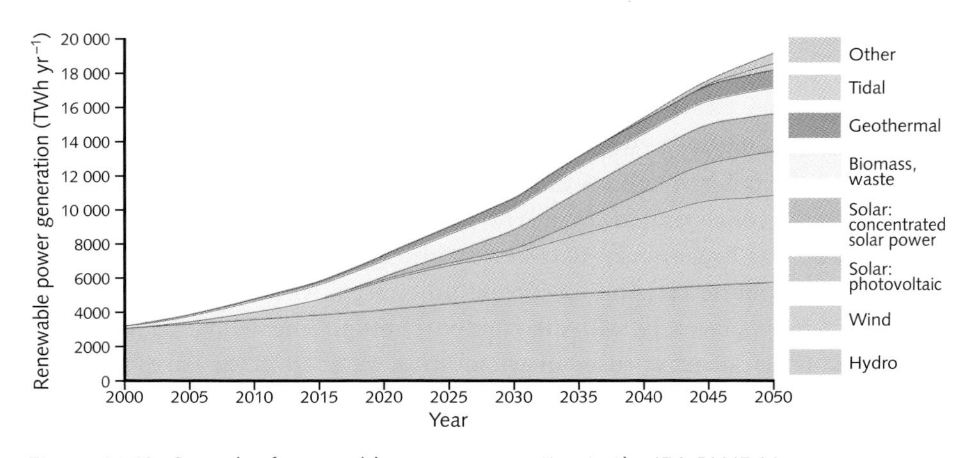

Figure 11.12 Growth of renewable power generation in the IEA BLUE Map energy scenario, 2000–2050.

project on the Yangtze River in China, generates about 18 GW of electricity. Two other large schemes, each of over 10 GW capacity, are in South America at Guri in Venezuela and at Itaipu on the borders of Brazil and Paraguay. It is estimated[36] that there is potential for further exploitation of hydroelectric capacity to two or three times the amount that has currently been developed, much of this undeveloped potential being in Africa, Asia and Latin America. Large schemes, however, can have significant social impact (such as the movement of population from the reservoir site), environmental consequences (for example, loss of land, of species and of sedimentation to the lower reaches of the river), and problems of their own such as silting up, which have to be thoroughly addressed before they can be undertaken.

But hydroelectric schemes do not have to be large; small hydroelectric sources are increasingly providing an important resource. Many units exist that generate a few kilowatts only to supply one farm or a small village. The attractiveness of small schemes is that they provide a locally based supply at modest cost. Substantial growth in 'small hydro' – much of it designed to run in-river – has occurred during the last decade or so, but only about 5% of the global potential resource of 150 or 200 GW has yet been exploited.

An important facility provided by some hydro schemes is that of pumped storage. Using surplus electricity available in off-peak hours, water can be pumped from a lower reservoir to a higher one. Then, at other times, by reversing the process, electricity can be generated to meet periods of peak demand. The efficiency of conversion can be as high as 80% and the response time a few seconds, so reducing the need to keep other generating capacity in reserve. Currently about 100 GW of pumped storage capacity is available worldwide but the potential is considered to be at least ten times that figure.

Biomass energy

Second in current importance as a renewable energy source is the use of biomass.[37] The annual global primary production of biomass of all kinds expressed in energy units is about 4500 EJ (= 107 Gtoe). About 1% of this is currently turned into energy mostly in developing countries – we have labelled it 'traditional biomass'. It has been estimated that about 6% of the total could become available from energy crops taking into account the economics of production and the availability of suitable land.[38] The energy so generated would represent about 75% of current world energy consumption, so in principle a large contribution from biomass could be made towards global energy needs. It is a genuinely renewable resource in that the carbon dioxide that is emitted when the biomass is burnt is turned back into carbon, through the process of photosynthesis, in the renewed biomass when it is grown again. The word biomass not only covers crops of all kinds but also domestic, industrial and agricultural dry waste material and wet waste material, all of which can be used as fuel for heating and to power electricity generators; some are also appropriate to use for the manufacture of liquid or gaseous fuels (see next section). Since biomass is widely distributed, it is particularly appropriate as a distributed energy source suitable for rural areas. For instance, in Upper Austria, with a population of 1.5 million, in 2003 14% of their total energy came from local biomass – planned to increase to 30% by 2010 and to continue to grow substantially thereafter.

In much of the developing world, most of the population live in areas where there is no access to modern or on-grid energy. They rely on 'traditional biomass' (fuelwood, dung, rice husks and other forms of biomass) to satisfy their needs for cooking and heating. About 10% of world energy originates from these sources, supplying over one-third of the world's population. Although these sources are in principle renewable, it is still important that they are employed efficiently, and a great deal of room for increased efficiency exists. For instance, a large proportion of each day is often spent in collecting firewood especially by the women, increasingly far afield from their homes.

The burning of biomass in homes causes serious health problems and has been identified by the World Health Organization as one of the most serious causes of illness and mortality especially amongst children. For instance, much cooking is still carried out on open fires with their associated indoor pollution and where only about 5% of the heat reaches the inside of the cooking pot. The introduction of a simple stove can increase this to 20% or with a little elaboration to 50%.[39] An urgent need exists for the large-scale provision of stoves using simple technology that is sustainable – although there is often considerable consumer resistance to their introduction. Other means of reducing fuelwood demand are to encourage alternatives such as the use of fuel from crop wastes, of methane from sewage or other waste material, or of solar cookers (mentioned again later on). From the existing consumption of 'traditional biomass' there is the potential to produce sustainable 'modern' energy services with much greater efficiency and much less pollution for the 2 billion or so people who currently rely on this basic energy source. A particular challenge is to set up appropriate management and infrastructure for the provision of these services in rural areas in developing countries (see box below).

Consider for instance the use of waste.[40] There is considerable public awareness of the vast amount of waste produced in modern society. The UK, for example, produces each year somewhat over 30 million tonnes of domestic solid waste, or about half a tonne for every citizen – a typical value for a country in the developed world. Even with major programmes for recycling some of it, large quantities would still remain. If it were all incinerated for power generation (modern technology enables this to be done with negligible air pollution) nearly 2 GW could be generated, about 5% of the UK's electricity requirement.[41] Uppsala in Sweden is an example of a city with a comprehensive district heating system, for which, before 1980, over 90% of the energy was provided from oil. A decision was then made to move to renewable energy and by 1993 energy from waste incineration and from other biomass fuel sources provided nearly 80% of what is required for the city's heating.

Biomass projects in rural areas in the developing world

In much of the developing world, most of the population live in areas where there is little or no access to electricity or modern energy services. There is large potential for creating local biomass projects to provide such services. Figure 11.13 shows a schematic of a modern biogas plant and examples are given of pilot projects,[42] all of which could be replicated many times.

Rural power production, India

India, still a predominantly agriculture-based country, produces approximately 400 million tonnes of agro waste every year. A fraction of this is used for cooking purposes and the balance is either burned or left to decompose. India also imports large quantities of fossil-fuel-based furnace oil to supply power and heat to millions of small-to large-scale urban/rural industrial units.

Linus Strategic Energy Solutions, an Indian company, are producing environmentally sustainable briquettes from agro waste suitable for use as a fuel in these industrial units. In addition to the reduction in costly and environmentally damaging fossil fuel use and the cash savings generated by the user, this cycle also has the potential to generate new sources of income for farmers supplying the biomass, new business opportunities for rural entrepreneurs processing and selling the briquettes and additional rural employment in the collection and processing of the agro waste.

Also in India, Decentralised Energy Systems India Private Limited are piloting the first independent power projects of around 100 kW capacity in rural India owned and operated by village community co-operatives. An example is a small co-operative in Baharwari, Bihar State, where a biomass gasification power plant is used as a source of electricity for local enterprises, for instance for pumping water in the dry season. Local income is thereby generated that enables villagers to expand their micro-industries and create more jobs – all of which in turn increases the ability of people to pay for improved energy services. A 'mutuality of interest' is created between biomass fuel suppliers, electricity users and plant operators.

Integrated biogas systems, Yunnan, China

The South–North Institute for Sustainable Development has introduced a novel integrated biogas system in the Baima Snow Mountains Nature Reserve, Yunnan Province. The system links a biogas digester, pig-sty, toilet and greenhouse. The biogas generated is used for cooking and replaces the burning of natural firewood, the 'greenhouse' pigsty increases the efficiency of pig-raising, the toilet improves rural environmental hygiene, and vegetables and fruits planted in the greenhouse increase the income of local inhabitants. Manure and other organic waste from the pigsty and toilet are used as the raw material for biogas generation which delivers about 10 kWh per day of useful energy (cf Figure 11.13). The operation of 50 such systems has considerably reduced local firewood consumption.

Biomass power generation and coconut oil pressing, the Philippines

The Community Power Corporation (CPC) has developed a modular biopower unit that can run on waste residue or biomass crops and can enable village-level production of coconut oil. CPC and local partners are using the modular biopower unit fuelled by the waste coconut shells to provide electricity to a low-cost mini-coconut-oil-mill (developed by the Philippines Coconut Authority and the University of Philippines), 16 of which are now operating in various Philippine villages. Furthermore, the biopower unit generates waste heat which is essential for drying the coconuts prior to pressing.

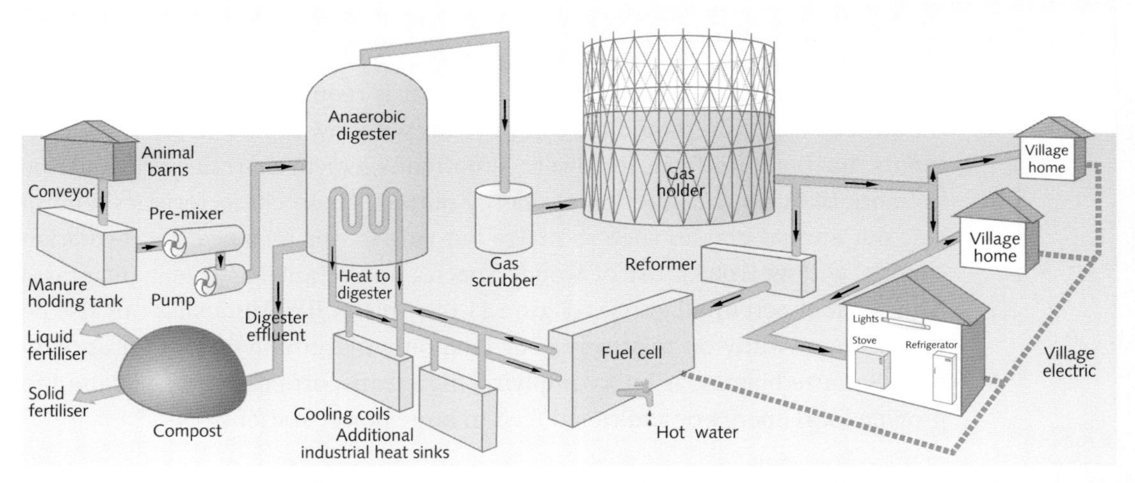

Figure 11.13 Schematic of digester for biogas plant for local supply (not to scale). The fuel cell for generating electricity anticipates the availability of more advanced fuel cells. In the meantime the reformer and fuel cell could be replaced by an internal combustion gas engine and a generating set.

But what about the greenhouse gas generation from waste incineration? Carbon dioxide is of course produced from it, which contributes to the greenhouse effect (see question 4 Chapter 3). However, the alternative method of disposal is landfill (most of the waste in the UK currently is disposed of by landfill). Decay of the waste over time produces carbon dioxide and methane in roughly equal quantities. Some of the methane can be collected and used as a fuel for power generation. However, only a fraction of it can be captured; the rest leaks away. Because methane is a much more effective greenhouse gas, molecule for molecule, than carbon dioxide, the leaked methane makes a substantial contribution to the greenhouse effect. Detailed calculations show that if all UK domestic waste were incinerated for power generation rather than landfilled, the net saving per year in greenhouse gas emissions would be equivalent to about 10 million tonnes of carbon as carbon dioxide.[43] Since this is about 5% of the total UK greenhouse gas emissions, we can infer that power generation from waste could be a significant contribution to the reduction in overall emissions.

Other wastes resulting from human or agricultural activity are wet wastes such as sewage sludge and farm slurries and manures. Bacterial fermentation in the absence of oxygen (anaerobic digestion) of these wastes produces biogas, which is mostly methane and which can be used as a fuel to produce energy (Figure 11.13) There is room for an increasing contribution from these sources. If the potential for power generation from agricultural and industrial waste was taken into account, the savings in emissions arising from domestic waste already mentioned could be approximately doubled.

Turning now to the use of crops as a fuel, the potential is large. Many different crops can be employed as biomass for energy production. However, because of the relatively low efficiency of conversion of solar energy to biomass, the amount of land required for significant energy production by this means is large – and it is important that land is not taken over that is required for food production. An ideal energy crop should have high yields with low inputs. In energy terms, inputs, for instance from fertilisers, crop management or transport, must be no more than a small fraction of energy output. These characteristics tend to rule out annual grasses such as maize but rule in, for instance, short-rotation coppice willow from a list of woody species and *Miscanthus* ('elephant grass') from a list of perennial grasses (Figure 11.14). Grasses like *Miscanthus* can also be grown successfully on relatively poor land only marginally useful for agriculture. Because biomass is bulky implying high transport costs, it is best used to provide local energy or additional feed to large power stations.

In the IEA BLUE Map scenario, global biomass use (including biofuels, see below) increases nearly fourfold by 2050 accounting for nearly one-quarter of total world primary energy. It is then by far the most important renewable energy source. About half of this will come from crop and forest residues and other waste, the other half from purpose-grown energy crops. These will require a land area equivalent to about half the land area currently under agriculture in Africa or 10% of the world's total.

Biofuels

Biofuels are currently produced from starch, sugar and oil feedstocks that include wheat, maize, sugar cane, palm oil and oilseed rape. The best-known example of their use comes from Brazil where since the 1970s large plantations of sugar cane have produced ethanol for use as a fuel mainly in transport, generating, incidentally, much less local pollution than petrol or diesel fuel from fossil sources. Residues from ethanol or sugar production are used to generate electricity to power the factories and to export to the grid, ensuring good efficiency for the total process both in terms of energy and of saving carbon emissions.

Decisions about the large-scale production of biofuels must be guided by thorough and comprehensive assessments that address their overall efficiency and overall contribution to the reduction of carbon emissions.[44] Also requiring careful assessment is the degree to which their use of land is competing with food crops (as for instance with the use of maize) or adding to deforestation of tropical forests (as for instance with some palm oil plantations) that itself contributes substantially to greenhouse gas emissions. Examples have recently come to light of adverse consequences (for instance on world food prices) arising from a lack of adequate assessment.

Energy is also available in cellulosic biomass – as is evident from a cattle's rumen which turns grass into energy. On a laboratory scale, this process can be replicated and biofuels produced from lignocellulose from grasses or woody material or from the residue from cereal or other crops. A strong focus of recent work is turning this into commercial large-scale production of biofuels especially from woody wastes or from grasses such as *Miscanthus* grown on marginal land where it is not competing with food crops. Already this is beginning to happen and the scenarios I have presented assume that these second-generation biofuels, as they are called, can be successfully developed on the scale required.[45]

Figure 11.14 *Miscanthus* and willow growing at the Institute for Grassland and Environmental Research (IGER), Aberystwyth University,UK.

Wind energy

Energy from the wind is not new. Two hundred years ago windmills were a common feature of the European landscape; for example, in 1800 there were over 10 000 working windmills in Britain. During the past few years they have again become familiar on the skyline especially in countries in Western Europe (for instance, Denmark, Germany, UK and Spain) and in western North America. Slim, tall, sleek objects silhouetted against the sky, they do not have the rustic elegance of the old windmills, but they are much more efficient. A typical wind energy generator installed during the last decade will have three-bladed propeller about 50 m in diameter and a rate of power generation in a wind speed of 12 m s^{-1} (43 km h^{-1}, 27 mph or Beaufort Force 6), of about 700 kW. On a site with an average wind speed of about 7.5 m s^{-1} (an average value for exposed places in many western regions of Europe) it will generate an average power of about 250 kW. The generators are often sited close to each other in wind farms that may include several dozen such devices. The size of the largest generators has grown steadily, roughly doubling every five years, the largest now being 5 MW to 6 MW units with rotor diameters of around 120 metres.

From the point of view of the electricity generating companies the difficulty with the generation of electrical power from wind is that it is intermittent. There are substantial periods with no generation at all. The generating companies can cope with this in the context of a national electricity grid that pools electrical power from different sources providing that the proportion from intermittent sources is not too large.[46] Some public concern about wind farms

Wind turbines.

arises because of loss of visual amenity. Offshore sites, that are not seen to possess the same amenity disadvantage and that generally provide stronger and steadier winds, are being increasingly used for large wind farms.

Rapid growth has occurred in many countries in the installation of wind generators for electricity generation over the past decade – a growth that continues unabated. Over 100 GW peak operating capacity has now (2008) been installed worldwide providing about 1% of global electricity supply. With this large growth, economies of scale have brought down the cost of the electricity generated so that it is competitive with the cost of electricity from fossil fuels. Because the power generated from the wind depends on the cube of the wind speed (a wind speed of 12.5 m s^{-1} is twice as effective as one of 10 m s^{-1}) it makes sense to build wind farms on the windiest sites available. Some of these are to be found in Western Europe where rapid growth in wind generation is occurring. In Denmark, for instance, nearly 20% of electricity is now generated by wind – increasingly from wind farms being built offshore. Substantial offshore wind energy generation is also planned for the UK. Developing countries are also making increased use of wind energy. For instance, India with 8 GW

Wind power on Fair Isle

A good example of a site where wind power has been put to good effect is Fair Isle, an isolated island in the North Sea north of the Scottish mainland.[48] Until recently, the population of 70 people depended on coal and oil for heat, petrol for vehicles and diesel for electricity generation. A 50-kW wind generator was installed in 1982 to generate electricity from the persistent strong winds of average speed over 8 m s^{-1} (29 km h^{-1} or 18 mph). The electricity is available for a wide variety of purposes; at a relatively high price for lighting and electronic devices and at a lower price controlled amounts are available (wind permitting) for comfort heat and water heating. At the frequent periods of excessive wind further heat is available for heating glasshouses and a small swimming pool. Electronic control coupled with rapid switching enables loads to be matched to the available supply. An electric vehicle has been charged from the system to illustrate a further use for the energy.

With the installation of the wind generator, which now supplies over 90% of the island's electricity, electricity consumption has risen about fourfold and the average electricity costs have fallen from 13p per kWh to 4p per kWh. A second wind turbine of 100 kW capacity was installed in 1996/7 to meet increasing demand and to improve wind capture.

of installed capacity ranks as fourth in the world in wind generation. By 2050 under the IEA BLUE Map scenario, 12% of global electricity is projected to be provided from wind energy.

Wind energy is also particularly suitable for the generation of electricity at isolated sites to which the transmission costs of electricity from other sources would be unacceptable. Because of the wind's intermittency, some storage of electricity or some back-up means of generation has to be provided as well. The installation on Fair Isle (see box) is a good example of an efficient and versatile system. Small wind turbines also provide an ideal means for charging batteries in isolated locations; for instance, about 100 000 are in use by Mongolian herdsmen. Wind energy is often also an ideal source for water pumps – over 1 million small wind machines are used for this purpose worldwide.[47]

In the longer term it can be envisaged that wind generation could expand into areas remote from direct electrical connection providing an effective means for energy storage can be developed (for instance, using hydrogen; more of that possibility later in the chapter).

Energy from the Sun: solar heating

The simplest way of making use of energy from the Sun is to turn it into heat. A black surface directly facing full sunlight can absorb about 1 kW for each square metre of surface. In countries with a high incidence of sunshine it is an effective and cheap means of providing domestic hot water, extensively employed

Solar water heating

The essential components of a solar water heater (Figure 11.15) are a set of tubes in which the water flows embedded in a black plate insulated from behind and covered with a glass plate on the side facing the Sun. A storage tank for the hot water is also required. A more efficient (though more expensive) design is to surround the black tubes with a vacuum to provide more complete insulation. Over 10 million households worldwide have solar hot water systems.[49]

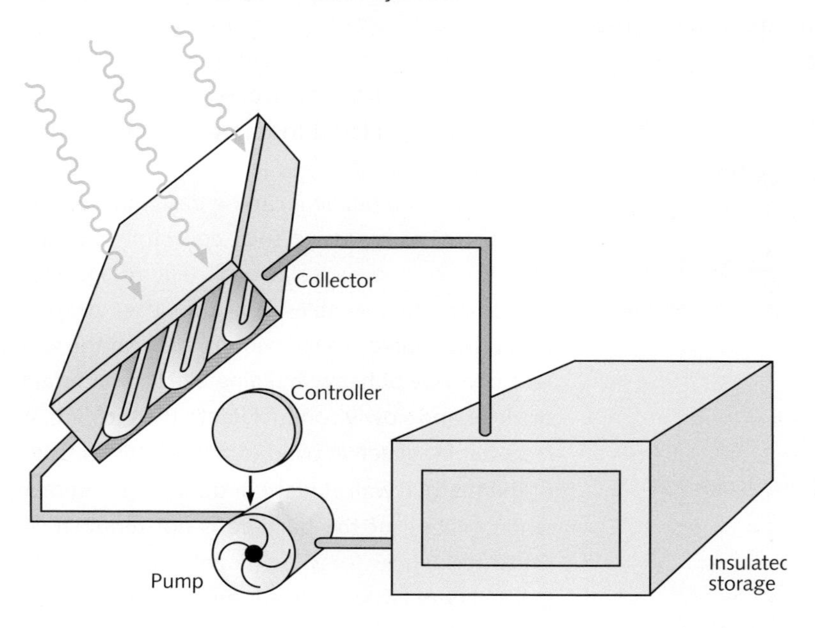

Figure 11.15 Design of a solar water heater: a solar collector connected to a storage tank through a circulating pump. Alternatively, if the storage is above the collector, the hot water will collect through gravity flow.

in countries such as Australia, Israel, Japan and the southern states of the USA (see box). In tropical countries, a solar cooking stove can provide an efficient alternative to stoves burning wood and other traditional fuels. Thermal energy from the Sun can also be employed effectively in buildings (it is called passive solar design), in order to provide a modest boost towards heating the building in winter and, more importantly, to provide for a greater degree of comfort and a more pleasant environment (see box).

Energy from the Sun: concentrating solar power

Solar energy can be converted into electricity either through its heat energy focused on to a boiler, for instance, to produce steam – known as concentrating solar power (CSP) – or by means of photovoltaic (PV) solar cells (see box). It is widely agreed that both possess the potential to be large contributors to global renewable energy. For instance, the current electricity needs of the United

Solar energy in building design

All buildings benefit from unplanned gains of solar energy through windows and, to a lesser extent, through the warming of walls and roofs. This is called 'passive solar gain'; for a typical house in the UK it will contribute about 15% of the annual space heating requirements. With 'passive solar design' this can relatively easily and inexpensively be increased to around 30% while increasing the overall degree of comfort and amenity. The main features of such design are to place, so far as is possible, the principal living rooms with their large windows on the south side of the house in the Northern hemisphere, with the cooler areas such as corridors, stairs, cupboards and garages with the minimum of window area arranged to provide a buffer on the north side. Conservatories can also be strategically placed to trap some solar heat in the winter.

The wall of a building can be designed specifically to act as a passive solar collector, in which case it is known as 'solar wall' (Figure 11.16).[50] Its construction enables sunlight, after passing through a double-glazed window, to heat the surface of a wall of heavy building blocks that retain the heat and slowly conduct it into the building. A retractable reflective blind can be placed in front of the thermal wall at night or during the summer when heating of the building is not required. A set of residences for 376 students at Strathclyde University in Glasgow in southwest Scotland has been built with a 'solar wall' on its south-facing side. Even under the comparatively unfavourable conditions during winter in Glasgow (the average duration of bright sunshine in January is only just over one hour per day) there is a significant net gain of heat from the wall to the building.

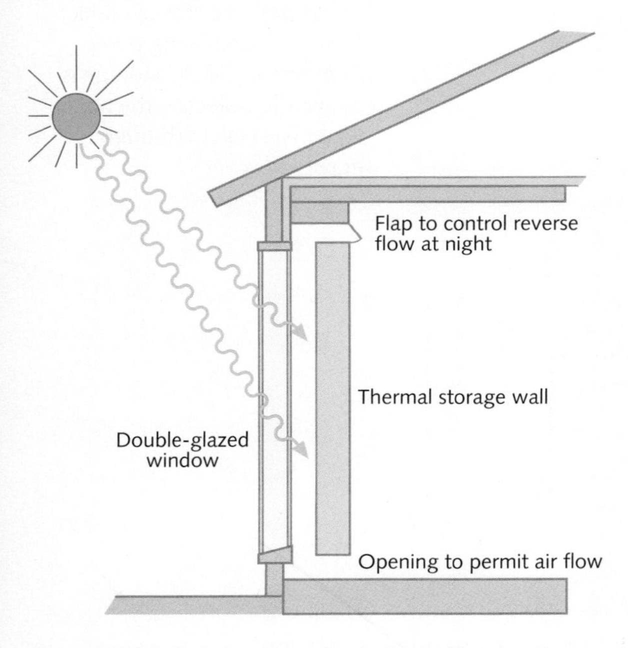

Figure 11.16 Construction of a 'solar wall', sometimes called a Trombe wall.

States could be generated from the solar energy falling on PV cells over an area of 400 km square or on CSP installations covering a somewhat smaller area. However, at the present time for large-scale electricity provision, neither is competitive in cost with conventional energy sources or with wind energy. Both require sufficient injection of investment for research and development to grow the economies of scale required to bring costs down to acceptable levels. I will address CSP and PV in turn.

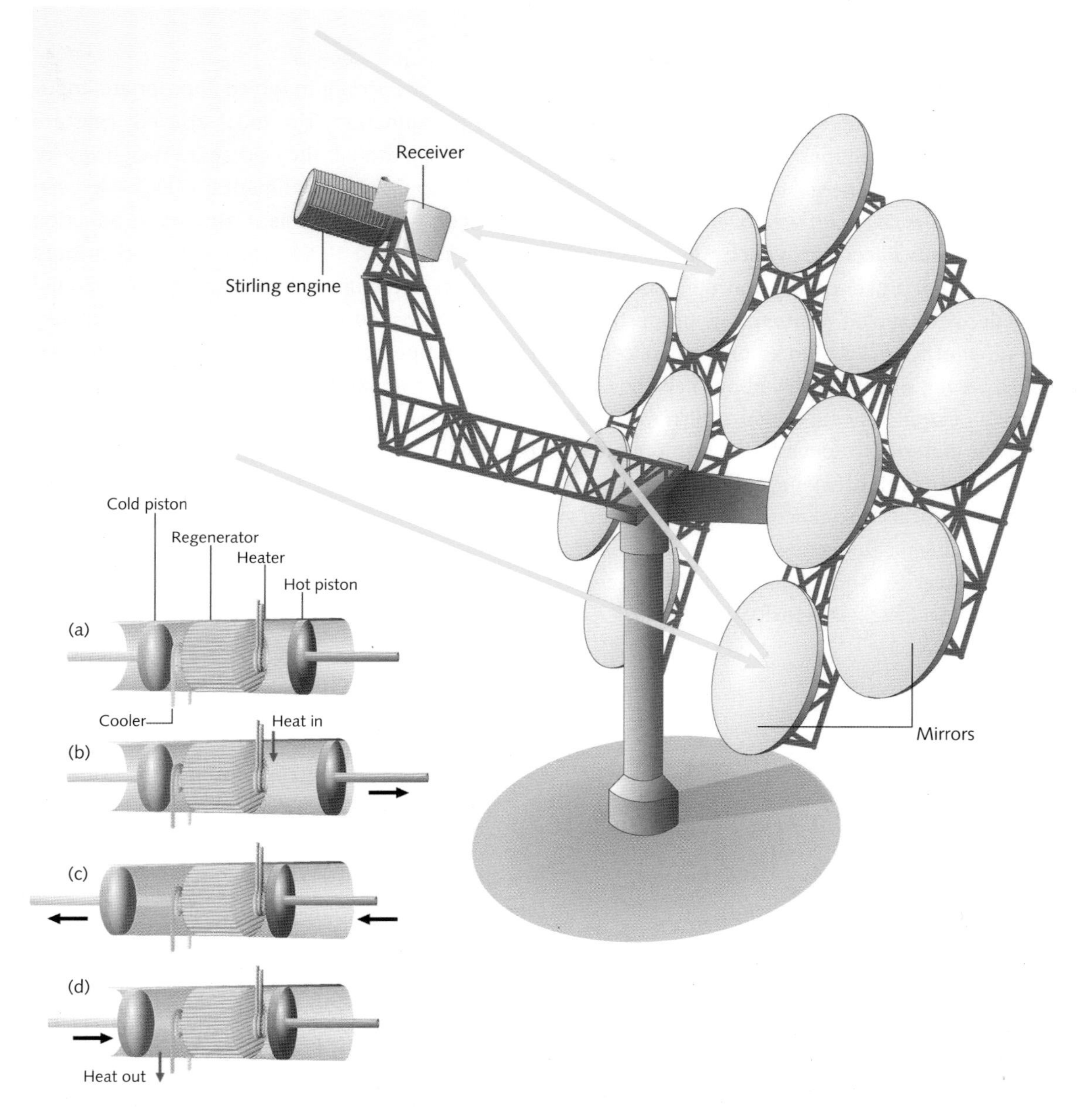

Figure 11.17 A concentrating solar power (CSP) system for electricity generation, that consists of a solar thermal array of a number of dish-shaped mirrors each focusing radiation on a receiver attached to a Stirling engine (see bottom left) that converts heat into electricity.

The photovoltaic solar cell

The silicon photovoltaic (PV) solar cell[51] consists of a thin slice of silicon into which appropriate impurities have been introduced to create what is known as a p–n junction. The most efficient cells are sophisticated constructions using crystalline silicon as the basic material; they possess efficiencies for conversion of solar energy into electricity typically of 15% to 20%; experimental cells have been produced with efficiencies well over 20%. Single crystal silicon is less convenient for mass production than amorphous silicon (for which the conversion efficiency is around 10%), which can be deposited in a continuous process onto thin films. Other alloys (such as cadmium telluride and copper indium diselenide) with similar photovoltaic properties can also be deposited in this way and, because they have higher efficiencies than amorphous silicon, are likely to compete with silicon for the thin-film market.[52] However, since typically about half the cost of a solar PV installation is installation cost, the high efficiency of single crystal silicon, which means a smaller size, remains an important factor. A number of new PV materials or devices are also under intensive investigation some of which are beginning to compete in terms of efficiency or cost.

Cost is of critical importance if PV solar cells are going to make a significant contribution to energy supply. This has been coming down rapidly. More efficient methods and larger-scale production are bringing the cost of solar electricity down to levels where it can compete with other sources. The decline in cost with increase in installed capacity over the last 25 years and a projection for the next 5 years is illustrated in Figure 11.18.

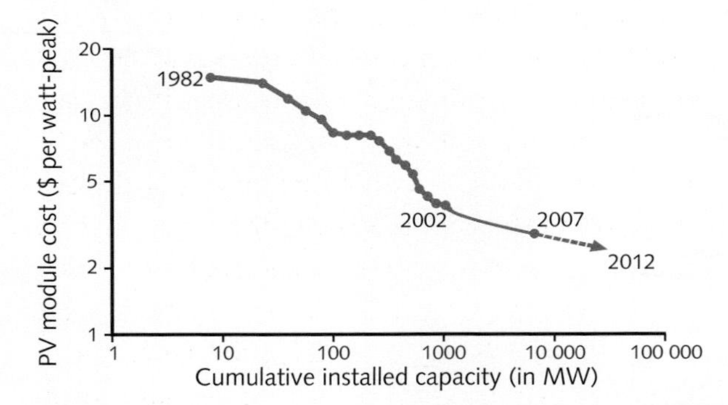

Figure 11.18 The increase in installed capacity and the falling cost of PV modules over the last 25 years and (dashed) projected into the future (data from 1982–2002 from Shell Renewables and from 2002–2007 from Energy Technology Perspectives, International Energy Agency 2008).

In CSP, to produce a sufficiently high temperature at the boiler, the solar energy has to be concentrated using mirrors (Figure 11.17). One arrangement employs trough-shaped mirrors aligned east–west which focus the Sun on to an insulated black absorbing tube running the length of the mirror. A number of such installations have been built, particularly in the USA, where such solar thermal installations provide over 350 MW of commercial electricity. Developments that are currently being pursued are of integration of solar and fossil fuel heat sources in combined cycle operation to enable continuous electricity provision throughout the day, and, in arid areas, of co-generation of power and heat for desalination for the delivery of fresh water.

Energy from the Sun: solar photovoltaics

Turning now to photovoltaic sources of electricity, solar panels on spacecraft have provided electrical power from the earliest days of space research 50 years ago. They now appear in a host of different ways in everyday life; for instance, as power sources for small calculators or watches or for lighting of public areas in remote places. Their efficiency for conversion of solar energy into electrical energy is now generally between 10% and 20%. A panel of cells of area 1 m^2 facing full sunlight will therefore deliver between 100 and 200 W of electrical power. A cost-effective way of mounting PV modules is on the surface of manufactured items or built structures rather than as free-standing arrays. In the fast-growing building-integrated-PV (BIPV) sector, the PV façade replaces and avoids the cost of conventional cladding. Installed on rooftops in cities, they provide a way for city dwellers to contribute renewably to their energy needs. Japan was the first to encourage rooftop solar installations and by 2000 had installed 320 MW capacity. Germany and the USA followed with large rooftop programmes, Germany with a target of 100 000 roofs that was met by 2003 and the USA with a target by 2010 of 1 million roofs. The cost of energy from solar cells has reduced dramatically over the past 20 years (see box) so that they can now be employed for a wide range of applications and providing the fall continues can also begin to contribute to the large-scale generation of electricity.

Solar energy schemes can be highly versatile in size or application. Small PV installations can provide local sources of electricity in rural areas especially in developing countries. About a third of the world's population have no access to electricity from a central source. Their predominant need is for small amounts of power for lighting, radio and television, refrigerators and air conditioning and for pumping water. The cost of PV installations for these purposes is now competitive with other means of generation (such as diesel units). Over the 20 years to the year 2000, over 1 million 'solar home systems' and 'solar

Local energy provision in Bangladesh

In a box earlier in the chapter on page 337, I outlined the components of a strategy for energy provision, one of which emphasised the value of local and distributed sources of energy as opposed to centralised sources in large units serving large grid networks.

An example of local provision is Grameen Shakti (meaning Rural Power) in Bangladesh – a subsidiary of Professor Yunus's famed Grameen Bank – that has developed an affordable solar home system (Figure 11.19) offered to rural communities through a soft credit facility.[54] From a small beginning in 1997, Grameen Shakti now powers over 135 000 homes. A biogas system is now also being offered, using poultry waste and cow dung to produce gas for cooking, ranging in cost from $US200 to $US1400 depending on its size – the larger one being appropriate for a cluster of homes. Training for local people who can be employed as technicians for installation, maintenance and operation of the systems is also provided. The availability of lighting and local energy is providing new business opportunities.

The availability of carbon credits through the Clean Development Mechanism is enabling the cost to be reduced bringing it within the range of some of the poorest people. Their aim is to have 1 million solar home systems and 1 million biogas systems by 2015.

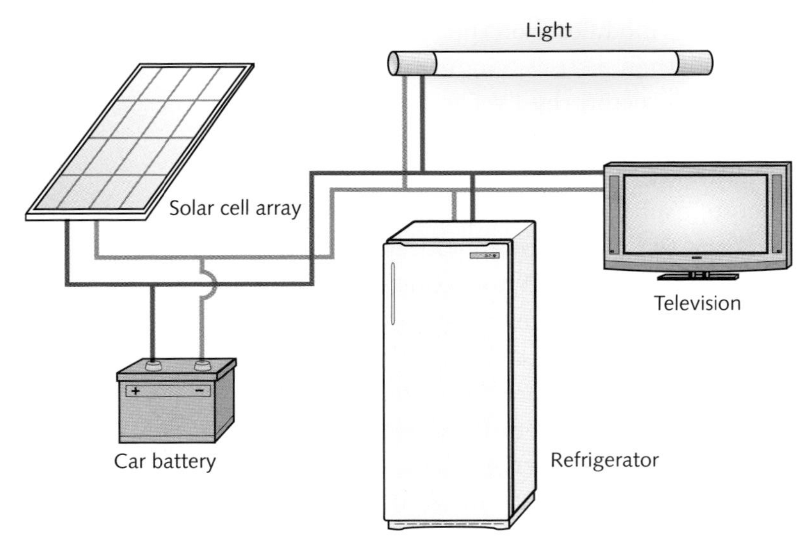

Figure 11.19 A simple 'solar home system' now being marketed in many countries in Africa, Asia and South America for a cost of a few hundred US dollars. An array of 36 solar cells, covering an area of 60 cm × 60 cm, provides around 40 W of peak power. This is sufficient to charge a car battery that can power fluorescent or LED lights, a few hours of radio and up to one hour of television per day. With more restricted use of these devices or with a larger solar array, a small refrigerator can be added to the system.

lanterns' were installed in Asia, Africa and South American countries.[53] Solar home systems provide typically 15–100 W from a solar array (Figure 11.19) and cost in the range of $US200–1200. Smaller 'solar lanterns' (typically 10 W) provide lighting only. Larger installations are required for public buildings. For instance small hospitals can benefit from a PV power source as small as 1–2 kW that, backed by batteries, can provide for lighting, refrigeration for vaccines, autoclave sterilisation, pumping for hot water (produced through a solar-thermal system) and radio. Many thousands of water pumps are now powered by solar PV and thousands of communities receive drinking water from solar-PV-powered purifiers/pumps. The potential for further growth and development of solar systems is clearly very large. For instance, mini-electrical grids powered by a combination of solar PV, wind, biomass and diesel are beginning to emerge especially in the remoter parts of China and India.

The total installed world capacity of PV grew from about 500 MW peak in 1998 to about 6000 MW peak in 2006, an increase of over 30% per year. With that rate of continued growth in both PV and CSP it should be possible to more than meet the projected growth of the contribution of solar energy to world energy supply by at least 1000-fold from today's levels to around 11% of global electricity production by 2050, as in the IEA BLUE Map scenario. In the short term, increased development of local installations is likely to have priority; later, with the expectation of a significant cost reduction (Figure 11.18), penetration into large-scale electricity generation will become more possible. Eventually, because of its simplicity, convenience and cleanliness, it is expected that electricity from solar PV sources will become one of the largest – if not the largest – of the world's energy sources.

Other renewable energies

We have so far covered the renewable energy sources for which there is potential for growth on a scale that can make a substantial contribution to overall world energy demand. We should also mention briefly other renewable energy technologies that contribute to global energy production and which are of particular importance in certain regions, namely geothermal energy from deep in the ground and energies from the tides, currents or waves in the ocean.

The presence of geothermal energy from deep down in the Earth's crust makes itself apparent in volcanic eruptions and less dramatically in geysers and hot springs. The temperature of the crust increases with depth and in favourable locations the energy available may be employed directly for heating purposes or for generating electrical power. Although very important in particular places, for instance in Iceland, it is currently only a small contributor (about 0.3%) to

Figure 11.20 A tidal stream turbine.

total world electricity; its contribution is estimated to rise by about a factor of 10 by 2050 in the IEA BLUE Map scenario.[55]

Large amounts of energy are in principle available in movements of the ocean; but in general they are not easy to exploit. Tidal energy is the only one currently contributing significantly to commercial energy production. It has the advantage over wind energy of being precisely predictable and of presenting few environmental or amenity problems. The largest tidal energy installation is a barrage across the estuary at La Rance in France; the flow from the barrage is directed through turbines as the tide ebbs so generating electricity with a capacity of up to 240 MW. Several estuaries in the world have been extensively studied as potential sites for tidal energy installations. The Severn Estuary in the UK, for instance, possesses one of the largest tidal ranges in the world and has the potential to generate a peak power of over 8000 MW and provide at least 6% of the total UK electricity demand. Other estuaries in the UK with tidal maxima at different times of day from the Severn could help to fill in the gaps when Severn power would not be available – so providing for a more continuous energy supply. Although the long-term cost of the electricity generated from the largest schemes could be competitive, the main deterrents to such schemes are seen to be the high capital upfront cost and the possibility of environmental impacts. However, the opportunity of harnessing such substantial quantities of long-term carbon-free power is now being taken seriously in the UK and in other countries such as China where also there are large tidal ranges.

Other proposals for tidal energy have been based on the construction of tidal 'lagoons' in suitable shallow regions offshore where there is a large tidal range.[56] Turbines in the lagoon walls would generate electricity as water flows in and out of the lagoons. Some of the environmental and economic problems of barrages built in estuaries might therefore be avoided.

The energy in tidal streams in coastal areas can be exploited in much the same way as wind energy from the atmosphere is harnessed (Figure 11.20). Although

Figure 11.21 A prototype of the Wave-Dragon – a device that generates ~7 MW of electrical power from the energy available in waves is in process of installation off the coast of southwest Wales, near Milford Haven. Waves break as they rise up a ramp facing them and enter a reservoir creating a small head. Energy is generated by running water down through turbines lower in the structure. The turbines are the only moving parts.

the speeds of water are lower than that of the wind, the greater density of sea water results in higher energy densities and requires smaller turbine diameters for similar power output. Substantial energy is also present in ocean waves. A number of ingenious devices have been designed to turn this into electrical energy (Figure 11.21)[57] and some are beginning to provide commercial power. However, because of the hostile ocean environment, early exploitation is comparatively costly. What is urgently needed for both tidal and wave energy is an adequate level of research, development and initial investment.

The waters around the coasts of Western Europe provide some of the best opportunities to exploit wave energy; for instance, tidal and wave energy together have the potential to provide up to 20% of the UK's electricity.

The support and financing of carbon-free energy

Energy free of carbon emissions on the scale required to meet any stabilisation scenario for carbon dioxide will only be realised if it is competitive in cost with energy from other sources. Under some circumstances renewable energy

Policy instruments

Action in the energy sector on the scale required to mitigate the effects of climate change through reduction in the emissions of greenhouse gases will require significant policy initiatives by governments in co-operation with industry. Some of these initiatives are the following:[61]

- putting in place appropriate institutional and structural frameworks;
- energy pricing strategies (carbon or energy taxes and reduced energy subsidies);
- reducing or removing other subsidies (e.g. agricultural and transport subsidies) that tend to increase greenhouse gas emissions;
- tradeable emissions permits (see Chapter 10, page 299);[62]
- voluntary programmes and negotiated agreements with industry;
- utility demand-side management programmes;
- regulatory programmes, including minimum energy efficiency standards (e.g. for appliances and fuel economy);
- stimulating R&D to make new technologies available;
- market pull and demonstration programmes that stimulate the development and application of advanced technologies;
- renewable energy incentives during market build-up;
- incentives such as provisions for accelerated depreciation or reduced costs for consumers;
- information dissemination for consumers especially directed towards necessary behavioural changes;
- education and training programmes;
- technological transfer to developing countries;
- provision for capacity building in developing countries;
- options that also support other economic and environmental goals.

sources are already competitive in cost, for instance in providing local sources of energy where the cost of transporting electricity or other fuel would be significant; some examples of this (such as Fair Isle in Scotland – see box above) have been given. However, when there is direct competition with fossil fuel energy from oil and gas, many renewable energies at the present compete only marginally. In due course, as easily recoverable oil and gas reserves begin to run out, those fuels will become more expensive enabling renewable sources to compete more easily. However, that is still some time away and for renewables to begin now to displace fossil fuels to the extent required, appropriate financial incentives must be introduced to bring about the change. Further to provide for carbon capture and storage from fossil-fuel power stations and for some energy-efficient measures, additional finance will also be necessary.

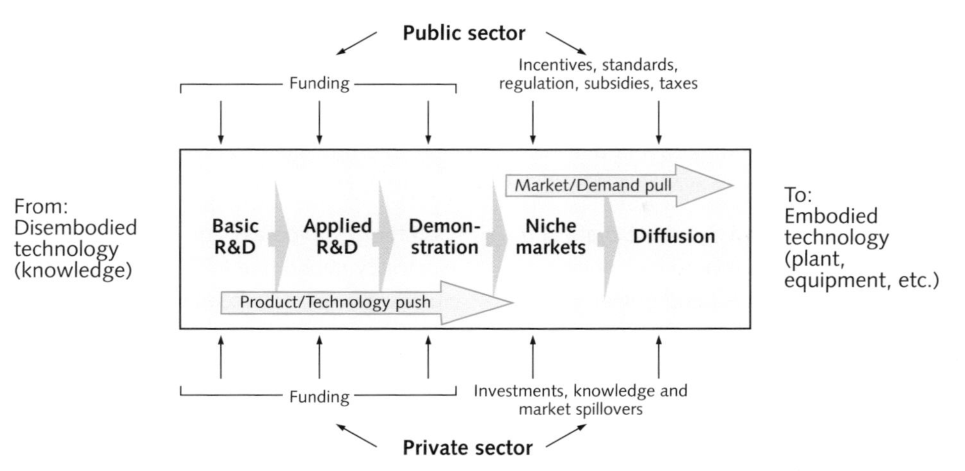

Figure 11.22 The process of technology development and its main driving forces.

As we saw in Chapter 9, the basis of such incentives would be the principle that the polluter should pay by the allocation of an environmental cost to carbon dioxide emissions. There are three main ways in which this can be done. Firstly, through a direct subsidy being provided by governments to carbon-free energy. Secondly, through the imposition of a carbon tax. Suppose, for instance, that through taxes or levies an additional cost of between $US25 and 50 per tonne of carbon dioxide (figures mentioned in the context of environmental costs towards the end of Chapter 9) were to be associated with carbon dioxide emissions, between 1 and 4 cents per kWh would be added to the price of electricity from fossil fuel sources– which could bring some renewables (for instance, biomass and wind energy) into competition with them.[58] It is interesting to note that in many countries substantial subsidies are attached to energy – worldwide they amount on average to the equivalent of more than $10 per tonne of carbon dioxide. A start with incentives would therefore be made if subsidies were removed from energy generated from fossil fuel sources (see box below).

A third way of introducing an environmental cost to fossil fuel energy is through tradeable permits in carbon dioxide emissions, as are being introduced under arrangements for the management of the Kyoto Protocol (Chapter 10, page 299). These control the total amount of carbon dioxide that a country or region may emit while providing the means for industries to trade permits for their allowable emissions within the overall total.

These fiscal measures are relatively easy to apply in the electricity sector. Electricity, however, only accounts for about one-third of the world's primary energy use. They also need to be applied to solid, liquid or gaseous fuels that

Table 11.2 Key mitigation technologies and practices by sector

Sector	Key mitigation technologies and practices currently commercially available	Key mitigation technologies and practices projected to be commercialised before 2030
Energy supply	Improved supply and distribution efficiency; fuel switching from coal to gas; nuclear power; renewable heat and power (hydropower, solar, wind, geothermal and bioenergy); combined heat and power; early applications of carbon capture and storage (CCS, e.g. storage of removed carbon dioxide from natural gas).	CCS for gas- biomass-and coal-fired electricity generating facilities; advanced nuclear power; advanced renewable energy, including tidal and wave energy, concentrating solar and solar PV.
Transport	More fuel-efficient vehicles; hybrid vehicles; cleaner diesel vehicles; biofuels; modal shifts from road transport to rail and public transport systems; non-motorised transport (cycling, walking); land-use and transport planning.	Second-generation biofuels; higher-efficiency aircraft; advanced electric and hybrid vehicles with more powerful and reliable batteries.
Buildings	Efficient lighting and daylighting; more efficient electrical applicances and heating and cooling devices; improved cooking stoves, improved insulation; passive and active solar design for heating and cooling; alternative refrigeration fluids, recovery and recycle of fluorinated gases.	Integrated design of commercial building including technologies such as intelligent meters that provide feedback and control; solar PV integrated in buildings.
Industry	More efficient end-use electrical equipment; heat and power recovery; material recycling and substitution; control of non-carbon dioxide gas emissions; and a wide array of process-specific technologies.	Advanced energy efficiency, CCS for cement, ammonia and iron manufacture; insert electrodes for aluminium manufacture,
Agriculture	Improved crop and grazing land management to increase soil carbon storage; restoration of cultivated peaty soils and degraded lands; improved rice cultivation techniques and livestock and manure management to reduce methane emissions; improved nitrogen fertiliser application techniques to reduce nitrous oxide emissions; dedicated energy crops to replace fossil fuel use; improved energy efficiency.	Improvements of crop yields.

Table 11.2 (Cont.)

Forestry/forests	Afforestation; reforestation; forest management; reduced deforestation; harvested wood product management; use of forestry products for bioenergy to replace fossil fuel use.	Tree species improvement to increase biomass productivity and carbon sequestration, Improved remote sensing technologies for analysis of vegetation/ soil carbon sequestration potential and mapping land use change.
Waste management	Landfill methane recovery; waste incineration with energy recovery; composting of organic waste; controlled waste water treatment; recycling and waste minimisation.	Biocovers and biofilters to optimise methane oxidation.

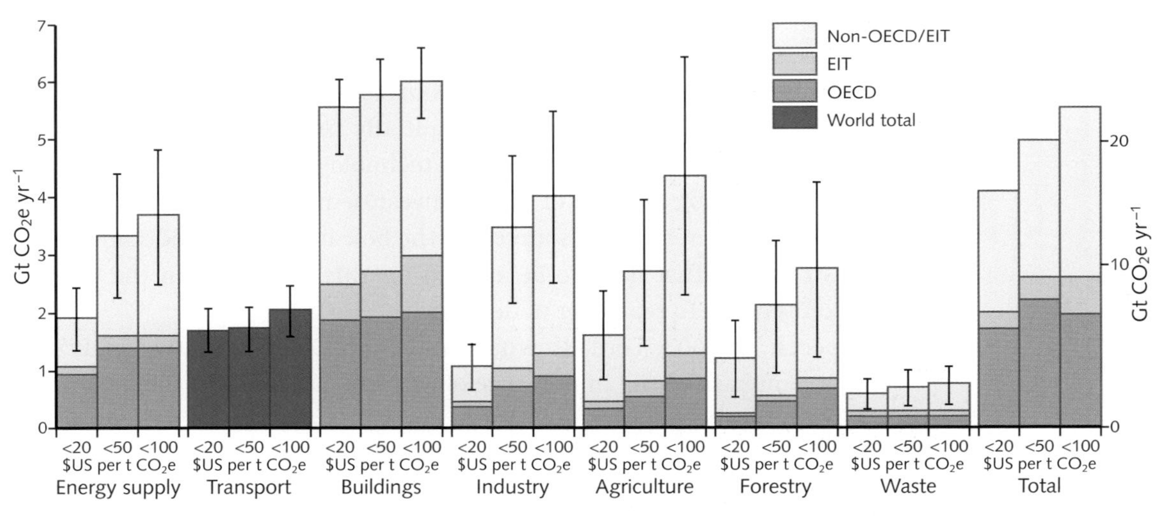

Figure 11.23 Estimates of the economic potential for global mitigation in 2030 for different sectors as a function of carbon price in terms of $US per tonne CO_2e. The ranges shown by the studies are indicated by vertical lines. Sectors used different baseline scenarios in between SRES B2 and A1B. Note that not all categories were included leading to an underestimation of the total economic potential of the order of 10–15%. Also electricity efficiency potentials were incorporated in the building and industry sectors. The three columns on the right combine the other columns into a total with a scale on the right-hand side.

are used for heating, industry and transport. It has already been mentioned that, currently, liquid fuels such as ethanol derived from biomass are significantly more expensive than those derived from fossil-fuel sources. Although there is an expectation that the processing of biomass will become more efficient – the rapid development of technologies in bioengineering will help – it is unlikely that the employment of biomass-derived fuels can occur on the scale required without the application of appropriate financial incentives.

There is a further crucial area where incentives are also required if renewable energy sources are going to come on stream sufficiently rapidly to meet the need. That is the area of research and development (R&D) – the latter is especially vital. Figure 11.22 illustrates how R&D fit into the normal process of technology development. Government R&D, averaged worldwide, currently runs at about $US10 billion per year or about 1% of worldwide capital investment in the energy industry of about $US1 trillion (million million) per year (about 3% of GWP). On average, in developed countries it has fallen by about a factor of 2 since the mid 1980s. In some countries the fall has been much greater, for instance in the UK where government-sponsored energy R&D fell by about a factor of 10 from the mid 1980s to 1998 when, in proportion to GDP, it was only one-fifth of that in the USA and one-seventeenth of that in Japan.[59] It is surprising – and concerning – that such falls in R&D have occurred at a time when the need to bring new renewable energy sources on line is greater than it has ever been. Energy R&D needs to be substantially increased and carefully targeted so as to enable promising renewable technologies to be introduced more quickly. An increasing fraction of capital investment in the energy industry is also needed for new renewable sources. In the box above are listed some of the policy instruments that need to be applied for this revolution in the way we generate our energy to really get under way.

In a speech in 2003, Lord Browne, the Group Chief Executive of British Petroleum, emphasised the importance of actively planning for the long term. After explaining the steps to be taken to combat change in the energy sector and the major investments that will be required he went on to say:[60]

> If such steps are to be taken, it is important to demonstrate the real value of taking a long-term approach which transcends the gap in time between the costs of investment and the delivery of the benefits. Political decisions are often taken on a very short-term basis and the challenge is to demonstrate the benefits of the actions which need to be taken for the long term ... The role of business is to transform the possibilities into reality. And that means being severely practical – undertaking very focused research and then experimenting with the different

possibilities. The advantage of the fact that the energy business is now global is that international companies can both access the knowledge around the world and can then apply it very quickly throughout their operations.

Mitigation technologies and potential in 2030

Table 11.2 summarises the various technologies and practices addressed in the last few sections (including also those relating to methane and forestry addressed in Chapter 10) and the possible contributions from different sectors both now and by 2030 to the required reductions in greenhouse gases.[63] Figure 11.23 shows a range of estimates from a number of studies for the mitigation of CO_2e emissions from the different sectors assuming different levels of the carbon price in terms of $US per tonne of CO_2e. Comparing these estimates with the reductions in 2030 shown in Figure 11.4 indicates that, for a carbon price in the range $US50–100 per tonne CO_2e, the required mitigation is achievable to lead to CO_2e stabilisation at 450 ppm CO_2e.

Technology for the longer term

This chapter has concentrated mostly on what can be achieved with available and proven technology during the next few decades. It is also interesting to speculate about the more distant future and what relatively new technologies may become dominant during the twenty-first century. In doing so, of course, we are almost certainly going to paint a more conservative picture than will actually occur. Imagine how well we would have done if asked in 1900 to speculate about technology change by 2000! Technology will certainly surprise us with possibilities not thought of at the moment. But that need not deter us from being speculative!

There is general agreement that a central component of a sustainable energy future is the fuel cell that with high efficiency converts hydrogen and oxygen directly into electricity (see box). In the fuel cell the electrolytic process of generating hydrogen and oxygen from water is reversed – the energy released by recombination of the hydrogen and oxygen is turned back into electrical energy. Fuel cells can have high-efficiency of 50–80% and they are pollution free; their only output other than electricity and heat is water. They offer the prospect of high-efficiency, small-scale power generation. They can be made in a large range of sizes suitable for use in transport vehicles or to act as local sources of electrical power for homes, for commercial premises or for many applications in industry. Much research and development has been put

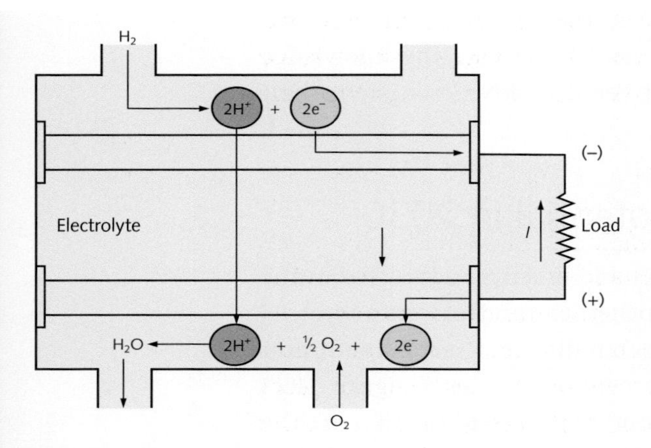

Figure 11.24 Schematic of a hydrogen–oxygen fuel cell. Hydrogen is supplied to the porous anode (negative electrode) where it dissociates into hydrogen ions (H^+) and electrons. The H^+ ions migrate through the electrolyte (typically an acid) to the cathode (positive electrode) where they combine with electrons (supplied through the external electrical circuit) and oxygen to form water.

Fuel cell technology

A fuel cell converts the chemical energy of a fuel directly into electricity without first burning it to produce heat.[64] It is similar to a battery in its construction. Two electrodes (Figure 11.24) are separated by an electrolyte which transmits ions but not electrons. Fuel cells possess high theoretical efficiency. Typical efficiencies in practice are in the range of 40–80%.

Hydrogen for fuel cells may be supplied from a wide variety of sources, from coal[65] or other biomass (see Note 31), from natural gas,[66] or from the hydrolysis of water using electricity generated from renewable sources such as wind power or solar photovoltaic (PV) cells (see box on page 364).

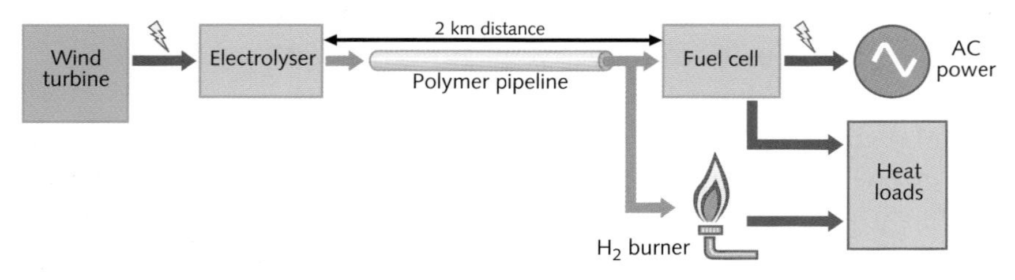

Figure 11.25 Remote area power system, employing hydrogen and fuel cell technology, supplying energy to a New Zealand farming community from a hill top wind turbine 2 km away. Electricity from the wind turbine is used to generate hydrogen by electrolysis of water. A polymer pipe 2 km long conveys the hydrogen to a fuel cell and/or hydrogen burner so providing heat and power to the farming community. The pipe not only provides a cheap way of transporting energy but also, by allowing pressure in the pipe to vary, provides a useful amount of hydrogen storage.

into fuel cells in recent years that has confirmed their potential as an important future technology. Although several technical challenges are yet to be resolved, there is an expectation that fuel cells will come into widespread use within the next decade.

Hydrogen for fuel cells can be generated from a wide variety of sources (see box). The most obvious renewable source is through the hydrolysis of water using electricity from PV cells exposed to sunlight or from wind turbines – an efficient process, over 80% of the electrical energy can be stored in the hydrogen. There are many regions of the world where sunshine or wind is plentiful and where suitable land not useful for other purposes would be readily available. The cost of electricity from PV or wind sources has been coming down rapidly (Figure 11.18) – a trend that will continue with technological advances and with increased scale of production. The IEA BLUE Map scenario assumes significant penetration of hydrogen fuel cell vehicles into the transport sector well before 2050.

Hydrogen is also important for other reasons. It provides a medium for energy storage and it can easily be transported by pipeline or bulk transport. An efficient local rural application is shown in Figure 11.25. For larger and more general applications, the main technical problem to be overcome is to find efficient and compact ways of storing hydrogen. Present technology (primarily in cylinders at high pressure) is bulky and heavy, especially for use in transport vehicles. A number of other possibilities are being explored. Other technologies for local energy storage, for instance, flywheels, super capacitors and superconducting magnetic energy storage (SMES), are also being actively explored.[67] As the

Power from nuclear fusion

When at extremely high temperatures the nuclei of hydrogen (or one of its isotopes, deuterium or tritium) are fused to form helium, a large amount of energy is released. This is the energy source that powers the Sun. To make it work on Earth, deuterium and tritium are used; from 1 kg, 1 GW can be generated for one day. The supply of material is essentially limitless and no unacceptable pollution is produced. A temperature of 100 million degrees Celsius is required for the reaction to occur. To keep the hot plasma away from the walls of the reaction vessel, it is confined by strong magnetic fields in a 'magnetic bottle' called a Tokamak. The challenges are to create effective confinement and a robust vessel.

Fusion power has been produced on Earth at levels up to 16 MW.[69] This has generated the confidence in a consortium of countries to build a new power-station-scale device called ITER capable of 500 MW with the object of demonstrating commercial viability. If this is successful, it is estimated that the first commercial plant could be in operation within 30 years.

drive towards energy efficiency becomes more vital, many of these will find appropriate applications.

Most of the technology necessary for a hydrogen energy economy is available now.[68] If its attractiveness from an environmental point of view were recognised as a dominant reason for its rapid development, a hydrogen economy could take off more rapidly than most energy analysts are currently predicting.

Iceland is a country that is in the forefront of the development of a hydrogen economy and aspires to be largely free of the use of fossil fuels by 2030–40. Much of its electricity already comes from hydroelectric or geothermal sources. The first hydrogen fuel station in Iceland was opened in April 2003 and several buses powered by fuel cells are its first customers.

Finally, in this section looking at the longer term, there is the possibility of power from nuclear fusion, the energy that powers the Sun (see box). If this can be harnessed, virtually limitless supplies of energy could be provided. The result of the next phase in this programme of work will be watched with great interest.

A zero carbon future

In Chapter 10 beginning on page 293, after an analysis of the increasing seriousness of the consequences of global warming as the temperature rises, arguments were put forward for setting a target for the increase of global average temperature of no greater than 2 deg C above its pre-industrial level. Inspection of the profile of CO_2 emissions in the IEA BLUE Map scenario published in *Energy Technology Perspectives* 2008 (Figures 11.4 and 10.3), which aims to meet this target, shows that the current increase in emissions, year on year, halts before 2015 after which emissions begin to fall substantially and continuously. As is eloquently stated by the IEA in their *World Energy Outlook* for 2008 (see box), to achieve such a profile will require extreme urgency and determination in applying the technologies and appropriate incentives included in the box on policy instruments on page 370 and in Table 11.2, Figure 11.23 and summarised also in Figures 11.26 and 11.27.

But will this be enough to meet the target of 2 °C? The following six assumptions and uncertainties were pointed out in Chapter 10:

1. Is a stabilisation level of 450 ppm CO_2e equivalent to the 2 degree target? It is in fact only a best estimate providing a 50% chance of success.
2. Because 20% of global greenhouse gas emissions currently arise from tropical deforestation, the IEA BLUE Map scenario for emissions from the energy sector will not get close to meeting the 2°C target unless a slowing of tropical

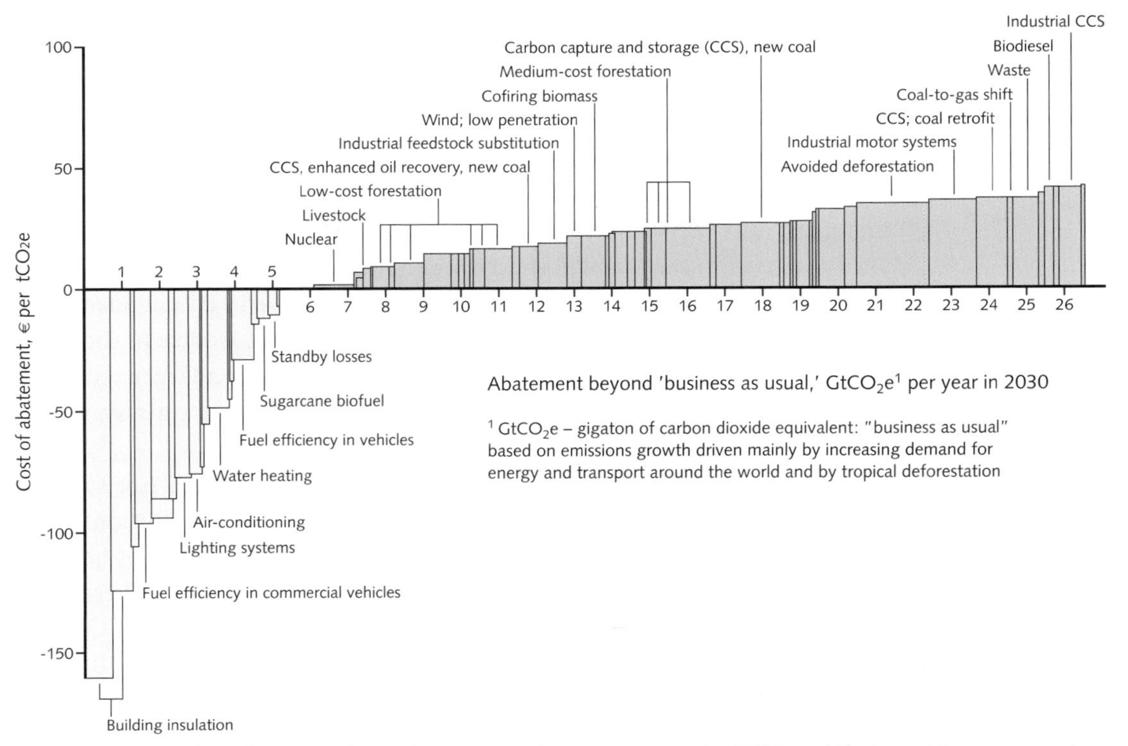

Figure 11.26 Technical options for reducing greenhouse gases up to 2030 and their cost in euros per tonne of carbon saved as estimated by workers in the Stockholm office of McKinsey & Company. Only some of the main options are labelled. Options below the line produce net savings, above the line net costs. The x-axis shows abatement for each of the options below 'business as usual' in $GtCO_2e$ per year. By 2030 they total 26 $GtCO_2e$ per year, enough to meet the 450 ppm stabilization curve shown in Figure 11.27.

Figure 11.27 Waymarks for annual global energy carbon di oxide emissions road map to 2050 showing International Energy Agency (IEA) Reference scenario (red) and a profile (green) aimed at targets of < 2°C temperature rise from pre-industrial and 450 ppm CO_2 stabilisation (cf Figure 10.3). The division between developed and developing countries from today until 2050 is a construction based on the developed countries' share, compared with that of developing countries, peaking earlier and reducing further e.g. by at least 80% by 2050..

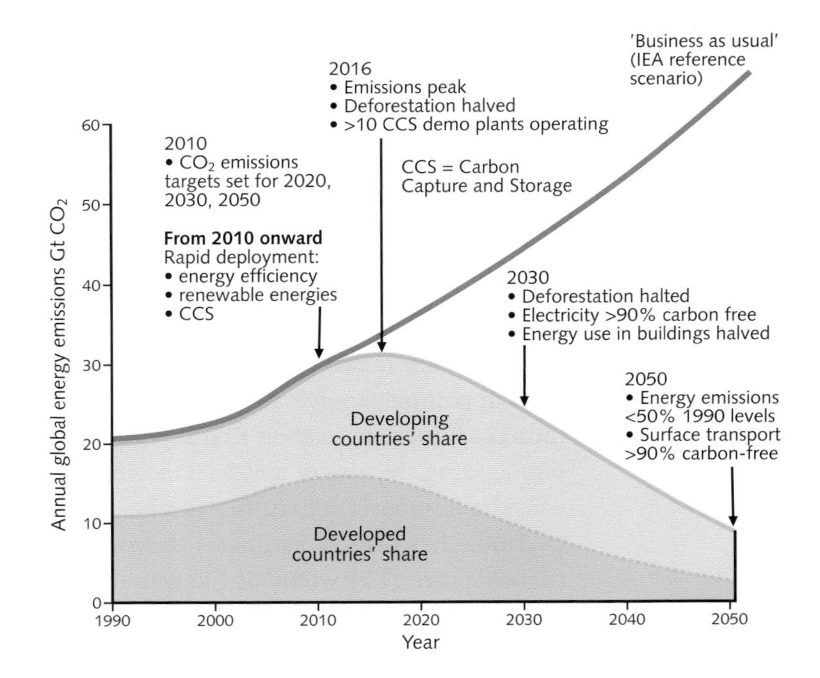

deforestation begins now with its complete halting over the next twenty or thirty years (Figure 10.3).

3. The uncertainty in climate-carbon-cycle feedback presented in Figure 10.3 shows that, even if tropical deforestation is halted completely, emissions reductions in the energy sector as in the IEA BLUE Map scenario do not adequately allow for that feedback. To allow for likely values of that feedback, the emissions reduction curve constructed in Figure 11.27 assumes a larger reduction (from 1990, of 60% in global emissions or 50% in energy sector emissions) than that provided in the IEA BLUE Map scenario.
4. Cooling from aerosols (Chapter 10 page 313–14) provides an offset to some of the present warming. As pressure to reduce atmospheric pollution grows and as the use of coal and oil is phased out, the concentration of aerosols could reduce in future years faster than most scenarios suggest. These aerosol reductions will have to be matched by matching reductions in CO_2 targets. But since reductions in aerosols take effect immediately (the lifetime of aerosols in the lower atmosphere is only a few days), these matching CO_2 reductions need to be anticipated. There is a need to prepare now for that anticipation, for instance, by researching the most cost effective means of removing substantial amounts of CO_2 from the atmosphere as mentioned on page 315.
5. It was assumed in Chapter 10 that some reduction of the concentration of other greenhouse gases, namely methane, nitrous oxide and halocarbons, from their 2005 levels would be possible and would compensate at least partially for the likely reductions in aerosols. Measures leading to these reductions need to be put in place but there is substantial uncertainty regarding their effectiveness.
6. Given only a 50% chance of meeting the 2°C target, what are the chances of avoiding higher global average temperature rises that would bring much more severe consequences? For instance (see Tables 7.1, 7.5 and 7.6), with a 4°C target, irreversible melting of some of the polar ice sheets becomes much more likely as does the possibility of changes in the large scale ocean circulation – in addition to much more severe impacts from extreme events. This is a point raised by the UK Climate Change Committee, the independent body appointed to give advice to the UK government about climate change targets and action, in their first report of December 2008.[70] In addition to recommending the aim of a 50% chance of no more than 2°C by 2100, they also consider it important to fulfil a further aim of less than 1% chance of 4 degrees by 2100. They present estimates showing that an emissions profile similar to that of Figure 11.27 would be likely to meet these criteria although the carbon dioxide equivalent concentration could reach at least 500 ppm by 2100 before

IEA World Energy Outlook 2008

In chapters 10 and 11, I have already referred extensively to the work of the International Energy Agency (IEA) and to their *Energy Technology Perspectives* published in June 2008 that outlines an energy future up to 2050 that aims at stabilization of atmospheric carbon dioxide consistent with a rise in global surface temperature of less than 2 °C. The IEA's annual *World Energy Outlook* (WEO) is the most important world publication in the energy field. The WEO published in November 2008 includes extensive analysis of greenhouse gas emissions from energy production. There follow some quotes from its Executive Summary, beginning with the opening sentences. The italics are theirs.

> The world's energy system is at a crossroads. Current global trends in energy supply and consumption are patently unsustainable – environmentally, economically, socially. But that can – and must – be altered; *there's still time to change the road we're on.* It is not an exaggeration to claim that the future of human prosperity depends on how successfully we tackle the two central energy challenges facing us today; securing the supply of reliable and affordable energy; and effecting a rapid transformation to a low-carbon, efficient and environmentally benign system of energy supply. What is needed is nothing short of an energy revolution.
>
> Preventing catastrophic and irreversible damage to the global climate ultimately requires a major decarbonisation of the world energy sources. The 15th conference of the Parties to be held in Copenhagen in November 2009, provides a vital opportunity to negotiate a new global climate change policy regime for beyond 2012 … The consequences for the global climate of policy inaction are shocking … The road from Copenhagen must be paved with more than good intentions.
>
> Any agreement will have to take into account the importance of a handful of major emitters. The five largest emitters of energy-related CO_2 – China, the United States, the European Union, India and Russia – together account for almost two-thirds of global CO_2 emissions … The contributions to emissions reductions made by China and the United States will be critical to reaching a stabilisation goal.
>
> The energy sector will have to play the central role in curbing emissions – through major improvements in efficiency and rapid switching to renewables and other low-carbon technologies such as carbon capture and storage (CCS) … Governments have to put in place appropriate financial incentives and regulatory frameworks that support both energy security and climate policy goals in an integrated way.
>
> The scale of the challenge in the 450 Policy Scenario … The technology shift, if achievable, would certainly be unprecedented in scale and speed of deployment. Increased public and private spending on research and development in the near term would be essential to develop the advanced technologies needed to make the 450 Policy Scenario a reality is immense.
>
> It is within the power of all governments, of producing and consuming countries alike, acting alone or together, to steer the world towards a cleaner, cleverer and more competitive energy system. Time is running out and the time to act is now.

Energy policy in the UK

A number of important reports concerned with energy policy have been published in the UK since the year 2000.

The first of these is *Energy in a Changing Climate* published in 2000 by the Royal Commission on Environmental Pollution (RCEP)[74] – an expert body that provides advice to government. It supported the concept of 'contraction and convergence' (Figure 10.5) as a basis for future international action to reduce greenhouse gas emissions and pointed out that application of this concept would imply a goal of 60% reduction in UK emissions of greenhouse gases by 2050. To achieve such a goal more effective measures are needed to increase energy efficiency (especially in buildings) and to encourage the growth of renewable energy sources especially by greatly increased R&D.

An *Energy Review* by the Policy and Innovation Unit (PIU) of the UK Cabinet Office,[75] published in 2002, provided input into an Energy White Paper, *Our Energy Future: Creating a Low Carbon Economy*, a policy statement by the UK government in 2003[76] that accepted the goal set by the RCEP of a 60% reduction in emissions by 2050. An estimate in the PIU review of the cost to the UK economy of realising the RCEP goal is expressed as a possible slowing in the growth of the UK economy of six months over the 50-year period.

A further Energy White Paper was introduced in 2007 and in November 2008, a Climate Change Bill was passed by Parliament that legislates mandatory targets of an 80% reduction of national CO_2e emissions from 1990 levels by 2050 and of 26% by 2020.[77] It also provides for revision of these targets should the Climate Change Committee, that is also set up by the Bill, deem it necessary. The challenge to the government and the country is the practical realization of these targets.

turning down in the 22nd century. However, they also show that emissions profiles that peak later than 2016 are unlikely to meet either of the criteria.[71] As mentioned in chapter 10, page 315, similar arguments about the danger of even smaller increases of global average temperature – more than around 1 °C – have led James Hansen to argue that action needs to be organised to remove carbon dioxide from the atmosphere to bring its concentrations back to around 350 ppm.[72]

Taking all these six points on board emphasises emphatically that fundamental to the scientific story I have presented is the move to a *zero carbon future* with no significant net anthropogenic emissions of greenhouse gases into the atmosphere. Over the past two decades as our scientific understanding has grown, it has been increasingly realised that we must make this move as quickly as possible. A path towards achieving large reductions has been demonstrated by the IEA and other bodies. It is achievable, affordable and will bring with it many co-benefits. The immediate challenge is to ensure that global CO_2 emissions peak well before 2020 and to begin now to work towards zero carbon by or even before 2050[73] – which implies even tougher action than that presented in Figure 11.27.

SUMMARY

This chapter has outlined the ways in which energy for human life and industry is currently provided. Growth in conventional energy sources at the rate required to meet the world's future energy needs will generate greatly increased emissions of greenhouse gases that will lead to unacceptable climate change. Such would not be consistent with the agreements reached at the United Nations Conference on Environment and Development at Rio de Janeiro in June 1992 when the countries of the world committed themselves to the action necessary to address the problems of energy and the environment. The objective of the Climate Convention agreed at that Conference requires that emissions of carbon dioxide be drastically reduced so that the concentration of carbon dioxide in the atmosphere is stabilised by the end of the twenty-first century. As the IEA have stated in their World Energy Outlook 2008, 'What is needed is nothing short of an energy revolution' on a global scale. Necessary areas of action are the following.

- Many studies have shown that in most developed countries improvements in energy efficiency of 30–50% or more can be achieved at little or no net cost and often with overall saving (see Figure 11.26). But industry and individuals will require not just encouragement, but incentives if the savings are to be realised.
- A long-term energy strategy needs to be formulated nationally and internationally, that addresses economic considerations alongside environmental and social ones that takes into account *inter alia* the need for development of local energy source as well as centralised ones and the requirement for energy security.
- Since no technology can provide a 'silver bullet' solution, all possibilities for low-carbon energy must be explored and appropriately developed so as to realise all effective contributions as rapidly as possible. Essential to this process will be much increased investment in Research/Development.
- To stem the rapid growth of carbon dioxide emissions from coal and gas fired power stations, carbon capture and storage (CCS) must be installed aggressively and urgently in new power stations and retrofitted where possible in existing stations.
- Much of the necessary technology is available for renewable energy sources (especially 'modern' biomass, wind and solar energy) to be developed and

implemented, so as to replace energy from fossil fuels. For this to be done on an adequate scale, an economic framework with appropriate incentives needs urgently to be set up. Policy options available include the removal of subsidies, carbon or energy taxes (which recognise the environmental cost associated with the use of fossil fuels) and tradeable permits coupled with capping of emissions.

- To meet the targets of 2 °C global average temperature rise and 450 ppm carbon dioxide equivalent stabilisation described in Chapter 10, rapid decarbonisation must take place in all sectors of energy generation and use, the aim being that before 2050 global electricity provision and most transport must be carbon free (Figure 11.27). The overall aim is for a zero carbon future to be realised as quickly as possible.
- Arrangements are urgently needed to ensure that technology is available for all countries (including developing countries through technology transfer) to develop their energy plans with high efficiency and to deploy renewable energy sources (for instance, local biomass, solar energy or wind generators) as widely as possible.
- World investment in the energy industry up to 2050 (including by consumers in capital equipment that consumes energy, e.g. in motor vehicles) in the IEA Reference scenario (i.e. business-as-usual) is estimated to be about $US250 trillion (million million) or about 6% of cumulative world GDP over the period. For the IEA Blue Map scenario the additional investment needs are estimated as $US45 trillion, an increase of 18% over the Reference scenario. The IEA also point out that, compared with the Reference scenario, the BLUE Map scenario will result in fuel savings over the period 2005 to 2050, amounting to about $US50 trillion – of the same order as the increased investment required.

These actions imply a technological revolution on a scale and at a rate of change much greater than any the world has yet experienced – a revolution that involves the whole world community working together in unprecedented cooperation both in bringing it about and enjoying its benefits. It demands clear policies, commitment and resolve on the part of governments, industries and individual consumers. Because of the long life time of energy infrastructure (e.g. power stations) and also because of the time required for the changes required to be realised, there is an inescapable urgency about the actions to be taken. As the World Energy Council pointed out over 15 years ago, 'the real challenge is to communicate the reality that the switch to alternative forms of supply will take many decades, and thus the realisation of the need, and commencement of the appropriate action, must be *now*' (their italics).[78] If that was true in 1993 it even more true *now* (my italics!).

QUESTIONS

1 Estimate how much energy you use per year in your home or your apartment. How much of this comes from fossil fuels? What does it contribute to emissions of carbon dioxide?
2 Estimate how much energy your car uses per year. What does this contribute to emissions of carbon dioxide?
3 Look up estimates made at different times over the last 30 years of the size of world reserves of coal, oil and gas. What do you deduce from the trend of these estimates?
4 Estimate annual energy saving for your country as a result of: (1) unnecessary lights in all homes being switched off; (2) all homes changing all light bulbs to low-energy ones; (3) all homes being maintained 1°C cooler during the winter.
5 Find out for your country the fuel sources that contribute to electricity supply. Suppose a typical home heated by electricity in the winter is converted to gas heating, what would be the change in annual carbon dioxide emissions?
6 Find out about the cost of heat pumps and building insulation. For a typical building, compare the costs (capital and running costs) of reducing by 75% the energy required to heat it by installing heat pumps or by adding to the insulation.
7 Visit a large electrical store and collate information relating to the energy consumption and the performance of domestic appliances: refrigerators, cookers, microwave ovens and washing machines. Which do you think are the most energy efficient and how do they compare with the least energy efficient? Also how well labelled were the appliances with respect to energy consumption and efficiency?
8 Consider a flat-roofed house of typical size in a warm, sunny country with a flat roof incorporating 50 mm thickness of insulation (refer to Table 11.1). Estimate the extra energy that would have to be removed by air conditioning if the roof were painted black rather than white. How much would this be reduced if the insulation were increased to a thickness of 150 mm?
9 Rework the calculations of total heating required for the building considered in Table 11.1 supposing insulation 250 mm thick (the Danish standard) were installed in the cavity walls and in the roof.
10 Look up articles about the environmental and social impact of large dams. Do you consider the benefits of the power generated by hydroelectric means are worth the environmental and social damage?

11 Suppose an area of 10 km^2 was available for use for renewable energy sources, to grow biomass, to mount PV solar cells or to mount wind generators. What criteria would determine which use would be most effective? Compare the effectiveness for each use in a typical area of your country.
12 What do you consider the most important factors that prevent the greater use of nuclear energy? How do you think their seriousness compares with the costs or damages arising from other forms of energy production?
13 In the IPCC 1995 Report, Chapter 19, you will find information about the LESS scenarios. In particular, estimates are provided, for different alternatives, of the amount of land that will be needed in different parts of the world for the production of energy from biomass. For your own country or region, find out how easily, on the timescale required, it is likely that this amount of land could be provided. What would be the likely consequences arising from using the land for biomass production rather than for other purposes?
14 In making arguments for a carbon tax would you attempt to relate it to the likely cost of damage from global warming (Chapter 9), or would you relate it to what is required to enable appropriate renewable energies to compete at an adequate level? Find recent cost information about different renewable energies and estimate the level of carbon tax that would enable there to be greater employment of different forms of renewable energy: (1) at the present time, (2) in 2020.
15 In discussing policy options, attention is often given to 'win–win' situations or to those with a 'double dividend', i.e. situations in which, when a particular action is taken to reduce greenhouse gas emissions, additional benefits arise as a bonus. Describe examples of such situations.
16 Of the policy options listed towards the end of the chapter, which do you think could be most effective in your country?
17 List the various environmental impacts of different renewable energy sources, biomass, wind, solar PV and marine (tidal and wave). How would you assess the seriousness of these impacts compared with the advantages to the environment of the contribution from these sources to the reduction of greenhouse gas emissions?
18 Discuss the advantages and disadvantages of local energy sources as opposed to centralised energy provision through large grid networks. Identify locations in the world known to you where local or centralised would be most appropriate.
19 Compare how the elements in the Kaya Identity – energy intensity, carbon intensity, GDP per capita and population–have changed in a sample group

of countries over the last 20 years. Comment on the differences between the countries and the possible reasons for them.

20 In a speech on Energy in Washington on 17 July 2008, Al Gore referred to striking examples of deliberate and rapid actions taken by US Presidents in the past – by Franklin Roosevelt in the Lease–Lend programme in 1941 and by John F. Kennedy who launched the Apollo project in 1961. He proposed that deliberate action should now be taken by the United States to tackle the problem of climate change and proposed, for instance, that a target for the US could be set of carbon-free electricity within the next 10 years. Reflect on the benefits to the world and to the United States of these particular actions in the past and compare them with the challenge of climate change now.

21 Suppose the net cooling from aerosols were halved from 2050 onwards and that greenhouse gases other than CO_2 were still at their 1990 levels, estimate the change in profile of further reductions of CO_2 that would be necessary to maintain the 450 ppm CO_2e stabilisation target. Does your answer require withdrawal of CO_2 already in the atmosphere between 2050 and 2100 (refer to Figure 10.3)? If so, make an estimate of how much and investigate what means might accomplish it.

FURTHER READING AND REFERENCE

Metz, B., Davidson, O., Bosch, P., Dave, R., Meyer, L. (eds) 2007. *Climate Change 2007: Mitigation of Climate Change. Contribution of Working Group III to the Fourth Assessment Report of the Intergovernmental Panel on Climate Change.* Cambridge: Cambridge University Press.

Technical Summary
Chapter 2 Framing issues (e.g. links to sustainable development, integrated assessment)
Chapter 3 Issues relating to mitigation in the long-term context
Chapter 4 Energy supply
Chapter 5 Transport and its infrastructure
Chapter 6 Residential and commercial buildings
Chapter 7 Industry
Chapter 10 Waste management
Chapter 11 Mitigation from a cross-sectoral perspective
Chapter 12 Sustainable development and mitigation

International Energy Agency. *World Energy Outlook 2007.*

Part A Global energy prospects
Part B China's energy prospects
Part C India's energy prospects

International` Energy Agency, *World Energy Outlook 2008* (published in November 2008, about 500 pages including many figures; contains detail upto 2030 of the IEA Reference scenario and of Scenarios designed to meet the challenge of the environment and climate change.)

International Energy Agency, *Energy Technology Perspectives 2008* (chapters on energy scenarios, different sectors and technologies)

Stern, N. 2006. *The Economics of Climate Change*. Cambridge: Cambridge University Press. The Stern Review: especially chapters in Part III on the economics of mitigation.

Boyle, G., Everett, R., Ramage, J. (eds.) 2003. *Energy Systems and Sustainability: Power for a Sustainable Future*. Oxford: Oxford University Press.

Scientific American, Special Issue on Energy's future beyond carbon, September 2006 (including novel technologies and a hydrogen future).

Monbiot, G. 2007 *Heat: How to Stop the Planets from Burning*. London: Allen Lane. A lively presentation of some of the technical, political and personal dilemmas.

NOTES FOR CHAPTER 11

1 1 toe = 11.7 MWh = 4.19×10^{10} J;
1 Gtoe = 1.9×10^{18} J = 1.9 EJ;
1 toe per day = 485 kW;
1 toe per year = 1.33 kW.

2 Report of G8 Renewable Energy Task Force, July 2001.

3 International Energy Agency. 2008. *World Energy Outlook 2008*. Paris: International Energy Agency.

4 International Energy Agency. 2008. *Energy Technology Perspectives*. Paris: International Energy Agency.

5 *Ibid*., p. 55ff.

6 *Ibid*., pp. 127ff.

7 *Ibid*., pp. 221ff.

8 Socolow, R.H., Pacala, S.W. 2006. *Scientific American*., **295**, 28–35.

9 See Chapter 9, page 272.

10 That government expenditure on energy R&D in the UK is now less than 5% of what it was 20 years ago provides an illustration of lack of commitment or urgency on the government's part.

11 IEA, *World Energy Outlook*, 2006, Table 14.6.

12 World Energy Council. 1993. *Energy for Tomorrow's World: The Realities, the Real Options and the Agenda for Achievement*. New York: World Energy Council, p. 122.

13 WEC, *Energy for Tomorrow's World*, p. 113.

14 Ways of achieving large reductions in all these sectors are described by von Weizacker, E., Lovins, A. B., Lovins, L. H. 1997. *Factor Four, Doubling Wealth: Halving Resource Use*. London: Earthscan.

15 More detail of heat pumps and their applications in Smith, P. F. 2003. *Sustainability at the Cutting Edge*. London: Architectural Press, pp. 45–50.

16 From National Academy of Sciences. 1992. *Policy Implications of Greenhouse Warming*. Washington, DC: National Academy Press, Chapter 21.

17 Smith, P.F. 2007. *Sustainability at the Cutting Edge*, second edition. Amsterdam: Elsevier.

18 *ibid*, pp. 135–7.

19 Smith, P.F. 2001. *Architecture in a Climate of Change*. London: Architectural Press.

20 See, for instance, von Weizacker, E., Lovins, A. B., Lovins, L. H. 1997. *Factor Four, Doubling Wealth: Halving Resource Use*. London: Earthscan, pp. 28–9.

21 For instance Roaf, S. *et al*. 2001. *Ecohouse: a Design Guide*. London: Architectural Press.

22 See www.zedfactory.com/bedzed/bedzed.html.

23 Note that the air transport fraction should be at least doubled to allow for the effects of increased high cloud mentioned in Chapter 3 on page 63.

24 See Mobility Report of World Business Council on Sustainable Development:www.wbcsd.ch.

25 More detail in Moomaw, W.R., Moreira 2001. Section 3.4, in Metz, J.R., Davidson, O., Swart, R., Pan, J., (eds.) 2001. *Climate Change 2001: Mitigation*.

26 From the Summary for policymakers, in Penner, J., Lister D., Griggs, D.J., Dokken, D.J., Mcfarland, M. (eds.) 1999. *Aviation and the Global Atmosphere. A Special Report of the IPCC*. Cambridge: Cambridge University Press.

27 For more detail on industrial emissions and possible reductions see *Energy Technology Perspectives*, IEA, Chapter 12.

28 From speech by Lord Browne, BP Chief Executive to the Institutional Investors Group, London, 26 November 2003.

29 For example, British Sugar with an annual turnover in 1992 of £700 million spent £21 million per year on energy. Through low-grade heat recovery, co-generation schemes and better control of heating and lighting, the spend on energy per tonne of sugar had been reduced by 41% from that in 1980 (example quoted in *Energy, Environment and Profits*. 1993. London: Energy Efficiency Office of the Department of the Environment).

30 See International Energy Agency, *Capturing CO_2*: www.ieagreen.org.uk; also Furnival S. 2006 Carbon capture and storage, *Physics World*, **19**, 24–27. Also see *IPCC Special Report*. Metz, B. *et al.* (eds.) 2005 *Carbon Dioxide Capture and Storage*. Cambridge: Cambridge University Press. Also available from www.ipcc.ch.

31 Carbonaceous fuel is burnt to form carbon monoxide, CO, which then reacts with steam according to the equation $CO + H_2O$ = carbon dioxide + H_2.

32 International Energy Agency, *Energy Technology Perspectives 2008*, pp. 134–5

33 See *Scientific American*, **295**, 52–9, 2006.

34 In countries such as the UK, there are substantial quantities of plutonium now in surplus from military programmes that could be used in nuclear power stations (and degraded in the process) – assisting with greenhouse gas reductions in the medium term and not adding to the proliferation problem. See Wilkinson, W. L. 2001. Management of the UK plutonium stockpile: the economic case for burning as MOX in new PWRs. *Interdisciplinary Science Reviews*, **26**, 303–6.

35 'Large' hydro applies to schemes greater than 10 MW in capacity; 'small' hydro to schemes smaller than 10 MW.

36 For more information see IEA, *Energy Technology Perspectives*, Chapter 12, from where the numbers for hydro potential quoted here have been taken.

37 See review by Loening, A. 2003. Landfill gas and related energy sources; anaerobic digesters; biomass energy systems. In *Issues in Environmental Science and Technology*, No. 19. Cambridge: Royal Society of Chemistry, pp. 69–88.

38 Moomaw and Moreira, Section 3.8.4.3.2 in Metz *et al.* (eds.) *Climate Change 2001: Mitigation*.

39 Twidell, J., Weir, T. 1986. *Renewable Energy Resources*. London: E. and F. Spon, p. 291.

40 See review by Loening A. 2003.

41 From *Report of the Renewable Energy Advisory Group*, Energy Paper No. 60. 1992. London: UK Department of Trade and Industry.

42 These projects are supported by the Shell Foundation (www.shellfoundation.org), a charity set up to promote sustainable energy for the Third World.

43 See *Incineration of Waste*. 1993. 17th Report of the Royal Commission on Environmental Pollution. London: HMSO, pp. 43–7.

44 *Sustainable Biofuels: Prospects and Challenges*, report by Royal Society of London 2008: www.royalsoc.org.

45 Tollefson, J. 2008. Not your father's biofuels. *Nature*, **451**, 880–3. From first to second generation biofuel technologies, International Energy Agency, IEA, paris 2008, for a discussion of the technology's status, also available at http://www.iea.org/Textbase/publications/free_new_desc.acp?PUBS_ID=2074

46 See Infield, D., Rowley, P. 2003. Renewable energy: technology considerations and electricity integration. *Issues in Environmental Science and Technology*, No. 19. Cambridge: Royal Society of Chemistry, pp. 49–68.

47 Martinot, E. *et al.* 2002. Renewable energy markets in developing countries. *Annual Review of Energy and the Environment*, **27**, 309–48.

48 Twidell and Weir, *Renewable Energy Resources*, p. 252.

49 Martinot *et al.* 2002.

50 Smith, P.F. 2001. *Architecture in a Climate of Change*. London: Architectural Press.

51 For a summary of current technology see IEA, *Energy Technology Perspectives*, IEA, Chapter 11.

52 See solar energy news feature in *Nature*, 2006, **443**, 19–24.

53 Martinot *et al.* 2002.

54 http://www.grameen-info.org/grameen/gshakti/index.html.

55 Renewables for Heating and Cooling, 2007, International Energy Agency, Paris, available at: http://www.iea.org/Textbase/publications/free_new_desc.asp/?PUBS_ID=1975

56 See www.tidalelectric.com.

57 See Boyle, G. 1996. *Renewable Energy Power for a Sustainable Future*. Oxford: Oxford University Press.

58 See Elliott, D. 2003. Sustainable energy: choices, problems and opportunities. *Issues in Environmental Science and Technology* No. 19. Cambridge: Royal Society of Chemistry, pp. 19–47.

59 *Energy: The Changing Climate*. 2000. 22nd Report of the Royal Commission on Environmental Pollution. London: Stationery Office, p. 81.

60 From speech by Lord Browne, BP Chief Executive, to the Institutional Investors Group, London, 26 November 2003.

61 Based on Summary for policymakers. Section 4.4, in Watson, R.T., Zinyowera, M.C., Moss, R.H. (eds.) 1996. *Climate Change 1995: Impacts Adaptations and Mitigations of Climate Change: Scientific–Technical Analyses. Contribution of Working Group II to the Second Assessment Report of the Intergovernmental Panel on Climate Change*. Cambridge: Cambridge University Press.

62 See Mullins, F. 2003. Emissions trading schemes: are they a licence to pollute? *Issues in Environmental Science and Technology* No.19. Cambridge: Royal Society of Chemistry, pp. 89–103.

63 The material for this section comes from Chapter 11, Section 11.3 (also summarised in the Summary for Policymakers) of Metz *et al.* (eds.) *Climate Change 2007: Mitigation*.

64 For a recent review see Eikerling, M. *et al.* 2007. *Physics World*, **20**, 32–6.

65 For hydrogen from coal see Liang-Shih Fan. 2007, *Physics World*, **20**, 37–41.

66 By reacting natural gas (methane CH_4) with steam through the reaction $2H_2O + CH_4 = CO_2 + 4H_2$.

67 See Swarup, R. 2007. *Physics World*, **20**, 42–5.

68 See *Scientific American*, **295**, 70–77, 2006.

69 McCraken, G., Stott, P. 2004. *Fusion, the Energy of the Universe*. New York: Elsevier/Academic Press.

70 The Committee on Climate Change, www.theccc.org.uk/reports, Inaugural Report December 2008, *Building a low-carbon economy – the UK's contribution to tackling climate change, Part 1, the 2050 target*.

71 The IEA *World Energy Outlook* published in December 2008 presents a 450 ppm scenario that differs from that in their *Energy Technology Perspectives* of June 2008. In particular, without extensive early retirement of existing power stations and other energy capital, it is not considered possible to achieve a peak of CO_2 emissions before 2020. Therefore, although still demanding, their new scenario weakens the possibility of achieving the 2 °C target and fails to meet the criteria set by the UK Committee on Climate Change.

72 James Hansen, Bjerknes Lecture at American Geophysical Union, 17 December 2008, www.columbia.edu/~jeh1/2008/AGUBjerknes_20081217.pdf.

73 see for instance, www.climatesafety.org; or www.zerocarbonbritain, published by Centre for Alternative Technology, Machynlleth, Wales.

74 www.rcep.org.uk

75 www.cabinet-office.gov.uk/innovation/2002/energy/report/index.htm

76 www.dti.gov.uk/energy/whitepaper/index.shtml

77 www.defra.gov.uk

78 From Energy for *Tomorrow's World: the Realities, the Real Options and the Agenda for Achievement*. WEC Commission Report. NewYork: World Energy Council, 1993, p. 88.

79 From *Energy for Tomorrow's World: the Realities, the Real Options and the Agenda for Achievement*. WEC Commission Report. NewYork: World Energy Council, 1993, p.88.

The global village

12

Claude Monet's views of the River Thames and the Houses of Parliament show the sun struggling to shine through London's smog-laden atmosphere (1904).

THE PRECEDING chapters have considered the various strands of the global warming story and the action that should be taken. In this last chapter I want first to present some of the challenges of global warming, especially those which arise because of its global nature. I then want to put global warming in the context of other major global problems faced by humankind.

Global warming – *global* pollution

A hundred years ago, the French painter Claude Monet spent time in London and painted wonderful pictures of the light coming through the smog. London was blighted by intense *local pollution* – from domestic and industrial chimneys around London itself. Thanks to the Clean Air Acts beginning in the 1950s, those awful smogs belong to the past – although London's atmosphere could be still cleaner.

Today, however, it is not just *local* pollution that is a problem but *global* pollution. Small amounts of pollution for which each of us are responsible are affecting everyone in the world. The first example to come to light was in the 1970s and early 1980s when it was realised that very small quantities of chlorofluorocarbons (CFCs) emitted to the atmosphere from leaking refrigerators, aerosol cans or some industrial processes resulted in significant degradation of the ozone layer. The problem was highlighted by the discovery of the *ozone hole* in 1985. Beginning in 1987, an international mechanism for tackling and solving this problem was established through the Montreal Protocol through which all nations contributing to it agreed to phase out their emissions of harmful substances. The richer nations involved also agreed to provide finance and technology transfer to assist developing countries to comply. A way forward for addressing global environmental problems was therefore charted.

The second example of global pollution is that of global warming, the subject of this book. Greenhouse gases that enter the atmosphere from the burning of fossil fuels, coal, oil and gas and from other human activities such as widespread deforestation are leading to damaging climate change. This is a very much larger and more challenging problem to tackle than that of ozone depletion; it strikes so much nearer to the core of human resources and activities – such as energy and transport – upon which our quality of life depends. However, we have been at pains to point out, tackle it we must – although, as we have also been keen to explain, we can do so in the realisation that abatement of the use of fossil fuels need not destroy or even diminish our quality of life; it should actually improve it!

Global pollution demands *global* solutions. These need to address human attitudes broadly, for instance those concerned with resource use, lifestyle, wealth and poverty. They must also involve human society at all levels of aggregation – international organisations, nations with their national and local governments, large and small industry and businesses, non-governmental organisations (e.g churches) and individuals.

Sustainability – also a *global* challenge

To bring in the broader context demanded by global solutions, in Chapters 8 and 9 the concept of *sustainable development* was introduced. Sustainability is a modern term that is widely used that takes into account our breadth of concern about the environment and the Earth's resources.[1] Just what do we mean by sustainability?

Imagine you are a member of the crew of a large spaceship on a voyage to visit a distant planet. Your journey there and back will take many years. An adequate, high-quality, source of energy is readily available in the radiation from the Sun. Otherwise, resources for the journey are limited. The crew on the spacecraft are engaged for much of the time in managing the resources as carefully as possible. A local biosphere is created in the spacecraft where plants are grown for food and everything is recycled. Careful accounts are kept of all resources, with especial emphasis on non-replaceable components. That the resources be *sustainable* at least for the duration of the voyage, both there and back, is clearly essential.

Planet Earth is enormously larger than the spaceship we have just been describing. The crew of Spaceship Earth at over 6 billion and rising is also enormously larger. The principle of sustainability should be applied to Spaceship Earth as rigorously as it has to be applied to the much smaller vehicle on its interplanetary journey. Professor Kenneth Boulding, a distinguished American economist, was the first to employ the image of Spaceship Earth. In a publication in 1966 he contrasted an 'open' or 'cowboy' economy[2] (as he called an unconstrained economy) with a 'spaceship' economy in which sustainability is paramount.[3]

There have been many definitions of sustainability. The simplest I know is 'not cheating on our children'; to that may be added, 'not cheating on our neighbours' and 'not cheating on the rest of creation'. In other words, not passing on to our children or any future generation an Earth that is degraded compared to the one we inherited, and also sharing common resources as necessary with our neighbours in the rest of the world and caring properly for the non-human creation.

Many things are happening in our modern world that are just not sustainable.[4] In fact, we are all guilty of cheating in the three respects I have mentioned. Table 12.1 lists five of the most important issues, briefly showing how they are all connected together and also linked to other major areas of human activity or concern. All these issues present enormous challenges.

To illustrate these connections let me use the example of tropical deforestation. Every year tropical forest is cut down or burnt equivalent approximately to

Table 12.1 Important sustainability issues

Issue	Linked to
Global warming and climate change	Energy, transport, biodiversity loss, deforestation
Land-use change	Biodiversity loss, deforestation, climate change, soil loss, agriculture, water
Consumption	Waste, fish, food, energy, transport, deforestation, water
Waste	Consumption, energy, agriculture, food
Fishing	Consumption, food

the area of the island of Ireland. Some is to harvest valuable hardwoods unsustainably; some is to create grassland to raise cattle to provide beef for some of the world's richest countries or to grow soya beans mostly to use as animal feed again for the rich countries of the world. This level of deforestation adds significantly to the atmospheric greenhouse gases carbon dioxide and methane so increasing the rate of human-induced climate change. It also alters the local climate close to the region where the deforestation is occurring. For instance, in the Amazon if current levels of deforestation continue, some of Amazonia could become much drier, even semi-desert, during this century. Further, when the trees go, soil is lost by erosion; again in many parts of Amazonia the soil is poor and easily washed away. Tropical forests are also rich in biodiversity. Loss of forests results in irreplaceable species loss.

A helpful measure of sustainability is provided by the idea of the Ecological Footprint and, for any given human community, the extent to which this Footprint exceeds the resources and the land available. It has for instance been estimated that three planets would be needed to provide the Footprint for everyone on Earth to possess the lifestyle of the richer countries.[5]

Not the only global problem

Global warming is not the only global problem. There are other issues of a global scale; we need to see global warming in their context. Four problems of particular importance impact on the global warming issue.

The first is population growth. When I was born there were about 2000 million people in the world. At the beginning of the twenty-first century there were 6000 million. During the lifetime of my grandchildren it is likely to rise to 8000 or 9000 million.[6] Most of the growth will be in developing countries; by 2020 they will contain over 80% of the world's people. These new people will all make

Deforestation.

demands for food, energy and work to generate the means of livelihood – all with associated implications for global warming.

The second issue is that of poverty and the increasing disparity in wealth between the developed and the developing world. The gap between the rich nations and the poor nations is becoming wider. The net flow of wealth in the world is from the poorer nations to the richer ones. Increasingly there are demands that more justice and equity be realised within the world's communities. The Prince of Wales has drawn attention to the strong links that exist between population growth, poverty and environmental degradation (see box above).

The third global issue is that of the consumption of resources, which in many cases is contributing to the problem of global warming. Many of the resources now being used cannot be replaced, yet we are using them at an unsustainable rate. In other words, because of the rate at which we are depleting them, we are seriously affecting their use even at a modest level by future generations.

Poverty and population growth

The Prince of Wales, in addressing the World Commission on Environment and Development on 22 April 1992, spoke as follows:[7]

> I do not want to add to the controversy over cause and effect with respect to the Third World's problems. Suffice it to say that I don't, in all logic, see how any society can improve its lot when population growth regularly exceeds economic growth. The factors which will reduce population growth are, by now, easily identified: a standard of health care that makes family planning viable, increased female literacy, reduced infant mortality and access to clean water. Achieving them, of course, is more difficult – but perhaps two simple truths need to be writ large over the portals of every international gathering about the environment: we will not slow the birth rate until we address poverty. And we will not protect the environment until we address the issue of poverty and population growth in the same breath.

Further, over 80% of resources are consumed by 20% of the world's population and to propagate modern Western patterns of consumption into the developing world is just not realistic. An important component of sustainable development, therefore, is *sustainable consumption*[8] of all resources.

The fourth issue is that of global security. Our traditional understanding of security is based on the concept of the sovereign state with secure borders against the outside world. But communications, industry and commerce increasingly ignore state borders, and problems like that of global warming and the other global issues we have mentioned transcend national boundaries. Security therefore also needs to take on more of a global dimension.

The impacts of climate change will pose a threat to security. One of the most recent wars has been fought over oil. It has been suggested that wars of the future could be fought over water.[9] The threat of conflict must be greater if nations lose scarce water supplies or the means of livelihood as a result of climate change. A dangerous level of tension could easily arise, with large numbers of environmental refugees as projected in Chapter 7. As has been pointed out by Admiral Sir Julian Oswald,[10] who has been deeply concerned with British defence policy, a broader strategy regarding security needs to be developed which considers *inter alia* environmental threats as possible sources of conflict. In addressing the appropriate action to combat such threats, it may be better overall and more cost-effective in security terms to allocate resources to the removal or the alleviation of the environmental threat rather than to military or other measures to deal head-on with the security problem itself.

Consumption of resources for material goods contributes to global warming. Pollution is a side issue of our consumption.

The challenge to all sections of community

In facing the challenge, it is important to recognise that the problems raised by global warming are not only global but long term; the timescales of climate change, of major infrastructure change in energy generation or transport or of major changes in programmes such as forestry are of the order of several decades. The programme of action must therefore be seen as both urgent and evolving, based on continuing scientific, technical and economic assessments. As the IPCC 1995 Report states, 'The challenge is not to find the best policy today for the next 100 years, but to select a prudent strategy and to adjust it over time in the light of new information.'[11]

In pursuing the challenge, I list below some of the particular responsibilities for different communities of expertise that generally transcend national boundaries.

- For the world's *scientists* the brief is clear: to narrow the uncertainties regarding the science of climate change and to provide better information especially about expected changes in climate extremes at the regional and local levels. Not only politicians and policymakers, but also ordinary people in all countries and at all levels of society, need the information provided in the clearest possible form. Scientists also have an important role in contributing to the research necessary to underpin the technical developments, for example in the energy, transport, forestry and agriculture sectors, required by the adaptation and mitigation strategies we have described.
- In the world of *politics*, it is over 20 years since Sir Crispin Tickell drew attention to the need for international action addressing climate change.[12] Since then, a great deal of progress has been made with the signing in Rio in 1992 of the Framework Convention on Climate Change and with the setting up of the Sustainable Development Commission in the United Nations. The challenges presented by the Convention to politicians and decision-makers, working both nationally and internationally, are, firstly, to achieve the right balance of development against environmental concern, that is to achieve sustainable development, and secondly, to find the resolve to turn the many fine words of the Convention into adequate, genuine and urgent action (including both adaptation and mitigation) regarding climate change.
- In addressing both adaptation and mitigation, the role of *technology* is paramount. The necessary technology is available. Urgently needed is adequate investment by both governments and industry in essential research and development and in the training of adequate numbers of technologists and engineers to carry it out. An important component of the strategy is the transfer of appropriate technology between countries, especially in the energy sector. This has been specifically recognised in the Climate Convention which in Article 4, paragraph 5 states:

The developed country Parties ... shall take all practical steps to promote, facilitate and finance, as appropriate, the transfer of, or access to, environmentally sound technologies and know-how to other Parties, particularly developing country Parties, to enable them to implement the provisions of the Convention.

- The responsibilities of *industry* must also be seen in the world context. It is the imagination, innovation, commitment and activity of industry that will do most to solve the problem. Industries that have a global perspective, working as appropriate with governments, need to develop a technical, financial and policy strategy to this end. The challenge of global warming must not be seen by industry as a threat but as a great opportunity; many companies from some of the largest to the smallest are now seriously and effectively taking sustainability and environmental considerations on board.[13]

- There are also new challenges for *economists* and *social scientists*; for instance, that of adequately representing environmental costs (especially including those 'costs' that cannot be valued in terms of money) and the value of 'natural' capital, especially when it is of a global kind – as mentioned in Chapter 9. There is the further problem of dealing fairly with all countries. No country wants to be put at a disadvantage economically because it has taken its responsibilities with respect to global warming more seriously than others. As economic and other instruments (for instance, taxes, subsidies, capping and trading arrangements, regulations or other measures) are devised to provide the incentives for appropriate action regarding global warming by governments or by individuals, these must be seen to be both fair and effective for all nations. Economists working with politicians and decision-makers need to find imaginative solutions that recognise not just environmental concerns but political realities.
- There is an important role for *communicators* and *educators*. Everybody in the world is involved in climate change so everybody needs to be properly informed – to understand the evidence for it, its causes, the distribution of its impacts and the action that can be taken to alleviate them. Climate change is a complex topic; the challenge to educators (including churches and other organisations involved in teaching) – also to *the media* –is to inform in ways that are understandable, comprehensive, honest and balanced.
- All countries will need to adapt to the climate change that applies in their region. For many developing countries this will not be easy because of increased floods, droughts or significant rise in sea level. Reductions in risks from disasters are some of the most important adaptation strategies. A challenge for *aid agencies* therefore is to prepare for more frequent and intense disasters in vulnerable countries; the International Red Cross has already taken the lead in this.[14]
- Finally, there is a challenge for everybody (see box). None of us can argue that there is nothing we can usefully do. Edmund Burke, a British parliamentarian of 200 years ago, said: *No one made a greater mistake than he who did nothing because he could do so little.*

There are many small efforts that individuals can make, which in numbers can help mitigate global warming. Compared with paper directly from a forest source, recycled paper reduces water consumption by nearly 60% and energy use by 40%. Air and water pollution are decreased by 74% and 35% respectively.

The conception and conduct of environmental research

While completing the writing of this last chapter for the third edition in 2004 I attended the opening of the Zuckerman Institute for Connective Environmental Research at the University of East Anglia – a centre devoted to interdisciplinary research on the environment. An opening lecture was given by William Clark, Professor of International Science, Public Policy and Human Development at Harvard University.[18] I was particularly struck by his remarks concerning the changes that are necessary in the way research is conceived and conducted if science (both natural and social) and technology are going to provide more adequate support to environmental sustainability. He pointed out the need to address all aspects of a problem both in the conception of the research and in its conduct and particularly emphasised the following four requirements:

What the individual can do

I have spelled out the responsibilities of experts of all kinds – scientists, economists, technologists, politicians, industrialists, communicators and educators. There are important contributions also to be made by ordinary individuals to help to mitigate the problem of global warming.[15] Some of these are to:

- ensure maximum energy efficiency in the home – through good insulation (see box on page 342) against cold in winter and heat in summer and by making sure that rooms are not overheated and that light is not wasted;
- as consumers, take energy use into account, e.g. by buying goods that last longer and from more local sources and buying appliances with high energy efficiency;
- support, where possible, the provision of energy from non-fossil-fuel sources; for instance, purchase 'green' electricity (i.e. electricity from renewable sources) wherever this option is available;[16]
- drive a fuel-efficient car and choose means of transport that tend to minimise overall energy use; for instance, where possible, walking or cycling; think before travelling by air;
- check, when buying wood products, that they originate from a renewable source;
- contribute to projects that reduce carbon dioxide emissions – this can be a way of compensating for some of the emissions to which we contribute, e.g. from aircraft journeys;[17]
- through the democratic process, encourage local and national governments to deliver policies which properly take the environment into account.

- An *integrative, holistic approach* that considers the interactions between multiple stresses and between various possible solutions. Such an approach also seeks to integrate perspectives from both the natural and the social sciences, so as to understand better the dynamical interplay by which environment shapes society and society in turn reshapes environment. And these various integrations must also be in a global context.
- A *goal of finding solutions not just of characterising problems*. There is a tendency amongst scientists to talk forever about problems but leave solutions to others. Applied research seeking solutions is just as challenging and worthy as so-called fundamental research identifying and describing the problems.
- *Ownership by both scientists and stakeholders*.[19] People are more prepared to change their behaviour or beliefs in response to knowledge that they have had a hand in researching or shaping.

- *Scientists must see themselves more as facilitators of social learning and less as sources of social guidance.* The problems faced in environmental research are such that solutions will only be reached after a long and iterative learning process in which many sectors of society as well as scientists must be included.

Two other qualities that need to govern our attitude to research that have often received emphasis in this book are those of *honesty* (especially accuracy and balance in the presentation of results) and *humility* (see, for instance, the fourth bullet in the last paragraph and the quotation from Thomas Huxley in Chapter 8, page 255). Together with the theme of *holism* from the last paragraph, they make up *3 Hs*, an alliteration that assists in keeping them all in mind.

The goal of environmental stewardship

Early in this chapter I invited you to imagine a voyage on Spaceship Earth. Let me leave you with two further metaphors to provide some perspective on sustainability especially as it is seen from the rich developed world.

The first metaphor represents unsustainability; I owe it to Professor Bob Goudzwaard[20] of the Free University of Amsterdam. He asks us to imagine our position in the developed world to be like that of a passenger in a comfortable seat on a high-speed train – a *train à grande vitesse* (TGV) – rushing along its tracks through the villages and countryside. Our *view from inside* the train seems stable, smooth and peaceful. Looking through the window beside us we perceive movement, the landscape seems to be moving backwards – staying behind. That, of course, is an illusion; the speed of the train provides for us our different but seemingly fixed frame of reference. Now imagine another position in relation to the TGV. We are standing in the open air just by the tracks as the train passes by. The *view from outside* is very different. The train is going fast, rushing by – perhaps too fast! We look anxiously ahead of the train to where some children seem to be trying to cross the tracks.

We, modern human beings, tend to take the inside view, from where the dynamic patterns of growth, consumption and progress seem completely normal. We are constantly stretching to increase the speed and dynamism of these technological and market-driven patterns. From this inside view, we see poor countries as underdeveloped and lagging behind. We also see the natural world and the environment as unable to move fast enough, too restraining – we want them out of the way. We want to speed on with our own conception of progress as if the landscape were not there.

A second metaphor – of positive sustainability – comes from the natural world itself – that of a tree. For the first part of its life a tree puts all its available resources into growth. It needs to grow upwards as fast as possible to reach the forest canopy to compete for its share of light that will bring more effective growth. But then the tree matures. It has grown to its full size. It has no drive to grow taller or wider. Its efforts and use of resources are now directed into a different activity, that of producing fruit. It is the fruit that will guarantee the future for subsequent generations of its species and also provide sustenance for a wide range of other species of both plants and animals – so enabling the tree to fulfil its purpose within the total ecosystem where it resides.

A point that is frequently made about such stewardship is that many of the actions that must be taken to combat global warming are good to do anyway because they will lead towards the sustainability that is essential if we are to live in a world where happiness and justice thrive – the sort of world that most people long to see. Seeing action on climate change as a catalyst for these other changes provides even more impetus for immediate and aggressive action.

In our modern world we tend to be obsessed with material goals: economic growth, the latest gadgets, more leisure and so on. But for our fulfilment as human beings we also desperately need challenges of a moral or spiritual kind. Caring for the Earth, its peoples and its future could provide a common purpose uniting the world's peoples in a cooperative enterprise bringing value far beyond the immediate tasks. There exist strong connections, that I drew out in Chapter 8, between our basic attitudes and beliefs and environmental concern. I presented a picture of humans as stewards or gardeners of the Earth. Many people in the world are already deeply involved in a host of ways in action towards greater sustainability. Such concern could, however, with benefit to all, be elevated to a much higher public and political level both nationally and internationally. In an article in *Time Magazine* at the time of the Summit on Sustainable Development in Johannesburg in 2002, Kofi Annan, the Secretary-General of the United Nations presented 'Competing Futures' in the following terms:[21]

> Imagine a future of relentless storms and floods; islands and heavily inhabited coastal regions inundated by rising sea levels; fertile soils rendered barren by drought and the desert's advance; mass migrations of environmental refugees; and armed conflicts over water and precious natural resources.

Then, think again – for one might just as easily conjure a more hopeful picture: of green technologies; liveable cities; energy-efficient homes, transport and industry; and rising standards of living for all the people not just a fortunate minority. The choice between these competing visions is ours to make.

QUESTIONS

1 List and describe the most important environmental problems in your country. Evaluate how each might be exacerbated under the type of climate change expected with global warming.
2 It is commonly stated that my pollution or my country's pollution is so small compared with the whole, that any contribution I or my country can make towards solving the problem is negligible. What arguments can you make to counter this attitude?
3 Speak to people you know who are involved with industry and find out their attitudes to local and global environmental concerns. What are the important arguments that persuade industry to take the environment seriously?
4 Al Gore, Vice-President of the United States in 1996–2000, has proposed a plan for saving the world's environment.[22] He has called it 'A Global Marshall Plan' paralleled after the Marshall Plan through which the United States assisted Western Europe to recover and rebuild after the Second World War. Resources for the plan would need to come from the world's major wealthy countries. He has proposed five strategic goals for the plan: (1) the stabilisation of world population; (2) the rapid creation and development of environmentally appropriate technologies; (3) a comprehensive and ubiquitous change in the economic 'rules of the road' by which we measure the impact of our decisions on the environment; (4) the negotiation of a new generation of international agreements, that must be sensitive to the vast differences of capability and need between developed and developing nations; (5) the establishment of a cooperative plan for educating the world's citizens about our global environment. Consider these five goals. Are they sufficiently comprehensive? Are there important goals that he has omitted?

5 How do you think governments can best move forward towards strategic goals for the environment? How can citizens be persuaded to contribute to government action if it involves making sacrifices, for example paying more in tax?
6 Can you add to the list in the box at the end of the chapter of contributions that the individual can make?
7 The Jubilee 2000 campaign has worked towards the cancellation of Third World debt possibly in return for appropriate environmental action. Discuss whether this is a good idea and how it might be made more successful.
8 Millions of people (especially children) die in the world's poorer countries because they lack clean water. It is argued by Professor Bjorn Lomborg that resources that might be used in reducing carbon dioxide emissions could be better used in making sure that everyone has access to clean water.[23] Do you agree with this argument? If so how could the result be realised in practice?
9 It has been suggested that anthropogenic climate change should be considered as a Weapon of Mass Destruction. Discuss the validity of this comparison.
10 Consider the requirements for the conception and conduct of research that are detailed on page 401. Do you consider that they could be components of a checklist against which research proposals might be judged? How far does the research in which you are engaged or do the research programmes with which you are connected fulfil these requirements?
11 Analyse and compare the three metaphors introduced in the chapter – Spaceship Earth, the TGV and the Tree. Are they helpful in providing perspective and vision? Define the ways in which they might be misleading.
12 In 2000, at the Millennium Summit, the UN agreed eight Millennium Goals with targets for achievement by 2015. Look up these goals.[24] Comment on how each of the goals might be connected with the issue of climate change. Look up the further commitments, relevant to the environment and climate change, made at the Earth Summit in Johannesburg in 2002.[25] To what extent are these adequate to meet the challenge of climate change? What are the prospects for the goals and commitments being achieved by 2015?

▶ FURTHER READING AND REFERENCE

Metz, B., Davidson, O., Bosch, P., Dave, R., Meyer, L. (eds.) 2007. *Climate Change 2007: Mitigation of Climate Change. Contribution of Working Group III to the Fourth Assessment Report of the Intergovernmental Panel on Climate Change*. Cambridge: Cambridge University Press.
Chapter 13 Policies, instruments and cooperative arrangements
IPCC AR4 2007 Synthesis Report
World Wildlife Fund 2006. *Living Planet Report*. www.footprintnetwork.org
Framework Convention on Climate Change (FCCC). www.unfccc.int

NOTES FOR CHAPTER 12

1 Sustainability not only concerns physical resources. It is also applied to activities and communities. Environmental sustainability is strongly linked to social sustainability – about sustainable communities and sustainable economics. *Sustainable development* is an all-embracing term. The Brundtland Report, *Our Common Future*, of 1987 provides a milestone review of sustainable development issues.

2 Michael Northcott in *The Moral Climate*, (London: Dorton, Longman and Todd, 2007), Chapter 4, calls it *frontier economics*.

3 Kenneth Boulding was Professor of Economics at the University of Colorado, sometime President of the American Economics Association and of the American Association for the Advancement of Science. His article, 'The economics of the coming Spaceship Earth' was published in 1966 in *Environmental Quality in a Growing Economy* pp. 77–82.

4 See, for instance, UNEP, *Global Environmental Outlook 3*, London: Earthscan, 2002 and Berry, R. J. (ed.) 2007. *When Enough Is Enough: A Christian Framework for Environmental Sustainability*. London: Apollos.

5 See World Wildlife Fund 2006. *Living Planet Report*: www.panda.org/livingplanet

6 The United Nations predicts a rise from 6.7 billion in 2006 to 9.2 billion by 2050.

7 HRH the Prince of Wales, in the First Brundtland Speech, 22 April 1992, published in Prins, G. (ed.) 1993. *Threats without Enemies*. London: Earthscan, pp. 3–14.

8 Many of the world's national academies of science led by the Royal Society in London have joined together in a report pointing this out. See Appendix B in *Towards Sustainable Consumption: A European Perspective*. 2000. London: Royal Society.

9 The former United Nations Secretary-General, Boutros Boutros-Ghali, has said that 'the next war in the Middle East will be fought over water, not politics'.

10 Oswald, J. 1993. Defence and environmental security, in Prins, *Threats without Enemies*.

11 *Synthesis of Scientific–Technical Information Relevant to Interpreting Article 2 of the UN Framework Convention on Climate Change*. 1995. Geneva: IPCC, p. 17.

12 Tickell, C. 1986. *Climatic Change and World Affairs*, second edition. Boston, Mass.: Harvard University Press.

13 For instance, two of the largest oil companies, Shell and British Petroleum are taking action to reduce their internal carbon dioxide emissions and are also putting strong investment into renewable energies. Lord Browne, former Chief Executive of BP, said in a speech in Berlin in 1997, 'No single company or country can solve the problem of climate change. It would be foolish and arrogant to pretend otherwise. But I hope we can make a difference – not least to the tone of the debate – by showing what is possible through constructive action.'

14 The International Red Cross/Red Crescent has set up a Climate Centre based in the Netherlands as a bridge between Climate Change and Disaster

Preparedness. The activities of the Centre are concerned with Awareness (information and education), Action (development of climate adaptation in the context of Disaster Preparedness programmes) and Advocacy (to ensure that policy development takes into account the growing concern about the impacts of climate change and utilises existing experience with climate adaptation and Disaster Preparedness).

15 Some useful websites: Sierra Club USA, www.sierraclub.org/sustainable.consumption/; Union of Concerned Scientists, www.ucsusa.org; Energy Saving Trust, www.est.org.uk; Ecocongregation, www.encams.org; Christian Ecology Link, www.christian-ecology.org.uk; John Ray Initiative, www.jri.org.uk.

16 With changes in the organisation of electricity supply companies in some countries, it is becoming possible to purchase electricity, delivered by the national grid, from a particular generating source, see for instance for the UK www.greenelectricity.org or www.good-energy.co.uk.

17 See for instance Climate Care website: www.climatecare.org.uk.

18 Clark, W.C. 2003. Sustainability science: challenges for the new millennium. An address at the official opening of the Zuckerman Institute for Connective Environmental Research, University of East Anglia, Norwich, UK, 4 September 2003. http://sustainabilityscience.org/ists/docs/clark_zicer_opening030904.pdf.

19 This is illustrated by the experience of the IPCC, as described on page 266.

20 See www.allofliferedeemed.co.uk/goudzwaard/BG111.pdf

21 Kofi Annan, *Time Magazine*, 26 August 2002.

22 Expounded in the last chapter of Gore, A. 1992. *Earth in the Balance*. Boston, Mass.: Houghton Mifflin Company.

23 Lomborg, B. (ed.) 2004. *Global Crises, Global Solutions*. Cambridge: Cambridge University Press.

24 UN Millennium Goals: www.un.org/millenniumgoals

25 Earth Summit 2002: www.earthsummit2002.org

Appendix 1

SI unit prefixes

Quantity	Prefix	Symbol
10^{12}	tera	T
10^{9}	giga	G
10^{6}	mega	M
10^{3}	kilo	k
10^{2}	hecto	h
10^{-2}	centi	c
10^{-3}	milli	m
10^{-6}	micro	µ
10^{-9}	nano	n

Chemical symbols

CFCs	chlorofluorocarbons	N_2	molecular nitrogen
CH_4	methane	N_2O	nitrous oxide
CO	carbon monoxide	NO	nitric oxide
CO_2	carbon dioxide	NO_2	nitrogen dioxide
H_2	molecular hydrogen	O_2	molecular oxygen
HCFCs	hydrochlorofluorocarbons	O_3	ozone
HFCs	hydrofluorocarbons	OH	hydroxyl radical
H_2O	water	SO_2	sulphur dioxide

Appendix 2
Acknowledgements for figures, photos and tables

Figures

1.1 From *World Climate News*, no. 16, July 1999. Geneva: World Meteorological Organization. A similar map is prepared and published each year. Data from Climate Prediction Center, NOAA, USA.

1.2 Figure 2.7 from Watson, R. *et al.* (eds.) 2001. *Climate Change 2001: Synthesis Report. Contribution of Working Groups I, II and III to the Third Assessment Report of the Intergovernmental Panel on Climate Change.* Cambridge: Cambridge University Press.

1.3 World Meteorological Organization 1990. From *The Role of the World Meteorological Organization in the International Decade for Natural Disaster Reduction*, World Meteorological Organization report no, 745. Geneva: World Meteorological Organization.

1.4 Adapted from: Canby, T.Y. 1984. El Niño's ill wind. *National Geographic Magazine*, pp. 144–83.

1.5 After Fig. 1.1 from IPCC 2007. *Climate Change 2007: Synthesis Report. Contribution of Working Groups I, II and III to the Fourth Assessment Report of the Intergovernmental Panel on Climate Change.* Geneva: IPCC.

2.2 After FAQ 1.3, Fig. 1 from Le Treut, H., Somerville, R., Cubasch, U., Ding, Y., Mauritzen, C., Mokssit, A., Peterson, T., Prather, M., 2007. Historical overview of climate change. In Solomon, S., Qin, D., Manning, M., Chen, Z., Marquis, M., Averyt, K. B., Tignor, M., Miller, H. L., (eds.) *Climate Change 2007: The Physical Science Basis. Contribution of Working Group I to the Fourth Assessment Report of the Intergovernmental Panel on Climate Change.* Cambridge: Cambridge University Press, p. 115.

2.5 Spectrum taken with the infrared interferometer spectrometer flown on the satellite Nimbus 4 in 1971 and described by Hanel, R.A. *et al.* 1971. *Applied Optics*, **10,** 1376–82.

2.7 After FAQ 1.1, Figure 1. Le Treut *et al.* In Solomon *et al.* (eds.) *Climate Change 2007: The Physical Science Basis.*

2.8 From Houghton, J.T. 2002. *The Physics of Atmospheres*, third edition. Cambridge: Cambridge University Press.

3.1 Figure 1.1 from Bolin, B., Sukumar, R. 2000. Global perspective. In Watson, R.T., Noble, I.R., Bolin, B., Ravindranath, N.H., Verardo, D.J., Dokken, D.J. (eds.) *Land Use, Land-use Change, and Forestry*, IPCC Special Report. Cambridge: Cambridge University Press, Chapter 1.

3.2 (a) After Fig. SPM 1 from Summary for Policymakers. In Solomon *et al.*. (eds.) *Climate Change 2007: The Physical Science Basis.*
b) After Fig. TS3 from Technical Summary, 2007. In Solomon *et al.* (eds.) *Climate Change 2007: The Physical Science Basis.*

3.3 After Fig. 2.3a from Foster, P., Ramaswamy, V., Artaxo, P., Berntsen, T., Betts, R., Fahey, D.W., Haywood, J., Lean, J., Lowe, D.C., Myhre, G., Nganga, J., Prinn, R., Raga, G., Schulz, M., Van Dorland, R., 2007. Changes in atmospheric constituents and radiative forcing. In Solomon *et al.* (eds.) *Climate Change 2007: The Physical Science Basis.*

3.4 Figure 3.4 from Prentice, I.C. *et al.* 2001. The carbon cycle and atmospheric carbon dioxide. Chapter 3 in Houghton, J.T., Ding, Y., Griggs, D.J., Noguer, M., van der Linden, P.J., Dai, X., Maskell, K., Johnson, C.A. (eds.) *Climate Change 2001: The Scientific Basis. Contribution of Working Group I to the Third Assessment Report of the Intergovernmental Panel on Climate Change.* Cambridge: Cambridge University Press.

3.5 (a) and (b) From Chris Jones, UK Meteorological Office.

3.6 After Fig. SPM1 from Summary for Policymakers. In Solomon *et al.* (eds.) *Climate Change 2007: The Physical Science Basis.*

3.7 (a) After Fig. 2.11 from Foster, P. *et al.* In Solomon *et al.* (eds.) *Climate Change 2007: The Physical Science Basis.* (b) From Hadley Centre Briefing, December 2005. *Climate Change and the Greenhouse Effect.* Exeter: UK Met Office, p. 19.

3.8 After Fig. 7.24 from Denman, K.L., Brasseur, G., Chidthaisong, A. Ciais, P., Cox, P.M., Dickinson, R.E., Hauglustaine, D., Heinze, C., Holland, E., Jacob, D., Lohman, U., Ramachandran, S., da Silva Dias, P.L., Wofsy, S.C., Zhang, X. 2007. Couplings between changes in the climate system and biogeochemistry. In Solomon *et al.* (eds.) *Climate Change 2007: The Physical Science Basis.*

3.9 After Fig. 7.20 from Denman *et al.* In Solomon *et al.* (eds.) *Climate Change 2007: The Physical Science Basis.*

3.10 After Cloud droplet radii in micrometers for ship track clouds. In King, M.D., Parkinson, C.L., Partington, K.C., Williams, R.G. (eds.) 2007. *Our Changing Planet.* Cambridge: Cambridge University Press, p. 70.

3.11 After Fig. SPM 2 from Summary for Policymakers, In Solomon *et al.* (eds.) *Climate Change 2007: The Physical Science Basis.*

3.12 After Global surface albedo. In King *et al.*, (eds). *Our Changing Planet,* p. 129.

4.1 (a) and (b) After FAQ 3.1, Fig. 1 from Trenberth, K.E., Jones, P.D., Ambenje, P., Bojariu, R., Easterling, D., Klein Tank A., Parker, D., Rahimzadeh, F., Renwick, J.A., Rusticucci, M., Sode, B., Zhai, P. 2007. Observations: Surface and atmospheric climate change. In Solomon *et al.* (eds.) *Climate Change 2007: The Physical Science Basis,* p. 253.

4.2 After Fig 3.8 from Trenberth *et al.* In Solomon *et al.* (eds.) *Climate Change 2007: The Physical Science Basis.*

4.3 After Fig 3.17 from Trenberth *et al.* In Solomon *et al.* (eds.) *Climate Change 2007: The Physical Science Basis.*

4.4 After Fig. SPM3 from Summary for Policymakers. In Solomon *et al.* (eds.) *Climate Change 2007: The Physical Science Basis.*

4.5 After Fig 6.10 from Jansen, E., Overpeck, J., Briffa, K.R., Duplessy, J.C., Joos, F., Masson-Delmotte, V., Olago, D., Otto-Bliesner, B., Peltier, W.R., Rahmstorf, S., Ramesh, R., Raynaud, D., Rind, D., Solomina, O., Villalba, R., Zhang, D. 2007. Paleoclimate. In Solomon *et al.* (eds.) *Climate Change 2007: The Physical Science Basis.*

4.6 (a) Adapted from Raynaud, D. *et al.* 1993. The ice core record of greenhouse gases. *Science*, **259**, 926–34. (b) After Fig. 6.3 from Jansen, *et al.* In Solomon *et al.* (eds.) *Climate Change 2007: The Physical Science Basis.*

4.7 Adapted from Broecker, W. S., Denton, G. H. 1990. What drives glacial cycles. *Scientific American*, **262**, 43–50.

4.8 Adapted from Professor Dansgaard and colleagues, Greenland ice core (GRIP) members. 1990. Climate instability during the last interglacial period in the GRIP ice core. *Nature*, **364**, 203–7.

4.9 Adapted from Dansgaard, W., White, J. W. C., Johnsen, S. J. 1989. The abrupt termination of the Younger Dryas climate event. *Nature*, **339**, 532–3.

5.1 From UK Meteorological Office.

5.3 After Fig. 1.4 from Le Treut *et al.* In Solomon *et al.* (eds.) *Climate Change 2007: The Physical Science Basis.*

5.4 From UK Meteorological Office.

5.5 From UK Meteorological Office.

5.6 After Milton, S., Meteorological Office, quoted in Houghton, J. T. 1991. The Bakerian Lecture 1991: The predictability of weather and climate. *Philosophical Transactions of the Royal Society A*, **337**, 521–71.

5.7 After Lighthill, J. 1986. The recently recognized failure in Newtonian dynamics. *Proceedings of the Royal Society A*, **407**, 35–50.

5.8 After Fig. 1.9 from Palmer, T., Hagedorn, R., 2006. *Predicability of Weather and Climate*. Cambridge: Cambridge University Press. p. 18.

5.9 After Fig 3.27 from Trenberth *et al.* In Solomon *et al.* (eds.) *Climate Change 2007: The Physical Science Basis.*

5.10 From Houghton, The Bakerian Lecture 1991. London: Royal Society.

5.11 After Fig. 9.19 from Hegerl, G. C., Zwiers, F. W., Braconnot, P., Gillet, N. P., Luo, Y., Marengo Orsini, J. A., Nicholls, N., Penner, J. E., Stott, P. A. 2007. Understanding and attributing climate change. In Solomon *et al.* (eds.) *Climate Change 2007: The Physical Science Basis.*

5.12 After Plate 19.3 from Palmer and Hagedorn, *Predicability of Weather and Climate.*

5.13 After FAQ 1.2, Fig. 1 from Le Treut *et al.* In Solomon *et al.* (eds.) *Climate Change 2007: The Physical Science Basis*. p. 104.

5.15 (a) After Net cloud radiative forcing. In King *et al.* (eds.), *Our Changing Planet*, p. 23. (b) After Comparison of observed longwave, shortwave and net radiation at the top of the atmosphere for the tropics. In King, *et al.* (eds.), *Our Changing Planet*, p. 24.

5.16 See Siedler, G., Church, J., Gould, J. (eds.) 2001. *Ocean Circulation and Climate.* London: Academic Press. Original diagram from Woods, J. D. 1984. The upper ocean and air sea interaction in global climate. In Houghton, J. T. 1985. *The Global Climate*. Cambridge: Cambridge University Press, pp. 141–87.

5.18 From UK Meteorological Office.

5.19 This diagram and information about modelling past climates is from Kutzbach, J. E. 1992. In Trenberth, K. E. (ed.) *Climate System Modelling*. Cambridge: Cambridge University Press.

5.20 From Hansen, J. *et al.* 1992. Potential impact of Mt Pinatubo eruption. *Geophysics Research Letters*, **19**, 215–18. Also quoted in Technical summary, in Houghton, J. T., Meira Filho, L. G., Callander, B. A., Harris, N., Kattenberg, A., Maskell, K. (eds.)

1996. *Climate Change 1995: The Science of Climate Change.* Cambridge: Cambridge University Press.

5.21 From Sarmiento, J. L. 1983. *Journal of Physics and Oceanography*, **13**, 1924–39.

5.22 After Fig. 9.5 from Hegerl, G. C. *et al.* In Solomon *et al.*. (eds.) *Climate Change 2007: The Physical Science Basis.*

5.23

5.24 From the Hadley Centre Report 2002. *Regional Climate Modelling System.* Exeter: UK Met Office, p. 4.

6.1 Figure 17 from Technical summary. In Houghton *et al.* (eds.) *Climate Change 2001: The Scientific Basis.*

6.2 Figure 18 from Technical summary. In Houghton *et al.* (eds.) *Climate Change 2001: The Scientific Basis.*

6.4 (a) After Fig. SPM 5 from Summary for Policymakers. In Solomon *et al.* (eds.) *Climate Change 2007: The Physical Science Basis.* (b) After Fig. TS 26 from Technical summary, *ibid.* (c) After Fig. 10.5 from Meehl, G. A., Stocker, T. F., Collins, W. D., Friedlingstein, P., Gaye, A. T., Gregory, J. M., Kitoh, A., Knutti, J. M., Murphy, J. M., Noda, A., Raper, S. C. B., Watterson, I. G., Weaver, A. J., Zhao, Z. C. 2007. Global climate projections, *ibid.*

6.5 After Fig. 10.7 from Meehl *et al.* In Solomon *et al.* (eds.) *Climate Change 2007: The Physical Science Basis.* Cambridge University Press.

6.6 After Fig. SPM 6 from Summary for policymakers. In Solomon *et al.* (eds.) *Climate Change 2007: The Physical Science Basis.*

6.7 After Fig. SPM 7 from Summary for policymakers. In Solomon *et al.* (eds.) *Climate Change 2007: The Physical Science Basis.*

6.8 After Fig. 2.32 from Folland C. K., Karl T. R. *et al.* 2001. Observed climate variability and change, Chapter 2 in Houghton *et al.* (eds.) *Climate Change 2001: The Scientific Basis*, p. 155.

6.9 From Pittock, A. B. *et al.* 1991. Quoted in Houghton, J. T., Callander, B. A., Varney, S. K. (eds.) *Climate Change 1992: The Supplementary Report to the IPCC Assessments.* Cambridge: Cambridge University Press, p. 120.

6.10 After Fig. 10.18 from Meehl *et al.* In Solomon *et al.* (eds.) *Climate Change 2007: The Physical Science Basis.*

6.11 From Palmer, T. N., Raisanen, J. 2002. *Nature*, **415**, 512–14.

6.12 After Fig. 9 from Burke, E. J., Brown, S. J., Christidis, N. 2006. Modelling the recent evolution of global drought and projections for the 21st century with the Hadley Centre climate model. *Journal of Hydrometeorology*, **7**, 1113–25.

6.13 From Hadley Centre Report 2002. *Regional Climate Modelling System.* Exeter: UK Met Office.

6.14 From Hadley Centre Briefing, 2005. *Climate Change and the Greenhouse Effect.* Exeter: UK Met Office.

6.15 After Fig. 2.17 from Foster *et al.* In Solomon *et al.* (eds.) *Climate Change 2007: The Physical Science Basis.*

7.1 After Fig. 5.21 from Bindoff, N. L., Willebrand, J., Artale, V., Cazenave, A., Gregory, J., Gulev, S., Hanawa, K., Le Quere, C., Levitus, S., Nojiri, Y., Shum, C. K., Talley, L. D., Unnikrishnan, A. 2007 Observation: Oceanic climate change and sea level. In Solomon *et al.* (eds.) *Climate Change 2007: The Physical Science Basis.*

7.2 After Fig. FAQ 5.1, Fig. 1 from Bindoff *et al.* In Solomon *et al.* (eds.) *Climate Change 2007: The Physical Science Basis.*

7.3 After Fig. TS 8 from Technical Summary. In Solomon *et al.* (eds.) *Climate Change 2007: The Physical Science Basis,* p. 41.

7.4 From Broadus, J.M. 1993. Possible impacts of, and adjustments to, sea-level rise: the case of Bangladesh and Egypt. In Warrick, R.A., Barrow, E.M., Wigley, T.M.L. (eds.) 1993. *Climate and Sea-Level Change: Observations, Projections and Implications.* Cambridge: Cambridge University Press, pp. 263–75; adapted from Milliman, J.D. 1989. Environmental and economic implications of rising sea level and subsiding deltas: the Nile and Bangladeshi examples. *Ambio,* **18**, 340–5.

7.5 From Maurits la Rivière, J.W. 1989. Threats to the world's water. *Scientific American,* **261**, 48–55.

7.6 Figure 11.4(a) from Shiklomanov, I.A., Rodda, J.C. (eds.) 2003. *World Water Resources at the Beginning of the Twenty-First Century.* Cambridge: Cambridge University Press.

7.7 After Fig. 3.2 from Kundzewicz, Z.W., Mata, L.J., Arnell, N.W., Doll, P., Kabat, P., Jimenez, B., Miller, K.A., Oki, T., Sen, Z., Shiklomanov, I.A., 2007. Freshwater resources and their management. In Parry, M., Canziani, O., Palutikof, J., van der Linden, P., Hansen, C. (eds.) *Climate Change 2007: Impacts, Adaptation and Vulnerability. Contribution of Working Group II to the Fourth Assessment Report of the Intergovernmental Panel on Climate Change.* Cambridge: Cambridge University Press, p. 178

7.8 After Fig. 3.8 from Kundzewicz *et al.* In Parry *et al.* (eds.) *Climate Change 2007: Impacts, Adaptation and Vulnerability.*

7.9 From the Hadley Centre Report 2002. *Regional Climate Modelling System.* Exeter: UK Met Office p. 5.

7.10 From Tolba, M.K., El-Kholy, O.A. (eds.) 1992. *The World Environment 1972–1992.* London: Chapman and Hall, p. 135.

7.11 (a) and (b) After Fig. 5.2 from Easterling, W.E., Aggarwal, P.K., Batima, P., Brander, K.M., Erda, L., Howden, S.M., Kirilenko, A., Morton, J., Soussana, J.F., Schmidhuber, J., Tubiello, F.N. 2007. Food, fibre and forest products. In Parry *et al.* (eds.) *Climate Change 2007: Impacts, Adaptation and Vulnerability.*

7.12 Illustrating key elements of a study of crop yield and food trade under a changed climate. From Parry, M. *et al.* 1999. Climate change and world food security: a new assessment. *Global Environmental Change,* **9**, S51–S67.

7.13 After Fig. 4.1 from Fischlin, A., Midgley, G.F., Price, J.T., Leemans, R., Gopal, B., Turley, C., Rounsevell, M.D.A., Dube, O.P., Tarazona, J., Velichko, A.A. 2007. Ecosystems, their properties, goods and services. In Parry *et al.* (eds.) *Climate Change 2007: Impacts, Adaptation and Vulnerability.*

7.14 Adapted from Gates, D.M. 1993. *Climate Change and its Biological Consequences.* Sunderland, Mass.: Sinauer Associates, p. 63. The original source is Delcourt, P.A., Delcourt, H.R. 1981. In Romans, R.C. (ed.) *Geobotany II.* New York: Plenum Press, pp. 123–65.

7.15 From Gates, *Climate Change and its Biological Consequences,* p. 63.

7.16 Data from Bugmann, H. quoted in Miko U.F. *et al.* 1996. Climate change impacts on forests. In Watson, R.T., Zinyowera, M.C., Moss, R.H. (eds.) 1996. *Climate Change 1995: Impacts, Adaptation and Mitigation of Climate Change: Scientific – Technical*

Analyses. Contribution of Working Group II to the Second Assessment Report of the Intergovernmental Panel on Climate Change. Cambridge: Cambridge University Press, Chapter 1.

7.17 After Fig. TS16 from Technical summary. In Parry *et al.* (eds.) *Climate Change 2007: Impacts, Adaptation and Vulnerability.*

7.18 After Fig. 8.2 from Turley, C., Blackford, J.C., Widdecombe, S., Lowe, D., Nightingale, P.D., Rees, A.P. 2006. In Schellnhuber, H.J. (ed.) *Avoiding Dangerous Climate Change.* Cambridge: Cambridge University Press, Chapter 8.

7.19 After Fig. TS13 from Technical summary. In Parry *et al.* (eds.) *Climate Change 2007: Impacts, Adaptation and Vulnerability.*

8.1 After Lovelock, J.E. 1988. *The Ages of Gaia*. Oxford: Oxford University Press, p. 203.

8.2 From Lovelock, *The Ages of Gaia*, p. 82.

9.1 Figure 13.2 from Mearns, L.O., Hulme, M. *et al.* 2001. Climate scenario development. In Houghton *et al.* (eds.) *Climate Change 2001: The Scientific Basis.*

9.2 European Space Agency.

9.3 After Fig. Box 13.2 from Stern, N. 2007. *The Stern Review: The Economics of Climate Change.* Cambridge: Cambridge University Press. p. 343.

10.1 After Fig. SPM3 from IPCC *Climate Change 2007: Synthesis Report.*

10.2 After Fig. SPM11 from IPCC *Climate Change 2007: Synthesis Report.*

10.3 From Jason Lowe and Chris Jones at the Hadley Centre, UK Meteorological Office.

10.4 After Fig SPM3a from Summary for policymakers. In Metz, B., Davidson, O., Bosch, P., Dave, R., Meyer, L. (eds.) *Climate Change 2007: Mitigation. Contribution of Working Group III to the Fourth Assessment Report of the Intergovernmental Panel on Climate Change.* Cambridge: Cambridge University Press.

10.5 From the Global Commons Institute, Illustrating their 'Contraction and Convergence' proposal for achieving stabilisation of carbon dioxide concentration.

11.1 Adapted and updated from Davis, G.R. 1990. Energy for planet Earth. *Scientific American*, **263**, 21–7. Information from Fig. TS13, Technical summary. In Metz *et al.* (eds.) *Climate Change 2007: Mitigation.*

11.2 After Fig. 4.4 from Sims, R.E.H., Schock, R.N., Adegbululgbe, A., Fenham, J., Konstantinaviciute, I., Moomaw, W., Nimir, H.B., Schlamadinger, B., Torres-Martinez, J., Turner, C., Uchiyama, Y., Vuori, S.J.V., Wamukonya, N., Zhang, X. 2007. Energy supply. In Metz *et al.* (eds.) *Climate Change 2007: Mitigation.*

11.3 After Fig. SPM2 from Summary for policymakers. In Metz *et al.* (eds.) *Climate Change 2007: Mitigation.*

11.4 (a) After Fig 2.1 from *Energy Technology Perspectives 2008*. Paris: International Energy Agency. (b) After Fig 2.2, *ibid.* (c) After Fig. 2.3 *ibid.*

11.5 Adapted from *Scientific American*, **295**, p. 30.

11.7 From Abuerdeen City Council.

11.9 After Fig. 5.4, from Kahn Riberio, S., Kobayashi, S., Beuthe, M., Gasca, J., Greene, D., Lee, D.S., Muromachi, Y., Newton, P.J., Plotkin, S., Sperling, D., Wit, R., Zhou, P.J. 2007. Transport and its infrastructure. In Metz *et al.* (eds.) *Climate Change 2007: Mitigation.*

11.10 After Fig. 5.3 from Kahn Ribero *et al.* In Metz *et al.* (eds.) *Climate Change 2007: Mitigation*

11.11 After Fig. 5.6, *World Energy Outlook 2007.* Paris: International Energy Agency.

11.12 After Fig. 2.18, *Energy Technology Perspectives 2008*. Paris: International Energy Agency.
11.14 Dr John Clifton-Brown, Aberystwyth University.
11.15 Adapted from Twidell, J., and Weir, T. 1986. *Renewable Energy Resources*. London: E. and F. Spon, p. 100.
11.16 From Smith, P.F. 2001. *Architecture in a Climate of Change*. London: Architectural Press.
11.17 Adapted from *Scientific American*, **295**, 64.
11.18 From Shell Renewables.
11.19 From Williams, N., Jacobson, K., Burris, H. 1993. Sunshine for light in the night. *Nature*, **362**, 691–2. For more recent information on solar home systems see Martinot, E. *et al.* 2002. Renewable energy markets in developing countries. *Annual Review of Energy and the Environment*, **27**, 309–48.
11.20 From the 22nd Report of the Royal Commission on Environmental Pollution. London: Stationery Office.
11.21 www.wavedragon.com.
11.22 Figure TS6 from Metz *et al.* (eds.) *Climate Change 2007: Mitigation*.
11.23 Figure SPM6 from Metz *et al.* (eds.) *Climate Change 2007: Mitigation*.
11.24 Adapted from Twidell and Weir, *Renewable Energy Resources*, p. 399.
11.25 A.I. Gardiner, Pilbrow E.N., Broome S.R. and McPherson A.E., 2008. *HyLink - a renewable distributed energy application for hydrogen*. Proc., 3rd Asia Pacific Regional International Solar Energy Society (ISES) Conference, and 46th Australia and New Zealand Solar Energy Society (ANZSES) Annual Conference, 25–28 November, Sydney. http://isesap08.com
11.26 From Enkvist, P. *et al.* 2007. A cost curve for greenhouse gas reduction. *The McKinsey Quarterly*, **1**, 35–45.

Photos

Chapter 1

Hurricane Wilma Reuters: GM1DWIALXUAA
Hurricane Mitch NASA/Goddard Space Flight Center Scientific Visualization Studio
The Great Flood of 1993 NASA/Goddard Space Flight Center Scientific Visualization Studio
El Niño event of 1997–8 NASA/Goddard Space Flight Center Scientific Visualisation Studio

Chapter 2

Earth Rise NASA
Ice, oceans, land surfaces, and clouds © Cloudzilla, 6 December 2007
Venus, Earth, and Mars Lunar and Planetary Institute

Chapter 3

Industrial activity © 2006 Rinderart, courtesy ImageVortex.com
Didcot power station © Rose Davies, 7 February 2007
Plankton bloom ESA
Rice paddy fields MET Office

Ozone depletion NASA
Aircraft contrails NASA

Chapter 4

Himalayan glaciers Image provided by Jeffrey Kargel, USGS/NASA JPL/AGU, through NASA's Earth Observatory
Nukuoro Atoll NASA
Ice core research BAS ref 10002006

Chapter 5

Supercell Floridalightening.com
Lewis Fry Richardson UK Meteorological office
Saharan dust storm NASA image created by Jesse Allen, Earth Observatory, using data provided courtesy of the MODIS Rapid Response Team
Snow and ice Scott Polar Research Institute
Mount Pinatubo USGS/Cascades Volcano Observatory/Dave Harlow

Chapter 6

CloudSat spacecraft NASA/JPL
Torrential rain and flooding Environmental Canada/Associated Press
Sea surface temperature NASA/Goddard Space Flight Center Scientific Visualization Studio

Chapter 7

African droughts PA photos
Arctic Ice NASA/Goddard Space Flight Center Scientific Visualization Studio
Bangladesh floods PA photos
Acid Rain Science Photo Library
Hurricane Katrina Space Imaging
US Coast Guardsman US Coast Guard

Chapter 8

The Earth NASA
Golden toad US Fish and Wildlife Service

Chapter 9

***The Garden of Eden*, Jan Brueghel** V&A Museum
Industrial chimneys Getty Images
The IPCC delegation IPCC, Geneva
Bangladesh radar image ESA
***Nature* diagnosing climate change**

Chapter 10

Amazon rainforest canopy iStockphoto
Bolivian rainforest NASA/Goddard Space Flight Center Scientific Visualization Studio
Afforestation Agricultural Research Service / US Department of Agriculture
Landfill site © D'Arcy Norman, 24 September 2006

Chapter 11

Solar PV US Air Force
Sleipner T gas platform *Physics World*, **19,** 25 (2006), Institute of Physics
The Earth at night NASA/Goddard Space Flight Center Scientific Visualization Studio
Wind turbines iStockphoto

Chapter 12

Deforestation iStockphoto
Recycled paper iStockphoto

Tables

3.1 Denman, K.L. and Brasseur, G. *et al.* 2007. Chapter 7 in Solomon *et al.* (eds.) *Climate Change 2007: The Physical Science Basis.*

3.2 Prather, M., Ehhalt, D. *et al.* 2001. Atmospheric chemistry and greenhouse gases. Chapter 4, in Houghton *et al.* (eds.), *Climate Change 2001: The Scientific Basis.*

4.1 Table SPM-1 from *IPCC 2001 Synthesis Report* with some updates from Table 3.8 in Trenberth, K.E., Jones, P.D. *et al.*, Chapter 3, in Solomon *et al.* (eds.) *Climate Change 2007: The Physical Science Basis.*

6.1 Data for 2005 and estimates of uncertainty in Figure 3.11. For 2050 and 2100 from Ramaswamy, V. *et al.* 2001: Radiative forcing of climate change. In Houghton *et al.* (eds.) *Climate Change 2001: The Scientific Basis*, Chapter 6, Tables 6.1 and 6.14.

6.2 Table SPM-1 from Summary for policymakers. In Houghton *et al.* (eds.) *Climate Change 2001: The Scientific Basis.* Updated from Trenberth, K.E., Jones, P.D. *et al.*, Chapter 3, Table 3.8 and Christensen, J.H., Hewitson, B. *et al.*, Chapter 11, Table 11.2, in Solomon *et al.* (eds.) *Climate Change 2007: The Physical Science Basis.*

7.1 Based on Table TS3 IPCC AR4 2007 WG 2 Technical Survey p. 66.

7.2 Table SPM-4 from IPCC AR4 Synthesis Report.

7.3 Data from Munich Re, presented in Figure 8.6 in Vellinga, P., Mills, E. *et al.* 2001. In McCarthy *et al.*, (eds.) *Climate Change 2001: Impacts.*

7.4 Data from Munich Re, presented in Table 8.3 in Vellinga, P., Mills, E. *et al.* 2001. In McCarthy *et al.*, (eds.) *Climate Change 2001: Impacts.*

7.5 Table 19.6 in Smith, J.B. *et al.* Vulnerability to climate change and reasons for concern: a synthesis. In McCarthy *et al.* (eds.) *Climate Change 2001: Impacts.*

7.6 IPCC AR4 Synthesis Report, Table SPM 3.

10.2 Table 6.7 from Ramaswamy, V. *et al.* 2001. In Solomon *et al.* (eds.) *Climate Change 2001: The Scientific Basis;* also Solomon *et al.* (eds.) *Climate Change 2007. The Physical Science Basis.*

10.3 Adapted from IPCC AR4 *Synthesis Report*, Table SPM 6.

11.2 Table SPM 6 Metz *et al.* (ed.) *Climate Change 2007: Mitigatiz.*

Glossary

aerosol(s) A collection of airborne solid or liquid particles with a typical size between 0.01 and 10 μm that reside in the atmosphere from periods of hours to days or months. They may be natural or anthropogenic in origin. They influence climate directly through absorbing or scattering radiation or indirectly by acting as cloud condensation nuclei

afforestation Planting of new forests on lands that historically have not contained forests

Agenda 21 A document accepted by the participating nations at UNCED on a wide range of environmental and development issues for the twenty-first century

albedo The fraction of light reflected by a surface, often expressed as a percentage. Snow-covered surfaces have a high albedo level; vegetation-covered surfaces have a low albedo, because of the light absorbed for *photosynthesis*

anthropic principle A principle which relates the existence of the Universe to the existence of humans who can observe it

anthropogenic effects Effects which result from human activities such as the burning of *fossil fuels* or *deforestation*

AOGCM Atmosphere–ocean coupled general circulation model

atmosphere The envelope of gases surrounding the Earth or other planets

atmospheric pressure The pressure of atmospheric gases on the surface of the planet. High atmospheric pressure generally leads to stable weather conditions, whereas low atmospheric pressure leads to storms such as cyclones

atom The smallest unit of an *element* that can take part in a chemical reaction. Composed of a nucleus which contains *protons* and *neutrons* and is surrounded by *electrons*

atomic mass The sum of the numbers of *protons* and *neutrons* in the nucleus of an *atom*

biodiversity A measure of the number of different biological species found in a particular area

biological pump The process whereby carbon dioxide in the *atmosphere* is dissolved in sea water where it is used for *photosynthesis* by *phytoplankton* which are eaten by *zooplankton*. The remains of these microscopic organisms sink to the ocean bed, thus removing the carbon from the *carbon cycle* for hundreds, thousands or millions of years

biomass The total weight of living material in a given area

biome A distinctive ecological system, characterised primarily by the nature of its vegetation

biosphere The region on land, in the oceans and in the *atmosphere* inhabited by living organisms

business-as-usual The scenario for future world patterns of energy consumption and *greenhouse gas emissions* which assumes that there will be no major changes in attitudes and priorities

C3, C4 plants Groups of plants which take up *carbon dioxide* in different ways in *photosynthesis* and are hence affected to a different extent by increased atmospheric carbon dioxide. Wheat, rice and soya bean are C3 plants; maize, sugarcane and millet are C4 plants

carbon cycle The exchange of carbon in various chemical forms between the *atmosphere*, the land and the oceans

carbon dioxide One of the major greenhouse gases. Human-generated carbon dioxide is caused mainly by the burning of fossil fuels and deforestation

carbon dioxide fertilisation effect The process whereby plants grow more rapidly under an atmosphere of increased carbon dioxide concentration. It affects *C3 plants* more than *C4 plants*

Celsius Temperature scale, sometimes known as the centigrade scale. Its fixed points are the freezing point of water (0 °C) and the boiling point of water (100 °C)

CFCs Chlorofluorocarbons; synthetic compounds used extensively for refrigeration and aerosol sprays until it was realised that they destroy ozone (they are also very powerful greenhouse gases) and have a very long lifetime once in the *atmosphere*. The Montreal Protocol agreement of 1987 is resulting in the scaling down of CFC production and use in industrialised countries

chaos A mathematical theory describing systems that are very sensitive to the way they are originally set up; small discrepancies in the initial conditions will lead to completely different outcomes when the system has been in operation for a while. For example, the motion of a pendulum when its point of suspension undergoes forced oscillation will form a particular pattern as it swings. Started from a slightly different position, it can form a completely different pattern, which could not have been predicted by studying the first one. The weather is a partly chaotic system, which means that even with perfectly accurate forecasting techniques, there will always be a limit to the length of time ahead that a useful forecast can be made

CHP Combined Heat and Power; provided when heat produced by a power station is utilised for distinct heating instead of being wasted

CIS Commonwealth of Independent States (former USSR)

climate sensitivity The global average temperature rise under doubled *carbon dioxide* concentration in the *atmosphere*

climate The average weather in a particular region

CO_2e or carbon dioxide equivalent concentration The concentration of carbon dioxide that would cause the same radiative forcing as a given mixture of carbon dioxide and other greenhouse gases

compound A substance formed from two or more elements chemically combined in fixed proportions

condensation The process of changing state from gas to liquid

convection The transfer of heat within a fluid generated by a temperature difference

coppicing Cropping of wood by judicious pruning so that the trees are not cut down entirely and can regrow

cryosphere The component of the climate system consisting of all snow, ice and permafrost on and beneath the surface of the Earth and ocean

Daisyworld A model of biological feedback mechanisms developed by James Lovelock (see also *Gaia hypothesis*)

DC Developing country: also Third World country

deforestation Cutting down forests; one of the causes of the enhanced *greenhouse effect*, not only when the wood is burned or decomposes, releasing *carbon dioxide*, but also because the trees previously took carbon dioxide from the *atmosphere* in the process of photosynthesis

deuterium Heavy *isotope* of hydrogen

drylands Areas of the world where precipitation is low and where rainfall often consists of small, erratic, short, high-intensity storms

ecosystem A distinct system of interdependent plants and animals, together with their physical environment

El Niño A pattern of ocean surface temperature in the Pacific off the coast of South America, which has a large influence on world *climate*

electron Negatively charged component of the *atom*

element Any substance that cannot be separated by chemical means into two or more simpler substances

environmental refugees People forced to leave their homes because of environmental factors such as drought, floods or sea level rise

EU European Union

evaporation The process of changing state from liquid to gas

FAO The United Nations Food and Agriculture Organization

feedbacks Factors which tend to increase the rate of a process (positive feedbacks) or decrease it (negative feedbacks), and are themselves affected in such a way as to continue the feedback process. One example of a positive feedback is snow falling on the Earth's surface, which gives a high *albedo* level. The high level of reflected rather than absorbed solar radiation will make the Earth's surface colder than it would otherwise have been. This will encourage more snow to fall, and so the process continues

fossil fuels Fuels such as coal, oil and gas made by decomposition of ancient animal and plant remains which give off *carbon dioxide* when burned

FSU Countries of the former Soviet Union

Gaia hypothesis The idea, developed by James Lovelock, that the *biosphere* is an entity capable of keeping the planet healthy by controlling the physical and chemical environment

geoengineering Artificial modification of the environment to counteract *global warming*

geothermal energy Energy obtained by the transfer of heat to the surface of the Earth from layers deep down in the Earth's crust

global warming The idea that increased *greenhouse gases* cause the Earth's temperature to rise globally (see *greenhouse effect*)

Green Revolution Development of new strains of many crops in the 1960s which increased food production dramatically

greenhouse effect The cause of *global warming*. Incoming *solar radiation* is transmitted by the *atmosphere* to the Earth's surface, which it warms. The energy is retransmitted as *thermal radiation*, but some of it is absorbed by *molecules* of *greenhouse gases* instead of being retransmitted out to space, thus warming the atmosphere. The name comes from the ability of greenhouse glass to transmit incoming solar radiation but retain some of the outgoing thermal radiation to warm the interior of the greenhouse. The 'natural' greenhouse effect is due to the greenhouse gases present for natural reasons, and is also observed for the neighbouring planets in the solar system. The 'enhanced' greenhouse effect is the added effect caused by the greenhouse gases present in the atmosphere due to human activities, such as the burning of *fossil fuels* and *deforestation*

greenhouse gas emissions The release of *greenhouse gases* into the atmosphere, causing *global warming*

greenhouse gases *Molecules* in the Earth's *atmosphere* such as *carbon dioxide* (CO_2), methane (CH_4) and CFCs which warm the atmosphere because they absorb some of the *thermal radiation* emitted from the Earth's surface (see *greenhouse effect*)

GtC Gigatonnes of carbon (C) (1 gigatonne = 10^9 tonnes). 1 GtC = 3.7 Gt carbon dioxide

GWP Global warming potential: the ratio of the enhanced *greenhouse effect* of any gas compared with that of *carbon dioxide*

heat capacity The amount of heat input required to change the temperature of a substance by 1 °C. Water has a high heat capacity so it takes a large amount of heat input to give it a small rise in temperature

hectopascal (hPa) Unit of atmospheric pressure equal to *millibar*. Typical pressure at the surface is 1000 hPa

hydrological (water) cycle The exchange of water between the *atmosphere*, the land and the oceans

hydropower The use of water power to generate electricity

IEA International Energy Agency, a body that acts as energy advisor to 27 countries belonging to the *OECD*. In particular, it addresses the 3Es of energy policy, energy security, economic development and environmental protection.

IPCC Intergovernmental Panel on Climate Change – the world scientific body assessing *global warming*

isotopes Different forms of an *element* with different atomic masses; an element is defined by the number of *protons* its nucleus contains, but the number of *neutrons* may vary, giving different isotopes. For example, the nucleus of a carbon atom contains six protons. The most common isotope of carbon is ^{12}C, with six neutrons making up an atomic mass of 12. One of the other isotopes is ^{14}C, with eight neutrons, giving an atomic mass of 14. Carbon-containing compounds such as *carbon dioxide* will contain a mixture of ^{12}C and ^{14}C isotopes. See also *deuterium, tritium*

latent heat The heat absorbed when a substance changes from liquid to gas (*evaporation*), for example when water evaporates from the sea surface using the Sun's energy. It is given out when a substance changes from gas to liquid (*condensation*), for example when clouds are formed in the *atmosphere*

Milankovitch forcing The imposition of regularity on climate change triggered by regular changes in distribution of *solar radiation* (see *Milankovitch theory*)

Milankovitch theory The idea that major ice ages of the past may be linked with regular variations in the Earth's orbit around the Sun, leading to varying distribution of incoming *solar radiation*

millibar (mb) Unit of atmospheric pressure equal to *hectopascal*. Typical pressure at the surface is 1000 mb

MINK Region of the United States comprising the states of Missouri, Iowa, Nebraska and Kansas, used for a detailed climate study by the US Department of Energy

mole fraction (or mixing ratio) The ratio of the number of moles of a constituent in a given volume to the total number of moles of all constituents in that volume. It differs from volume mixing ratio (expressed for instance in ppmv, etc.) by the corrections for non-ideality of gases, which is significant relative to measurement precision for many *greenhouse gases*

molecule Two or more *atoms* of one or more *elements* chemically combined in fixed proportions. For example, atoms of the elements carbon (C) and oxygen (O) are chemically bonded in the proportion one to two to make molecules of the compound *carbon dioxide* (CO_2). Molecules can also be formed of a single element, for example ozone (O_3)

monsoon Particular seasonal weather patterns in sub-tropical regions which are connected with particular periods of heavy rainfall

neutron A component of most atomic nuclei without electric charge, of approximately the same mass as the *proton*

OECD Organization for Economic Cooperation and Development; a consortium of 30 countries (including the members of the European Union, Australia, Canada, Japan and the USA) that share commitment to democratic government and market economy

optical depth The fraction of a particular radiation incident on the top of the atmosphere that reaches a given level in the atmosphere is given by $\exp(-T)$ where T is the optical depth

ozone hole A region of the *atmosphere* over Antarctica where, during spring in the southern hemisphere, about half the atmospheric ozone disappears

palaeoclimatology The reconstruction of ancient *climates* by such means as ice-core measurements. These use the ratios of different *isotopes* of oxygen in different samples taken from a deep ice 'core' to determine the temperature in the *atmosphere* when the sample *condensed* as snow in the clouds. The deeper the origin of the sample, the longer ago the snow became ice (compressed under the weight of more snowfall)

parameterisation In climate models, this term refers to the technique of representing processes in terms of an algorithm (a process of step by step calculation) and appropriate quantities (or parameters)

passive solar design The design of buildings to maximise use of *solar radiation*. A wall designed as a passive solar energy collector is called a solar wall

photosynthesis The series of chemical reactions by which plants take in the Sun's energy, *carbon dioxide* and water vapour to form materials for growth, and give out oxygen. Anaerobic photosynthesis takes place in the absence of oxygen

phytoplankton Minute forms of plant life in the oceans

ppb parts per billion (thousand million) – measurement of mixing ratio (see *mole fraction*) or concentration

ppm parts per million – measurement of mixing ratio (see *mole fraction*) or concentration

Precautionary Principle The principle of prevention being better than cure, applied to potential environmental degradation

primary energy Energy sources, such as *fossil fuels*, nuclear or wind power, which are not used directly for energy but transformed into light, useful heat, motor power and so on. For example, a coal-fired power station which generates electricity uses coal as its primary energy

proton A positively charged component of the atomic nucleus

PV Photovoltaic: a solar cell often made of silicon which converts *solar radiation* into electricity

radiation budget The breakdown of the radiation which enters and leaves the Earth's *atmosphere*. The quantity of *solar radiation* entering the atmosphere from space on average is balanced by the *thermal radiation* leaving the Earth's surface and the atmosphere

radiative forcing The change in average net radiation at the top of the *troposphere* (the lower *atmosphere*) which occurs because of a change in the concentration of a *greenhouse gas* or because of some other change in the overall climate system. Cloud radiative forcing is the change in the net radiation at the top of the troposphere due to the presence of the cloud

reforestation Planting of forests on lands that have previously contained forests but that have been converted to some other use

renewable energy Energy sources which are not depleted by use, for example *hydropower*, PV solar cells, wind power and *coppicing*

respiration The series of chemical reactions by which plants and animals break down stored foods with the use of oxygen to give energy, *carbon dioxide* and water vapour

sequestration Removal and storage, for example, *carbon dioxide* taken from the *atmosphere* into plants via *photosynthesis*, or the storage of carbon dioxide in old oil or gas wells

sink Any process, activity or mechanism that removes a *greenhouse gas*, *aerosol* or precursor of a greenhouse gas or aerosol from the atmosphere

solar radiation Energy from the Sun

sonde A device sent into the *atmosphere* for instance by balloon to obtain information such as temperature and *atmospheric pressure*, and which sends back information by radio

stewardship The attitude that human beings should see the Earth as a garden to be cultivated rather than a treasury to be raided. (See also *sustainable development*)

stratosphere The region of the *atmosphere* between about 10 and 50 km altitude where the temperature increases with height and where the ozone layer is situated

sustainable development Development which meets the needs of the present without compromising the ability of future generations to meet their own needs

thermal radiation Radiation emitted by all bodies, in amounts depending on their temperature. Hot bodies emit more radiation than cold ones

thermodynamics The First Law of thermodynamics expresses that in any physical or chemical process energy is conserved (i.e. it is neither created nor destroyed). The Second Law of thermodynamics states that it is not possible to construct a device which only takes heat energy from a reservoir and turns it into other forms of energy or which only delivers the heat energy to another reservoir at a different temperature. The Law further provides a formula for the maximum efficiency of a heat engine which takes heat from a cooler body and delivers it to a hotter one

thermohaline Large-scale density-driven circulation in the circulation (THC) ocean caused by differences in temperature and salinity

transpiration The transfer of water from plants to the *atmosphere*

tritium Radioactive *isotope* of hydrogen, used to trace the spread of radioactivity in the ocean after atomic bomb tests, and hence to map ocean currents

tropical cyclone A storm or wind system rotating around a central area of low *atmospheric pressure* and occurring in tropical regions. They can be of great strength and are also called hurricanes and typhoons. Tornadoes are much smaller storms of similar violence

troposphere The region of the lower *atmosphere* up to a height of about 10 km where the temperature falls with height and where convection is the dominant process for transfer of heat in the vertical

UNCED United Nations Conference on Environment and Development, held at Rio de Janeiro in June 1992, after which the United Nations Framework Convention on Climate Change was signed by 160 participating countries

UNEP United Nations Environmental Programme – one of the bodies that set up the *IPCC*

UNFCCC United Nations Framework Convention on Climate Change with 192 member countries was agreed at the *UNCED* in 1992

UV Ultraviolet radiation

watt Unit of power

WEC World Energy Council – an international body with a broad membership of both energy users and the energy industry

wind farm Grouping of wind turbines for generating electric power

WMO World Meteorological Organization – one of the bodies that set up the *IPCC*

Younger Dryas event Cold climatic event that occurred for a period of about 1500 years, interrupting the warming of the Earth after the last ice age (so called because it was marked by the spread of an Arctic flower, *Dryas octopetala*). It was discovered by a study of palaeoclimatic data

zooplankton Minute forms of animal life in the oceans

Index